T0306134

THE NEUROSCIENCE OF SLEEP

THE NEUROSCIENCE OF SLEEP

EDITORS-IN-CHIEF

ROBERT STICKGOLD
Beth Israel Deaconess Medical Center, and
Harvard Medical School, Boston, MA, USA

MATTHEW P. WALKER
University of California, Berkeley, CA, USA

AMSTERDAM • BOSTON • HEIDELBERG • LONDON • NEW YORK • OXFORD
PARIS • SAN DIEGO • SAN FRANCISCO • SINGAPORE • SYDNEY • TOKYO
Academic Press is an imprint of Elsevier

ELSEVIER

ACADEMIC
PRESS

Academic Press is an imprint of Elsevier
32 Jamestown Road, London NW1 7BY, UK
30 Corporate Drive, Suite 400, Burlington, MA 01803, USA
525 B Street, Suite 1900, San Diego, CA 92101-4495, USA

Notice
No responsibility is assumed by the publisher for any injury and/or damage to persons or property as a matter of products liability, negligence or otherwise, or from any use or operation of any methods, products, instructions or ideas contained in the material herein, Because of rapid advances in the medical sciences, in particular, independent verification of diagnoses and drug dosages should be made

British Library Cataloguing in Publication Data
A catalogue record for this book is available from the British Library

Library of Congress Catalog Number: 2009932252

ISBN: 978-0-12-375073-0

For information on all Elsevier publications
visit our website at books.elsevier.com

Printed in the United States of America
Transferred to Digital Printing, 2011

TABLE OF CONTENTS

CONTRIBUTORS

S Ancoli-Israel
University of California at San Diego, San Diego, CA, USA

J S Antrobus
The City College of New York, New York, NY, USA

C Ballas
University of Pennsylvania School of Medicine, Philadelphia, PA, USA

S Banks
University of Pennsylvania School of Medicine, Philadelphia, PA, USA

R Basheer
Boston VA Healthcare System and Harvard Medical School, Brockton, MA, USA

M Bentivoglio
University of Verona, Verona, Italy

M Boly
University of Liège, Liège, Belgium

G Buzsáki
Rutgers University, Newark, NJ, USA

S M Caples
Mayo Clinic, Rochester, MN, USA

M A Carskadon
Brown University, and E. P. Bradley Hospital, Providence, RI, USA

A-M Chang
Harvard Medical School, Boston, MA, USA

J J Chrobak
University of Connecticut, Storrs, CT, USA

L Churchill
Washington State University, Pullman, WA, USA

C Cirelli
University of Wisconsin–Madison, Madison, WI, USA

M A Cramer-Bornemann
Hennepin County Medical Center, Minneapolis, MN, USA; and University of Minnesota Medical School, Minneapolis, MN, USA

T T Dang-Vu
University of Liege and Centre Hospitalier Universitaire, Liege, Belgium

M Desseilles
University of Liege and Centre Hospitalier Universitaire, Liege, Belgium

A Destexhe
Centre National de la Recherche Scientifique (CNRS), Gil-sur-Yvette, France

D F Dinges
University of Pennsylvania School of Medicine, Philadelphia, PA, USA

S P A Drummond
University of California at San Diego, La Jolla, CA, USA

S P A Drummond
University of California at San Diego and Veterans Affairs San Diego Healthcare System, San Diego, CA, USA

D J Earnest
Texas A&M University Health Science Center, College Station, TX, USA

M G Frank
University of Pennsylvania, Philadelphia, PA, USA

P M Fuller
Harvard Medical School, Boston, MA, USA

E Garcia-Rill
University of Arkansas for Medical Sciences, Little Rock, AR, USA

A Germain
University of Pittsburgh School of Medicine, Pittsburgh, PA, USA

C Guilleminault
Stanford University Sleep Medicine Program,
Stanford, CA, USA

H L Haas
Heinrich-Heine-Universität, Düsseldorf, Germany

H C Heller
Stanford University, Stanford, CA, USA

J A Hobson
Harvard Medical School, Boston, MA, USA

B E Jones
McGill University, Montreal, QC, Canada

K Kristensson
Karolinska Institutet, Stockholm, Sweden

J M Krueger
Washington State University, Pullman, WA, USA

D Kuiken
University of Alberta, Edmonton, AB, Canada

S Laureys
University of Liege and Centre Hospitalier
Universitaire, Liege, Belgium

S Laureys
University of Liège, Liège, Belgium

J A Lesku
Max Planck Institute for Ornithology,
Seewiesen, Germany

R Levin
Ferkauf Graduate School of Psychology, Bronx,
NY, USA

L Lin
Stanford University, Palo Alto, CA, USA

J Lu
Harvard Medical School, Boston, MA, USA

M W Mahowald
Hennepin County Medical Center, Minneapolis, MN,
USA; and University of Minnesota Medical School,
Minneapolis, MN, USA

P Maquet
University of Liege and Centre Hospitalier
Universitaire, Liege, Belgium

D Martinez-Gonzalez
Max Planck Institute for Ornithology,
Seewiesen, Germany

R W McCarley
Boston VA Healthcare System and Harvard Medical
School, Brockton, MA, USA

B S McKenna
San Diego State University, University of California
at San Diego, and Veterans Affairs San Diego
Healthcare System, San Diego, CA, USA

J T McKenna
Boston VA Healthcare System and Harvard Medical
School, Brockton, MA, USA

S C Mednick
University of California at San Diego, La Jolla,
CA, USA

E Mignot
Stanford University, Palo Alto, CA, USA

J D Minkel
University of Pennsylvania School of Medicine,
Philadelphia, PA, USA

G Moonen
University of Liège, Liège, Belgium

J M Mullington
Harvard Medical School, Boston, MA, USA

M Nicolelis
Duke University Medical Center, Durham, NC, USA,
and École Polytechnique Fédérale de Lausanne,
Lausanne, Switzerland

T Nielsen
Université de Montréal, Montreal, QC, Canada

S Nishino
Stanford University School of Medicine, Stanford,
CA, USA

E F Pace-Schott
Harvard Medical School, Boston, MA, USA

P Peigneux
University of Liege, Liege, Belgium

R Pelayo
Stanford University Sleep Medicine Program,
Stanford, CA, USA

N C Rattenborg
Max Planck Institute for Ornithology,
Seewiesen, Germany

D M Rector
Washington State University, Pullman,
WA, USA

S Ribeiro
Edmond and Lily Safra International Institute of
Neuroscience of Natal, and Universidade Federal
do Rio Grande do Norte, Natal, Brazil

M Rissling
University of California at San Diego, San Diego,
CA, USA

C H Schenck
Hennepin County Medical Center, Minneapolis,
MN, USA; and University of Minnesota Medical
School, Minneapolis, MN, USA

M Schredl
Central Institute of Mental Health,
Mannheim, Germany

T J Sejnowski
Salk Institute for Biological Studies and
University of California at San Diego, La Jolla,
CA, USA

O Selbach
Heinrich-Heine-Universität, Düsseldorf, Germany

O A Sergeeva
Heinrich-Heine-Universität, Düsseldorf, Germany

A Sirota
Rutgers University, Newark, NJ, USA

V K Somers
Mayo Clinic, Rochester, MN, USA

M Steriade
Université Laval, Laval, QC, Canada

R Stickgold
Beth Israel Deaconess Medical Center, and
Harvard Medical School, Boston, MA, USA

R Szymusiak
University of California, Los Angeles, CA, USA

L Tarokh
Brown University, Providence, RI, USA

G Tononi
University of Wisconsin–Madison, Madison,
WI, USA

S C Veasey
University of Pennsylvania School of Medicine,
Philadelphia, PA, USA

M P Walker
University of California, Berkeley, CA, USA

E J Wamsley
Harvard Medical School, Boston, MA, USA

A Zadra
Université de Montréal, Montréal, QC, Canada

P C Zee
Northwestern University Feinberg School of
Medicine, Chicago, IL, USA

PREFACE

It is striking that sleep, which occupies between a quarter and a third of each of our lives, remains so poorly understood. To put this ignorance in perspective, both REM sleep and the structure of DNA were discovered in 1953. Yet while the discovery of DNA's structure led to the development of the entire field of molecular biology and spawned the biotechnology industry, sleep researchers still struggle to identify the nature, mechanisms, and functions of sleep. As recently as the late 1990's, one noted sleep researcher, Allan Hobson, quipped that the only known function of sleep was to cure sleepiness.

There are legitimate reasons for this slow progress. For one thing, there does not seem to be a single mechanistic pathway to the generation or termination of sleep. Rather, sleep is regulated by a highly complex collection of interlocking mechanisms ranging from the circadian rhythm of the suprachiasmatic nucleus, in turn controlled by a series of conserved genes that also control the rhythms of fruit flies, to frontal cortical arousal systems, and psychological processes such as anxiety and depression. While current research focuses on the hypothalamus as the conduit through which all these regulatory factors act, the mechanisms controlling the actual production of sleep remain elusive.

This is not that surprising, given that sleep is, at it core, defined not as a physiological state, but as a state of consciousness. Human sleep can be defined as a state in which conscious awareness of the external world is lost, replaced either by an internally generated world of thoughts, feelings and hallucination, or, at other times, by a true absence of consciousness. While we can study the physiological correlates of this altered state of consciousness, we remain at a fundamental loss when we it comes to understanding how this altered state is generated.

Despite these drawbacks, the recent decades have produced a wealth of information and knowledge about the biology of sleep. The circadian control of sleep is now well described at genetic, molecular, and cellular levels. The hypothalamic regulation of sleep onset, and the reciprocal interactions that regulate the REM-NREM cycle have been described at the physiological level, and more and more at the molecular. The changes in neuromodulation and oscillatory brain patterns during sleep continue to be elucidated. In addition, sleep medicine has grown as a discipline, with narcolepsy now well described from the genetic and molecular levels through to new treatments. The nature of REM sleep behavior disorder has been clarified, and its relationship to Parkinson's disease is becoming clearer. Apnea is now a treatable disorder.

The functions of sleep remain less clear. The last decade has helped identify roles for sleep in memory consolidation, immune function, and endocrine and thermal regulation, but other functions remain to be confirmed, and the detailed mechanisms of these known functions remain unknown. Finally, dreaming is still almost as obscure a black box as it was half a century ago. The mechanisms that initiate and control the dream process are only guessed at, and it remains difficult to know how to even pose the question of possible functions for dreams. Still, dream research continues to make inroads, and there is hope that the next decade will bring exciting new insights.

The goal of this book, and the ordered chapters contained within, is to traverse the rich and complex descriptive levels of the sleeping brain. We travel the distance from molecules all the way to mind. The journey considers the basic brain mechanisms regulating and controlling sleep states, the functional benefits of sleep at both a brain and body level, and the detrimental consequence of sleep loss across an array of clinical disorders, and ends with a discussion of dreaming.

Sections I–IV: We begin with a historical perspective on sleep research, and the time-line of our current knowledge. The subsequent series of chapters examines the different neurobiological approaches to characterizing and defining sleep, from genetic (molecules), cellular (brain nuclei), systems (networks) level, and whole-brain perspectives, and how these canonical signatures change across the lifespan. Next, we consider the rich and often perplexing clues that come from appreciating the evolution of sleep, and its changes across develoment. Returning to regulation, we consider the elegant and sophisticated subcortical systems that orchestrate sleep and control its varied stages. Arousal states that arise from, or closely resemble, sleep are then reviewed, sharpening our understanding of how sleep differs from these related states.

Sections V–VII: Over the next several sections, we widen the aperture of our review. First, we consider the interplay between sleep and biological rhythms, specifically those of the circadian system. We start by describing the neural mechanisms and genetic determinants of the endogenous circadian rhythm. Next, the physiological and psychological implications of the circadian cycle are discussed. The impact of circadian rhythms on domains of alertness and sustained task performance is then examined, emphasizing conditions of sleep deprivation, and how dysfunction of circadian control can be understood at a human genetic level. Finally, recent efforts to pharmacologically manipulate sleep and sleepiness are reviewed.

Section VIII: Everything up to this point leads us to perhaps the most fundamental of questions – why do we sleep? More specifically, what are the functions of sleep? Appreciating that sleep likely supports not one but many different functions, a number of exciting hypotheses, including some recent ones, are discussed, and defended empirically. At a body level, these include the role of sleep in modulating endocrine function, regulating immune system integrity and optimally controlling thermoregulation. At a central brain level, the role of sleep in supporting a number of neurocognitive processes will be examined, including the benefits of sleep for learning, memory and brain plasticity.

Section X: With the neurobiology and functions of sleep under our belt, we conclude with a consideration of dreaming ("sleep mentation"). Different theories of dream function are discussed. This involves consideration of the interplay between REM and NREM, the association between dreaming and prior waking experience, and how these two [waking and dream consciousness] can be understood symbiotically. The consequence of dysfunctional sleep mentation is then considered, including the case of nightmares and clinical syndromes such as post-traumatic stress disorder.

In all, we have attempted to create a book that can serve many audiences. For the student, its articles cover the full range of sleep research, provide a firm base in the science of sleep, and offer insights into the directions in which the field is going. For the researcher and clinician, it provides an up-to-date summary of the current state of sleep research at both the basic and clinical level. While not delving deeply into clinical issues, the chapters on sleep disorders provide a wealth of references to more detailed discussion in this field. Finally, for the dilettante, each chapter is designed to function as a self-contained unit, independent of other chapters, allowing one to dip in here and there, exploring the science and mysteries of sleep at one's whim.

Robert Stickgold and Matthew P. Walker
Editors

INTRODUCTION

History of Sleep Research

R Pelayo and C Guilleminault, Stanford University
Sleep Medicine Program, Stanford, CA, USA

When the history of sleep research and sleep medicine began is a matter of debate, but modern sleep medicine may be related to development of sleep research tools. In 1929, J-H Berger recorded the electrical activity of the human brain and demonstrated differences in these rhythms during wakefulness and sleep. He called the signals that he recorded 'electroencephalogram' (EEG). At Harvard University, Loomis et al. studied sleep onset and overnight sleep, and they published the results of their investigation in several reports between 1937 and 1939, dividing sleep into five states (A–E). Blake et al. provided information on sleep subdivision. The description of specific EEG rhythms during sleep was the beginning of a new era in sleep research.

However, the overall development of neurosciences at the beginning of the twentieth century had other impacts on the clinical understanding of sleep. Constatin Von Economo probably provided the most insight on central nervous system structures involved in the control of sleep. While investigating the viral pandemic that began at the end of World War I and killed many people, he made important observations on patients affected by 'encephalitis lethargica.' He correlated the clinical symptoms of insomnia and continuous lethargy with the presence of two different sites of inflammatory lesions. Lesions in the preoptic area and anterior hypothalamus led to insomnia and hyperkinetic movements, whereas lesions of the posterior hypothalamus led to sleepiness, continuous lethargy, and eye movement abnormalities. The role of the ventrolateral preoptic nucleus in sleep induction has been demonstrated, as has the role of the hypocretin peptides along with the identification of the hypocretin-containing neurons in the posterolateral regions of the hypothalamus. However, in the early 1920s Von Economo had found that the anterior hypothalamic lesions were often extended to the diencephalon (basal ganglia), whereas the posterior hypothalamic lesions were associated with lesions in the periaqueductal gray matter with frequent extension to the oculomotor nuclei, which thus explained the oculomotor symptoms. This led Von Economo to speculate that the anterior hypothalamus contained a sleep-promoting area and that an area spanning from the posterior wall of the third ventricle to the third cranial nerve was involved in activity promoting wakefulness. He also speculated that the syndrome of narcolepsy involved this posterior hypothalamus area (a fact also mentioned by CP Symonds in 1926, based on Von Economo's findings), as reported again very recently.

In 1949, Moruzzi and Magoun made an important discovery that had a major impact on further electrophysiological research on sleep that began to attract increasingly more researchers. They showed that the brain stem netlike core of neurons, the reticular formation, had ascending pathways, leading to the discovery of the 'ascending reticular activating system.' Stimulation of this neuronal network induced long-lasting and widespread cortical activation, marked by replacement of cortical slow waves with fast activity. They also showed that lesions of the reticular formation, but not those of sensory pathways, produced a loss of cortical activation. The lesions with most marked effect were located in the oral pontine and midbrain reticular formation and the posterior hypothalamic–subthalamic regions where ascending pathways reach into the forebrain. Additional, smaller lesions in the midbrain tegmentum and caudal diencephalons showed that dissociation between cortical activation and behavioral arousal of wakefulness could be obtained.

Deficiency of cortical activation without prevention of behavioral responsiveness to sensory stimulation is obtained with lesions of central midbrain tegmentum, whereas lesions of the ventral tegmentum and hypothalamus lead to unresponsiveness without loss of cortical activation.

Following these initial studies, increasingly more lesions at different levels were performed, leading to a delimitation of structures responsible for activation of the cortex, more numerous than initially considered, including not only the reticular formation but also the posterior hypothalamus–subthalamus and basal forebrain systems (nucleus basalis of Meynert). Transection of the brain stem behind the oral pontine tegmentum by Batini and other collaborators of Moruzzi resulted in a complete insomnia.

The understanding of sleep was greatly changed by the discovery of what has been called rapid eye movement (REM) sleep, paradoxical sleep, or desynchronized sleep. It is related to the investigations by Nathaniel Kleitman on rest and activity cycles. Eugene Aserinsky, a graduate student of Kleitman, was collecting data on napping infants and designed a data sheet with 5 min periods with indication of 'periods of motility' or 'absence of motility.' Aserinsky and Kleitman obtained data suggesting the presence of an 'eye motility rhythm' and decided to verify it in adults. In order to monitor the eye movements during the night, they decided to measure eye movements using electro-oculography (EOG). While monitoring these electro-oculograms, they noted for the first time the presence of two very different types of movements: movements with bursts different from slow eye movements seen at sleep onset. Initially thought to be artifacts, direct observations of the phenomenon indicated by the recordings confirmed the reality of the movement. The investigators also noted that simultaneously during these specific eye movements, irregular breathing and heart rate accelerations were seen. Based on their observations, they assumed that there was a 'lightening' of sleep with these movements and that they may represent dreaming. Aserinsky and Kleitman had no continuous EEG recording but woke up their subjects during periods of occurrence and absence of eye movements, obtaining very different findings as reported in 1953. It was with another student, William C Dement, that Kleitman monitored EEG continuously to further investigate the association of EOG and EEG. In 1957, Dement and Kleitman reported the cyclical variation of EEG during sleep and its relation to eye movements, body motility, and dreaming. They found that the cyclical variation from a period of eye movement to the successive one was between 90 and 100 min, and that there was a regular re-occurrence of rapid eye movements and dreaming. However, at that time, sleep was still considered as only one state. They called these periods with rapid eye movements 'emergent stage 1' sleep. After reading Dement's 1958 report on the occurrence of low-voltage, fast EEG patterns during behavioral sleep in cats, French researcher Michel Jouvet replicated the findings and indicated that the cat also presented a muscle atonia in association with the rapid eye movement and the characteristic EEG pattern. In 1962, Jouvet used the term 'paradoxical sleep,' with the 'paradox' being that although the studied cat had an EEG that resembled wakefulness, behaviorally the animal remained asleep and unresponsive. With Francois Michel, Jouvet also reported that this new 'state' was associated with the presence of large electrical potentials that were first seen in the pons but appeared a few milliseconds later and were able to be recorded most easily in the lateral geniculate nucleus and, with further but limited delay, in the occipital cortex, leading to the term ponto-geniculo-occipital spikes. Sleep was divided by the Lyon, France, team into two states: slow wave sleep and paradoxical (also called dysynchronized or REM) sleep.

Following these discoveries, under the initiative of WC Dement, a small group of researchers formed the Association of Psychophysiological Study of Sleep. This small group invited other North American, European, and Asian researchers to participate in its research and to built a 'common sleep scoring system.' By consensus, these researchers agreed on a final product that had a tremendous impact on human sleep research. An atlas was created that could be used as a reference to score sleep in a similar way throughout the world. This atlas also included recommendations on how to monitor sleep as well as how and what to tabulate. This atlas, edited by Alan Rechtschaffen and Anthony Kales, was published in 1968.

These different efforts opened the field for systematic investigation of pathology of sleep in humans. However, there had previously been publications on specific disorders during sleep. In 1685, Willis gave the best description of restless leg syndrome, long before the 1945 publication of Ekbom that gave the name to the syndrome. An associated syndrome was noted with small movements of the big toe that occurred at regular intervals during sleep. This motor manifestation was questioned, and called 'nocturnal myoclonus' and erroneously thought to be an epileptic equivalent by Charles P Symonds in 1953; it was renamed periodic limb movement syndrome after demonstration of its linkage to restless leg syndrome (RLS) by Lugaresi and colleagues. Jean Baptiste Gelineau described the narcolepsy syndrome in 1880. He named the syndrome that he considered as an independent entity. If the terminology 'narcolepsy' began in 1880, Caffe in 1862, as cited by Gelineau, had already described a 'sleeping sickness' that plagued a subject with short attacks of sleepiness. The patient also presented 'hallucinations' that had him hospitalized in a mental hospital. Also, Westphal in 1877 published a report on a patient with sleeping attacks associated with episodes of loss of motor and language abilities. He also indicated that the mother was also affected by the same problem. What would be called the 'Pickwickian syndrome' by William Oster was first reported in the nineteenth century by Robert Caton in England. He reported on 8 February 1861 to the Clinical Society of London about a 37-year-old poulterer with excessive sleepiness: "The moment he sat down on his chair sleep came on and even when standing and walking he would sink into sleep constantly while serving customers in shop, sleep would come as he stood by the counter, he would wake and find himself holding in his hand the duck or chicken which he had been selling to a customer a quarter of an hour before, the customer having in the meantime departed." Caton goes on to describe that "when in sound sleep a very peculiar state of the glottis is observed, a spasmodic closure entirely suspending respiration (1889). The thorax and abdomen are seen to heave from fruitless contractions of the inspiratory and expiratory muscles; their efforts increase in violence for about a minute or a minute and a half, the skin mean time becoming more and more cyanosed, until at last, when the condition to an on-looker is most alarming, the glottic obstruction yields, a series of long inspirations and expirations follows and cyanosis disappears. This acute dyspneic attack does not awaken the patient. The night nurses stated that these attacks go on throughout the night." The usage of polygraphic recording during sleep led to a new understanding of syndromes during sleep.

The first 'polysomnogram' monitored during nocturnal sleep was probably by Bulow and Ingward, who studied breathing during sleep in normal subjects and in 1961 published a report stating that there was a clear change of breathing pattern during sleep onset, with the presence of a short respiratory pause and a change of carbon dioxide tension that occurred much earlier than the notion of the 'apnea threshold' and also a change in arterial CO2 set point when a transition from wakefulness to sleep was noted. Influenced by these researchers, Werner Gerardy in Heidelberg modified an EEG machine to monitor respiration and pulse rate. He performed the recording of an obese patient hospitalized for headaches and who presented episodes of 'periodic breathing' with intermittent absence of airflow and association with change of heart rate. Thereafter, Gerardy et al. monitored another 'Pickwickian' patient, noted a similar phenomenon, and reported their findings in 1960. They reported the first sleep recording of a Pickwikian patient and first demonstration of apnea during sleep. These different reports, followed by those of Gastaut et al. and others on Pickwickian subjects, were still the realm of neurologists and clinical neurophysiologists.

However, in 1970, Christian Guilleminault, who had benefited from the advances in understanding in sleep physiology, pharmacology, and biochemistry, believed that the existence of three different states of being – wakefulness, nonrapid eye movement (NREM) sleep, and REM sleep – with many important neuronal changes implied that the control of vital functions would be affected by passage from one state to another and that pathologies may be seen. For Guilleminault there was a different dimension: the states of sleep that led to different controls as those seen during wakefulness. The existence of these different controls during the two sleep states led to special clinical presentations, required treatment aimed at the dysfunction occurring during the sleep

states, and justified the development of what he called 'sleep medicine.' He was given limited funding in 1970 by the Societe Medicale des Hopitaux de Paris for investigation of sleep medicine patients in La Salepétriere Hospital, Paris. The skepticism in France led him to accept the invitation from William C Dement and Stanford University to work there in 1972.

Also in 1972, Elio Lugaresi and Pierre Sadoul organized the first international meeting on 'hypersomnia with periodic breathing' focusing on the abnormalities of breathing during sleep in obese subjects. In 1973, Guilleminault et al. reported sleep apnea and insomnia, and in 1975 the definition of an 'obstructive sleep apnea syndrome' was given. It was followed by the publication of obstructive sleep apnea syndrome in children and a presentation of the state of the art on sleep apnea syndrome in a monograph titled *Sleep Apnea Syndromes* that included most researchers involved in the study of the syndrome. Colin Sullivan, in the pulmonary physiology laboratory in Sydney, designed equipment to apply continuous air pressure to avoid the collapse of the upper airway. Using an air compressor and an airtight mask, Sullivan and colleagues succeeded in eliminating the sleep-related obstruction in a severely affected patient. After observing the results obtained with their prototype, they not only began treating patients in the hospital but also made major advances with regard to the possibility of a permanent nonaggressive treatment of obstructive sleep apnea (OSA) with application of continuous positive airway pressure (CPAP) through nasal mask at home, as reported in 1981.

The search for new treatments for OSA was not limited to the creation of nasal CPAP. Dr. Ikematsu, an otolaryngologist in Japan, developed a surgical procedure performed on the soft palate. The procedure was aimed at eliminating loud chronic snoring that was disturbing to spouses. Fujita, an American-Japanese otolaryngologist from Detroit, extended the procedure under the name uvulopalatopharyngoplasty (UPPP) in 1981. Because the procedure was not completely successful in several cases, this led to a search for other anatomy-oriented treatment modalities. The development of what became known as the 'Stanford surgical protocol' based on work by Nelson Powell, Robert Riley, and Christian Guilleminault began in 1981 and initially involved only mandibular geniotubercle advancement, but later involved a 'phase I' and a 'phase II' surgical approach, with phase II consisting of bimaxillary advancement that could provide results similar to those noted with nasal CPAP treatment. To avoid performing orthognathic surgery, efforts were made to use dental-related devices to protrude the tongue forward during sleep. Samelson and Cartright reported objective results obtained with the 'tongue retaining device.' Following these initial trials, different types of dental appliances were designed to help patients with the syndrome of mild to moderate severity or residual problems post-UPPP.

The understanding that OSA was, commonly, a familial syndrome, often with involvement of anatomical and particularly mild changes in craniofacial features, led to systematic investigation of nonsyndromic children not only in families of adult patients with OSA but also in children with chronic snoring. Orthodontic techniques were explored. Rapid maxillary distraction (RMD), a well-defined orthodontic procedure, was thus first used in adults by Cistulli et al. for mild OSA. It was applied thereafter to prepubertal children with residual OSA post-adenotonsillectomy and narrow hard palate by the Rome and Stanford teams. Also, mandibular distraction osteogenesis with or without RMD on the hard palate was introduced for treatment of teenagers. Finally, investigation of children and, later, adults with more sophisticated monitoring sensors allowed the description of an abnormal breathing pattern that may not be associated with the defined sleep apnea and hypopnea syndrome: the upper airway resistance syndrome.

The first international meeting on narcolepsy was held in 1975 at Montpellier, France, at which the modern definition of the syndrome was developed. In 1983, Fuji and Honda reported a dominant association between a specific group, HLA (DR2), and the presence of narcolepsy in Japanese people. A strong HLA association between a certain HLA typing and narcolepsy was quickly confirmed. Neely and colleagues investigated different ethnic groups, particularly African Americans, and did not find an association with DR2, and the Stanford group showed that the best and most common HLA association was with HLA DQB1–0602.

The discovery of a canine model for narcolepsy cataplexy led to many pharmacological studies and allowed the recognition that the narcoleptic Doberman had a genetic defect on a newly discovered brain system – the hypocretin/orexin system. This discovery was nearly simultaneously associated with the finding that mice with knockout hypocretin receptor presented cataplectic attacks. In 2000, Nishino et al. showed that human narcoleptics presented an absence of measurable hypocretin/orexine levels in cerebrospinal fluid, and it was found that genetic defect was very rare in human narcolepsy. Also in 2000, Peyron et al. and Thannickal et al. showed destruction of hypocretin cells in the posterolateral region of the hypothalamus (the same region as identified by Von Economo in the brains of human narcoleptics) was the most common cause of narcolepsy and identified the presence of primary and secondary narcolepsy. The cause of the primary narcolepsy, which is most common, is unknown.

After the association between RLS and periodic leg movement had been made by Lugaresi et al. in 1968, the presentation of periodic limb movement as an isolated syndrome was made by Guilleminault et al. in 1978 and extended in 1982 by Coleman et al. In 1982, Akpinar reported the beneficial effect of dopamine agonist in the treatment of RLS – a finding largely confirmed over time. However, the underlying defect associated with RLS is still elusive, despite the clear demonstration of a large group of familial cases and the identification in Quebec families of specific chromosomic loci associated with RLS occurrence.

Abnormal Behavior during Sleep

NREM Sleep Parasomnia

Broughton and Gastaut coined the term 'disorders of arousal' to qualify these clinical presentations and dissociate them from an epileptic disorder occurring during stages 3 and 4 of NREM sleep, usually during the first one-third of the night. Fisher et al. emphasized the presence of 'hypersynchronous' delta waves, which were later called 'delta bursts' or 'delta sequence' by others. The existence of a familial pattern of sleepwalking was noted by Kales et al., but data call into question whether the familial pattern is related directly to sleepwalking or it is related to an underlying cause of behavior. Analysis of sleep EEG patterns associated with NREM sleep parasomnia has used new approaches, particularly the investigation of the cyclic alternating pattern (CAP) initially described by Terzano et al. in adults. This analysis shows the presence of an abnormal CAP rate in patients, indicative of instability of NREM sleep. It is most often related to other underlying unrecognized and untreated sleep disorders, usually sleep-disordered breathing, particularly upper airway resistance syndrome or, rarely, RLS.

REM Behavior Disorder

Experimental lesions performed in cats on the brain stem induced an 'active' REM sleep: cats presented motor activity similar to dreaming (i.e., acting as if hunting or playing). They 'act out' their dreams due to the absence of REM sleep motor atonia. In humans, Japanese teams observed similar findings with delirium. It was also observed in patients with a brain stem neurodegenerative disease, olivo-ponto-cerebellar degeneration, by Quera-Salva and Guilleminault. Finally, Schenck and Mahowald reported a series of 'idiopathic cases' and called the disorder 'REM behavior disorder' (RBD). Initially thought to be 'idiopathic,' it progressively appeared to be associated with other neurological disorders – not only narcolepsy, a syndrome of dissociated REM sleep, but also very prominently with Parkinson's disease, multiple system atrophy, and Lewy's dementia. The notion has emerged that RBD is more commonly associated with synucleopathy, raising the issue of idiopathic RBD being the first clinical indication of the presence of Lewy bodies located mostly in the lower brain stem before caudal extension of lesions.

Sleep investigations uncovered another neurodegenerative disorder: fatal familial insomnia (FFI). The genetic nature of the disorder, its direct linkage to a prion disease, and its inexorable evolution toward the complete disappearance of sleep and then death were identified by the continuous research led by Elio Lugaresi and involving both the Gambetti team in Cleveland and many researchers in Bologna. FFI is associated with the development of very severe dysautonomia, complete insomnia before death, severe dementia, and bilateral symmetrical degeneration of the thalamus.

Sleep–wake is part of chronobiology, and the history of sleep disorders is very much intertwined with the understanding of chronobiology and its defect leading to dyschronosis or chronobiologic disorders. Identification of genes such as CLOCK and PERIOD (or PER) led to research on circadian rhythms and sleep pathologies. Two syndromes, 'sleep phase delay' and 'sleep phase advance' syndromes, were described. The term sleep phase delay syndrome was introduced by Weitzman and colleagues and by Czeisler et al., who reported the first treatment attempt called 'chronotherapy.' The discovery of a very large family in Utah presenting phase advance syndrome led to the identification of a specific genetic marker for this syndrome. Advances in the treatment of dyschronosis have been based on the investigation of the role of light and dark cycles and placement of sleep in this cycle.

Conclusion

In 1994, the Center for Sleep Research of the National Heart, Lung, and Blood Institute of the National Institutes of Health was created after a major Congress-supported effort by a special committee chaired by WC Dement. The American Sleep Disorders Association and its successor, the American Academy of Sleep Medicine, have had their efforts recognized in accreditation of training centers in sleep medicine, and the accreditation process was handed over to the American Council on General Medical Education in 2005. The Sleep Medicine Board examination will be administered under the auspices of the American Board of Medical Specialties beginning in 2007. Sleep medicine is a living history.

See also: Behavior and Parasomnias (RSBD); Cataplexy; Narcolepsy; PET Activation Patterns; Phylogeny and Ontogeny of Sleep; Sleep Apnea; Sleep Architecture; Sleeping Sickness.

Further Reading

Bulow K and Ingvar DH (1961) Respiration and state of wakefulness in normals, studied by spirography, capnography, and EEG: A preliminary report. *Acta Physiologica Scandinavia* 51: 230–238.

Czeisler CA, Richardson GS, Coleman RM, et al. (1981) Chronotherapy: Resetting the circadian clocks of patients with delayed sleep phase insomnia. *Sleep* 4: 1–21.

Ekbom KA (1945) Restless legs. *Acta Medica Scandinavia* 158 (supplement): 1–123.

Gastaut H, Lugaresi E, Berti-Ceroni G, and Coccagna G (eds.) (1968) *The Abnormalities of Sleep in Man*. Bologna: Aulo Gaggi.

Guilleminault C and Dement WC (1978) *Sleep Apnea Syndromes*. New York: Liss.

Guilleminault C, Dement WC, and Passouant P (eds.) (1976) *Narcolepsy*. New York: Spectrum.

Guilleminault C, Eldridge FL, and Dement WC (1973) Insomnia with sleep apnea: A new syndrome. *Science* 181: 856–858.

Jouvet M (1962) Research on the neural structures and responsible mechanisms in different phases of physiological sleep. *Archives Italiennes de Biologie* 100: 125–206.

Jouvet M and Michel F (1959) Corrélation électromyographiques du sommeil chez le chat décortique et mésencephalique chronique. *Comptes Rendus de la Société de Biologie (Paris)* 153: 29–45.

Juji T, Satake M, Honda Y, and Doi Y (1984) HLA antigens in Japanese patients with narcolepsy. All the patients were DR2 positive. *Tissue Antigens* 24: 316–319.

Kleitman N (1939) *Sleep and Wakefulness*. Chicago, IL: University of Chicago Press (Revised in 1963).

Lavie P (2003) *Restless Night*. pp. 29–45. New Heaven, CT: Yale University Press.

Loomis AL, Harvey FN, and Hobart G (1935) Further observations on the potential alpha-rhythms of the cerebral cortex during sleep. *Science* 2: 198.

Lugaresi E, Medori R, Montagna P, et al. (1986) Fatal familial insomnia and dysautonomia with selective degeneration of thalamic nuclei. *New England Journal of Medicine* 315: 997–1003.

Moruzzi G and Magoun HW (1949) Brain stem reticular formation and activation of the EEG. *Electroencephalography and Clinical Neurophysiology* 1: 455–473.

Peyron C, Faraco J, Rogers W, et al. (2000) A mutation in a case of early onset narcolepsy and a generalized absence of hypocretin peptides in human narcoleptic brains. *Nature Medicine* 6: 991–997.

Sadoul P and Lugaresi E (eds.) (1972) *Hypersomnia with Periodic Breathing: A Symposium, Bulletin de Physiopathologie Respiratoire* 8: 967–1200.

Schenck CH, Bundlie SR, Ettinger MG, and Mahowald MW (1986) Chronic behavioral disorders of human REM sleep: A new category of parasomnia. *Sleep* 9: 293–308.

Sullivan CE, Issa FG, Berthon-Jones M, and Eves L (1981) Reversal of obstructive sleep apnoea by continuous positive airway pressure applied through the nares. *Lancet* 1: 862–865.

Von Economo C (1930) Sleep as a problem of localization. *Journal of Nervous and Mental Disease* 71: 249–259.

DEFINITIONS AND
DESCRIPTIONS OF SLEEP

Sleep Architecture

E F Pace-Schott, Harvard Medical School, Boston, MA, USA

Human Sleep

Human sleep is a behavioral state that alternates with waking, relative to which it is characterized by a heightened threshold to sensory input, attenuation of motor output, characteristic changes in central and peripheral physiology, and diminished conscious awareness. It is during this state of altered consciousness that dreaming occurs. Two major types of sleep, rapid eye movement (REM) and non-rapid eye movement (NREM), are characteristic of all mammals and alternate with regular, species-specific periods, indicating that they are timed by a reliable oscillator. In adult humans, this period is 90–120 min, resulting in approximately four or five cycles per night, as shown in **Figure 1**.

Polysomnographic Recording of Sleep

Since the late 1960s, sleep stages have been visually scored using the standardized electroencephalographic (EEG), electrooculographic (EOG), and electromyographic (EMG) criteria of the 1968 manual developed by Alan Rechtschaffen and Anthony Kales. EEG, EOG, and EMG are collectively termed polysomnography (PSG). **Figure 2** shows the sites of placement of electrodes for recording these three measures, which constitute oscillating field potentials from the underlying cortex (EEG), relative distance of the electronegative retina from the electrode (EOG), and muscle tone (EMG). In recent years, these electrophysiological descriptors of sleep stage have been expanded to include results from functional neuroimaging (e.g., functional magnetic resonance imaging), quantitative EEG (e.g., spectral analysis), and magnetoencephalography (MEG). Although these new technologies are rapidly expanding our understanding of sleep neurophysiology, the PSG criteria of Rechtschaffen and Kales remain the gold standard by which the progression of sleep stages (or 'sleep architecture') is routinely characterized in clinical and research sleep labs. (Note, however, that the American Academy of Sleep Medicine published a substantive update of the Rechtschaffen and Kales manual in 2007.) The following sections first characterize sleep architecture in terms of the traditional PSG measures and then describe cellular neurophysiology relevant to the control of the sleep–wake

and REM–NREM cycles and, lastly, findings on regional brain activation in sleep.

Sleep Architecture

Before commencing with a description of sleep architecture, it is important to define three terms used to characterize the EEG: frequency, amplitude, and morphology. EEG measures changes in electrical potential on the scalp that, in turn, reflect fluctuation of membrane potentials integrated across large numbers of cortical neurons beneath the recording electrodes. Such fluctuations or 'brain waves' are rhythmic, and their features reflect varying degrees of synchrony in membrane potential changes among neurons in these large aggregates. Frequency refers to the rapidity of fluctuation in electrical potential between successive peaks (or successive troughs) of measured EEG brain waves and is measured in cycles per second or Hertz (Hz). Amplitude refers to the magnitude of electrical potential change between such waves' peaks and troughs and is measured in (micro) volts. Morphology refers to the shape of a particular EEG wave or group of waves, reflecting both their frequency and amplitude, which can assume characteristic forms such as the sleep spindle or K-complex described later.

The initiation of NREM sleep is gradual and is characterized by a slowing in frequency of brain waves in the EEG from predominance of beta (14–25 Hz) and gamma (25–80 Hz) frequencies seen with eyes open (**Figure 3(a)**) to alpha (8–13 Hz) with eyes closed (**Figure 3(b)**) and finally to a predominance of theta rhythm (4–7 Hz) in stage 1 NREM sleep (**Figure 3(c)**). The subsequent increasing depth of NREM sleep through stages 2–4 can be generally characterized as a progressive decrease in the predominant EEG frequency along with a progressive increase in average wave amplitude (**Figures 3(d), (e) and (g)**). In stage 1 NREM, characteristic slow eye movements are recorded in the EOG and muscle tone (EMG) is diminished relative to waking (**Figure 4**). Stage 1 is succeeded by stage 2 sleep, with an EEG characterized by further decreased mean frequency and increased amplitude of brain waves (**Figure 3(d)**). Stage 2 EEG is also characterized by two distinctive wave forms: sleep spindles, which are intermittent clusters of spikes at sigma frequency (12–14 Hz) with characteristic waxing and waning morphology, and the K-complex, which is a large negative (upward) deflection followed by a positive (downward) deflection (**Figure 3(d)**). In stage 3 sleep, sleep spindles decrease, EEG frequency further slows, and the amplitude of brain waves increases.

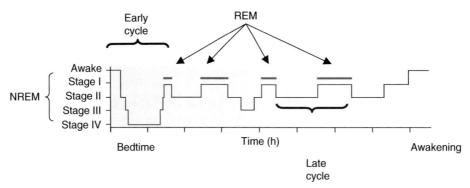

Figure 1 A night of sleep shown in the form of a 'hypnogram' in which the vertical axis shows depth of NREM sleep, red bars show REM, and the 90–120 min periodicity of each sleep cycle can be estimated as the duration from the end of one REM period to the end of the next one. Note that early in the night NREM is deeper and occupies more of the sleep cycle, whereas toward the end of the night, NREM sleep is lighter and more time is spent in REM. Adapted from Pace-Schott EF and Hobson JA (2002) The neurobiology of sleep: Genetic mechanisms, cellular neurophysiology and subcortical networks. *Nature Reviews Neuroscience* 3: 591–605, with permission.

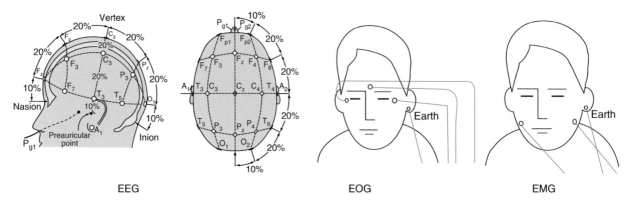

Figure 2 Placement of electrodes for polysomnography (PSG) that includes electroencephalography (EEG), electrooculography (EOG), and electromyography (EMG).

Very high-amplitude delta waves (1–3 Hz) occur in the deepest NREM sleep, stages 3 and 4, which are also known as slow-wave sleep (SWS) or delta sleep. A slower oscillatory rhythm of less than 1 Hz, the 'slow oscillation,' has been described in cats and humans. In healthy young adults between ages 20 and 40 years, stage 1 NREM occupies approximately 2–5% of total sleep, stage 2 occupies 45–55%, stage 3 occupies 3–8%, and stage 4 occupies 10–15%.

REM sleep, discovered in 1953 by Eugene Aserinsky and Nathaniel Kleitman, has also been termed 'paradoxical' sleep because, by contrast with NREM, frequencies in the EEG resemble those of waking but muscle tone reaches its minimum level (**Figure 3(f)**). This EMG minimum in REM results from central, brain-stem-mediated inhibition of spinal alpha motor neurons ('REM atonia'). Because of its wake-like EEG (**Figures 3(f)** and **3(g)**), REM has also been termed 'active' sleep. Also, because of its low-amplitude, high mixed-frequency waves, it has been termed 'desynchronized' sleep, in contrast with

NREM (especially SWS), in which high-amplitude waves reflect synchronous activity of cortical neurons. As its name suggests, saccadic rapid eye movements (REMs) occur during REM sleep (**Figure 4**). These REMs, which occur in singlets and clusters, reflect the fact that unlike skeletal muscles, oculomotor muscles are not atonic in REM. REMs are believed to reflect underlying, central phasic (brief) neural events that can also be measured as peripheral muscle twitches (that briefly overcome REM atonia), middle-ear muscle movements, and periorbital potentials. Such events, in humans, are likely related to ponto-geniculo-occipital waves – phasic potentials recorded sequentially in the pons, thalamic lateral geniculate body, and the occipital cortex that are a characteristic electrophysiological feature of REM in the cat. Unlike REM's phasic events, its tonic (persisting), wake-like EEG is believed to reflect cortical activation by the cholinergic ascending arousal systems described later. An additional EEG feature of REM sleep, the 'sawtooth wave,' is illustrated in **Figure 3(f)**.

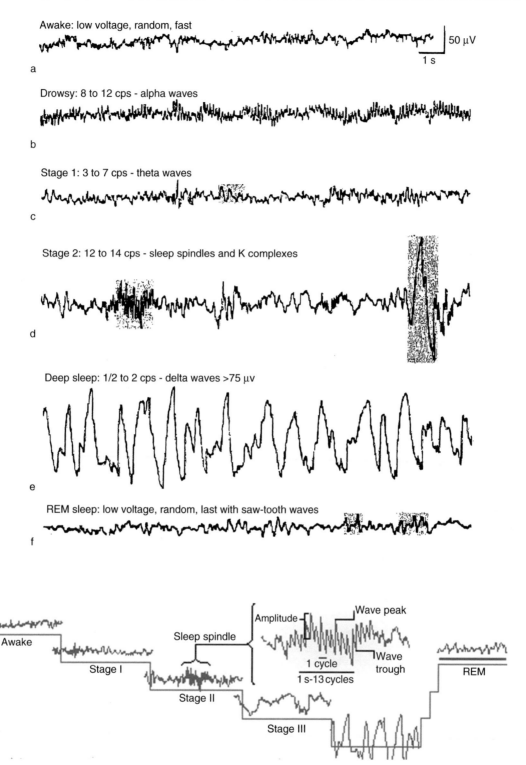

Figure 3 Depth of sleep is characterized by the frequency and amplitude of EEG waveforms with slower frequency and higher amplitude indicative of greater depth. The dominant EEG frequencies waking and sleep stages: (a) waking with eyes open (beta); (b) waking with eyes closed (alpha); (c) stage 1 NREM sleep (theta); (d) stage 2 NREM sleep (theta frequency with sleep spindle and K-complex wave forms); (e) stages 3 and 4 NREM sleep collectively termed 'slow-wave sleep' (delta); (f) REM sleep (wake-like EEG, characteristic saw-tooth waves); (g) a single sleep cycle from early in the night showing the complete series of stages from waking to NREM stages 1–4 and back through these stages to REM. (a–f) Hobson JA and Pace-Schott EF (2003) Sleep, dreaming, and wakefulness. In: Squire L, Bloom F, McConnell S, Roberts J, Spitzer N, and Zigmond M (eds.) *Fundamental Neuroscience*, 2nd edn., ch. 42. San Diego: Academic Press, with permission from Elsevier. (g) Pace-Schott EF and Hobson JA (2002) The neurobiology of sleep: Genetic mechanisms, cellular neurophysiology and subcortical networks. *Nature Reviews Neuroscience* 3: 591–605, with permission.

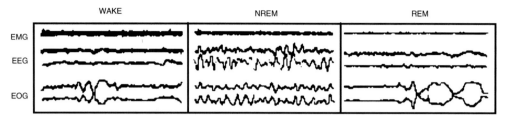

Figure 4 The combined EEG, EOG, and EMG of waking, NREM, and REM sleep. Note the decline in muscle tone from wake to NREM to REM, the similarity of waking and REM EEG, and the rapid eye movements in the EOG for which REM sleep is named. Reproduced from Hobson JA and Pace-Schott EF (2003) Sleep, dreaming, and wakefulness. In: Squire L, Bloom F, McConnell S, Roberts J, Spitzer N, and Zigmond M (eds.) *Fundamental Neuroscience*, 2nd edn., ch. 42. San Diego: Academic Press, with permission from Elsevier.

Autonomic and Peripheral Physiological Changes in Sleep

In NREM sleep relative to waking, heart rate, blood pressure, and respiration rate decline (**Figure 5**), sympathetic outflow of the autonomic nervous system decreases, and parasympathetic activity increases. Parasympathetic activity, as measured by heart rate variability indices, becomes closely linked to changes in the delta power band. In REM, heart rate, blood pressure, and respiration rate again rise but, as noted previously, muscle tone remains actively suppressed. In REM, homeostatic responses are attenuated. For example, respiration becomes insensitive to blood CO_2 and thermoregulation is largely absent. Additional peripheral features of REM include penile erections and clitoral engorgement. In the transition from NREM to REM sleep, sympathetic activity increases and parasympathetic activity decreases, but the parasympathetic:sympathetic ratio remains high as evidenced by pupillary miosis.

Neurophysiological Basis of Sleep Stage Changes

Control of sleep onset from the waking state is believed to be a function of the hypothalamus. Clifford Saper and colleagues have shown that there is a population of neurons in the anterior hypothalamus that, unlike much of the brain, shows increased metabolic activity during sleep (**Figure 6**). These neurons are concentrated in the ventrolateral preoptic area (VLPO) and produce both the inhibitory amino acid γ-aminobutyric acid (GABA) and the inhibitory neuropeptide galanin. At sleep onset, these neurons become active and inhibit ascending arousal systems of the brain stem, posterior hypothalamus, and basal forebrain (described later), thereby effecting the transition from the waking state to NREM sleep.

Once NREM is initiated, its delta and slow ($<1\,\text{Hz}$) oscillatory rhythms, as well as its spindle and K-complex wave forms are generated by reciprocally interconnected thalamic and cortical neural

circuits that have been extensively investigated and described by the late Mircea Steriade and colleagues. When the influence of subcortical ascending arousal systems decreases following sleep onset, thalamocortical circuits shift into an oscillatory mode during which transmission of sensory information to the cortex is blocked. Intrinsic oscillations of NREM represent the combined influence of two neuronal mechanisms: (1) the inhibitory influence of neurons of the thalamic reticular nucleus – a thin sheet of GABAergic neurons surrounding the thalamic periphery – on thalamic neurons projecting to the cortex, and (2) an oscillation within cortical neurons involving alternation between a prolonged hyperpolarization phase and a shorter depolarized phase with rapidly repeated spiking. This latter rhythm, generated within cortical networks, is the neuronal basis of the $<1\,\text{Hz}$ slow oscillation in the EEG and has been shown in animal models to exert an organizing or grouping influence on other intrinsic NREM sleep rhythms such as sleep spindles and the K-complex.

During waking, NREM oscillatory rhythms are prevented due to neuromodulation of thalamocortical circuits and cortex by ascending projections >from arousal-promoting systems of the brain stem, hypothalamus, and basal forebrain (**Figure 6**). Such neuromodulators include (1) acetylcholine from the midbrain–pons (mesopontine) junction in the brain stem, as well as from the basal forebrain (e.g., nucleus basalis); (2) the monoamines including norepinephrine (NE) and serotonin (5-HT) from the locus coeruleus and dorsal raphe, respectively, in the mesopontine brain stem dopamine from the midbrain periaqueductal gray, and histamine from the posterior hypothalamus; and (3) the neuropeptide orexin from the lateral hypothalamus. In contrast, during REM sleep, the intrinsic rhythms of NREM are prevented mainly by influence on the thalamus and basal forebrain from the cholinergic (acetylcholine) subcomponent of these systems originating in the mesopontine brain stem, including the laterodorsal tegmental and pedunculopontine nuclei.

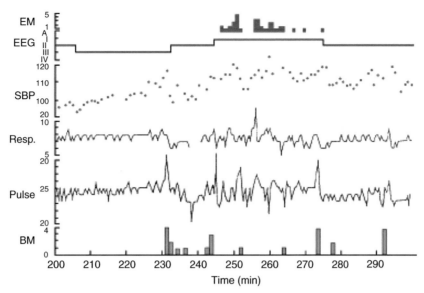

Figure 5 Peripheral physiological changes accompanying one sleep cycle. The EM trace shows the number of rapid eye movements per minute indicating the occurrence of REM sleep. The EEG trace is a hypnogram as in **Figure 1**. The SBP trace is systolic blood pressure, the Resp trace is respiration rate, the Pulse trace is heart rate, and the BM trace is the number of body movements per minute (movements increase at the beginning and end of REM). Reproduced from Hobson JA and Pace-Schott EF (2003) Sleep, dreaming, and wakefulness. In: Squire L, Bloom F, McConnell S, Roberts J, Spitzer N, and Zigmond M (eds.) *Fundamental Neuroscience*, 2nd edn. San Diego: Academic Press, with permission from Elsevier.

Homeostatic and Circadian Control of Sleep Onset

A two-process model of sleep propensity, proposed by Alexander Borbely, suggests that homeostatic mechanisms (process S) interact with circadian factors (process C) to bring about sleep–wake transitions. Process S refers to a drive to sleep that increases as a function of the duration spent awake, whereas process C refers to sleep propensity varying as a function of time of day. The physiological basis of process S is believed to be substances that build up in the brain as a function of time awake ('endogenous somnogens'), and the purine nucleoside adenosine is currently the most widely suggested candidate for such a substance, although others have also been proposed (e.g., interleukin-1, prostaglandin D1, and growth hormone releasing hormone). Robert McCarley and colleagues have suggested that the specific site at which adenosine acts to promote sleep is in the basal forebrain, although its somnogenic effects have been reported at other subcortical sites as well.

Sleep propensity due to process C is an example of a wide variety of physiological circadian rhythms whose 'master clock' resides in the suprachiasmatic nucleus (SCN) of the anterior hypothalamus. Cells of the SCN contain molecular clocks that maintain a near-perfect 24-h periodicity via interlocking positive and negative feedback control of transcription and translation of circadian genes. These SCN cells can become entrained to the ambient photoperiod by

light impinging on retinal ganglion cells that contain photoreceptive pigments such as melanopsin and which communicate monosynaptically with the SCN via the retinohypothalamic tract. Clifford Saper and colleagues have shown that the SCN, in turn, projects to various hypothalamic nuclei with circuits through subparaventricular and dorsomedial nuclei able to influence sleep–wake transitions via projections to the VLPO. Secretion of the pineal sleep hormone melatonin is also controlled by the SCN via a multisynaptic pathway and provides an internal signal of circadian time, as well as feedback to the SCN via its melatonin receptors.

Recent Discoveries Regarding the Brain in Sleep

The advent of new technologies in brain imaging has resulted in a greatly increased understanding of the behavioral and electroencephalographic features of human sleep. Positron emission tomography (PET) studies indicate that global cerebral oxygen utilization (detected using $H_2^{15}O$) and glucose metabolism (detected using ^{18}fluorodeoxyglucose) is decreased during NREM sleep compared to waking. Relative to waking, glucose metabolism is decreased by approximately 11% in stage 2 NREM and approximately 40% in SWS. fMRI studies have shown similar declines with deepening NREM sleep in the cerebral blood oxygen level-dependent signal that is measured using this technique. In contrast, during

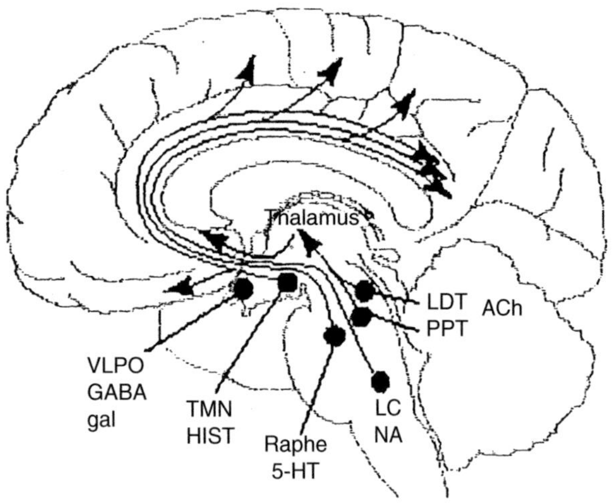

Figure 6 Major components of sleep inducing and wake-inducing neuromodulatory systems. To induce sleep, projections from the gamma-aminobutyric acid (GABA) and galanin (gal) producing cells of the ventrolateral preoptic nucleus (VLPO) of the anterior hypothalamus send inhibitory signals to each of the ascending arousal systems of the brain stem, basal forebrain, and hypothalamus. These arousal systems include: the tuberomammillary nucleus (TMN) of the posterior hypothalamus that produce histamine (HIST), serotonin (5-HT) producing cells of the dorsal raphe (Raphe), acetylcholine (ACh) producing cells of the laterodorsal tegmental (LDT) and pedunculopontine tegmental (PPT) nuclei and norepinephrine (NA) producing cells of the locus coeruleus (LC). Other ascending arousal systems not shown include the perifornical lateral hypothalamus that produces orexin, dopamine producing cells of the midbrain periaqueductal gray, and cholinergic projections from the basal forebrain (nucleus basalis, diagonal band of Broca, medial septum) to the limbic system and the cortex. Reproduced from Saper CB, Chou TC, and Scammell TE (2001) The sleep switch: Hypothalamic control of sleep and wakefulness. *Trends in Neurosciences* 24: 726–731, with permission from Elsevier.

REM, global cerebral energy metabolism increases so that it is equal to or even greater than that of waking.

Deactivation of frontal cortices is one of the first physiological signs of human sleep. This has been observed whether activation is measured by EEG, MEG, PET, or fMRI. Functional imaging studies have also shown deactivation of the thalamus to be an early event in sleep onset. As NREM sleep deepens, more posterior cortices also decrease their neuronal activity, as do subcortical areas including the hypothalamus, basal ganglia, and hippocampus. In addition to their earlier deactivation in NREM, frontal areas may also display a greater need for recovery sleep than more posterior regions. Alexander Borbely and colleagues have shown that slow-wave activity (SWA), measured as the spectral power in delta frequencies, can serve as a marker of homeostatic sleep pressure. SWA is inversely related to cortical activation, is greater in frontal than in parietal and occipital

regions in the first NREM episode of the night, and spreads in a frontal to posterior direction within the cortex. In addition to being the first part of the cortex to 'go to sleep,' frontal areas also appear to be the last part to wake up because their reactivation lags behind that of the rest of the brain.

After the transition from NREM to REM, much of the lateral frontal cortex remains less active than during waking. However, PET studies have shown that blood flow and glucose metabolism increase during REM in subcortical brain regions such as the pons, midbrain, thalamus, amygdala, hypothalamus, and basal ganglia, as well as limbic-related cortices such as parahippocampal cortex, temporal pole, anterior insula, caudal orbitofrontal cortex, subcallosal anterior cingulate, and medial prefrontal cortex (e.g., Brodmann areas 24, 32, and 10). Eric Nofzinger and colleagues have termed these midline structures an 'anterior paralimbic REM activation area' and have shown that their reactivation during REM following low activity in NREM can be as great or greater than their activation during waking.

Summary

Sleep, consisting of alternating REM and NREM states, is a universal behavior of all mammals. In humans, this cycle has a periodicity of 90–120 min and consists of characteristic substages of NREM characterized by brain waves of increasing amplitude and decreasing frequency as NREM deepens. The oscillating EEG patterns of NREM reflect synchronous activity in thalamocortical circuits. REM sleep is characterized by a unique, paradoxical combination of a fast, wake-like EEG but the least muscle tone of any stage. Characteristic peripheral and autonomic changes accompany stages of the sleep cycle with higher, more variable respiration, pulse, and blood pressure during REM than during NREM. Sleep propensity is determined by the interaction of a circadian rhythm ('process C') that is controlled, like other such rhythms, by the biological clock in the suprachiasmatic nucleus and a sleep homeostatic 'process S' that increasingly drives the need for sleep the longer one is awake. Waking is maintained by the ascending arousal systems of the brain stem, hypothalamus, and basal forebrain that send projections containing wake-promoting neuromodulators such as serotonin, norepinephrine, acetylcholine, histamine, orexin, and dopamine to the thalamus, limbic system, and cerebral cortex. During sleep, inhibitory projections from sleep-active sites in the anterior hypothalamus suppress these same arousal systems. In REM, mainly one component of these wake-active systems, that of acetylcholine, activates the forebrain. Functional neuroimaging studies have revealed that NREM sleep is associated with widespread cortical deactivation relative to waking that is particularly pronounced in frontal areas. During REM, midline limbic and paralimbic areas are selectively activated while much of the lateral cortex remains in the less activated state characteristic of NREM.

See also: Circadian Rhythms in Sleepiness, Alertness, and Performance; Gene Expression; Hippocampal–Neocortical Dialog; History of Sleep Research; Network Reactivation; PET Activation Patterns; Sleep: Development and Circadian Control; Sleep Oscillations and PGO Waves.

Further Reading

Borbely AA and Achermann P (2005) Sleep homeostasis and models of sleep regulation. In: Kryger MH, Roth T, and Dement WC (eds.) *Principles and Practice of Sleep Medicine,* 4th edn., pp. 405–417. Philadelphia: Saunders.

Braun AR, Balkin TJ, Wesensten NJ, et al. (1997) Regional cerebral blood flow throughout the sleep–wake cycle. *Brain* 120: 1173–1197.

Carskadon MA and Dement WC (2005) Normal human sleep: An overview. In: Kryger MH, Roth T, and Dement WC (eds.) *Principles and Practice of Sleep Medicine,* 4th edn., pp. 13–23. Philadelphia: Saunders.

Hobson JA (1989) *Sleep.* New York: Scientific American Library.

Hobson JA, Pace-Schott EF, and Stickgold R (2000) Dreaming and the brain: Toward a cognitive neuroscience of conscious states. *Behavioral and Brain Sciences* 23: 793–842.

Kaufmann C, Wehrle R, Wetter TC, Holsboer F, Auer DP, and Pollmacher T (2006) Brain activation and hypothalamic functional connectivity during human non-rapid eye movement sleep: An EEG/fMRI study. *Brain* 129: 655–667.

Kryger MH, Roth T, and Dement WC (eds.) (2005) *Principles and Practice of Sleep Medicine,* 4th edn. Philadelphia: Saunders.

Lydic R and Baghdoyan HA (eds.) (1999) *Handbook of Behavioral State Control: Molecular and Cellular Mechanisms,* Boca Raton, FL: CRC Press.

Mallick BN and Inoue S (eds.) (1999) *Rapid Eye Movement Sleep.* New York: Dekker.

Maquet P (2000) Functional neuroimaging of normal human sleep by positron emission tomography. *Journal of Sleep Research* 9: 207–231.

Maquet P, Ruby P, Maudoux A, et al. (2005) Human cognition during REM sleep and the activity profile within frontal and parietal cortices: A reappraisal of functional neuroimaging data. *Progress in Brain Research* 150: 219–227.

Nofzinger EA (2005) Functional neuroimaging of sleep. *Seminars in Neurology* 25: 9–18.

Ohayon MM, Carskadon MA, Guilleminault C, and Vitiello MV (2004) Meta-analysis of quantitative sleep parameters from childhood to old age in healthy individuals: Developing normative sleep values across the human lifespan. *Sleep* 27: 1255–1273.

Opp MR, Kilduff TS, Marcus CL, Armitage R, and Nofzinger EA (2005) *SRS Basics of Sleep Guide.* Westchester, IL: Sleep Research Society.

Pace-Schott EF and Hobson JA (2002) The neurobiology of sleep: Genetic mechanisms, cellular neurophysiology, and subcortical networks. *Nature Reviews Neuroscience* 3: 591–605.

Saper CB, Scammell TE, and Lu J (2005) Hypothalamic regulation of sleep and circadian rhythms. *Nature* 437: 1257–1263.

Steriade M (2006) Grouping of brain rhythms in corticothalamic systems. *Neuroscience* 137: 1087–1106.

Behavior and Parasomnias (RSBD)

M W Mahowald, M A Cramer-Bornemann, and C H Schenck, Hennepin County Medical Center, Minneapolis, MN, USA; and University of Minnesota Medical School, Minneapolis, MN, USA

Introduction

There are many different parasomnias. They may be divided into two broad categories: primary, which represent abnormalities of the sleep states *per se*, and secondary due to activity of other organ systems, which occur primarily or predominately during sleep. The primary parasomnias can be divided with regard to the parent state: nonrapid eye movement (NREM) sleep, rapid eye movement (REM) sleep, or a miscellaneous category not respecting sleep states. The secondary parasomnias can be categorized by the organ system involved. The most common primary parasomnias are the disorders of arousal, REM sleep behavior disorder, and rhythmic movement disorder. Common secondary parasomnias are nocturnal seizures and, less common, sleep-related expiratory groaning.

Primary Sleep Parasomnias

The concept that sleep and wakefulness are not invariably mutually exclusive states, and that the various state-determining variables of wakefulness, NREM sleep, and REM sleep may occur simultaneously or oscillate rapidly, is key to the understanding of the primary sleep parasomnias. The admixture of wakefulness and NREM sleep would explain confusional arousals (sleep drunkenness), automatic behavior, or microsleeps. The tonic and phasic components of REM sleep may become dissociated, intruding or persisting into wakefulness, thus explaining cataplexy, wakeful dreaming, lucid dreaming, and the persistence of motor activity during REM sleep (REM sleep behavior disorder).

Parasomnias are most apt to appear as the brain becomes reorganized across states; therefore, they often occur during the transition periods from one state to another. In view of the large number of neural networks, neurotransmitters, and other state-determining substances which must be recruited synchronously for full state declaration and the frequent transitions among states during the wake–sleep cycle, it is surprising that errors in state declaration do not occur more frequently.

In addition to the phenomenon of state dissociation, in which two states of being overlap or occur simultaneously, there are likely additional underlying physiologic phenomena which contribute to the appearance of complex motor behaviors during sleep. These include activation of locomotor centers during sleep, sleep inertia upon arousal, and sleep state instability.

NREM Parasomnias

Disorders of Arousal

The disorders of arousal are the most impressive and most common of the NREM sleep parasomnias. These share common features: They tend to arise from the deepest stages of NREM sleep (slow-wave sleep (stages 3 and 4 of NREM sleep)), thus usually occurring in the first third of the sleep cycle (and rarely during naps); and they are common in childhood, usually decreasing in frequency with increasing age. Disorders of arousal may be primed by febrile illness, alcohol, prior sleep deprivation, physical activity, emotional stress, or medications. Such factors should be considered as priming events in susceptible individuals, and not causal. Numerous other sleep disorders which result in arousals (obstructive sleep apnea, nocturnal seizures, or periodic limb movements) may also trigger disorders of arousal.

Persistence of these behaviors beyond childhood or their development in adulthood have erroneously been taken as an indication of underlying psychopathology. Numerous studies have dispelled this myth, indicating that significant psychopathology is usually not present in adults with disorders of arousal.

Disorders of arousal occur on a broad spectrum ranging from confusional arousals, through somnambulism (sleepwalking), to sleep terrors. Some take the form of 'specialized' behaviors, such as sleep-related eating and sleep-related sexual activity– without conscious awareness.

Confusional Arousals

These are often seen in children and are characterized by movements in bed, occasionally thrashing about, or inconsolable crying. 'Sleep drunkenness' is probably a variation on this theme. The prevalence of confusional arousals in adults is approximately 4%.

Sleepwalking

Sleepwalking is prevalent in childhood (1–17%), peaking at 11 or 12 years of age, and is far more common in adults (nearly 4%) than generally acknowledged. Sleepwalking may be either calm or agitated, with varying degrees of complexity and duration.

Sleep Terrors

The sleep terror is the most dramatic disorder of arousal. It is frequently initiated by a loud, blood-curdling scream associated with extreme panic, followed by prominent motor activity such as hitting the wall and running around or out of the bedroom, or even out of the house, resulting in bodily injury or property damage. A universal feature is inconsolability, and attempts at consolation are fruitless and may serve only to prolong or even intensify the confusional state. Complete amnesia for the activity is typical but may be incomplete. Sleep terrors are much more prevalent in adults than generally acknowledged (4% or 5%). Although usually benign, these behaviors may be violent, resulting in considerable injury to the victim or others or damage to the environment, occasionally with forensic implications.

Treatment of disorders of arousal is often not necessary. Reassurance of their typically benign nature, lack of psychological significance, and the tendency to diminish over time is often sufficient. The tricyclic antidepressants and benzodiazepines may be effective, and they should be administered if the behaviors are dangerous to person or property or extremely disruptive to family members. Nonpharmacologic treatment such as hypnosis is recommended for long-term management. Anticipatory awakening has been reported to be effective in treating sleepwalking in children. The avoidance of priming factors such as drugs, alcohol, and sleep deprivation is also important.

REM Sleep Parasomnias

REM Sleep Behavior Disorder

The most common REM sleep parasomnia is the REM sleep behavior disorder (RBD). One of the defining features of REM sleep is active paralysis of all somatic musculature (with the exception of the diaphragm to permit respiration during REM sleep). In experiments reported in 1965, bilateral lesions of pontine regions adjacent to the locus coeruleus in cats caused the absence of the expected atonia associated with REM sleep, allowing the cats to demonstrate prominent motor behaviors during REM sleep (oneiric behaviors). Patients with RBD lack the paralysis of REM sleep, permitting the acting out of dream mentation, often with violent or injurious results. The prevalence of RBD is approximately 0.5%.

The presenting complaint is that of vigorous sleep behaviors usually accompanying vivid striking dreams. These behaviors may result in repeated injury, including ecchymoses, lacerations, and fractures. The potential for injury to self or bed partner raises interesting and difficult forensic medicine issues. Two striking demographic characteristics of RBD are that it predominately affects males (approximately 90%) and usually begins after the age of 50 years. RBD in humans occurs in both an acute and a chronic form.

Until recently, most reported cases of acute transient RBD fell in the toxic/metabolic category, with the best-studied conditions being the withdrawal states – most commonly involving ethanol. A growing cause of acute RBD appears to be iatrogenic, particularly associated with tricyclic antidepressants and serotonin-specific and noradrenergic-specific reuptake inhibitors – both widely prescribed categories of antidepressants.

The chronic form is most often either idiopathic or associated with neurological disorders. Each basic category of neurological disease (vascular, neoplastic, toxic/metabolic, infectious, degenerative, traumatic, congenital, and idiopathic) could be expected to manifest this disorder. Careful longitudinal evaluation has demonstrated that the majority of patients with what initially appears to be 'idiopathic' RBD will eventually develop neurodegenerative disorders, most notably the synucleinopathies (Parkinson's disease, multiple system atrophy, including olivoponto-cerebellar degeneration and the Shy–Drager syndrome, or dementia with Lewy body disease). RBD may be the first manifestation of these conditions, and it may precede any other manifestation of the underlying neurodegenerative process by more than 10 years. With such a high percentage of patients with RBD eventually developing other manifestations of neurodegenerative diseases, it has been proposed that the term idiopathic RBD be replaced by 'cryptogenic' RBD – the implication being that if followed long enough, the underlying neurologic condition will declare itself.

Systematic longitudinal study of patients with such neurological syndromes indicates that RBD and REM sleep without atonia may be far more prevalent than previously suspected. Although the exact prevalence of RBD in extrapyramidal disease is unknown, subjective reports indicate that 25% of patients with Parkinson's disease have behaviors suggestive of RBD or sleep-related injurious behaviors, and polysomnographic studies found RBD in up to 47% of patients with Parkinson's disease with sleep complaints. In one large series of patients with multiple system atrophy, 90% were found to have REM sleep without atonia and 69% had clinical RBD.

There is a higher incidence of RBD in patients with narcolepsy. This association could be expected, because both narcolepsy and RBD are characterized by 'state boundary control' abnormalities. Furthermore, tricyclic antidepressants, SSRIs, and monoamine oxidase inhibitors, often prescribed to treat cataplexy associated with narcolepsy, can trigger or exacerbate RBD in this population.

Neuroimaging studies indicate dopaminergic abnormalities in RBD. Single-photon emission computed tomography (SPECT) studies have found reduced striatal dopamine transporters, and decreased striatal dopaminergic innervation has been reported. Decreased blood flow in the upper portion of the frontal lobe and pons has been reported, as has functional impairment of brain stem neurons. Positron emission tomography and SPECT studies have revealed decreased nigrostriatal dopaminergic projections in patients with multiple system atrophy and RBD. Decreased blood flow in the upper portion of the frontal lobe and pons has been found in one magnetic resonance imaging and SPECT study. Whole brain perfusion changes in patients with RBD are similar to those seen in Parkinson's disease. Impaired cortical activation as determined by electroencephalographic spectral analysis in patients with idiopathic RBD supports the relationship between RBD and neurodegenerative disorders.

There are a number of fascinating variations on RBD (beyond the scope of this article), all likely representing abnormalities of state boundary control. These include status dissociatus, the parasomnia overlap syndrome, and agrypnia excitata.

Clonazepam 0.5–1.0 mg at bedtime is effective in approximately 90% of cases, with little evidence of tolerance or abuse. Melatonin, often at doses of 3–12 mg at night, may also be effective. Pramipexole or levodopa may be effective. The other essential therapeutic intervention concerns environmental safety.

Rhythmic Movement Disorder

Rhythmic movement disorder (RMD), formerly termed jactatio capitis nocturna, refers to a group of behaviors characterized by stereotyped movements (rhythmic oscillation of the head or limbs and head banging or body rocking during sleep) seen most frequently in childhood. Its persistence into adulthood is not uncommon. It may be familial, and it is not the manifestation of any underlying psychiatric or psychological condition. RMD may arise from either REM or NREM sleep and may occur in the transition from wake to sleep. Significant injury from repetitive pounding may result. The etiology of RMD is unknown, and no systematic studies of pharmacologic or behavioral treatment have been reported.

Numerous other conditions may perfectly mimic the primary sleep parasomnias. These include obstructive sleep apnea, nocturnal seizures, psychogenic dissociative disorders, or malingering.

Isolated, often bizarre, sleep-related events may be experienced by perfectly normal people, and most do not warrant further extensive or expensive evaluation. The initial approach to the complaint of unusual sleep-related behavior is to determine whether further evaluation is necessary. The patient should be queried regarding the exact nature of the events. Because many of these episodes may be associated with partial or complete amnesia, additional descriptive information from a bed partner or other observer may prove invaluable. Home videotapes of the clinical event may be quite helpful. In general, indications for formal evaluation of parasomnias include behaviors that are potentially violent or injurious, are extremely disruptive to other household members, result in the complaint of excessive daytime sleepiness, or are associated with medical, psychiatric, or neurological symptoms or findings.

Secondary Sleep Parasomnias

Nocturnal Seizures

Sleep and epilepsy are common bedfellows. In most cases, epilepsy is highly state dependent: NREM sleep promotes seizures, whereas REM sleep is a relatively antiepileptic state. This state-dependent vulnerability to seizures reflects the dramatic reorganization of the entire central nervous system as it moves across the three states of being: wakefulness, NREM sleep, and REM sleep. There is extensive overlap and potential for confusion between other parasomnias and nocturnal seizures, particularly nocturnal frontal lobe epilepsy.

Sleep-Related Expiratory Groaning (Catathrenia)

Groaning during sleep has been termed 'catathrenia.' The behaviors occur intermittently during either REM or NREM sleep and are characterized by prolonged, often very loud, socially disruptive groaning sounds during expiration. It is poorly understood and awaits further definition and therapeutic studies. There is no known effective treatment.

Conclusion

Evaluation of problematic behaviors arising from the sleep period is best performed by experienced, multidisciplinary teams of sleep medicine professionals in centers with extensive polysomnographic monitoring capability.

See also: Cataplexy; Narcolepsy; PET Activation Patterns; Sleep Apnea; Sleep Architecture; Sleep Oscillations and PGO Waves; Sleeping Sickness.

Further Reading

Albin RL, Koeppe RA, Chervin RD, et al. (2000) Decreased striatal dopaminergic innervation in REM sleep behavior disorder. *Neurology* 55: 1410–1412.

Cramer Bornemann MA, Mahowald MW, and Schenck CH (2006) Parasomnias. Clinical features and forensic implications. *Chest* 130: 605–610.

Derry CP, Davey M, Johns M, et al. (2006) Distinguishing sleep disorders from seizures. Diagnosing bumps in the night. *Archives of Neurology* 63: 705–709.

Gagnon J-F, Postuma RB, Mazza S, et al. (2006) Rapid-eye-movement sleep behaviour disorder and neurodegenerative diseases. *Lancet Neurology* 5: 424–432.

Gagnon J-F, Postuma RB, and Montplaisir J (2006) Update on the pharmacology of REM sleep behavior disorder. *Neurology* 67: 742–747.

Hurwitz TD, Mahowald MW, Schenck CH, et al. (1991) A retrospective outcome study and review of hypnosis as treatment of adults with sleepwalking and sleep terror. *Journal of Nervous and Mental Disease* 179: 228–233.

Jouvet M and Delorme F (1965) Locus coeruleus et sommeil paradoxal. *Comptes Rendus Des Seances de la Societe de Biologie et de ses Filiales* 159: 895–899.

Kohyama J, Masukura F, Kimura K, and Tachibana N (2002) Rhythmic movement disorder: Polysomnographic study and summary of reported cases. *Brain and Development* 24: 33–38.

Maaza S, Soucy JP, Gravel P, et al. (2006) Assessing whole brain perfusion changes in patients with REM sleep behavior disorder. *Neurology* 67: 1618–1622.

Mahowald MW (2004) Parasomnias. *Medical Clinics of North America* 88: 669–678.

Mahowald MW and Schenck CH (2000) Parasomnias: Sleepwalking and the law. *Sleep Medicine Reviews* 4: 321–339.

Mahowald MW and Schenck CH (2001) Evolving concepts of human state dissociation. *Archives Italiennes de Biologie* 139: 269–300.

Mahowald MW and Schenck CH (2005) NREM sleep parasomnias. *Neurologic Clinics* 23: 1077–1106.

Nightingale S, Orgill JC, Ebrahim IO, et al. (2005) The association between narcolepsy and REM behavior disorder (RBD). *Sleep Medicine* 6: 253–258.

Ohayon M, Guilleminault C, and Priest RG (1999) Night terrors, sleepwalking, and confusional arousal in the general population: Their frequency and relationship to other sleep and mental disorders. *Journal of Clinical Psychiatry* 60: 268–276.

Ohayon MM, Caulet M, and Priest RG (1997) Violent behavior during sleep. *Journal of Clinical Psychiatry* 58: 369–376.

Pressman MR (2007) Factors that predispose, prime, and precipitate NREM parasomnias in adults: Clinical and forensic implications. *Sleep Medicine Reviews* 11: 5–30.

Schenck CH and Mahowald MW (1990) Polysomnographic, neurologic, psychiatric, and clinical outcome report on 70 consecutive cases with REM sleep behavior disorder (RBD): Sustained clonazepam efficacy in 89.5% of 57 treated patients. *Cleveland Clinic Journal of Medicine* 57(supplement): S9–S23.

Schenck CH and Mahowald MW (1994) Review of nocturnal sleep-related eating disorders. *International Journal of Eating Disorders* 15: 343–356.

Schenck CH and Mahowald MW (2005) REM sleep parasomnias. *Neurologic Clinics* 23: 1107–1126.

Shapiro CM, Trajanovic NN, and Fedoroff JP (2003) Sexsomnia – A new parasomnia? *Canadian Journal of Psychiatry* 48: 311–317.

Shouse MN, da Silva AM, and Sammaritano M (1996) Circadian rhythm, sleep, and epilepsy. *Journal of Clinical Neurophysiology* 13: 32–50.

Tobin JD Jr. (1993) Treatment of somnambulism with anticipatory awakening. *Journal of Pediatrics* 122(3): 426–427.

Vetrugno R, Provini F, Plazzi G, et al. (2001) Catathrenia (nocturnal groaning): A new type of parasomnia. *Neurology* 56: 681–683.

Sleep Oscillations and PGO Waves

M Steriade, Université Laval, Laval, QC, Canada

Behavioral and Electrographic Signs of Sleep Stages

Sleep Onset

The popular view that behavioral quiescence is the sign that heralds sleep may be valid for the full-blown state of sleep but not for the preparatory period during which many animal species display complex motor behaviors directed to find a safe home for sleep. The defining signs of the period when one falls asleep are changes in brain electrical activity (electroencephalogram (EEG)) that are associated with long periods of inhibition in neurons located in the thalamus, a deep structure in the brain and a gateway for most sensory signals in their route to the cerebral cortex. The consequence is that the incoming messages are blocked in the thalamus and the cerebral cortex is deprived of information from the outside world.

In contrast with the rapid awakening from sleep, the period of falling asleep is quite long. Thus, sleep onset may not be exclusively attributable to neuronal mechanisms, which operate on short timescales. It is conceivable that the mechanisms of falling sleep will only be revealed with concerted studies on sleep humoral factors (acting on longer timescale) and on their actions on neurons in critical brain areas. Currently, little is known about the effects of sleep-promoting chemical substances on neurons. The list of putative sleep humoral factors is quite long. Some of them are (1) adenosine, which exerts a tonic inhibitory control on both upper brain stem and basal forebrain cholinergic activating (awakening) neurons; (2) muramyl peptides, which have a chemical structure that is similar to that of serotonin, a neurotransmitter that was implicated in the generation of early sleep stages; and (3) prostaglandin D_2, whose receptors are found in the anterior hypothalamus, thought to be a critical zone for promoting sleep.

The concepts postulating that sleep is a passive phenomenon due to closure of cerebral gates leading to brain deafferentation or, alternatively, an active phenomenon promoted by inhibitory mechanisms arising in some cerebral areas have long been considered as opposing views. However, neurons with inhibitory influences (located in the anterior hypothalamus) act on neurons that exert excitatory influences on the brain (located in the posterior hypothalamus and upper brain stem reticular core), and the final outcome of these relations is the disconnection of the forebrain. Therefore, the two mechanisms (active and passive) of sleep onset are probably successive steps within a chain of events, and they are complementary rather than opposed.

Slow-Wave and Rapid Eye Movement Sleep

Sleep consists of an alternation between two distinct stages. The early stage is characterized by large-scale synchronization of low-frequency EEG activity (0.5–15 Hz), which is generated by the cerebral cortex and thalamus. Synchronization is a state in which two or more oscillators display the same frequency because of mutual influences. The notion of synchronization supposes the coactivation as well as concerted inhibitory processes in a large number of neurons. The summation of neuronal activities is sufficiently large to be recorded with gross electrodes, such as scalp electrodes when recording the human EEG. Because of the prevalent low-frequency waves, this stage is termed slow-wave sleep. This electrophysiological feature is associated with behavioral quiescence and suspension of conscious processes related to the external world because the transmission of signals is inhibited in the thalamus, even before these signals reach the cerebral cortex, the highest site of consciousness. However, not all mental processes are obliterated during slow-wave sleep because the brain oscillations that occur during this stage give rise to synaptic plasticity that underlies processes through which information is stored, such as consolidation of memory traces acquired during the state of wakefulness. Also, a peculiar form of dreaming mentation is present during slow-wave sleep.

The second stage of sleep is associated with rapid eye movements (REMs) and the sign that distinguishes it from the other two states of vigilance (slow-wave sleep and waking) is the largely suppressed motor output (muscular atonia) due to inhibition of spinal motor neurons. In addition, spiky potentials (ponto-geniculo-occipital (PGO) waves), related to ocular saccades, can be recorded from the thalamus and cortex and are regarded as the physiological correlate of dreaming. This state was also termed active because the level of brain alertness is similar to, or even higher than, the state of wakefulness. (This should not imply, however, that slow-wave sleep is inactive from the standpoint of brain functions since neuronal recordings have shown that cortical neurons fire during this stage at almost the same level as during waking.) Thus, REM sleep is a brain-active state associated with paralysis of limbs' muscles. In other words, motor commands from a highly excitable cerebral cortex cannot be executed because of the inhibition of motor neurons in the spinal cord. Although REM sleep seems very similar to wakefulness because of intense brain activity, some differences also exist between these two states, mainly due to more effective inhibitory processes in waking and the virtual silence of monoaminergic neurons. This might explain the psychological differences between the logical thought in waking and the bizarre, illogical, and emotional content of dreaming in REM sleep.

Sleep Oscillations: Sites of Generation and Cellular Mechanisms

Three Major Oscillations Define Slow-Wave Sleep

The brain rhythms that characterize slow-wave sleep are the slow oscillation (0.5–1 Hz) that appears from the onset of full-blown sleep; spindles (7–15 Hz), whose amplitudes are highest during stage 2 of slow-wave sleep; and delta waves (1–4 Hz), which appear in stages 3 and 4, when their great amplitudes may obscure spindles.

The slow oscillation was first described using intracellular recordings from different neuronal types in anesthetized animals and, in the same article, was also detected in EEG recordings during natural slow-wave sleep in humans in which cyclic groups of delta waves at 1–4 Hz recurred with a slow periodicity of 0.4 or 0.5 Hz. The distinctness between delta and slow oscillations also came from human studies showing that the typical decline in delta activity (1–4 Hz) from the first to the second sleep episode was not present at frequencies characteristic for the slow oscillation (range, 0.55–0.95 Hz). The cortical nature of the slow oscillation was demonstrated by its survival in the cerebral cortex after large thalamic lesions and its absence in the thalamus of decorticated animals. This rhythm was recorded in all major types of neocortical neurons, including pyramidal-shaped and local circuit inhibitory neurons. The slow oscillation is composed of a prolonged depolarizing ('up') phase followed by a long-lasting hyperpolarizing ('down') phase. The hyperpolarizations of cortical neurons occur from the onset of sleep and distinguish sleep from the steady depolarization seen during waking (**Figure 1**). The depolarizing phase consists of excitatory postsynaptic potentials (EPSPs), some intrinsic excitatory currents, and also inhibitory postsynaptic potentials (IPSPs) that reflect the action of synaptically coupled local

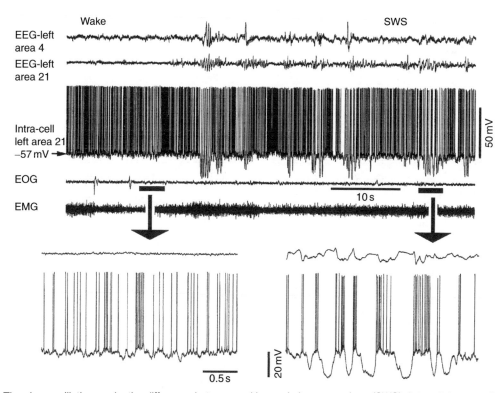

Figure 1 The slow oscillation marks the difference between waking and slow-wave sleep (SWS). Intracellular recording of regular-spiking neuron from left cortical area 21 of chronically implanted cat, together with depth EEG from left areas 4 and 21, electrooculogram (EOG), and electromyogram (EMG). Transition from waking to SWS. Two epochs, one in waking and the other in SWS, are marked by horizontal traces (below EOG) and expanded below (arrows). Note the occurrence of episodic cyclic hyperpolarizations of the slow oscillation since the very onset of SWS, as indicated by EEG.

circuit GABAergic neurons. The hyperpolarizing (silent) phase is not produced by inhibitory interneurons but is due to disfacilitation (removal of synaptic, mainly excitatory, inputs) in intracortical and thalamocortical networks and also to some K^+ currents.

Sleep spindles are waxing-and-waning waves, grouped in sequences that recur every 2–5 s. They are generated in the thalamus, even in the absence of the cerebral cortex, but corticothalamic neurons have a decisive role in the near simultaneity of spindles over widespread thalamic and cortical territories. The pacemaker of spindles is the thalamic reticular nucleus, which consists exclusively of inhibitory (GABAergic) neurons. In the absence of this nucleus, spindles are not seen in the remaining thalamus and cortex, but this oscillation is present in the deafferented reticular nucleus. The cellular mechanisms underlying spindles are as follows. At each spindle wave, thalamic reticular neurons fire spike bursts that impose IPSPs on their targets, thalamocortical neurons. Following IPSPs, thalamocortical neurons discharge postinhibitory rebound spike bursts that are transmitted to cortex, where they rhythmically excite cortical neurons and thus give rise to spindles. The rebound is an intrinsic property of thalamic neurons and depends on a Ca^{2+}-mediated conductance, which is inactive during waking and REM sleep when neurons are depolarized and is de-inactivated (uncovered) during slow-wave sleep when thalamocortical neurons are hyperpolarized.

Some delta waves arise in the cortex without thalamic participation. Another type of delta potentials is generated in thalamocortical neurons through interplay between two of their intrinsic properties.

Coalescence of low-frequency corticothalamic rhythms by the slow oscillation   Unlike rhythms that appear within distinct frequency bands and are generated in restricted neuronal circuits of simplified experimental preparations (e.g., isolated thalamic or cortical slices maintained *in vitro* and decorticated or thalamectomized animals *in vivo*), the normal brain does not generally display separate oscillations during slow-wave sleep. Rather, it displays a coalescence of the slow oscillation with other sleep rhythms (spindles and delta) as well as with faster (beta and gamma) rhythms that are superimposed on the depolarizing phase of the slow oscillation (**Figure 2**). The circuitry and neuronal mechanisms that account for the grouping of low-frequency and fast-frequency rhythms by the slow oscillation are discussed later.

During the depolarizing phase of the slow oscillation, the synchronous firing of neocortical neurons impinges upon thalamic reticular pacemaking neurons, thus creating conditions for formation of spindles,

which are transferred to thalamocortical neurons and up to cortex, at which level spindles shape the tail of the slowly oscillatory cycle (**Figure 2**, middle panel in 'slow + spindle'). This connectivity explains why a cycle of the slow oscillation is followed by a brief sequence of spindles in thalamocortical neurons and in the cortical EEG (**Figure 2**, left panel in 'slow + spindle'), as seen with intracellular recordings in animal studies as well as with EEG recordings in human slow-wave sleep. The sequence of grapho-elements consisting of an ample surface-positive transient, corresponding to the excitation in deeply lying cortical neurons, followed by a slower, surface-negative component and eventually a few spindle waves, represents the combination between the slow and spindle oscillations. It is termed the K-complex and is a reliable sign for stage 2 of human sleep, but it is apparent in all stages of slow-wave sleep. These investigations indicate that K-complexes are the expression of the spontaneously occurring, cortically generated slow oscillation, although K-complexes can also be evoked by sensory stimuli during sleep.

Why fast rhythms are also observed during slow-wave sleep   The unexpected association between a slow sleep rhythm and fast oscillations that are conventionally regarded as defining the electrical activity of waking and REM sleep is explained by the voltage dependency of fast oscillations. Indeed, cortical and thalamic neurons generate beta and gamma oscillations at relatively depolarized values of the membrane potential. Thus, fast rhythms are sustained during the steady depolarization of cortical neurons during waking and REM sleep, selectively appear over the depolarizing phase of the slow sleep oscillation, and are absent during the hyperpolarizing phase of the slow oscillation (**Figure 2**). The association between slow oscillation and fast rhythms has also been reported in human sleep.

Fast Oscillations during REM Sleep and Wakefulness

Beta and gamma waves (20–80 Hz) occur spontaneously during REM sleep and waking and are evoked by intense attention, conditioned responses, tasks requiring fine movements, or sensory stimuli. Beta and gamma rhythms can interchangeably be termed fast because neurons may pass from beta to gamma oscillation in very short periods of time (0.5–1 s), with slight depolarization of cortical neurons. Studies in humans have also shown that there is no precise cutoff between the beta and gamma bands since these activities may fluctuate simultaneously, as indicated by increased activities within both beta and gamma frequency bands (20–40 Hz) during cognitive processes implicating memory. The synchronized fast

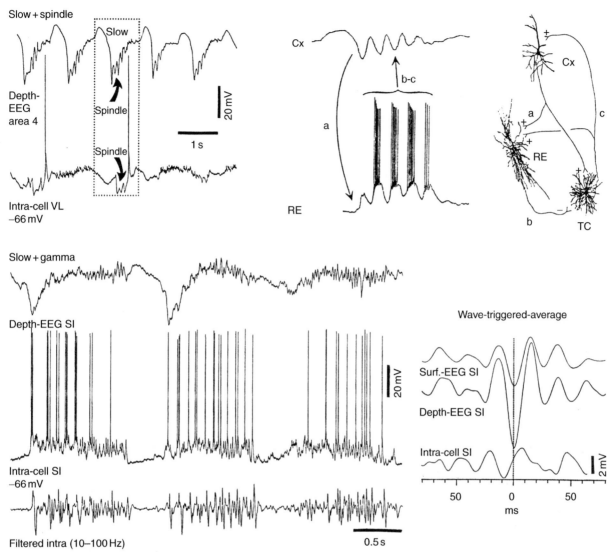

Figure 2 Coalescence of slow oscillation with spindle and gamma rhythms. Intracellular recordings from cortical and thalamic neurons in cats. Slow + spindle, combined slow oscillation and spindle. (Left) Depth EEG from cortical area 4 and intracellular recording of thalamocortical neuron from ventrolateral (VL) nucleus. The excitatory component (negative depth EEG wave, downward deflection) of the slow cortical oscillation (0.9 Hz) is followed by a sequence of spindle waves at 10 Hz (arrows). One typical cycle of these two combined rhythms is indicated by the dotted box; note the inhibitory postsynaptic potentials in the VL neuron leading to a postinhibitory rebound. (Right) Top and bottom traces represent field potential from the depth of association cortical area 5 and intracellular recording from thalamic reticular (RE) neuron. In neuronal circuits (far right), synaptic projections are indicated with small letters, corresponding to the arrows at left, which indicate the time sequence of the events. The depolarizing phase of the field slow oscillation (depth-negative, downward deflection) in the cortex (Cx) travels through the corticothalamic pathway (a) and triggers in the thalamic reticular nucleus (RE) a spindle sequence that is transferred to thalamocortical cells (TC) of the dorsal thalamus (b) and thereafter back to the cortex (c), where it shapes the tail of the slow oscillatory cycle (middle panel). Slow + gamma, fast activity (40 Hz) crowns the depolarizing phase of the slow oscillation. Three traces depict depth EEG waves from primary somatosensory cortex (S1), intracellular recording from S1 neuron, and filtered intracellular trace (between 10 and 100 Hz). Note the fast waves (40 Hz) during the depolarizing phase of the slow sleeplike oscillation and the absence of such fast waves during hyperpolarization.

rhythmic activity led to hypotheses postulating that linkages between spatially distributed oscillatory elements in the visual cortex may be the bases for pattern recognition function. However, beta and gamma activities are also present in the spontaneous activity of neurons and EEG, and these oscillations have been recorded under deep anesthesia when cognitive processes are suspended.

The best candidates for generation and synchronization of fast activities are fast rhythmic-bursting neurons. These neurons are both corticothalamic excitatory and local circuit inhibitory neurons, which may serve synchronization of fast activities within corticothalamic circuits and also be part of the role played by cortical inhibitory neurons in these oscillations. The thalamus and neocortex display coherent beta and gamma rhythms. At variance with the long-range synchrony of the slow sleep oscillation, fast rhythms are synchronized over restricted cortical territories and within specific circuits between thalamocortical and neocortical areas.

Synaptic Plasticity Induced by Sleep Oscillations

Far from being epiphenomena of intrinsic properties and synaptic operations in cortical and thalamic neuronal networks, with little or no functional significance, sleep oscillations lead to synaptic plasticity and are implicated in memory consolidation. Earlier views that considered sleep to be associated with widespread inhibition throughout the cortex and subcortical structures, which would lead to abolition of cognitive and conscious events, have been challenged by intracellular recordings during natural slow-wave sleep demonstrating rich spontaneous firing of neocortical neurons. These data refute the assumption that cortical neurons are inactive in this state. Although external signals are blocked at the thalamic level during slow-wave sleep, mainly because thalamocortical neurons are inhibited during spindle oscillations, the intracortical dialog is maintained and the responsiveness of cortical neurons to cortical volleys is even increased during slow-wave sleep. In humans too, this cortically evoked response is stronger in slow-wave sleep than in waking. That the neocortex is active during slow-wave sleep suggests a reorganization/specification of neuronal circuits. This view is supported by studies using indicators of neuronal activities during slow-wave sleep in humans, revealing more marked changes in those neocortical areas that are implicated in memory tasks and decision making during wakefulness.

Then, spontaneously occurring brain rhythms may lead during sleep to increased responsiveness and plastic changes in the strength of connections among neurons, a mechanism through which information is stored. Next, the results of animal experiments and human studies on synaptic plasticity are briefly presented.

The experimental model of sleep spindles is the sequence of augmenting responses, defined as thalamically evoked cortical potentials that grow in size during the first stimuli at a frequency of approximately 10 Hz, which mimics the initially waxing pattern of spindle waves. In the intact brain, augmenting responses evoked by rhythmic thalamic stimulation are characterized in cortical neurons by an increase in the secondary depolarization, at the expense of the primary EPSP. Synaptic plasticity is not only seen by progressively enhanced amplitudes of responses during the pulse train but also by persistence of self-sustained

potentials, with the same pattern and frequency as those of responses during the prior stimulation period (**Figure 3(a)**). Stimulation of the corticothalamic projection with pulse trains at 10 Hz results in evoked responses and also, after protracted stimulation, in 'spontaneously' occurring spike bursts whose form and rhythmicity are similar to those of evoked responses, as if the repetition of volleys was imprinted in the 'memory' of the corticothalamic network (**Figure 3(c)**). Short- and medium-term (5–30 min) neuronal plasticity can occur inside cortical circuitry, even in the absence of the thalamus (**Figure 3(b)**).

Thus, besides their role in cortical disconnection through inhibition of incoming messages in the thalamus, sleep spindles are also operational in important cerebral functions. During spindles, rhythmic and synchronized spike bursts of thalamic neurons depolarize the dendrites of neocortical neurons, which is associated with massive Ca^{2+} entry that may provide an effective signal to efficiently activate Ca^{2+}/calmodulin-dependent protein kinase II, implicated in synaptic plasticity of excitatory synapses.

Similar phenomena, with Ca^{2+} entry in dendrites and somata of cortical neurons, occur with rhythmic spike trains associated with oscillations in the slow (0.5–1 Hz) or delta (1–4 Hz) frequency bands, during later stages of slow-wave sleep. The hypothesis that the slow oscillation is responsible for the consolidation of memory traces acquired during the state of wakefulness is supported by data showing that slow and delta oscillations are implicated in the cortical plasticity of the developing cortex.

These experimental data are congruent with human studies demonstrating that the overnight improvement of discrimination tasks requires some steps, including those in early slow-wave sleep stages. Also, procedural memory formation may be associated with oscillations during early sleep stages. The early part of night sleep favors retention of declarative memories, whereas the late part of sleep favors retention of nondeclarative memories.

At variance with the commonly used notion of global brain processes in slow-wave sleep are two major findings reported in humans subjects: slow-wave sleep activity increases 2 h after a motor learning task, and the enhancement is expressed locally in parietal association areas that receive converging visual and proprioceptive inputs relevant to spatial attention and skilled actions.

The increased cortical activity that accounts for consolidation of memory traces during slow-wave sleep also explains the presence of dreaming mentation during this state. In slow-wave sleep, dreaming is rational and repetitive, in contrast with the vivid, illogical, and emotional perceptions during REM

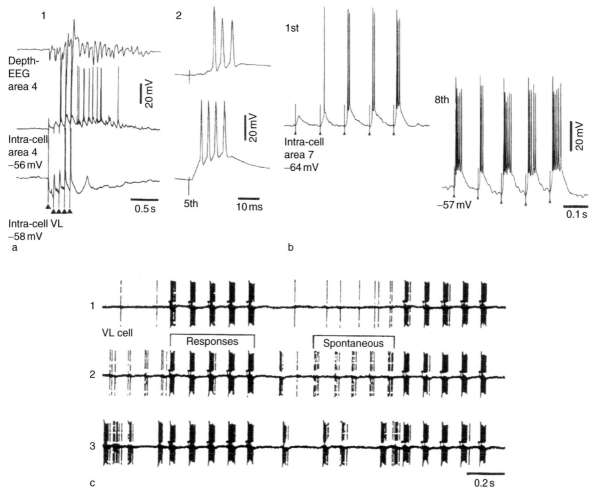

Figure 3 Synaptic plasticity in thalamocortical and intracortical systems, and 'memory' of electrical responses in corticothalamic system, induced by low-frequency stimuli mimicking sleep spindles. (a) Dual simultaneous intracellular recordings from cortical and thalamocortical (TC) neurons in cat (top trace is depth EEG from area 4). (1) Pulse train (five stimuli at 10 Hz; arrowheads) applied to the thalamic ventrolateral (VL) nucleus produced augmenting responses in the cortical neuron, whereas simultaneously recorded VL neuron displayed hyperpolarization. (2) Expanded fifth response; the augmented response in cortical neuron followed the rebound spike burst of the VL neuron. Note the self-sustained oscillatory activity at 10 Hz in the cortical neuron after cessation of thalamic stimuli, despite persistent hyperpolarization in the VL neuron. (b) Intracellular responses of cat area 7 bursting cortical neuron to repetitive callosal stimulation (10 Hz). The thalamus ipsilateral to the recorded neuron was extensively lesioned using kainic acid. The intracortical augmenting responses to the first and eighth pulse trains are illustrated. Note the depolarization by approximately 7 mV and the increased number of action potentials within bursts after repetitive stimulation. (c) Extracellular recording of the VL neuron in brain stem-transected cat. Motor cortex stimulation with pulse trains at 10 Hz (stimuli are marked by dots). (1) The pattern of responses in the thalamic VL neuron in early stages of rhythmic pulse trains. (2 and 3) Responses at later stages of stimulation. Note the appearance of spontaneous spike bursts resembling the evoked ones as a form of 'memory' in the corticothalamic circuit.

sleep. The brain is never 'empty,' and mental activity is present during all stages of normal sleep.

Ponto-Geniculo-Occipital Waves

PGO waves are sharp field potentials that are generally recorded in the visual thalamic lateral geniculate (LG) nucleus and occipital cortex, where they appear in clusters of up to six waves, closely related to gaze direction in dream imagery. Although the term PGO indicates their presence in the visual pathway, these potentials appear in many thalamic nuclei and cortical areas, outside the visual system. This widespread occurrence is due to the generalized thalamic projections of PGO generators, which are cholinergic neurons located in the pedunculopontine tegmental (PPT) nucleus and an adjacent nucleus at the mesopontine junction. Five distinct neuronal types have been recorded within cholinergic nuclei, whose discharges are temporally related to PGO field potentials in the visual thalamic nucleus.

The major input sources of PPT neurons are in adjacent areas of the upper brain stem reticular formation, hypothalamic areas, and cerebral cortex. Impulses in the PGO generators can be triggered by any of these inputs through direct excitation or postinhibitory rebound excitation. Then, PPT neurons can be regarded as the final link in the brain stem–thalamic path that generates PGO waves. Because PGO potentials are thought to represent 'the stuff dreams are made of' and in view of the emotional nature of dreaming mentation in REM sleep, hypothalamic and forebrain structures that store some of the emotionally charged information may be most effective in driving the PGO brain stem neuronal generators.

In naturally sleeping cats – the species of choice for experimental studies because they display sleep patterns similar to those of humans – PGO waves precede other signs of REM sleep by approximately 30–90 s, and they appear during the last period of slow-wave sleep. Thus, there is a transitional period between EEG-synchronized and EEG-activated (REM) sleep that is called the pre-REM period, during which PGO waves appear over the background of a fully synchronized EEG. Thalamic neurons are hyperpolarized during the pre-REM period, when the sleep EEG is still fully synchronized, whereas they are depolarized during REM sleep. These two states (pre-REM and fully developed REM sleep) generate different PGO-related responses of LG neurons to brain stem cholinergic inputs, which influence the signal-to-noise ratio in the visual channel – that is, the ratio between the neuronal activity related to the PGO signal and the background firing of the same neuron. During pre-REM, the activity of thalamic LG cells starts with a short spike burst coinciding with a high-amplitude PGO wave and continues with a train of single spikes. By contrast, during full-blown REM sleep, the rate of LG cells' spontaneous firing is much higher than in pre-REM; LG cells do not fire spike bursts but, rather, single action potentials; and the amplitudes of PGO waves are much lower than during the pre-REM state. The peri-PGO histograms of LG neuronal activities show that the signal-to-noise ratio reaches values of approximately 7 during the pre-REM epoch, whereas the ratio values during REM sleep are approximately 2.

Why all these details? Because PGO waves are commonly regarded as the physiological correlate of dreaming, the greater signal-to-noise ratio in the LG-cortical channel during the pre-REM epoch than during REM sleep suggests that the vivid imagery associated with dreaming sleep may appear before fully developed REM sleep, during a period of apparent slow-wave sleep. Discussing dreaming from experiments in cats is justified in view of the behavioral repertoire typical for dreaming mentation that can be elicited in cats after having prevented muscular atonia

by lesions of certain brain stem reticular structures. The idea that PGO waves with greater amplitudes during the pre-REM stage may reflect more vivid imagery during that epoch than even during REM sleep corroborates earlier data showing that after interrupting sleep immediately after the occurrence of the first PGO wave (in the pre-REM stage) and eliminating approximately 30 s of the slow-wave sleep stage that precedes REM sleep, the increased time of the REM sleep rebound was due to phasic events (PGO waves) rather than the loss of REM sleep *per se*. This observation fits in well with data on dream reports from the last epoch of EEG-synchronized sleep, which are indistinguishable from those obtained from REM sleep awakenings.

PGO waves are thought to play a role in development and structural maturation of the brain. PGO waves also exert specific excitatory effects on complex neurons in the visual cortex, and their deprivation may modify the expression of long-term potentiation in immature animals.

See also: Gene Expression; Hippocampal–Neocortical Dialog; Network Reactivation; PET Activation Patterns; Sleep Architecture; Sleep-Dependent Memory Processing.

Further Reading

Achermann P and Borbély A (1997) Low-frequency (<1 Hz) oscillations in the human sleep EEG. *Neuroscience* 81: 213–222.

Contreras D, Destexhe A, Sejnowski TJ, and Steriade M (1996) Control of spatiotemporal coherence of a thalamic oscillation by corticothalamic feedback. *Science* 274: 771–774.

Contreras D and Steriade M (1995) Cellular basis of EEG slow rhythms: A study of dynamic corticothalamic relationships. *Journal of Neuroscience* 15: 604–622.

Gais S, Plihal W, Wagner U, and Born J (2000) Early sleep triggers memory for early visual discrimination skills. *Nature Neuroscience* 3: 1335–1339.

Hobson JA and Pace-Schott EF (2002) The cognitive neuroscience of sleep: Neuronal systems, consciousness and learning. *Nature Reviews Neuroscience* 3: 679–693.

Huber R, Ghilardi MF, Massimini M, and Tononi G (2004) Local sleep and learning. *Nature* 430: 78–81.

Maquet P, Degueldre C, Delfiore G, et al. (1997) Functional neuroanatomy of human slow wave sleep. *Journal of Neuroscience* 17: 2807–2812.

Massimini M, Ferrarelli F, Huber R, Esser SK, Singh H, and Tononi G (2005) Breakdown of cortical effective connectivity during sleep. *Science* 309: 2228–2232.

Massimini M, Huber R, Ferrarelli F, and Tononi G (2004) The sleep slow oscillation as a traveling wave. *Journal of Neuroscience* 24: 6862–6870.

Mölle M, Marshall L, Gais S, and Born J (2002) Grouping of spindle activity during slow oscillations in human non-REM sleep. *Journal of Neuroscience* 22: 10941–10947.

Plihal W and Born J (1997) Effects of early and late nocturnal sleep on declarative and procedural memory. *Journal of Cognitive Neuroscience* 9: 534–547.

Steriade M (2003) *Neuronal Substrates of Sleep and Epilepsy.* Cambridge, UK: Cambridge University Press.

Steriade M and Llinás RR (1988) The functional states of the thalamus and the associated neuronal interplay. *Physiological Reviews* 68: 649–742.

Steriade M and McCarley RW (2005) *Brain Control of Wakefulness and Sleep.* New York: Springer.

Steriade M, Nuñez A, and Amzica F (1993) A novel slow (<1 Hz) oscillation of neocortical neurons *in vivo*: Depolarizing and hyperpolarizing components. *Journal of Neuroscience* 13: 3252–3265.

Steriade M, Paré D, Bouhassira D, Deschênes M, and Oakson G (1989) Phasic activation of lateral geniculate and perigeniculate neurons during sleep with ponto-geniculo-occipital spikes. *Journal of Neuroscience* 9: 2215–2229.

Steriade M and Timofeev I (2003) Neuronal plasticity in thalamocortical networks during sleep and waking oscillations. *Neuron* 37: 563–576.

Steriade M, Timofeev I, and Grenier F (2001) Natural waking and sleep states: A view from inside neocortical neurons. *Journal of Neurophysiology* 85: 1969–1985.

Stickgold R, James L, and Hobson JA (2000) Visual discrimination learning requires sleep after training. *Nature Neuroscience* 3: 1237–1238.

PET Activation Patterns

T T Dang-Vu and M Desseilles, University of Liege
and Centre Hospitalier Universitaire, Liege, Belgium
P Peigneux, University of Liege, Liege, Belgium
S Laureys and P Maquet, University of Liege and
Centre Hospitalier Universitaire, Liege, Belgium

Introduction

Positron emission tomography (PET) shows the distribution in the body (or in the brain, in the context of
neuroimaging) of compounds labeled with positron-
emitting isotopes. In the field of sleep research, two
compounds have mainly been used to assess the global
and regional cerebral activity during the different sleep
stages: [^{18}F]fluorodeoxygucose, which is a marker of
glucose metabolism, and oxygen-15-labeled water
($H_2^{15}O$), which is an indirect marker of blood flow. In
the framework of sleep research, PET images are
acquired in combination with simultaneous recordings
of brain electrical activity by electroencephalography
for sleep staging. After data transformation and interindividual standardization, statistical procedures allow
the comparison of regional cerebral activity between
two conditions (e.g., a specific sleep stage and wakefulness) or the examination of correlation patterns
between regional activity distribution in a specific condition and measurements of relevant physiological
events (e.g., sleep oscillations within a specific sleep
stage). The resulting functional brain maps have contributed important information about human sleep
physiology, assessed in the light of animal experimental
data. Sleep functions have also been investigated with
this technique. In particular, a role for sleep in learning
and memory has been evidenced using PET recording
during posttraining sleep associated with behavioral
measurements.

Sleep Stages and Sleep Oscillations

PET studies have shown that global and regional
patterns of brain activity during wakefulness are
remarkably different from those obtained during
sleep (i.e., rapid eye movement (REM) sleep and
non-REM sleep).

Non-Rapid Eye Movement Sleep

Non-REM sleep is a state of synchronization of the
electroencephalogram (EEG), along with the production of specific oscillations within thalamo-cortical
networks: spindles, delta waves, and slow oscillations.

When compared to wakefulness and REM sleep,
non-REM sleep is characterized by a decrease in
global cerebral blood flow and regional cerebral
blood flow (rCBF). The largest decreases in rCBF are
observed in a set of subcortical and cortical areas,
including the dorsal pons, mesencephalon, thalami,
basal ganglia, basal forebrain, prefrontal cortex, anterior cingulate cortex, and precuneus (**Figure 1(b)**).

A lower activity in the brain stem and thalamus
was expected from animal data of non-REM
sleep-generation mechanisms; a decreased firing rate
in the brain stem induces the sequential alternation of
long hyperpolarization and short depolarization
patterns in thalamic neurons, which leads to the
formation of non-REM sleep rhythms (spindles,
delta oscillations, and slow) among thalamocortical
networks. Due to the low time resolution of the PET
technique (i.e., one scan is the activity averaged over a
period of time ranging from 40 to 90 s) and because
the hemodynamic influences of hyperpolarization
predominate over those of depolarization phases,
brain areas where non-REM sleep rhythms are most
expressed appear deactivated in PET studies.

At the cortical level, the pattern of deactivation is
not homogeneously distributed. Indeed, the least
active areas in non-REM sleep are located in associative cortices, especially the ventromedial prefrontal
cortex (VMPF), which includes the orbitofrontal
and anterior cingulate cortices. The VMPF is also
one of the most active brain areas during the awake
resting state and is involved in important cognitive
processes such as action monitoring and decision
making. In contrast, the primary cortices were the
least deactivated cortical areas during non-REM
sleep. This specific segregation of cortical activity
remains poorly understood, although some hypotheses
have been proposed, for instance, (1) that associative
areas might be more profoundly influenced by non-
REM sleep rhythms than primary cortices because
they are the most active cerebral areas during wakefulness and (2) that sleep intensity is homeostatically
related to prior waking activity at the regional level.

The precuneus is another cortical area that displays
a reduced activity during non-REM sleep in PET
studies. It is a region particularly active in wakefulness, during which it is involved in visual mental
imagery processes, explicit memory retrieval, and
consciousness. The precuneus is also deactivated during other states of decreased consciousness such as
pharmacological sedation, hypnotic states, and vegetative states. The role of the precuneus during sleep
still remains unclear. Its decreased activity during

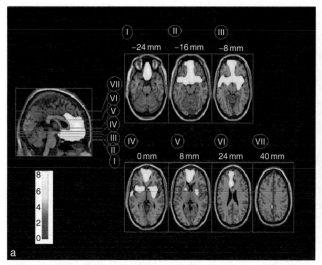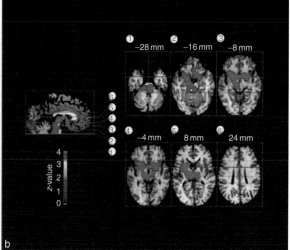

Figure 1 Functional neuroanatomy of normal human non-REM sleep, as assessed by $H_2^{15}O$ PET: (a) brain areas in which rCBF decreases as a function of delta power during non-REM sleep (stages 2–4); (b) brain areas in which rCBF decreases during non-REM sleep as compared to wakefulness and REM sleep. In (a), image sections are displayed on different levels of the z-axis, as indicated at the top of each picture. The color scale indicates the range of z-values for the activated voxels, superimposed on a canonical T1-weighted MRI image. Displayed voxels are significant at $p < 0.05$ after correction for multiple comparisons. Note that different analyses (correlation (a) vs. subtraction (b)) gave rise to strikingly similar results in terms of the regional blood flow distribution. MRI, magnetic resonance imaging; PET, positron emission tomography; rCBF, regional cerebral blood flow; REM, rapid eye movement. Adapted from Maquet P, Degueldre C, Delfiore G, et al. (1997) Functional neuroanatomy of human slow wave sleep. *Journal of Neuroscience* 17(8): 2807–2812. Copyright 1997 by the Society for Neuroscience.

non-REM sleep might reflect a homeostatic compensation of a high waking activity.

The basal forebrain and the basal ganglia (mostly the striatum) have also been found to be consistently deactivated during non-REM sleep in PET sleep studies. The basal forebrain is a functionally and structurally heterogeneous structure in which a majority of neurons is involved in cortical activation during wakefulness and REM sleep. Its deactivation during non-REM sleep may therefore reflect a lower activity of these arousal-promoting neurons. The role of the basal ganglia, and especially the striatum, in sleep regulation remains speculative, however. Two hypotheses have been put forward. First, the striatum receives massive afferent inputs from the frontal cortex and the thalamus, which are also deactivated during non-REM sleep. These structures are most likely to participate to the formation of non-REM sleep rhythms by oscillating synchronously between long phases of hyperpolarization and bursts of discharges. Due to the fronto- and thalamo-striatal connections, basal ganglia neurons may likewise oscillate following these non-REM sleep-rhythm sequential patterns and, thus, appear deactivated at the macroscopic level. According to the second proposal, the striatum may also send projections to the pedunculopontine tegmental nucleus (PPT) in the brain stem and induce the disinhibition of this activating structure, subsequently leading to

cortical arousal during wakefulness. In this perspective, the decreasing activity in the striatum during non-REM sleep may be related to a reduced propensity to arousal as well.

PET studies have not merely compared the activity between non-REM sleep and other stages of sleep or wakefulness. Another way to describe brain activity during this sleep stage was to search for the neural correlates of non-REM sleep oscillations (spindles and delta waves) by looking for brain areas in which rCBF values correlate with the EEG activity of interest (i.e., power density in the sigma or delta frequency band). Using this approach, spindle activity (12–15 Hz) has been shown to correlate negatively with rCBF in the thalamus, meaning that the higher the power density within the spindle frequency range on EEG recordings, the lower the thalamic activity. This result is in line with spindle-generation mechanisms in mammals, which are dominated by the cyclic repetition of hyperpolarization and spike bursts in the thalamic neurons. Delta activity (1.5–4 Hz) correlates negatively with rCBF in the VMPF, basal forebrain, striatum, and precuneus (**Figure 1(a)**). The resulting map is very similar to the brain map of the regions less activated during non-REM sleep compared to REM sleep and wakefulness (**Figure 1(b)**), which emphasizes the notion that delta activity is a prominent feature of non-REM sleep. A major difference, however, is the absence of significant correlation between delta and thalamus activity, whereas the thalamus is markedly deactivated during non-REM sleep compared to other sleep stages or wakefulness. This discrepancy can be explained taking into account that two types of delta activity have been described in animals: a stereotyped delta rhythm, whose generation depends on intrinsic properties of thalamocortical neurons, and a cortical polymorphous delta rhythm, which persists after extensive thalamectomy. Therefore, the delta correlation map might preferentially reflect the brain areas involved in the generation of cortical delta waves during non-REM sleep. The physiology of these cortically generated delta oscillations, and their relationship with the slow rhythm, is still poorly understood.

It should be emphasized here that deactivation patterns found with PET studies do not imply that these brain areas remain idle during non-REM sleep. As already stated, non-REM sleep oscillations are produced by the recurrent and sequential alternation of hyperpolarization and depolarization phases in the thalamic and cortical neurons. The latter are characterized by bursts of neuronal firing temporally organized by the non-REM sleep slow oscillation. PET is insensitive to these bursts because it averages brain activity over long periods, during which the effects on regional brain function of prolonged hyperpolarization periods exceed those of shorter depolarization phases. This issue should be addressed in future studies using techniques with higher spatial and temporal resolution, such as combined EEG-functional magnetic resonance imaging (fMRI), that will provide activation patterns closer to the genuine non-REM sleep physiology, dominated by synchronous and low-frequency oscillations.

Rapid Eye Movement Sleep

REM sleep is characterized by a desynchronized EEG with high-frequency, low-amplitude activity, a major muscle atonia, spontaneous rapid eye movements, and an intense oneiric mentation. In contrast to non-REM sleep, PET studies showed that the global cerebral blood flow during REM sleep is sustained to a level comparable to wakefulness. At the regional level, several brain areas even display a higher rCBF during REM sleep compared to wakefulness and/or non-REM sleep: the pontine tegmentum, thalamus, amygdala, hippocampus, anterior cingulate cortex, temporal–occipital areas, and the basal forebrain. Conversely, regional decreases in activity are also found during REM sleep, in the dorsolateral prefrontal cortex (DLPF), posterior cingulate gyrus, precuneus, and inferior parietal cortex (**Figure 2**).

Activation of the pontine tegmentum, thalamic nuclei, and basal forebrain is in agreement with REM sleep-generation mechanisms in animals. REM sleep is generated by cholinergic processes arising from brain stem structures, located in the pedunculopontine tegmentum and laterodorsal tegmentum, which mediate widespread cortical activation via a dorsal pathway innervating the thalamus and a ventral pathway innervating the basal forebrain.

Consistent activations during REM sleep are also found in limbic and paralimbic structures, including the amygdaloid complexes, hippocampal formation, and anterior cingulate cortex. There is evidence that the amygdala plays a role in REM sleep modulation. For instance, carbachol (cholinergic agonist) injections in the central nucleus of amygdaloid complexes enhance REM sleep duration, and a shift from non-REM sleep to REM sleep is observed in rats after serotonin injection into the amygdala. The amygdala also appears to modulate important physiological features of REM sleep. For example, recent PET data suggest that the large variability in heart rate during REM sleep could be explained by a prominent influence of the amygdaloid complexes. Both amygdala and hippocampal formation are also critical for memory systems and may thus participate in the processing of memory traces during REM sleep.

PET studies showed higher rCBF values in temporal–occipital areas during REM sleep compared to wakefulness. These areas include the inferior temporal cortex and the fusiform gyrus, which are extrastriate cortices belonging to the ventral visual stream. Functional interactions between the primary visual cortex (striate cortex) and the extrastriate cortex during REM sleep were also assessed in a PET study. This study showed that extrastriate cortex activation is significantly correlated with striate cortex deactivation during REM sleep, whereas their activities are positively correlated during wakefulness. This result has been interpreted as supporting the hypothesis that REM sleep allows internal information processing (between extrastriate areas and their paralimbic projections, both activated during REM sleep) in a closed system dissociated from interactions with the environment (via striate cortex and prefrontal cortex, both deactivated during REM sleep).

Regional deactivations during REM sleep are mostly located in the DLPF (inferior and middle frontal gyrus), precuneus, posterior cingulate cortex, and a part of the parietal cortex (temporal–parietal region and inferior parietal lobule). In contrast, the activity in the superior parietal lobe and in the superior and medial prefrontal cortex is similar to waking level. The reasons of these cortical deactivations are still unclear. Animal data show that the cortical areas less

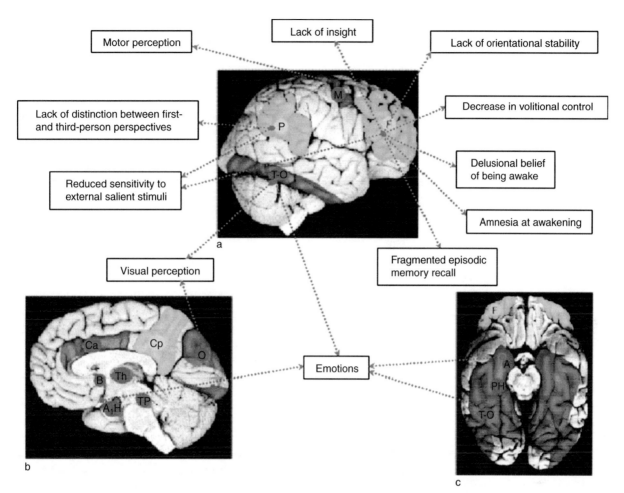

Figure 2 Schematic representation of the functional neuroanatomy of normal human REM sleep: (a) lateral view; (b) medial view; (c) ventral view. Regions in which there is a relative increase in neural activity during REM sleep are colored in red; regions in which neural activity decreases are colored in blue. Green arrows illustrate the proposed relationships between the activity of specific brain areas and several dreaming features, which may be accounted for by regional patterns of activity during REM sleep. A, amygdala; B, basal forebrain; Ca, anterior cingulate gyrus; Cp, posterior cingulate gyrus and precuneus; F, prefrontal cortex (middle, inferior, and orbito-frontal cortices); H, hypothalamus; M, motor cortex; P, parietal cortex (inferior parietal lobule); PH, parahippocampical gyrus; O, occipital-lateral cortex; REM, rapid eye movement; Th, thalamus; T-O, temporal–occipital extrastriate cortex; TP, pontine tegmentum. Reproduced from Schwartz S and Maquet P (2002) Sleep imaging and the neuro-psychological assessment of dreams. *Trends in Cognitive Science* 6(1): 23–30, with permission from Elsevier.

active during REM sleep (the inferior parietal and DLPF) receive only few inputs from the amygdala, whereas areas more active during REM sleep (the anterior cingulate and right parietal operculum) receive rich amygdalar inputs, suggesting that amygdala may modulate the pattern of cortical activity during REM sleep. This hypothesis is also supported by PET results showing that functional interactions between the amygdala and the temporal–occipital cortices are different in the context of REM sleep than in non-REM sleep or wakefulness. The amygdalo-cortical network during REM sleep might contribute, in particular, to the selective processing of emotionally relevant memories.

In animals, rapid eye movements during REM sleep are closely related to the occurrence of ponto-geniculo-occipital (PGO) waves. PGO waves are bioelectrical phasic potentials occurring during the transition from non-REM sleep to REM sleep or during REM sleep itself. They are observed in many parts of the animal brain, but they are most easily recorded in the pons, lateral geniculate bodies of the thalamus, and occipital cortex, hence their name. Animal data also suggest that PGO waves might have important functional roles, such as a facilitation of brain plasticity. In humans, a PET study found correlations during REM sleep, but not during wakefulness, between spontaneous eye movements and rCBF in the occipital cortex and lateral geniculate bodies of the thalamus, supporting the hypothesis that PGO-like activities are present during REM sleep in humans as well as in animals.

Neuroimaging and Dreams

Dreaming is experienced every night by many humans as multisensory mental representations occurring spontaneously during sleep, often organized in a narrative manner. Dreams are more often associated with REM sleep (dream reports are present in 90% of the episodes in subjects awakened during a REM sleep period). Dream during REM sleep is classically considered the canonical dream with several characteristics: hallucinoid imagery, narrative structure, bizarreness, hyperemotionality, delusional acceptance, and deficient memory of its content. Dream reports are usually less organized, poorer, and shorter during non-REM sleep.

At the neuroimaging level, it has been proposed that the PET activation patterns during REM sleep might be interpreted in the light of dreams. Indeed, brain functional segregation during REM sleep, as already described, may account for several typical dreaming features (**Figure 2**). We present a few examples here.

Dream reports differentially involve the sensory modalities. Visual components are nearly always present, auditory components are present in 40–60% of dreams, movement and tactile sensations are present in 15–30%, and smell and taste components are present in less than 1%. The activation of the associative posterior (temporal–occipital) cortices may be related to these perceptual aspects of dreams, consistently dominated by visual and auditory elements. Accordingly, a cessation of visual dream imagery was reported in some patients with temporal–occipital lesions.

Dream content is also characterized by the prominence of emotions especially negative emotions such as fear and anxiety. Responses to threatening stimuli or stressful situations are known to be modulated by the amygdala during wakefulness. The high limbic, and especially amygdalar, activity during REM sleep may therefore underpin the emotional intensity occurring during dreams. Moreover, PET data have shown functional interactions between the amygdala and temporal–occipital cortices during REM sleep, and fMRI studies during wakefulness have found positive relationships between the emotional intensity of visual stimuli and both amygdalar and inferotemporal cortex activity. Together, these data suggest that emotional experience during dreams may engage

specific brain networks encompassing the amygdala and temporal–occipital cortices rather than a single brain area.

The bizarreness of dream reports is another recurrent hallmark. The cortical patterns of hypoactivity during REM sleep may contribute to these aspects. Indeed, the prefrontal regions deactivated during REM sleep overlap with the areas involved in the selection of stimulus–response associations according to contextual signals, past events, and internal goals. The decreased activity of prefrontal regions consequently impairs the efficiency of this integrative system during REM sleep. This may, for instance, account for the inability of the dreamer to integrate information of a whole episode, leading to an oneiric content in which characters, times, and places are fused, incongruous, and discontinuous. It may also explain the decrease in volitional control and the dreamer's failure to organize mental representation toward a well-identified internal goal. Prefrontal inactivity during REM sleep may, then, explain why the dreamer is unable to control the flow of dream events.

These hypoactive areas also encompass frontal areas participating in the processing of episodic memory, that is, the ability to encode and recollect personally experienced events set in a particular spatiotemporal context. Prefrontal areas are involved in the monitoring of episodic memory retrieval by checking the accuracy and completeness of the processed information. The impairment of these areas during REM sleep might explain why, although 65% of dream reports contain residues of recent waking activity, only 1.4% of them are considered to represent a replay of full memory episodes. During dreams, episodic elements might be reactivated in a fragmented fashion, but the deactivation of the prefrontal cortex prevents the various details of past events from being integrated into an identifiable life episode. Impairment in these areas might also explain the well-known amnesia on awakening that prevents most people from accurately remembering the events experienced during the dream episode.

Overall, PET studies have brought interesting hypotheses about human dream organization. However, these assumptions remain largely speculative and partial, especially because combined dream and functional imaging data are still very sparse. In future studies, the use of scales to quantify and categorize the dream narrative in terms of different perceptual, emotional, or bizarre elements should provide crucial explanatory variables to model neuroimaging data. This combination of systematic neuropsychological assessments of dreams with neuroimaging data might greatly improve our insight into dreaming mechanisms by bringing genuine functional maps of the dreaming brain.

Sleep and Memory

The idea that sleep is involved in the processing of information that should be memorized dates back to the theory of memory consolidation a century ago. Consolidation here refers to the processing of memory traces during which the traces may be reactivated, analyzed, and gradually incorporated into long-term memory. In addition to anecdotic similarities between brain areas activated in canonical maps of normal human sleep and learning-related areas, other PET studies have supported a role for sleep in learning and memory. These studies suggest that neuronal activity patterns observed during a learning episode are reinstated during posttraining sleep. These reactivations allow for the adaptation of intercellular connection strengths between the elements of the network and the incorporation of the new experience into long-term memory. Both REM sleep and non-REM sleep are differentially involved in these processes.

During REM sleep, but not during non-REM sleep, the premotor and visual cortical areas engaged in the implicit learning of a procedural motor task are reactivated in subjects previously trained, compared to nontrained subjects. Such reactivations do not occur if the subjects are trained in a random version of the same task, showing that processing during sleep is not merely due to intense task exposure but rather relies on the sequential content of the material. These data speak for the reprocessing during REM sleep of procedural memory traces acquired during previous wakefulness.

During non-REM sleep, but not during REM sleep, the hippocampal areas activated during a spatial learning task are reactivated in subjects previously trained, compared to nontrained subjects. This result suggests a reprocessing during non-REM sleep of recent spatial memory traces acquired during previous wakefulness. Moreover, this study demonstrated a significant correlation between rCBF increases in the hippocampal areas during non-REM sleep and an overnight gain in behavioral performance, supporting the hypothesis that this offline reprocessing is related to plastic changes underlying a subsequent improvement in performance.

PET studies have thus contributed to a large amount of evidence supporting a role of sleep in learning and memory. These results are in agreement with behavioral data suggesting that REM sleep and non-REM sleep differentially modulate the consolidation of procedural and spatial/episodic memories, respectively, in the model called the dual process hypothesis. However, other behavioral data support, and brain-imaging results do not oppose to, the sequential hypothesis, in which the ordered succession of non-REM sleep and REM sleep are necessary for the consolidation of memory traces, whatever the memory system. Furthermore, memory systems and mechanisms are complex and heterogeneous, and it remains unclear whether sleep influences all memory systems. Behavioral data already suggest that not all memories need sleep to consolidate. More studies are needed to refine the characterization of the precise contribution of each sleep stage in the processing of the different categories of memory traces.

Conclusion and Perspectives

PET studies have contributed to the understanding of human sleep neurophysiology by describing the functional neuroanatomy of sleep stages, by proposing brain correlates of dreaming features, and by showing cerebral reactivations during sleep of regions involved in prior learning. Yet our knowledge about normal human sleep remains fragmentary. These seminal studies only pave the way for future works that will benefit from the development of multimodal techniques such as combined EEG-fMRI. This neuroimaging method is technically more challenging in sleep studies, but it also has a number of advantages, including better spatial and temporal resolutions. This should help in describing more accurately the dynamics of brain activity throughout the sleep–wake cycle and especially in relation to the sleep physiological events, such as sleep oscillations. Future studies should also continue to characterize the specific extent of sleep's contribution to the learning and memory processes, which remains a topic of intense debates. There is no doubt that functional neuroimaging studies bring us significant insights into sleep functions (e.g., the relationships between sleep and brain plasticity) and sleep physiology, which is a prerequisite to the investigation of sleep disorders.

See also: Autonomic Dysregulation During REM Sleep; History of Sleep Research; REM/NREM Differences in Dream Content; Sleep and Consciousness; Sleep Architecture; Sleep Oscillations and PGO Waves; Sleep-Dependent Memory Processing; Theories of Dream Function.

Further Reading

Braun AR, Balkin TJ, Wesensten NJ, et al. (1998) Dissociated pattern of activity in visual cortices and their projections during human rapid eye movement sleep. *Science* 279(5347): 91–95.

Dang-Vu TT, Desseilles M, Albouy G, et al. (2005) Dreaming: A neuroimaging view. *Swiss Archives of Neurology and Psychiatry* 156(8): 415–425.

Dang-Vu TT, Desseilles M, Laureys S, et al. (2005) Cerebral correlates of delta waves during non-REM sleep revisited. *Neuroimage* 28(1): 14–21.

Dang-Vu TT, Desseilles M, Peigneux P, and Maquet P (2006) A role for sleep in brain plasticity. *Pediatric Rehabilitation* 9(2): 98–118.

Maquet P (2000) Functional neuroimaging of normal human sleep by positron emission tomography. *Journal of Sleep Research* 9(3): 207–231.

Maquet P (2001) The role of sleep in learning and memory. *Science* 294(5544): 1048–1052.

Maquet P, Degueldre C, Delfiore G, et al. (1997) Functional neuroanatomy of human slow wave sleep. *Journal of Neuroscience* 17(8): 2807–2812.

Maquet P, Laureys S, Peigneux P, et al. (2000) Experience-dependent changes in cerebral activation during human REM sleep. *Nature Neuroscience* 3(8): 831–836.

Maquet P, Peters J, Aerts J, et al. (1996) Functional neuroanatomy of human rapid-eye-movement sleep and dreaming. *Nature* 383(6596): 163–166.

Maquet P, Ruby P, Maudoux A, et al. (2005) Human cognition during REM sleep and the activity profile within frontal and parietal cortices: A reappraisal of functional neuroimaging data. *Progress in Brain Research* 150: 219–227.

Maquet P, Smith C, and Stickgold R (2003) *Sleep and Brain Plasticity*. Oxford: Oxford University Press.

Peigneux P, Laureys S, Fuchs S, et al. (2001) Generation of rapid eye movements during paradoxical sleep in humans. *Neuroimage* 14(3): 701–708.

Peigneux P, Laureys S, Fuchs S, et al. (2004) Are spatial memories strengthened in the human hippocampus during slow wave sleep? *Neuron* 44(3): 535–545.

Rauchs G, Desgranges B, Foret J, and Eustache F (2005) The relationships between memory systems and sleep stages. *Journal of Sleep Research* 14(2): 123–140.

Schwartz S and Maquet P (2002) Sleep imaging and the neuropsychological assessment of dreams. *Trends in Cognitive Sciences* 6(1): 23–30.

Hippocampal–Neocortical Dialog

J J Chrobak, University of Connecticut, Storrs, CT, USA
A Sirota and G Buzsáki, Rutgers University, Newark, NJ, USA

Introduction

Understanding how large groups of neurons interact across brain structures is important for understanding brain function. During both awake and asleep brain states there are structured interactions between groups of neurons (ensembles) in the hippocampus and groups of neurons in the neocortex. Ensembles fire together in predictable patterns during hippocampal theta and sharp waves in both the awake and sleeping brain, as well as during neocortical sleep spindles, delta waves, and slow oscillations. How are the ensemble patterns observed in the hippocampus related to ensemble patterns observed in the neocortex, and do they differ in the awake versus asleep brain?

Two prominent hippocampal activity patterns (theta and sharp waves) occur during specific behavioral states (**Figure 1**). The theta pattern occurs during active exploration of the environment, with a number of neurons throughout the hippocampal system discharging collectively every 50–100 ms. Theta is considered an 'online' state in which information about the 'world' is channeled from primary sensory pathways through the neocortical associative cortices (e.g., perirhinal and parahippocampal cortices) and then fed into the circuitry of the hippocampus. Theta pattern also occurs during rapid eye movement (REM) sleep.

During moments of immobility and quiet resting, sharp waves (SPWs) dominate the hippocampus, with large numbers of hippocampal neurons discharging in a fast-frequency (200 Hz) 'ripple' volley in relation to the slow (~100 ms) SPWs. Sharp waves are considered an output state, when the hippocampus reengages neocortical circuits, perhaps providing a comment, an index, or simply a replay about the content of prior theta-related input. This pattern has been suggested to serve an 'offline' consolidation process. Similar SPW patterns also dominate the hippocampus during slow-wave sleep (SWS). Do the ensemble firing patterns during SPWs in awake and sleep states support different functions? Defining the similarities and differences across awake and sleep states may shed light on their significance as well as on the function of distinct sleep states. More importantly, it is necessary to uncover the mechanism by which these 'replay' patterns are coordinated with the ongoing neocortical activity, to ensure information transfer between the two structures.

It has been often suggested that sleep and particular sleep stages (e.g., REM) play an important role in memory, perhaps consolidating important information or erasing trivial information gathered by the waking brain. According to several models, memory involves an initial encoding stage followed by a longer period of consolidation, with some models suggesting that ensemble interactions during REM and SWS contribute to memory consolidation. Several research groups have demonstrated that neurons within the hippocampus that fire together during exploration fire together and in temporally related sequences during both REM and SWS. Could these structured interactions during sleep support memory consolidation processes? What are the mechanisms that give rise to these interactions?

Information about the spatial extent and temporal synchronization of hippocampal and neocortical networks during both asleep and awake brain states may offer insight into the mechanism and function of large-scale integration and segregation of ensemble activity. Do the coordinated discharges of hippocampal neurons during SPWs actually influence neocortical neurons? What influence does the concerted action of neocortical neurons during sleep rhythms (spindles, delta waves, slow oscillation) have on hippocampal ensemble activity? Sleep spindles and delta waves reflect a concerted discharge of thalamic, thalamocortical, and neocortical neurons, and several findings now link the occurrence of SPWs to the synchronization of neocortical neurons during sleep spindles and delta waves. Thus, fairly large networks of hippocampal, thalamic, and neocortical neurons have the potential to synchronize their activity during SPWs. Do these coordinated interactions support the specific transfer of information between hippocampal and neocortical cell assemblies?

We suggest that structured interactions among neurons during awake and sleep serve as building blocks for systems-level functions. The dialogs between the hippocampus and neocortex during SPWs, during both awake and sleep states, may play an important role in consolidating hippocampal-dependent memory traces in neocortical stores. A protracted consolidation process, taking place over days/months, could then support the observation that hippocampal insults do not affect memories acquired at some point prior to insult (equally days/months/years). The logic and evidence in

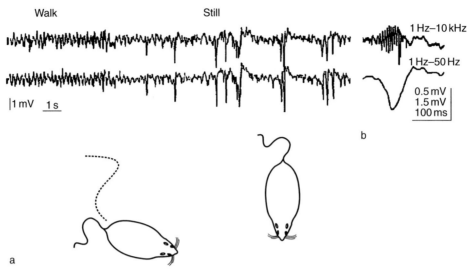

a

b

Figure 1 Behavior-dependent macroscopic (electroencephalogram) states in the hippocampus. (a) Extracellular recordings from the CA1 stratum radiatum of the left (top) and right (bottom) hippocampus during the transition from exploration walking to being still. Note regular theta waves during walking, and large, negative sharp waves during immobility. Note also that sharp waves are bilaterally synchronous. (b) A single sharp wave with simultaneously recorded fast-field oscillation from the CA1 pyramidal layer at a faster time scale. Adapted from Buzsaki G (1989) Two-stage model of memory formation: A role for 'noisy' brain states. *Neuroscience* 31: 551–570, with permission from Elsevier.

support of a role for SPWs and neocortical oscillations in hippocampal-dependent memory, and their role in hippocampal–neocortical dialogs during SWS, are overviewed in the following sections.

The Basics: A Two-Stage Model of Memory Trace Formation

The neuronal activity of the hippocampal formation (hippocampus, subiculum, and entorhinal cortex) is organized around two population patterns, theta and SPW. Theta and SPW patterns seem to serve companion processes. The theta pattern occurs during exploratory behavior and REM sleep. During theta, hippocampal circuits receive rhythmic input from neurons within the entorhinal cortex. The entorhinal cortex transmits information from neocortical association areas (e.g., perirhinal, parahippocampal) to the hippocampus. Thus, theta synchronizes the input pathways into the hippocampus from the neocortex.

In contrast, SPWs occur during consummatory behavior and SWS. During SPWs, the output neurons of the hippocampal formation participate in organized population bursts. During each burst, SPW neurons in the CA1, subiculum, and deep layers of the entorhinal cortex fire fast-frequency population volleys at ~200 Hz. The fast-frequency discharge is accompanied by a fast-frequency (~200 Hz) field potential referred to as a 'ripple' near the perisomatic region. The ripple field potential reflects synchronized

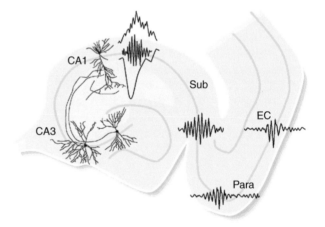

Figure 2 Self-organized burst of activity in the hippocampal CA3 region produces a field potential in the dendritic layer of CA1 and a short-lived fast-frequency field oscillation (200-Hz ripple) within stratum pyramidale, as well as a phase-related discharge of the neurons. The hippocampal output, in turn, produces similar sharp wave–ripple complexes in the subiculum (Sub), parasubiculum (Para), and deep layers of the entorhinal cortex (EC). Adapted from Chrobak JJ and Buzsaki G (2005) Synaptic plasticity and self-organization in the hippocampus. *Nature Neuroscience* 11: 1560–1567, with permission.

inhibitory input to the soma from local γ-aminobutyric acid (GABA)ergic neurons (e.g., basket cells).

Each SPW is a transient (40–100 ms) population volley that involves the discharge of roughly 40 000 to 80 000 neurons in the CA3–CA1–subicular complex–entorhinal cortex axis within a 50- to 200-ms time frame. Both SPWs and their associated fast-frequency ripples (**Figure 2**) are present in the hippocampus of all mammals investigated, including humans. Several features of this fast-frequency population volley make this network pattern an excellent candidate for altering synaptic connectivity. First, the time window of the SPW population burst roughly corresponds to the time constant of the N-methyl-D-aspartate (NMDA) receptor, through which Ca^{2+}, the key ion for the induction of synaptic plasticity, enters the neuron. Second, the fast-frequency network output (~200 Hz) is the ideal frequency range for initiating long-term potentiation (LTP). Third, during the SPW burst there is a three- to fivefold gain of network excitability, creating an ideal condition for synaptic potentiation. Last, the SPW engages the neurons that make up the output network of the hippocampus (CA1, subiculum, and entorhinal cortex). This network logically would transfer information back to neocortical structures, and SPW-related discharge has been observed in a number of hippocampal/entorhinal target structures, including the perirhinal cortex, amygdala, ventral striatum, and prefrontal cortices.

Hippocampal Replay of Waking Patterns during Sleep

Neuronal activity during sleep may promote memory-related synaptic plasticity more effectively than does neuronal activity during awake states. Why? First, neurons, free from the influence of sensory-driven input, can discharge independently and perhaps randomly, thereby erasing synaptic modifications brought about by the awake brain. The expected result could be to refresh or wipe clean the synaptic modifications associated with previous experience. This would provide a fresh 'tabula rasa' every morning, ready for information from the new day. Alternatively, neuronal pathways used and modified in the awake brain could be replayed repeatedly, and the repeated activation could be used to enhance important/relevant synaptic modifications. The molecular processes supporting synaptic potentiation are quite protracted (hours/days), thus perhaps necessitating repeated instantiation of the patterns supporting potentiation.

Because SPWs occur during immobility, during consummatory behaviors, and, mostly, during SWS, it was inferred that SPWs during SWS could play an important role in memory consolidation. The repeated replay of ensemble patterns (representations) related to recent ensemble patterns could provide a dynamic network mechanism for the slow, offline, transfer of hippocampal computations to the neocortex. Some specific predictions of the model are that (1) neurons do not randomly participate in SPW volleys; (2) synaptic connectivity, firing rates, and population patterns

during SWS remain similar unless specifically altered by awake experience; and (3) SPWs induce synaptic modifications in the output networks of the hippocampal formation.

Several physiological experiments support the conjectures and predictions of the two-stage model of memory consolidation. Ensembles of neurons discharging in spatial specific (place cells) environments continue to fire together at a high rate during subsequent SWS and REM sleep episodes (**Figure 3**). This replay of neuronal patterns has been observed in both rats and monkeys and has also been observed during SPWs during awake immobility as well as during SWS. Neurons 'tied' together by sensory experience(s) in the awake world tend to discharge together whenever the probability of their discharge is increased. In most of these studies, no specific 'learning' experience took place. Place (spatial)-specific neurons fired as a rat navigated on a linear track or circular environment, and subsequently discharged together some time later, during both awake states and during subsequent sleep (both REM sleep and SWS). Typically, the probability of two or more neurons discharging together is increased during sleep directly following an experience, as compared to sleep before spatial exploration. In addition to showing that neurons fire together, these experiments demonstrate that the specific sequence of place cells that discharge as a rat navigates an environment replay later in the same or related sequences. The degree to which these sequential patterns are maintained over time (hours to days)

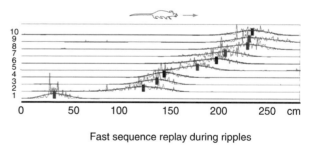

Fast sequence replay during ripples

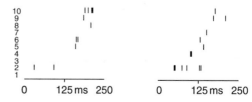

Figure 3 Temporally compressed replay of learned neuronal sequences. Top: Spatiotemporal firing rate changes of 10 CA1 place units on an elevated track. Bottom: Representative examples of firing patterns of the same units during two ripple events of nonrapid eye movement sleep. Note the similarity between ripple-related spike sequences and the order of place-cell activation on the track. Adapted from Lee AK and Wilson MA (2002) Memory of sequential experience in the hippocampus during slow wave sleep. *Neuron* 36: 1183–1194, with permission from Elsevier.

has been somewhat equivocal and may vary as a function of experience and/or novelty.

The latter can be examined by comparing firing rates and coactivation of neuronal groups in sleep episodes interrupted by performance in a well-trained task, such as food foraging in an open field and running in a wheel. The discharge frequency of concurrently recorded pyramidal neurons remains remarkably stable across several sleep/wake cycles. Neurons with high firing rates during wheel running (associated with theta) continue to fire at a high rate during drinking or awake immobility. Similarly, fast-firing neurons in SWS sustain their relatively high rates during REM sleep. When firing rates of neuronal populations are compared in two sleep episodes, interrupted by a wheel-running session or foraging in an open field, discharge frequencies of individual cells are robustly correlated. Thus, coactivation patterns during sleep episodes preceding and following an awake session are very strongly correlated. Several studies suggest that an underlying anatomical or anatomical/physiological matrix governs the propensity of neurons to fire, and the probability that they fire together is simply altered by awake experience. Any recent alterations in the probability of firing are maintained across subsequent sleep episodes.

All told, these findings demonstrate that sensory-related firing patterns during awake states are correlated with similar patterns during subsequent periods of immobility and during sleep states. Do these correlated patterns demonstrate a unique role for sleep patterns in maintaining ensemble representations of prior events? No, they do not. The correlated patterns demonstrate that neurons that fire together in a relatively stable network generally fire together when any member of the ensemble discharges. Important caveat in these studies is that causality is derived from the difference between sleep sessions before and after exploration, while many studies demonstrate considerable nonspecific difference of network activity patterns between these seemingly equivalent states. While subtle modifications, small perturbations, in the stability of a relatively hard-wired network could support mnemonic representations and be used for recognition and retrieval of prior events, the described studies do not demonstrate that they do. One possible way to attribute causality to sleep–wake correlated patterns is to manipulate or perturb the system.

Novel Experience Alters Firing Patterns in Subsequent Sleep

Can synaptic change associated with a specific experience persist in such a way that changes could be detected in subsequent sleep? Few studies have examined the role of learning on ensemble firing patterns. Exposure to a novel situation can result in long-lasting changes in the firing patterns of hippocampal pyramidal neurons. The correlated firing rates of individual neurons during exploration of a novel environment and subsequent sleep episode are stronger than are the correlated rates between exploration and the preceding sleep episode. Importantly, firing rates observed in the sleep episodes before and after exposure to novelty are poorly correlated (**Figure 4**). Exploration in a novel environment has a similar effect on correlated activity among neurons, with neuronal pairs exhibiting high awake correlations continuing to display high correlation during subsequent sleep episodes.

A more direct support for the 'replay' of learned neuronal patterns comes from experiments in which, instead of two neurons firing together, three or more neurons (up to six) exhibit sequential temporal relationships during awake and subsequent sleep. Typically neurons involved in the sequences were both identical and replayed much faster during sleep than in the waking animal. The dynamics of each ripple indicate that tens of thousands of hippocampal output neurons discharge as a population volley at 200 Hz. In summary, specific neuronal sequences during SWS are determined by prior sensory experience and by the dynamics of the SPW ripple process.

Interactions between Neocortex and Hippocampus during Sleep

Though the hippocampus is critical to the encoding of new memories, it is fairly well understood that the hippocampus plays a time-limited role in memory consolidation. After an ill-defined time period (days, weeks, and months), most memories can be retrieved in the absence of the hippocampus. Thus, a gradual translation or translocation of neural patterns to neocortical targets is presumed to mediate 'long-term' memory storage. A number of models posit that the hippocampus plays a role in bringing ensembles of cortical neurons together to form a linked 'chain' of information related to a particular memory. The role of the hippocampus is to activate each link in the chain, allowing for memory retrieval, and/or to forge the links such that the entire chain can become hippocampal independent. In the absence of hippocampal input, as occurs in human temporal lobe amnesia or experimental animals, the links in the chain cannot readily be forged or retrieved. With an intact hippocampus, the repeated replay of hippocampal patterns during offline states, during REM or SPWs in SWS, may forge the neocortical links such that memories can eventually be retrieved without the hippocampus.

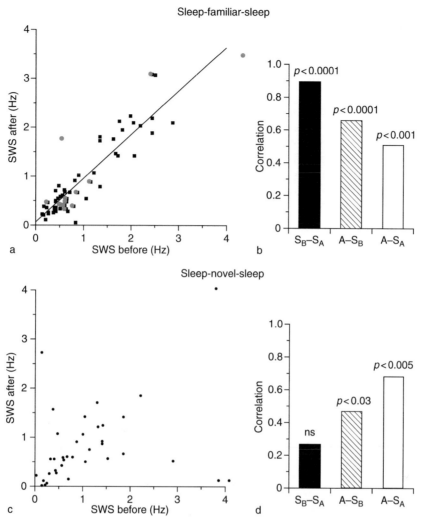

Figure 4 Comparison of discharge frequency of pyramidal cells in two successive slow-wave sleep (SWS) episodes, interrupted either (a) by a wheel running session (familiar) or (c) by novel exploration. Mean correlation values for 'sleep before' vs. 'sleep after' (S_B–S_A), 'awake' vs. 'sleep before' (A–S_B), and 'awake' vs. 'sleep after' (A–S_A) sessions in the well-trained running wheel task (b) and in the novel environment (d). Note high and low correlations between successive sleep sessions in the familiar and novel environments, respectively. Adapted from Hirase H, Leinekugel X, Czurko A, et al. (2001) Firing rates of hippocampal neurons are preserved during subsequent sleep episodes and modified by novel awake experiences. *Proceedings of the National Academy of Sciences of the United States of America* 98: 9386–9390, with permission.

The SPWs seem like an ideal candidate mechanism for the gradual offline strengthening of neocortical patterns related to memory, regardless of how the process is conceptualized.

Deep slow-wave sleep is characterized by widespread synchronized oscillatory patterns, defined primarily by spatially coherent delta waves during which spiking activity of nearly all principal cells and interneurons in the neocortex is suspended. The intervals between the silent periods define slow oscillations (0.3–1.5 Hz). The relatively abrupt shifts in population firing and silence states are reflected by the bimodal (depolarized and hyperpolarized or UP and DOWN) distribution of the membrane potential in neocortical, entorhinal, and subicular neurons. The UP state is often associated with K-complexes and thalamocortical sleep spindles (12–18 Hz).

Several studies have described a concurrence of hippocampal SPWs in association with neocortical sleep spindles and delta waves. The excitatory front of the UP cortical states can spread to the hippocampus via the entorhinal cortex, as reflected by the active sinks in the dentate molecular layer and stratum lacunosum-moleculare of the CA1 region, the main targets of the entorhinal cortex. As a result of the strong neocortical drive, dentate granule cells and other neurons of the dentate area fire coherently with the input from the entorhinal neurons during the slow oscillation. The rest of the hippocampus is less affected, mainly due to the strong feed-forward inhibitory circuits between granule cells and CA3 pyramidal cells. Nevertheless, the majority of the output of CA1 pyramidal cells also increases their discharge rate in phase with the cortical UP state, likely driven by direct perforant path input. Important in the present context, a majority of SPW/ripple complexes are also associated with the enhanced neocortical activity during slow-wave sleep: the probability of occurrence of SPWs/ripples is highest at and after the onset of the cortical UP state. Furthermore, SPWs are modulated by single cycles by neocortical spindles (**Figure 5**), suggesting a more direct role for neocortical inputs in triggering SPWs.

In summary, cortical slow oscillations temporally coordinate the occurrence of K-complexes, thalamocortical spindles, and hippocampal SPWs. Such periodic 'rebooting' of neuronal activity in widespread neocortical, paleocortical, and hippocampal networks should have important functional consequences. Could specific subtle patterns of neocortical inputs 'select' the ensemble participants in each specific SPW? Is the neocortex querying the hippocampus about the relationship between ensemble members as instantiated in the neocortical input? The possible scenario of such a mechanism is as follows: After each DOWN state, the neocortex self-organizes its global activity from locally generated patterns. The DOWN–UP transition, perhaps involving thalamocortical spindle, triggers organized firing patterns of cortical neurons, which, in turn, can lead to activation of specific subpopulations of hippocampal neurons. As a result of the neocortex-biased selection, the activated hippocampal neurons can generate SPW-related synchronous output. Because the SPW is a punctuate event, whereas the UP state and sleep spindle are temporally protracted, the hippocampal output can be directed to the still-active neocortical assemblies. We hypothesize that the temporal coordination of neocortical and hippocampal strong bursting patterns favors the conditions in which the hippocampal output may assist in modifying neocortical synaptic connectivity.

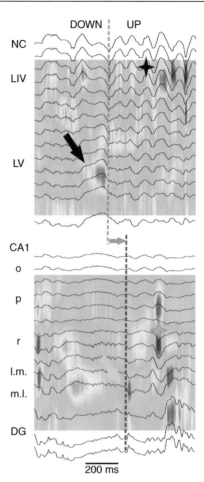

Figure 5 Spread of neocortical activity into the hippocampus during slow-wave sleep. Current-source density and superimposed local field potential traces of simultaneously recorded events in the neocortex (NC; top) and the hippocampus (bottom). A delta wave, seen as the red source in layer 5 (LV; black arrow), is followed by a sleep spindle (black star). DOWN–UP state transition is indicated by the orange dashed line. In the hippocampus, the entorhinal cortex-mediated DOWN–UP transition is reflected by a phase reversal of the local field potential and a sink (blue) in the molecular layer (white arrow). The horizontal arrow denotes the neocortico-dentate time lag (~100 ms); gray star denotes sharp wave in CA1 stratum radiatum; o, stratum oriens; p, pyramidal layer; r, radiatum; l.m., lacunosum-moleculare; m.l., molecular layer; DG, dentate gyrus. Adapted from Isomura Y, Sirota A, Ozen S, et al. (2006) Integration and segregation of activity in entorhinal-hippocampal subregions by neocortical slow oscillations. *Neuron* 52: 871–882, with permission from Elsevier.

Conclusions

The collective behavior of neurons (e.g., theta and gamma oscillations, SPWs and related 200 Hz ripples, and sleep spindles) offers a link between complex cognitive functions and behaviors governed by the brain and the biophysical, synaptic properties of individual neurons. Many of these collective behaviors can be observed in both the awake and sleeping brain. Understanding the differences between these collective patterns in the awake and sleeping brain should provide further insight into function. The current assumption about this function is that experience-initiated changes of synaptic connectivity are consolidated during sleep by way of transient, reciprocal, and strongly synchronous interactions between the neocortex and hippocampus.

See also: PET Activation Patterns; Sleep Architecture; Sleep Oscillations and PGO Waves; Sleep-Dependent Memory Processing.

Further Reading

Born J, Rasch B, and Gais S (2006) Sleep to remember. *Neuroscientist* 12(5): 410–424.

Buzsaki G (1989) Two-stage model of memory formation: a role for 'noisy' brain states. *Neuroscience* 31: 551–570.

Chrobak JJ and Buzsaki G (2005) Synaptic plasticity and self-organization in the hippocampus. *Nature Neuroscience* 11: 1560–1567.

Foster DJ and Wlson MA (2006) . *Nature* 440: 680–683.

Hirase H, Leinekugel X, Czurko A, et al. (2001) Firing rates of hippocampal neurons are preserved during subsequent sleep episodes and modified by novel awake experiences. *Proceedings of the National Academy of Sciences of the United States of America* 98: 9386–9390.

Isomura Y, Sirota A, Ozen S, et al. (2006) Integration and segregation of activity in entorhinal–hippocampal subregions by neocortical slow oscillations. *Neuron* 52: 871–882.

Kudrimoti HS, Barnes CA, and McNaughton BL (1999) Reactivation of hippocampal cell assemblies: Effects of behavioral state, experience, and EEG dynamics. *Journal of Neuroscience* 19: 4090–4101.

Lee AK and Wilson MA (2002) Memory of sequential experience in the hippocampus during slow wave sleep. *Neuron* 36: 1183–1194.

Nadasdy Z, Hirase H, Czurko A, et al. (1999) Replay and time compression of recurring spike sequences in the hippocampus. *Journal of Neuroscience* 19: 9497–9507.

Peigneux P, Laureys S, Fuchs S, et al. (2004) Are spatial memories strengthened in the human hippocampus during slow wave sleep? *Neuron* 44: 535–545.

Sejnowski TJ and Destexhe A (2000) Why do we sleep? *Brain Research* 886(1–2): 208–223.

Siapas AG and Wilson MA (1998) Coordinated interactions between hippocampal ripples and cortical spindles during slow-wave sleep. *Neuron* 21: 1123–1128.

Sirota A, Csicsvari J, Buhl D, et al. (2002) Communication between neocortex and hippocampus during sleep in rodents. *Proceedings of the National Academy of Sciences of the United States of America* 100: 2065–2069.

Steriade M and Timofeev I (2003) Neuronal plasticity in thalamocortical networks during sleep and waking oscillations. *Neuron* 37: 563–576.

Tononi G and Cirelli C (2006) Sleep function and synaptic homeostasis. *Sleep Medicine Reviews* 10(1): 49–62.

Wilson MA and McNaughton BL (1994) Reactivation of hippocampal ensemble memories during sleep. *Science* 265(5172): 676–679.

Network Reactivation

S Ribeiro, Edmond and Lily Safra International Institute of Neuroscience of Natal, and Universidade Federal do Rio Grande do Norte, Natal, Brazil
M Nicolelis, Duke University Medical Center, Durham, NC, USA, and École Polytechnique Fédérale de Lausanne, Lausanne, Switzerland

Introduction

The persistence of neural activity following a given stimulus or behavior is a widespread phenomenon in vertebrates. Simple introspection suffices to experience the phenomenon, which occurs during waking as well as during sleep. Network reactivation occurs, for instance, when a song heard by chance on the radio lingers in one's auditory imagination for days thereafter. Such reverberation seems to reflect a particularly robust encoding of the song memory, causing an involuntary recollection that runs on top of the heavy background of sensory interference that typically characterizes waking. When occurring during sleep, network reactivation gives rise to dreams as well as nonvivid thoughts ('mentation'), depending on the specific sleep state considered.

The Canadian psychologist Donald Hebb was probably the first to point out that patterns of neuronal activity present during memory encoding remain in the brain after stimulus cessation, maintaining the freshly acquired information 'alive' until structural cellular changes, able to transform a short-lived reverberatory trace into a long-lasting memory, have time to occur. In the past 20 years, evidence has accumulated suggesting that network reactivation is a critical step for learning, being related to many of its idiosyncrasies. It has also been demonstrated that network reactivation is enhanced during sleep, offering a plausible explanation for the beneficial role of sleep in memory consolidation. In addition, network reactivation elucidates the persistence of salient waking experiences during dreaming, a phenomenon labeled 'day residue' by Sigmund Freud. To date, network reactivation has been detected with a variety of techniques from the cellular to the systems level, including the transcriptional upregulation of activity-dependent genes, single neuron activity recorded with microwires, and large-scale neural activity measured by electroencephalographic waves and cerebral blood flow. This article reviews the properties and functions of network reactivation.

Feedback and Rhythms

The mechanistic investigation of network reactivation dates back to the pioneering studies conducted by Lorente de Nó on the dynamics of electrical excitation in hardwired neuronal loops. The famous disciple of Santiago Ramón y Cajal was interested in understanding how activity induced in a given neuron would be passed on to other cells so as to eventually return to the stimulus origin by way of recurrent connections. Based on much anatomy and some physiology, Lorente de Nó proposed that feedback loops comprising multiple neurons echo activation upon stimulation of a single neuron, creating depolarization cycles that only dissipate after several laps take place.

The notion of small-circuit network reactivation enthused neuroscientists in the twentieth century because recursive processes such as this could in principle implement clocks, pacemakers, and memory building blocks. This enthusiasm led to the discovery of specialized neuronal loops that generate recurrent traveling waves within and across brain structures. The combination of recursive architecture, local inhibitory neurons, and the modulatory input of different neurotransmitters (e.g., acetylcholine) gives rise to rhythms of variable duration which are characteristic of global brain states such as waking, sleep, and dreaming. Different substates occur within the major states, such as very long episodes of sustained regional oscillation, including theta rhythm (4–9 Hz in different species), which is composed of sinusoidal oscillations that strongly engage the hippocampus during alert waking as well as during rapid eye movement (REM) sleep, the dreaming phase of sleep in humans. Other rhythms include short-lived oscillations in the range of milliseconds to seconds, such as the cortical spindles (10–14 Hz) that are particularly abundant during stage 2 sleep, constituting its defining characteristic. Stage 2 sleep is one of the substates of slow-wave sleep (SWS), the nondreaming phase of sleep in humans.

The span of brain oscillations deemed behaviorally relevant goes from very slow, whole brain waves (<1 Hz) to very fast ripples (~200 Hz) that engage small groups of hippocampal neurons. All these oscillations depend on the interaction of specific neural circuits, such as the septohippocampal axis for theta rhythm, the thalamocortical loop for spindles, and intrahippocampal networks for ripples. Oscillations may coexist in time and space, phase locking at specific moments so as to optimize communication among brain areas. For instance, slow oscillations of the membrane potential in the cerebral cortex constrain neurons to 'upstates' and 'downstates.' The slow oscillatory upstate represents the depolarizing phase of the slow oscillation, leading to discrete bursts of cortical activity that are followed by hippocampal sharp wave/ripples complexes, preferentially triggered by the upstate.

Reactivation, Reverberation, and Replay

Postulated by Donald Hebb more than 50 years ago, the term 'reverberation' was replaced in the past few decades by 'reactivation' and 'replay,' but neither term fully accounts for the general phenomenon in question. Although network reactivation decreases substantially during waking, it does not disappear completely when animals are awake. This was demonstrated by electrophysiological studies in rodents and can easily be verified by introspection. For most people, it suffices to remain quiet under reduced sensory stimulation in order for mnemonic images to appear. The available data indicate that salient experience causes sustained neural reverberation rather than discrete reactivation, in the sense that memory traces are continuously detectable during subsequent periods across all behavioral states in a state-dependent manner. Reverberating patterns of neural activity associated with past experience are largely, but not completely, masked during waking by incoming sensory inputs.

Another related term that became well known in the early 1990s is memory replay, which implies a high-fidelity repetition of past patterns of neural activity. However, the accuracy of neuronal reverberation is still controversial. In the rodent hippocampus, the path run by the animal during wakefulness can be very accurately re-created based on the spatiotemporal neuronal firing patterns observed during sleep. In contrast, researchers found low values for firing rate correlations among neurons in the cerebral cortex, suggesting that cortical reactivation implements a low-fidelity emulation of memories rather than a perfect replay. This is probably related to the fact that many portions of the mammalian cerebral cortex are dedicated to the joint representation of multiple perceptions or acts. As a consequence, single cells are recruited into many different neuronal groups, multiplexing information in a way that likely blurs the detectable network reactivation of any discrete engram. A notable counterpoint is found in songbirds: The nucleus RA, single output of the telencephalic song circuitry, displays a highly accurate replay of song-related firing patterns (**Figure 1(d)**). This remarkably rare case of high-fidelity replay derives from the very specialized neural processing carried out by this

motor nucleus, namely the descending activation of a single, unique, and invariant sequential engram: the bird's own song.

A Link with Memory

Despite the importance of brain rhythms to explain major differences among behavioral states, nowadays it is recognized that the kind of network reactivation that underlies the recall of fine-grain memories arises from the interaction of much more complex neural circuits, in which the reverberation of neuronal activity does not derive trivially from a preexisting anatomical loop. Beginning in the 1970s, the investigation of primates performing working memory tasks showed that neurons in the prefrontal cortex show persistent firing rate changes during delay periods that separate stimulation from behavioral response. Currently, there is a consensus that network reactivation in the prefrontal cortex and other associative areas of intricate connectivity is directly implicated in the temporary storage of stimulus-related information required for correct behavioral responses. These so-called 'memory neurons' can remain active for many seconds after stimulus removal, keeping information about the past available for future decision making. In these cases, the experimental task shaped the neuronal behavior, uncovering neurons able to reverberate information on voluntary demand. This kind of delayed goal-directed reactivation is extremely important in nature for hunters and hunted alike. However, it does not account for all the time we spend in spontaneous network reactivation, either day- or night dreaming.

Network Reactivation Increases during Sleep

A major breakthrough in the search to identify the mechanisms underlying spontaneous network reactivation resulted from an experiment on the poststimulus sleep activity of hippocampal neurons performed by Constantine Pavlides and Jonathan Winson at the Rockefeller University in the late 1980s. Their elegant study took advantage of the fact that pyramidal hippocampal neurons in the rat tend to fire very selectively according to the spatial position of the animal in a given environment, hence the name 'place cells.' Using chronic electrode implants in the hippocampus, the experimenters first identified and recorded pairs of neurons with nonoverlapping place fields – that is, places in which the neurons fired the most. Then, rats were restricted from entering the place field of either cell overnight. Finally, animals were confined for 10–15 min within the place field of one of the cells so as to strongly stimulate one neuron while suppressing the activity of the yoked one. By

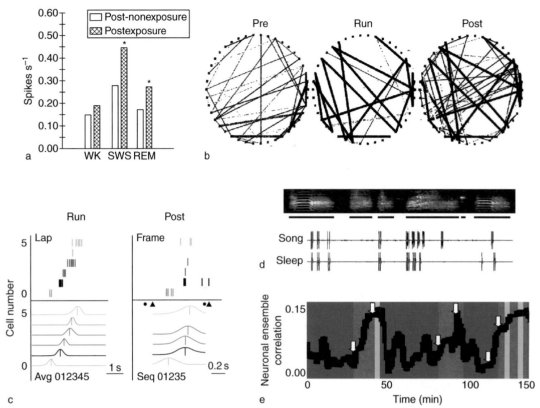

Figure 1 Neuronal reactivation in rats and birds. (a) Hippocampal place cells are reactivated during SWS and REM sleep after waking (WK) exposure to their corresponding place fields. *$p < 0.05$. (b) For hippocampal place cells, pairwise neuronal correlations established during waking exposure to a circular maze (RUN) are preserved during subsequent SWS (POST), while being absent from prior SWS (PRE). In each circle, neurons are represented by dots over the perimeter, and correlations are represented as lines whose thickness is proportional to correlation strength. (c) SWS replays neuronal firing sequences observed during RUN in the hippocampus. Notice the temporal compression of the neuronal patterns during SWS. Shown are firing patterns during a single running lap on the left–right trajectory (left) or during an equivalent sleep 'frame' period (right). One neuron per row, one spike per tick. (d) Premotor neurons in nucleus RA of a zebra finch accurately replay singing-specific activity during sleep. Shown are raw traces of neuronal activity (900 ms) recorded during singing (premotor activity) and sleep (spontaneous activity). A color spectrograph of the song that the bird sang is shown on top, with horizontal bars indicating different song syllables. (e) To estimate similarity with past patterns of neuronal ensemble activity produced by the exploration of novel objects in rats, neuronal activity patterns were compared with extensive recordings of neuronal action potentials using a template-matching algorithm. As indicated by the white arrows, neuronal ensemble correlations increase gradually between waking (blue) and sleep (red for SWS and green for REM sleep). (a) Adapted from Pavlides C and Winson J (1989) Influences of hippocampal place cell firing in the awake state on the activity of these cells during subsequent sleep episodes. *Journal of Neuroscience* 9(8): 2907–2918. (b) Adapted from Wilson MA and McNaughton BL (1994) Reactivation of hippocampal ensemble memories during sleep. *Science* 265(5172): 676–679. (c) Adapted from Ji D and Wilson MA (2007) Coordinated memory replay in the visual cortex and hippocampus during sleep. *Nature Neuroscience* 10(1): 100–107. (d) Adapted from Dave AS and Margoliash D (2000) Song replay during sleep and computational rules for sensorimotor vocal learning. *Science* 290(5492): 812–816. (e) Adapted from Ribeiro S, Gervasoni D, Soares ES, et al. (2004) Long-lasting novelty-induced neuronal reverberation during slow-wave sleep in multiple forebrain areas. *PLoS Biology* 2(1): 126–137.

quantifying the neuronal firing rates during and after spatial confinement across the wake–sleep cycle, Pavlides and Winson found that changes in firing rates recorded during waking persist during ensuing SWS and REM sleep (**Figure 1(a)**). In other words, place cells exposed to their place fields prior to sleep showed increased firing rates during SWS and REM sleep compared to place cells that were not exposed to their place fields.

In 1994, Matthew Wilson and Bruce McNaughton showed that pairs of neurons recorded in the hippocampus carry into subsequent SWS the temporal firing

relationships observed when animals run a circular path beforehand (**Figure 1(b)**). Follow-up of these discoveries by different research groups established that reactivation is strongest and most stable during SWS, becomes more variable during REM sleep, and decreases during waking. The difference between sleep states probably arises from the overall low activation of most of the brain during SWS, in contrast with the elevated neural activation that characterizes REM sleep, which is so high as to be considered waking-like, hence the alias 'paradoxical sleep' preferred by most European neuroscientists. Given the consistent increase in network reactivation during SWS, its higher variance during REM sleep, and the short duration of total REM sleep in comparison with total SWS (~1:4 in rats and humans), one must conclude that SWS plays the major role in network reactivation, although dreaming sleep (i.e., REM sleep) also participates in the process.

Reactivation and Learning

The great importance ascribed to network reactivation by neuroscientists stems from its putative link with memory consolidation. Although the early breakthroughs in the field were mostly the result of electrophysiological investigation of rodent neurons, it was the survey of human brains with electroencephalography, positron emission topography, and functional magnetic resonance imaging that established a firm link between learning and network reactivation (**Figure 2(a)**). Postacquisition reverberation of regional brain activity during sleep has been shown to be proportional to memory acquisition and to quantitatively predict learning. Such reverberation is reflected in local increases of blood oxygenation as well as the amplitude of low-frequency electroencephalographic waves (delta band, <4 Hz) (**Figure 2(b)**). If learning coincides with an increase in slow oscillatory activity, the converse is also true: The application of 0.75 Hz pulses by transcranial magnetic stimulation during SWS enhances the retention of hippocampus-dependent

memories in humans, but similar stimulation at 5 Hz does not produce the effect.

Network reactivation during sleep seems to favor the anatomical reorganization of memory traces, an increasingly recognized hallmark of memory consolidation. In the case of spatial memory tasks, which depend on the hippocampus for initial encoding, sleep favors a displacement of the activation elicited by task performance, from the hippocampus to striatal networks. Indeed, different lines of evidence suggest that network reactivation during sleep is directly implicated in the temporally graded propagation of memories from the hippocampus to cortical areas, a phenomenon made famous by patient H.M., who developed anterograde amnesia for declarative memories after bilateral removal of the hippocampus. Likewise, sleep may promote engram propagation in the case of implicit memory tasks, which initially involve the cerebellum but progressively engage the basal ganglia. To summarize, network reactivation during sleep seems to play a rather important role in the strengthening and relocation of memory traces within the brain.

Characterization of Reactivation Properties

One possible explanation for the increase in network reactivation during sleep is that endogenous brain activation, freed from sensory interference during

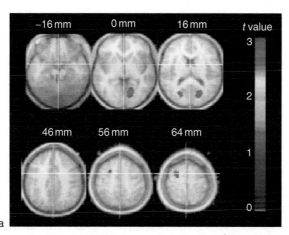 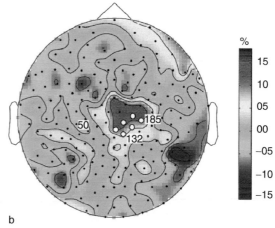

Figure 2 Regional brain reactivation in humans. (a) Brain regions reactivated during REM sleep in subjects previously trained in a visuomotor task. Shown are brain maps of regional cerebral blood flow as measured by positron emission tomography. The activated regions comprise the cuneus and adjacent striate striate cortex bilaterally, the left premotor cortex, and the mesencephalon. Maps show six different brain levels (from 16 mm below to 64 mm above the bicommissural plane), superimposed on the average MRI image of the sleeping subjects. (b) The amplitude of slow waves recorded with electroencephalography increases locally in subjects trained in a motor task. Shown is the topographic distribution of the percentage change in the amplitude of slow waves between the task condition and a control task, void of motor learning. White dots indicate the cluster of six electrodes showing increased amplitude of slow waves after motor learning. (a) Adapted from Maquet P, Laureys S, Peigneux P, et al. (2000) Experience-dependent changes in cerebral activation during human REM sleep. *Nature Neuroscience* 3(8): 831–836. (b) Adapted from Huber R, Ghilardi MF, Massimini M, and Tononi G (2004) Local sleep and learning. *Nature* 430: 78–81.

off-line states, flows preferentially through previously activated synapses. In principle, this would suffice to cause neuronal firing patterns originally produced during waking to reverberate significantly above chance levels during sleep. For instance, telencephalic reactivation may be driven by the pontine waves that are particularly abundant during REM sleep. To date, experience-dependent network reactivation during sleep has been observed in rodents, macaques, humans, and songbirds, indicating a very widespread biological phenomenon. Sleep promotes the coordinated reverberation of waking-acquired activity patterns across large groups of neurons in the hippocampus, primary sensory and associative cortices, basal ganglia, and even primary sensory thalamic relays. The synchronization of neuronal firing by slow cortical oscillations seems to generate discrete activity frames between the hippocampus and the cerebral cortex, promoting coordinating interactions across these brain regions during the membrane potential upstate (i.e., the depolarizing phase of the slow oscillation).

It is unclear whether reverberation is favored by novel stimuli, which are more salient for perception, or by well-trained stimuli/behaviors, which are deeply engrained by habit. Results favoring either option coexist in the literature. There is evidence that major changes in firing rate patterns last longer in the cerebral cortex than in the hippocampus, but more research is needed to elucidate this issue. The duration of network reactivation is also under intense debate. Whereas several studies point to fast-fading effects lasting less than 1 h, some experiments have detected signs of mnemonic reverberation for up to 48 h after stimulus removal.

It has been proposed that network reactivation during sleep occurs under different regimes of time compression or expansion, perhaps allowing for more efficient mnemonic processing. Some studies indicate that the experience-dependent reverberation of neuronal firing patterns is hastened during SWS, presumably because the activity bursts characteristic of that state (e.g., hippocampal ripples) promote the fast replay of entire neuronal firing sequences (**Figure 1(c)**). Similarly, there is some evidence suggesting that network reactivation is slowed during REM sleep, but a consensus regarding this issue is yet to be achieved. Another point of contention, with controversial evidence, is whether sleep also harbors antireverberation – that is, patterns of neuronal ensemble activity statistically more dissimilar from the waking reference patterns than expected by chance. In principle, the novelty-induced reverberation and antireverberation of neuronal firing patterns could play balancing roles in the delineation of specific memory traces, embossed in the synaptic landscape of the forebrain.

Reactivation at the Molecular Level

The very wide prevalence of network reactivation within the brain and across species suggests that it derives from extremely conserved elements of the vertebrate nervous system, such as the major anatomical connections and the molecular backbone shared by all neurons. As discussed previously, the overall plan of the vertebrate brain comprises several circuits prone to recursive activity, such as the thalamocortical loop. The conspicuous rhythms implemented by these circuits are ridden by finer-grain reactivation at the level of single neurons. Such neuronal reactivation, rather than large-scale rhythms, constitutes the most likely substrate for the recall of specific memories.

The acquisition of a single new memory involves activity-dependent molecular mechanisms able to store information at the synaptic level from many minutes to a few hours. Most of these mechanisms depend critically on the rise of calcium, either permeated through membrane channels (e.g., NMDA voltage-gated channels) or as a result of increased intracellular release. Elevated calcium levels activate phosphorylating enzymes (kinases) that trigger major changes in the internal biochemical milieu of the cell. Network reactivation immediately after memory acquisition seems to require, for instance, the phosphorylation of calcium/calmodulin kinase II (CaMKII), an autocatalytic enzyme that can by itself sustain local changes in synaptic strength. In this case, the substrate of reactivation is not an anatomical loop but, rather, a molecular one since CaMKII maintains memory by persistent self-phosphorylation.

Whereas the initial stabilization of recently acquired memories relies on spatially restricted synaptic mechanisms such as CaMKII activation, it is well established that the long-lasting persistence of mnemonic traces requires *de novo* protein synthesis. Indeed, calcium-dependent kinase activation triggers signaling cascades that couple membrane depolarization to genomic regulation inside the cell nucleus, leading to the transcriptional upregulation of several immediate early genes (IEGs) required for the long-term maintenance of memory. One of these genes is the activity-regulated cytoskeleton-associated protein (Arc), a calcium-dependent IEG that interacts with glutamatergic AMPA receptors, actin, and CaMKII to promote synaptic remodeling. Arc mRNA is transported to dendrites for local translation, leading to retrograde (presomatic) effects. Another activity-dependent gene regulated by calcium is zif-268 (also known as egr-1, krox-24, NGFI-A, and ZENK), which encodes a transcription factor with binding sites on the promoters of hundreds of different genes. The zif-268 protein is thought to modulate synaptic remodeling by controlling the expression of genes directly involved in synaptic function, such as synapsins, the most abundant protein constituent of synapses, involved in the regulation of synaptic vesicle release. The transcriptional control of synapsin levels by the zif-268 protein illustrates how neuronal depolarization can prompt anterograde (postsomatic) synaptic remodeling. Such a mechanism could, in principle, disseminate synaptic changes downstream of long neuronal chains, effectively propagating memory traces across separate circuits.

The large-amplitude neural oscillations typical of SWS promote marked periodic fluctuations of calcium levels in activated synapses, as indicated by extracellular calcium measurements during sleep. As a consequence, it has been hypothesized that SWS is concomitant with the activation of multiple calcium-dependent kinases, with a role in memory formation. This would result in a pretranscriptional amplification of synaptic changes encoding novel memory traces during SWS, as suggested by recent data (**Figure 1(e)**, white arrows). Although experimental evidence linking all the elements of these calcium-dependent metabolic cascades to sleep-dependent network reactivation is missing, results indicate that REM sleep triggers the activation/upregulation of several calcium-dependent molecules required for neuronal plasticity and memory

formation, such as zif-268, arc, cyclic AMP response element-binding protein, and brain-derived nerve growth factor (**Figure 3(a)**). The sleep-dependent transcriptional upregulation of the zif-268 gene seems to also reveal memory trace relocation from the hippocampus to the cerebral cortex (**Figure 3(b)**). The activity-dependent anterograde response caused by the zif-268 gene represents a putative molecular mechanism for the propagation of synaptic changes across neuronal circuits, leading to memory relocation.

The fact that neuronal activity can be traced to molecular markers of plasticity establishes a strong parallel between the cellular and systems levels, offering support for the Hebbian notion that memory requires two consecutive and distinct steps, namely network reactivation for short-term recall and synaptic remodeling for long-term storage. The current results suggest that SWS and REM sleep play distinct and complementary roles in memory consolidation, with memory recall (network reactivation) occurring mainly during SWS and memory storage (plasticity-related gene) sparked off during REM sleep. Such a mechanism fulfills early conceptual notions of a two-step process for memory consolidation during sleep and is in line with evidence that SWS and REM sleep have synergistic effects on human procedural learning and developmental plasticity.

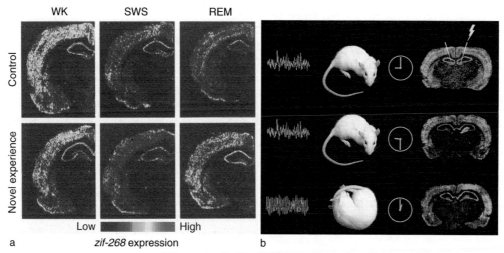

Figure 3 Molecular reactivation. (a) The mRNA levels of the activity-dependent gene *zif-268*, a transcription factor with an important role in memory consolidation, are upregulated during REM sleep in rats previously exposed to novel experience but not in control animals kept in their habitual home cages. Shown are autoradiograms of brain sections subjected to radioactive *in situ* hybridization. (b) The unilateral induction of hippocampal long-term potentiation during waking leads to marked unilateral gene upregulation in the hippocampus 30 min after high-frequency electrical stimulation (yellow arrow). Four hours later, during REM sleep, zif-268 is downregulated in the hippocampus but strongly induced in extrahippocampal regions such as the cerebral cortex and the amygdala, indicating that REM sleep disengages the hippocampus and engages extrahippocampal regions. Traces on the left represent hippocampal local field potentials typical of waking (top and middle) and REM sleep (bottom). (a) Reproduced from Ribeiro S, Goyal V, Mello CV, and Pavlides C (1999) Brain gene expression during REM sleep depends on prior waking experience. *Learning & Memory* 6(5): 500–508, with permission from Gold spring Harbor Laboratory press. (b) Adapted from Ribeiro S, Mello CV, Velho T, Gardner TJ, Jarvis ED, and Pavlides C (2002) Induction of hippocampal long-term potentiation during waking leads to increased extrahippocampal zif-268 expression during ensuing rapid-eye-movement sleep. *Journal of Neuroscience* 22(24): 10914–10923.

Dreams, Waking, and Sleep Mentation

In adult humans, REM sleep is almost always accompanied by dreaming, as can be easily demonstrated by waking experimental subjects during REM sleep and requesting dream reports. During human REM sleep, a selected set of forebrain areas are reactivated, including portions of the hypothalamus, amygdala, septum, and ventral striatum, as well as the anterior cingulate, orbitofrontal, entorhinal, and insular cortices. The dorsolateral prefrontal cortex, however, shows reduced activity. Given the role of prefrontal cortex activity in the executive control of behavior, its reduction during REM sleep has been proposed to underlie the poor logical concatenation and reduced self-awareness of dreaming in comparison with waking.

Dreams may have first evolved as a collateral effect of extended REM sleep, a characteristic mammalian trait. Pet owners know that cats and dogs seem to act out dreams during sleep. More controlled evidence of dreaming in nonhuman mammals was obtained by lesions of the brain stem nuclei that promote muscle atonia during REM sleep. Cats with such lesions sleep quietly through SWS, but upon entering REM sleep they become suddenly agitated by vigorous species-specific behaviors, such as meowing and pouncing. How are dreams generated, and what is their biological function?

As noted by Sigmund Freud, dreams often involve elements of the experience of the preceding day(s), the day residue. Although high-fidelity memory reactivation must be at the roots of the cognitive functions of sleep and dreams, it does not account entirely for the symbolic complexity that characterizes the oneiric narrative. After all, it is not common to dream about the exact repetition of waking scenes. On the contrary, most human dreams are characterized by the intrusion of illogical elements, leading to unforeseen associations. Human dreams are subjective narratives composed of familiar and unfamiliar beings, things, and places interacting around a self-representation of the dreamer that mostly observes an unfolding plot. Dreams vary in intensity, ranging from confused and faint impressions to complex time-evolving narratives with vivid imagery and surprising turns. Although dreams tend to be dominated by visual images, they can also involve combinations of auditory, olfactory, tactile, gustatory, motor, vestibular, and linguistic modalities. Dreams can sometimes be extremely pleasant or just the opposite, but they are usually characterized by a mix of emotions. Dreams are also hyperassociative, linking characters, places, and actions in bizarre ways. Dreams can also anticipate events of the coming day(s), particularly when subjects undergo extreme anxiety and expectation. A good example is provided by the dreams of students during the night before taking difficult exams; these dreams often contain detailed anticipatory simulations of the expected challenges, in content and/or context.

Freud also postulated that dream narratives fulfill wishes (or anti-wishes) of the awake subject, simulating the realization (or frustration) of specific desires. Whereas the prevalence of overt wish fulfillment in the dreams of normal adults is low, it is common in young children. Children living in war zones have a much higher incidence of nightmares, a result that has been interpreted as evidence that the function of dreams is to simulate environmental threats. Based on the existence of blissful dreams, other neuroscientists have expanded the threat-simulation theory to include the simulation of rewarded behaviors as a dream function. The notion of dreaming as behavior simulation leading to either reward or punishment is supported by the neuropsychological investigation of dreamless patients. These patients, who often conserve REM sleep, represent rare clinical cases with lesions of mesocorticolimbic pathways connecting reward centers with the thalamus, striatum, and cerebral cortex. Although REM sleep is spared in these cases, the loss of dreaming is accompanied by subjective reports of sleep disturbance, as assessed by qualitative interviews. This suggests a role for dreaming in sleep maintenance, which is another Freudian postulate.

The current evidence indicates that dreams reflect the intense and noisy network reactivation that takes place during extended REM sleep, with the cerebral cortex lightened up at high neuronal firing rates and oscillatory frequencies predominantly above 30 Hz (gamma band). The qualitative feeling of 'quasi-reality' in dreams derives from the fact that network reactivation during REM sleep occurs at an intensity and oscillatory spectrum comparable to those of waking. Such high levels of excitation cause the reverberating memories that comprise dreams to appear bright and vivid to the self-representation, and variations in neural activity levels explain the dynamic range of vividness that characterizes normal dreaming. Likewise, the simultaneous occurrence of neuronal reverberation in multiple forebrain sites during REM sleep explains the wide variation in dream 'sensory' modality. There is also a compelling relation between the higher variability of network reactivation during REM sleep and the fragmentation, condensation, and bizarreness of dreams. Far from being random, dream narratives highlight waking events according to how recent, novel, and behaviorally significant they were (i.e., dreams are directed by the anxieties and preoccupations of the dreamer). Dreams seldom occur during SWS. Instead, the subjective experience of SWS consists of low-intensity but coherent thoughts resembling waking reasoning. The coherence of such 'mentation' likely reflects the high stability of neuronal reverberation during SWS. Similarly, the lack of intense imagery during SWS probably reflects the decreased cortical activity and slow oscillations that characterize network reactivation in this state.

Concluding Remarks

Network reactivation is a fundamental property of neuronal circuits that occurs at several levels, ranging from molecular and cellular mechanisms to the consequent systemic and behavioral manifestations. Reactivation is based on the existence of recurrent biological loops, from self-activating enzymes to large networks of correlated neurons distributed over multiple brain regions, with various degrees of duration. Although reactivation is favored by the interruption of sensory and motor interference provided by sleep, data suggest that it occurs nonstop during waking as well, spanning one's entire lifetime.

See also: Gene Expression; Hippocampal–Neocortical Dialog; History of Sleep Research; PET Activation Patterns; REM/NREM Differences in Dream Content; Sleep Architecture; Sleep Oscillations and PGO Waves; Sleep-Dependent Memory Processing; Theories of Dream Function.

Further Reading

Dave AS and Margoliash D (2000) Song replay during sleep and computational rules for sensorimotor vocal learning. *Science* 290(5492): 812–816.

Hebb DO (1949) *The Organization of Behavior: A Neuropsychological Theory.* New York: Wiley.

Huber R, Ghilardi MF, Massimini M, and Tononi G (2004) Local sleep and learning. *Nature* 430: 78–81.

Ji D and Wilson MA (2007) Coordinated memory replay in the visual cortex and hippocampus during sleep. *Nature Neuroscience* 10(1): 100–107.

Llinas RR and Pare D (1991) Of dreaming and wakefulness. *Neuroscience* 44(3): 521–535.

Maquet P, Laureys S, Peigneux P, et al. (2000) Experience-dependent changes in cerebral activation during human REM sleep. *Nature Neuroscience* 3(8): 831–836.

Marshall L, Helgadottir H, Molle M, and Born J (2006) Boosting slow oscillations during sleep potentiates memory. *Nature* 444(7119): 610–613.

Nádasdy Z, Hirase H, Czurkó A, Csicsvari J, and Buzsáki G (1999) Replay and time compression of recurring spike sequences in the hippocampus. *Journal of Neuroscience* 19(21): 9497–9507.

Orban P, Rauchs G, Balteau E, et al. (2006) Sleep after spatial learning promotes covert reorganization of brain activity. *Proceedings of the National Academy of Sciences of the United States of America* 103(18): 7124–7129.

Pavlides M and Winson J (2006) Influences of hippocampal place cell firing in the awake state on the activity of these cells during subsequent sleep episodes. *Journal of Neuroscience* 9(8): 2907–2918.

Ribeiro S, Goyal V, Mello CV, and Pavlides C (1999) Brain gene expression during REM sleep depends on prior waking experience. *Learning & Memory* 6(5): 500–508.

Ribeiro S, Gervasoni D, Soares ES, et al. (2004) Long-lasting novelty-induced neuronal reverberation during slow-wave sleep in multiple forebrain areas. *PLoS Biology* 2(1): 126–137.

Ribeiro S, Mello CV, Velho T, Gardner TJ, Jarvis ED, and Pavlides C (2002) Induction of hippocampal long-term potentiation during waking leads to increased extrahippocampal zif-268 expression during ensuing rapid-eye-movement sleep. *Journal of Neuroscience* 22(24): 10914–10923.

Ribeiro S and Nicolelis MAL (2004) Reverberation, storage and postsynaptic propagation of memories during sleep. *Learning & Memory* 11(6): 686–696.

Wang XJ (2001) Synaptic reverberation underlying mnemonic persistent activity. *Trends in Neurosciences* 24(8): 455–463.

Wilson MA and McNaughton BL (1994) Reactivation of hippocampal ensemble memories during sleep. *Science* 265(5172): 676–679.

Winson J (1985) *Brain and Psyche.* New York: Anchor Press.

Gene Expression

C Cirelli and G Tononi, University of
Wisconsin–Madison, Madison, WI, USA

The Molecular Correlates of Sleep May Offer Clues to Its Functions

All animal species studied to date sleep, and sleep may serve multiple functions, which may differ in different species. It is possible, however, that sleep also has a core function that is conserved from invertebrates to mammals. If this is the case, that function is most likely a cellular one because flies and humans share most pathways for intercellular and intracellular signaling, from membrane receptors and ion channels to nuclear transcription factors, but differ significantly in the number, anatomy, and complexity of brain circuits. The idea that the functions of sleep may ultimately relate to cellular and molecular aspects of neural function is not new. Giuseppe Moruzzi argued that "sleep concerns primarily not the whole cerebrum, nor even the entire neocortex, but only those neurons or synapses, and possibly glial cells, which during wakefulness are responsible for the brain functions concerned with conscious behavior." Moruzzi suggested that neural cells or synapses supporting waking conscious activity undergo plastic changes that make sleep necessary, although he did not speculate about the mechanisms. Others have suggested that sleep may help restore brain energy metabolism, may be needed to maintain the synaptic efficacy of the neural circuits not frequently used during wakefulness, or may favor a generalized synaptic downscaling following waking-induced synaptic potentiation. It has also been suggested that the rich spontaneous activity of neocortical neurons during slow-wave sleep may consolidate memory traces acquired during wakefulness. Several of these hypotheses are not mutually exclusive, and sleep may actually favor different cellular processes.

The analysis of the molecular correlates of sleep and wakefulness may help to understand the benefits that sleep brings at the cellular level. Specifically, the identification of the genes whose expression changes in the brain between sleep and wakefulness may clarify if, and why, brain cells need sleep and why their functions are impaired if they are prevented from doing so during sleep deprivation. In the past 15 years, subtractive hybridization, mRNA differential display, and microarrays have emerged as new powerful methods in transcriptomics, the study of the complete set of RNA transcripts produced by the genome at any one time. Microarrays are currently the most comprehensive and sensitive method to detect genome-wide expression changes, and they have allowed the identification of hundreds of transcripts whose expression is modulated by behavioral state. In addition, transcriptomics analysis has also identified functional categories of genes whose expression varies during sleep, wakefulness, and after different periods of sleep deprivation. At least some of these gene categories are conserved from invertebrates to mammals.

Table 1 lists the studies that have used transcriptomics approaches to identify genes whose transcript (mRNA) levels vary as a function of sleep, wakefulness, and/or sleep deprivation. It is clear from this table that the main focus has been on the cerebral cortex, although other brain regions have been studied, including cerebellum, hypothalamus, and brain stem. The cerebral cortex is of major interest because it is responsible for the cognitive defects observed after sleep deprivation, and it is at the center of most hypotheses concerning the functions of sleep.

The following general conclusions can be drawn from the studies listed in **Table 1**, and they are discussed further in this article:

- Hundreds of genes in the brain change their expression due to a change in behavioral state.
- Some genes increase their expression during sleep, whereas others do so during waking and sleep deprivation.
- Sleep and waking genes belong to different functional categories.
- Three functional categories of waking genes are conserved from flies to mammals.
- The noradrenergic system plays a role in the induction of some waking genes.
- The expression of several genes changes after chronic, but not acute, sleep loss.

Hundreds of Genes in the Brain Change Their Expression between Sleep and Waking

A first important finding of transcriptomics studies is that gene expression in the brain changes extensively as a function of behavioral states, sleep, spontaneous wakefulness, and short-term (a few hours) or long-term (several days) sleep deprivation.

Table 1 Studies that use transcriptomics approaches to identify molecular correlates of sleep (S), wakefulness (W), and sleep deprivation (SD)

Study[a]	Method used/no. of tested transcripts	Experimental details	Major findings
1	Subtractive hybridization (~4000 transcripts)	24 h SD (forced locomotion) vs. time-matched controls; rat forebrain	Four transcripts ↓ by SD, six transcripts ↑ by SD, all unknown
2	mRNA differential display (~9000 transcripts)	3 h S, W, SD (novel objects); rat cerebral cortex	One unknown transcript ↑ in S; 11 transcripts ↑ in W-SD, including c-fos, NGFI-A, and several mitochondrial genes
3	mRNA differential display (~9000 transcripts) + cDNA arrays (Clontech, 1176 transcripts)	8 h S, W, SD (novel objects); rat cerebral cortex	12 transcripts ↑ in S (all unknown except membrane protein E25); 44 transcripts ↑ in W-SD, including IEGs and heat shock proteins
4, 5	cDNA arrays (Clontech, 1176 + 588 +140 transcripts)	6 h SD (gentle handling), 6 h SD + 4 h recovery S vs. time-matched controls; mouse cerebral cortex (changes must be >100%); four array replicas	IEGs and heat shock proteins ↑ during SD Grp94 ↑ in recovery sleep
6	Affymetrix GeneChips, >24 000 probe sets, >15 000 reliably detected	8 h S, W, SD (novel objects); rat cerebral cortex (changes must be >20%); three to five array replicas; two statistical methods; qPCR; estimated false positives ~20%, based on qPCR confirmation of 52 transcripts (using independent groups of animals)	106 known transcripts and 155 ESTs ↑ in S; 95 known transcripts and 395 ESTs ↑ in W-SD
7	mRNA differential display (~5000 transcripts) + Affymetrix GeneChips, ~8000 probe sets, ~5000 reliably detected	8 h SD with or without prior LC lesion (DSP-4); rat cerebral cortex (changes must be >20%); three array replicas; two statistical methods; estimated false positives <20%, based on qPCR confirmation of 32 transcripts (using independent groups of animals)	20% (16/95) of the known transcripts ↑ in W-SD downregulated after cortical depletion of norepinephrine
8	Affymetrix GeneChips, 18 955 probe sets, ~9400 reliably detected	9 h S, W, SD (forced locomotion); fly head (changes must be >50% in two fly lines); four array replicas; one statistical method; estimated false positives <20%, based on qPCR confirmation of 19 transcripts (using independent groups of animals)	12 transcripts ↑ in S and 121 transcripts ↑ in W-SD
9	Affymetrix GeneChips, 1322 probe sets, ~600 reliably detected	6 h SD (gentle handling), 6 h SD + 2 h recovery S vs. time-matched controls; rat cerebral cortex, hypothalamus, basal forebrain (changes must be >50%); two array replicas	
10	Affymetrix GeneChips, 26 261 probe sets, ~17 000 reliably detected	1-week SD (disk-over-water method) compared to yoked controls, 9 h W, and 9 h SD; rat cerebral cortex (changes must be >20%); three to eight array replicas; one statistical method; estimated false positives <5%, based on qPCR confirmation of 15 transcripts (using independent groups of animals)	75 transcripts ↑ and 28 transcripts ↓ after long-term SD

[a]Studies: 1, Rhyner TA, Borbely AA, and Mallet J (1990) Molecular cloning of forebrain mRNAs which are modulated by sleep deprivation. *European Journal of Neuroscience* 2: 1063–1073; 2, Cirelli C and Tononi G (1998) Differences in gene expression between sleep and waking as revealed by mRNA differential display. *Brain Research: Molecular Brain Research* 56: 293–305; 3, Cirelli C and Tononi G (2000) Gene expression in the brain across the sleep–wakefulness cycle. *Brain Research* 885: 303–321; 4, Terao A, Steininger TL, Hyder K, et al. (2003) Differential increase in the expression of heat shock protein family members during sleep deprivation and during sleep. *Neuroscience* 116: 187–200; 5, Terao A, Greco MA, Davis RW, Heller HC, and Kilduff TS (2003) Region-specific changes in immediate early gene expression in response to sleep deprivation and recovery sleep in the mouse brain. *Neuroscience* 120: 1115–1124; 6, Cirelli C, Gutierrez CM, and Tononi G (2004) Extensive and divergent effects of sleep and wakefulness on brain gene expression. *Neuron* 41: 35–43; 7, Cirelli C and Tononi G (2004) Locus coeruleus control of state-dependent gene expression. *Journal of Neuroscience* 24: 5410–5419; 8, Cirelli C, LaVaute TM, and Tononi G (2005) Sleep and wakefulness modulate gene expression in *Drosophila*. *Journal of Neurochemistry* 94: 1411–1419; 9, Terao A, Wisor JP, Peyron C, et al. (2006) Gene expression in the rat brain during sleep deprivation and recovery sleep: An Affymetrix GeneChip study. *Neuroscience* 137: 593–605; 10, Cirelli C, Faraguna U, and Tononi G (2006) Changes in brain gene expression after long-term sleep deprivation. *Journal of Neurochemistry* 98(5): 1632–1645.
ESTs, Expressed sequence tags; IEGs, immediate early genes.

Five percent or more of the transcripts tested in the brain show changes in gene expression according to behavioral state. Interestingly, a similar percentage of genes (1–10%) change because of differences in circadian time. In the cerebral cortex of rats, out of the 15 459 transcripts we tested, 808 (5.2%) were affected by time of day independently of behavioral state, and 752 (4.9%) were affected by sleep and wakefulness independently of time of day. These data indicate that, at least in the rat, day/nighttime and sleep/wakefulness influence gene expression in the cerebral cortex to a similar extent. Peripheral tissues such as liver and heart also show significant changes in gene expression as a function of circadian time. Whether this is also true for sleep-dependent and wakefulness-dependent genes is unknown because none of the array studies published so far has focused on peripheral tissues.

Some Genes Increase Their Expression during Sleep, Whereas Others Do So during Waking and Sleep Deprivation

We found that in the rat cerebral cortex, 95 known genes and 395 expressed sequence tags (ESTs; i.e., short transcribed sequences whose function is unknown) show higher expression after 8 h of spontaneous wakefulness or sleep deprivation, whereas 106 known genes and 155 ESTs show increased expression after 8 h of sleep. In the fly head, for which we used the same experimental protocol previously applied to rats (three experimental groups, including 8 h of sleep, 8 h of spontaneous wakefulness, and 8 h of total sleep deprivation), we identified 121 waking transcripts, including both known genes and ESTs, and 12 sleep transcripts, mostly unknown. Finally, we compared gene expression in the cerebral cortex of long-term sleep-deprived rats and sleeping controls and found that 226 transcripts are upregulated after long-term sleep deprivation and 113 after sleep. Thus, it seems that if all transcribed sequences are considered (including ESTs), there are more genes whose expression is upregulated during waking than during sleep. However, hundreds of genes are also upregulated during sleep, confirming the results of intracellular recording studies that show that sleep is far from being a quiescent state of global inactivity. In general, the largest fold changes are observed for waking genes. However, this is mainly due to the fact that waking increases the expression of transcription factors (e.g., c-fos) and effector genes (e.g., Arc) that behave as 'immediate early genes,' whose mRNA and protein levels may increase severalfold relative to sleep. Most waking and sleep genes, instead, change their

expression by 20–50%, a fold change similar to that seen for most genes whose expression varies because of circadian time.

Sleep and Waking Genes Belong to Different Functional Categories

Sleep and waking transcripts belong to different functional categories and thus may favor different cellular processes. Most genes whose expression increases during wakefulness (or short-term sleep deprivation) code for proteins involved in energy metabolism, synaptic excitatory transmission, transcriptional activity, acquisition of new information, and the response to cellular stress. Sleep genes, instead, code for proteins involved in protein synthesis, synaptic downscaling and/or memory consolidation, lipid metabolism, and in membrane trafficking and maintenance.

Three Functional Categories of Waking Genes Are Conserved from Flies to Mammals

Three functional categories of genes are consistently induced in several animal species (fruit flies, mice, rats, and hamsters) during waking and short-term sleep deprivation relative to sleep. The first one includes genes involved in energy metabolism, including mitochondrial genes such as cytochrome oxidase (CO) I and IV. One study in rats also found an increase in CO enzymatic activity after 3 h of wakefulness. In mice, 6 h of sleep deprivation also affects the mRNA levels of genes involved in glycogen metabolism (glycogen synthase and glycogen phosphorylase) and increases the activity of glycogen synthase, possibly to counterbalance the rapid depletion, at the transition from sleep to wakefulness, of astrocytic glycogen stores. Eight hours of spontaneous waking or sleep deprivation also induces the expression in the rat cerebral cortex of *Glut1*, the gene coding for one of the more widely expressed glucose transporters in the central nervous system. The upregulation during waking of genes involved in energy metabolism is perhaps not surprising because glucose metabolism in many brain regions is higher in wakefulness than in non-rapid eye movement (NREM) sleep, and the brain relies almost exclusively on glucose as its energy substrate. These changes, however, do not seem to persist when sleep deprivation is protracted for 24 h or several days. This is consistent with imaging studies in animals, normal human subjects, and fatal familial insomnia patients showing no change or a decrease in cerebral metabolic rate after prolonged sleep loss relative to normal wakefulness.

Thus, the rapid upregulation of mitochondrial genes, *Glut1*, and glycogen-related genes may represent a mechanism by which the brain responds to the immediate increase in energy requirement during spontaneous waking or after a few hours of sleep deprivation, whereas prolonged sleep loss is associated with reduced energy demands.

A second group of genes consistently upregulated during wakefulness and downregulated during sleep includes heat shock proteins, chaperones, and, in general, genes involved in the response to cellular stress. Cellular stress is the cell's reaction to any adverse environmental condition that perturbs cellular homeostasis, with potential macromolecular damage (i.e., damage to proteins, DNA, RNA, and lipids). Wakefulness and short-term deprivation are associated with higher mRNA levels of small heat shock proteins with actin-stabilizing properties, such as HSP27 and αB crystallin; molecular chaperones (HSP70 and HSP60); and chaperones in the endoplasmic reticulum, such as BiP and Grp94. BiP (immunoglobulin heavy chain-binding protein) shows the largest and most consistent upregulation in the brain of awake animals, including mice, rats, hamsters, and sparrows, as well as in the head of fruit flies. BiP assists in the folding and assembly of newly synthesized glycoproteins and secretory polypeptides and increases in *BiP* expression have been classically described after heat shock and ischemia, when unfolded proteins accumulate in the endoplasmic reticulum. This unfolded protein response is also induced by other conditions that adversely affect the function of the endoplasmic reticulum, such as anoxia, glucose deprivation (BiP is also called glucose-regulated protein 78), and depletion of Ca^{2+} stores in the endoplasmic reticulum. Importantly, the unfolded protein response also triggers a slowing down of protein synthesis. In the rat cerebellum, 8 h of spontaneous wakefulness or sleep deprivation is associated with an increase in the mRNA levels of *PEK*. *PEK* codes for the EIF2a kinase that mediates the inhibition of protein synthesis during the unfolded protein response. Moreover, in the mouse cerebral cortex, 6 h of sleep deprivation is also associated with signs of slowing down of protein synthesis, as indicated by the phosphorylation of key components of the translational machinery and by changes in the ribosomal profile.

A third group of genes consistently upregulated during wakefulness and downregulated during sleep includes transcripts sensitive to membrane depolarization and synaptic activity. In the rat cerebral cortex, they include *RGS2*, *Homer/Vesl*, tissue-type plasminogen activator, casein kinase 2, cyclooxygenase 2, *cpg2*, and connexin 30. Moreover, several genes involved in activity-dependent neural plasticity and long-term potentiation, such as *Arc*, *BDNF*, *Homer/Vesl*, and *NGFI-A*, have been consistently found to be upregulated in the cerebral cortex of awake or short-term sleep-deprived rats. At least some of the same genes are induced during wakefulness and/or short-term sleep deprivation in mice, hamsters, and sparrows. In the fly head, many of the depolarization-sensitive genes involved in synaptic plasticity as identified by Guan et al. in 2005 are also waking-related genes, including *Astray*, *Kek2*, *stripe*, and *Hr38*. The fly wakefulness-related gene *stripe* is a member of the Egr family of transcription factors. In the rat brain, several members of this family, including *NGFI-A* (*Egr-1*), *krox 20* (*Egr-2*), and *NGFI-C* (*Egr-4*), are induced during wakefulness. *Hr38*, another fly wakefulness-related gene, is the *Drosophila* ortholog of the mammalian NGFI-B family of orphan nuclear receptors. *NGFI-B* is a wakefulness-related gene in the rat cerebral cortex. Thus, it seems that in both mammals and flies several plasticity-related genes involved in synaptic growth and potentiation are specifically induced during wakefulness.

The observation that genes involved in long-term potentiation and the acquisition of new memories are specifically induced during wakefulness should not be too surprising. After all, we learn when we are awake and able to adapt to the environment, and not when we are asleep. However, if sleep is not involved in learning, is there a role for it in other aspects of synaptic plasticity? Well-controlled experiments show that sleep may favor the performance of some, but not all, tasks. The strongest and more consistent data relate to procedural tasks and suggest an important role for NREM sleep or a combination of NREM and REM sleep. What remains unknown, however, are the mechanisms by which sleep may enhance performance. Animal and human studies have shown that brain circuits activated during the execution or the acquisition of certain tasks are 'reactivated' during the subsequent sleep period. The presence of this memory trace, as shown by electrophysiological recordings and imaging studies, could reflect the fact that sleep replays the waking experience, and by doing so it somehow allows the specific consolidation of the learned tasks. However, it is also possible that this trace is the inevitable consequence of the synaptic potentiation that occurs during learning, and that sleep enhances performance by actually allowing a generalized downscaling of synapses. The first and most important effect of this downscaling is thought to be the decrease in the metabolic cost of synaptic activity, which increases during waking due to synaptic potentiation. A second effect of the generalized but proportional decrease in synaptic weight is an increase in the signal-to-noise ratio, which could improve the performance of some learned tasks.

If sleep is associated with a generalized depression of synaptic strength, molecular markers of synaptic depression should be increased during sleep relative to waking. Studies have identified and validated several specific and direct markers of synaptic strength, with the best established among them being the density, composition, and phosphorylation state of the GluR1 subunit of the glutamatergic AMPA receptors. We are currently measuring the number and phosphorylation state of synaptic GluR1 in the rat cerebral cortex after 6 h of sleep or sleep deprivation. Preliminary results show that in several cortical areas, sleep is associated with a relative dephosphorylation of GluR1 at serine 845, a finding consistent with the hypothesis that sleep is associated with synaptic downscaling.

The Noradrenergic System Plays a Role in the Induction of Some Waking Genes

One of the mechanisms that underlie the widespread changes in cortical gene expression between sleep and wakefulness is the activity of the noradrenergic system of the locus coeruleus (LC), whose neurons project diffusely over the entire brain. LC cells are tonically active during wakefulness, reduce their firing rate during NREM sleep, and cease firing during REM sleep. Moreover, LC activity increases phasically in response to salient events and in relation to the decision to act. In a series of experiments, rats were subjected to unilateral or bilateral LC lesions to deplete one or both sides of the brain of norepinephrine. In these animals, the raw electroencephalogram (EEG) and its power density spectrum were not significantly affected by the lesion. However, all cortical areas depleted of norepinephrine showed a marked decrease, during wakefulness, of the expression of plasticity-related genes such as *Arc*, *BDNF*, *NGFI-A*, and *P-CREB* as well as of stress response genes such as heat shock proteins and *BiP*. By contrast, the transcript for the translation elongation factor 2 was the only known sleep-related transcript whose expression increased after LC lesion (**Figure 1**). In a complementary experiment in mice, the activity of the LC of one side was increased using a conditional transgenic approach. This manipulation resulted in an increased ipsilateral expression of *NGFI-A* in cortical and subcortical target areas. Thus, LC activity during wakefulness modulates neuronal transcription to favor synaptic potentiation and memory acquisition and to counteract cellular stress, whereas LC inactivity during sleep may play a permissive role to enhance brain protein synthesis.

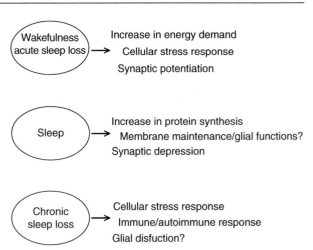

Figure 1 Schematic representation of the major functional categories of genes upregulated in the rat cerebral cortex after 3–8 h of spontaneous waking or sleep deprivation (acute sleep loss), 1 week of total sleep deprivation (chronic sleep loss), and 8 h of spontaneous sleep.

The Expression of Several Genes Changes after Chronic, but Not Acute, Sleep Loss

In an array study, we analyzed the expression of more than 26 000 transcripts in the cerebral cortex of long-term sleep-deprived rats. Animals were kept awake for 7 days using the disk-over-water method. This method uses minimal stimulation to enforce chronic sleep deprivation in the sleep-deprived rat, while it simultaneously applies to the control rat the same stimulation but without severely limiting its sleep. The sleep-deprived and the control rats are each housed on one side of a divided horizontal disk suspended over a shallow tray of 2 or 3 cm deep water. EEG and electromyogram are continuously recorded to detect sleep states. When the experimental rat falls asleep, the disk is automatically rotated at low speed, awakening the rat and forcing it to walk opposite to disk rotation to avoid being carried into the water. The yoked control rat receives the same physical stimulation because it is on the same disk. However, whereas sleep is severely reduced (by 70–80%) in the sleep-deprivation rat, the control rat can sleep whenever the sleep-deprived rat is spontaneously awake and eating, drinking, or grooming, and thus its sleep is reduced by only 25–40%.

We compared long-term sleep-deprived rats with rats spontaneously awake or sleep deprived for only 8 h and identified 75 transcripts upregulated and 28 transcripts downregulated after chronic sleep loss. Most transcripts specifically increased or decreased their expression due to the long-term sleep deprivation and not because of the disk stimulation, because they were not affected in the yoked controls. A large group of genes upregulated after prolonged sleep loss codes for immunoglobulins, including two autoantibodies. Another group includes stress response genes such as the macrophage inhibitor factor-related- protein MRP14, expressed in microglia. Arylsulfotransferase, an enzyme involved in the catabolism of catechol-amine, is also upregulated after chronic sleep deprivation. These genes are either not induced (antibodies and MRP14) or induced to a lesser extent (arylsulfotransferase) after acute sleep loss, suggesting that short-term and long-term sleep deprivation differ to some extent at the molecular level. We also compared long-term and short-term sleep deprivation to sleep and found that only a minority of genes were similarly up- or downregulated in the two sleep- deprivation conditions. Among the genes whose expression was most significantly downregulated in long-term sleep-deprived animals relative to sleeping animals were myelin constituents, including the most abundant structural protein component of myelin – the proteolipid protein. These findings suggest that sustained sleep loss may trigger a generalized inflammatory and stress response in the brain. Some glial cells (e.g., microglia) may help to protect neurons against this cellular insult, but others (e.g., oligodendrocytes) may also suffer some of its negative consequences. It will be important in the future to test whether sleep loss may be detrimental to the maintenance of cellular membranes and specifically to myelin.

Genes Upregulated during Sleep

Figure 1 shows the sleep-related transcripts identified so far in the rat cerebral cortex. Unfortunately, only one study in mammals was designed to identify genes upregulated after several (8) hours of sleep relative to both spontaneous waking and sleep deprivation to rule out effects of stress and circadian time. Another study in flies used the same experimental conditions used in rats, but for technical reasons the entire fly head, rather than the brain only, was used to hybridize the arrays. This most likely decreased the ability to detect state-dependent genes. Only 12 genes were identified as sleep-related in flies, and the function of most of them is unknown. Thus, the extent to which sleep genes are conserved across species is yet to be determined.

Two sleep genes in the rat cerebral cortex code for the translation elongation factor 2 and the initiation factor 4AII, and they are involved in protein synthesis. The mRNA levels of the elongation factor EF2 are also increased during sleep in the brain of Djungarian hamsters. A positive correlation between sleep and protein synthesis has been suggested by previous studies, and evidence suggests that waking and sleep

deprivation are less conducive to protein synthesis than sleep. Whether sleep favors protein synthesis globally or whether it enhances the synthesis of specific classes of proteins is unclear.

Other sleep genes identified in the rat cerebral cortex are involved in the consolidation of long-term memory as well as in synaptic depression and depotentiation. Among them are calmodulin-dependent protein kinase IV, calcineurin, FK506 binding protein 12, inositol 1,4,5-trisphosphate receptor, and amphiphysin II. As mentioned previously, the mechanism by which sleep enhances memory and performance is still debated, but these molecular data are compatible with the idea that sleep benefits the brain by producing a global synaptic downscaling.

Another large group of sleep-related transcripts in the rat cerebral cortex is involved in membrane trafficking and maintenance. Some of these transcripts are important for the synthesis/maintenance of membranes in general and of myelin in particular, including oligodendrocytic genes coding for myelin structural proteins (e.g., *MOBP, MAG, plasmolipin,* and *CD9*), myelin-related receptors, and enzymes. Transcripts with higher expression in sleep also code for enzymes involved in the synthesis and transport of cholesterol, a major constituent of myelin and other membranes and an important factor in regulating synaptic efficacy. As mentioned previously, long-term sleep deprivation decreases the expression of the gene coding for proteolipid protein, a major myelin component. Intriguingly, among the few known sleep genes in *Drosophila* are the gene *anachronism* (*ana*) and three genes involved in lipid metabolism (*CG8756, CG9009,* and *CG11407*). Ana is a glycoprotein secreted by glial cells, and it controls the timing of neuroblast development. Whether *ana* plays any role in membrane trafficking and synaptic plasticity remains to be determined.

Conclusions

As it is apparent from the small number of studies in **Table 1**, transcriptomics analysis is still a work in progress, and it would be inappropriate to draw firm conclusions from what has been published to date. Continuous improvement in the array technology will certainly favor more progress in this area. For instance, the current arrays can assess the expression of 20 000–30 000 transcripts, but this is still far from a true 'whole genome' analysis. Low expression genes are often difficult to detect, especially in large and heterogeneous tissue samples such as most brain areas. Moreover, current arrays focus on the coding portion of the genome, but new 'tiling' arrays, in which probe sets are laid down at regular intervals across the entire genome, are becoming commercially available. This will permit the identification of new transcriptionally active regions of the genome that have not been tested in more conventional arrays. Moreover, proteomics, the study of the complete repertoire of proteins that contribute to a cell's physiological phenotype, is a natural next step in the study of molecular correlates of behavioral states.

See also: Hippocampal–Neocortical Dialog; Network Reactivation; PET Activation Patterns; Phylogeny and Ontogeny of Sleep; Sleep and Waking in Drosophila; Sleep Architecture; Sleep Deprivation and Brain Function; Sleep Oscillations and PGO Waves.

Further Reading

Cirelli C, Faraguna U, and Tononi G (2006) Changes in brain gene expression after long-term sleep deprivation. *Journal of Neurochemistry* 98: 1632–1645.

Cirelli C, Gutierrez CM, and Tononi G (2004) Extensive and divergent effects of sleep and wakefulness on brain gene expression. *Neuron* 41: 35–43.

Cirelli C, LaVaute TM, and Tononi G (2005) Sleep and wakefulness modulate gene expression in *Drosophila*. *Journal of Neurochemistry* 94: 1411–1419.

Cirelli C, Pompeiano M, and Tononi G (1996) Neuronal gene expression in the waking state: A role for the locus coeruleus. *Science* 274: 1211–1215.

Cirelli C and Tononi G (1998) Differences in gene expression between sleep and waking as revealed by mRNA differential display. *Brain Research: Molecular Brain Research* 56: 293–305.

Cirelli C and Tononi G (2000) Gene expression in the brain across the sleep–wakefulness cycle. *Brain Research* 885: 303–321.

Cirelli C and Tononi G (2004) Locus coeruleus control of state-dependent gene expression. *Journal of Neuroscience* 24: 5410–5419.

Guan Z, Saraswati S, Adolfsen B, and Littleton JT (2005) Genome-wide transcriptional changes associated with enhanced activity in the *Drosophila* nervous system. *Neuron* 48: 91–107.

Moruzzi G (1972) The sleep–waking cycle. *Ergebnisse der Physiologie* 64: 1–165.

Naidoo N, Giang W, Galante RJ, and Pack AI (2005) Sleep deprivation induces the unfolded protein response in mouse cerebral cortex. *Journal of Neurochemistry* 92: 1150–1157.

Nikonova EV, Vijayasarathy C, Zhang L, et al. (2005) Differences in activity of cytochrome C oxidase in brain between sleep and wakefulness. *Sleep* 28: 21–27.

Petit JM, Tobler I, Allaman I, Borbely AA, and Magistretti PJ (2002) Sleep deprivation modulates brain mRNAs encoding genes of glycogen metabolism. *European Journal of Neuroscience* 16: 1163–1167.

Rhyner TA, Borbely AA, and Mallet J (1990) Molecular cloning of forebrain mRNAs which are modulated by sleep deprivation. *European Journal of Neuroscience* 2: 1063–1073.

Taishi P, Sanchez C, Wang Y, Fang J, Harding JW, and Krueger JM (2001) Conditions that affect sleep alter the expression of molecules associated with synaptic plasticity. *American Journal of Physiology – Regulatory, Integrative and Comparative Physiology* 281: R839–R845.

Terao A, Greco MA, Davis RW, Heller HC, and Kilduff TS (2003) Region-specific changes in immediate early gene expression in response to sleep deprivation and recovery sleep in the mouse brain. *Neuroscience* 120: 1115–1124.

Terao A, Steininger TL, Hyder K, et al. (2003) Differential increase in the expression of heat shock protein family members during sleep deprivation and during sleep. *Neuroscience* 116: 187–200.

Terao A, Wisor JP, Peyron C, et al. (2006) Gene expression in the rat brain during sleep deprivation and recovery sleep: An Affymetrix GeneChip study. *Neuroscience* 137: 593–605.

Tononi G and Cirelli C (2006) Sleep function and synaptic homeostasis. *Sleep Medicine Reviews* 10: 49–62.

ONTOGENY AND PHYLOGENY OF SLEEP

Phylogeny and Ontogeny of Sleep

**J A Lesku, D Martinez-Gonzalez, and
N C Rattenborg**, Max Planck Institute for Ornithology,
Seewiesen, Germany

Introduction

Sleep is a prominent yet enigmatic animal behavior.
Despite the apparent ubiquity of sleep and associated
vulnerability that (in part) defines sleep, the functions
of sleep remain elusive. The seemingly simplest
approach to determining sleep's functions would be
to compare animals that sleep with those that do not
and then determine whether a relationship exists
between various traits thought to be functionally
involved in sleep and the presence or absence of
sleep. However all animals adequately studied sleep
in one form or another making such comparisons
currently impossible. Nevertheless, of the 30 or more
animal phyla, detailed sleep information is known
for only two: Chordata (includes vertebrates) and
Arthropoda (includes insects). Thus, there is an
opportunity for truly sleepless animals to be discovered in the future. For example, sponges (phylum:
Porifera) are nearest to the base of the Metazoan
phylogenetic tree and lack a nervous system. Based
on the generally accepted belief that a plastic nervous
system is the biological target benefiting from sleep,
sponges should not sleep.

Determining whether sleep evolved many independent times or only once early in the evolution of
animals is of fundamental importance in contemporary comparative sleep research. With the application
of genetic approaches to the study of sleep, it has
become important to establish homology between
sleep in mammals and sleep in invertebrates, such as
the fruit fly (*Drosophila melanogaster*), where gen-
omes generally have less redundancy and are amena-
ble to manipulation. Establishing homology between
sleep in fruit flies and sleep in mammals would reinforce the usefulness of invertebrate models for determining the molecular correlates of human sleep. Even
without homology at the molecular level, comparative studies can enhance our understanding of sleep.
For instance, evolutionary convergence (i.e., distantly
related species that independently evolved similar
sleep-related traits) can provide insight into sleep by
revealing overriding principles otherwise obscured by
nonessential traits specific to one lineage.

The lack of a highly resolved cladogram of sleep
across Animalia has not impeded comparative analyses of sleep. Indeed, many studies have tried to
determine the relationships between the time species
spend asleep and various constitutive, physiological,
and ecological traits. Historically, this approach has
been applied only in mammals, in which the electrophysiological correlates of sleep have been identified
and quantified in a large number of species; however,
a recent study provides insight into the correlates of
sleep in birds as well. Although the strengths of functional hypotheses derived from comparative analyses
are necessarily limited given the correlational nature
of the data, the evolutionary patterns gleaned from
such analyses, especially those that employ modern
phylogenetic statistical techniques, provide a framework for the development of experimentally testable
hypotheses.

As with comparing sleep among taxonomic groups,
changes in sleep occurring during early development
can provide insight into the functions of sleep. For
example, changes in sleep duration, intensity, or the
relative proportion of the two sleep states in mammals and birds over early ontogeny might be functionally linked to the concurrent development of the
central nervous system. In this article, we review
current knowledge of sleep in various taxonomic
groups and over early ontogeny, and we relate important patterns to existing hypotheses for the functions
of sleep. By doing so, we demonstrate the contribution that comparative sleep research has made to our
greater understanding of why we sleep.

Behavioral Definition of Sleep

Sleep is foremost an animal behavior broadly characterized by quiescence with reduced responsiveness to
stimuli. Sleep can be distinguished from other quiescent states (e.g., hibernation) by rapid reversibility to
wakefulness with sufficient stimulation and homeostatic regulation; that is, sleep shows a compensatory
increase in intensity or time following sleep loss.
Sleeping animals often retreat to a species-specific
location and assume a characteristic posture. Although the occurrence of concurrent sleep and swimming in some aquatic mammals is an exception to the
quiescence criterion, these criteria apply to the vast
majority of animals studied, ranging from insects to
mammals and birds.

Mammalian Sleep

Behavioral sleep is associated with specific changes in
brain activity. Placental and marsupial (or therian)
mammals exhibit two basic states of sleep: slow-wave
sleep (SWS), also known as non-rapid eye movement

(non-REM) sleep, and REM sleep. In nonhuman animals, the term SWS usually refers to all non-REM sleep, whereas SWS refers only to stages 3 and 4 of non-REM sleep in humans. SWS is characterized by an electroencephalogram (EEG) of low-frequency, high-amplitude activity arising from the large-scale synchronous slow oscillations of neurons in the neocortex. SWS is homeostatically regulated with SWS-related slow-wave activity (SWA) (typically 0.5–4.0 Hz EEG power density) reflecting the intensity of SWS, as arousal thresholds are correlated with the amount of SWA and SWA increases as a function of prior time awake. REM sleep is characterized by an EEG of high-frequency, low-amplitude activity similar to wakefulness, with a hippocampal theta rhythm (between 4 and 9 Hz or higher depending on the species) observed in some mammals, and an atonic electromyogram. Heart and respiratory rate are irregular during REM sleep, and thermoregulatory mechanisms that rely on motor control are diminished. REM sleep does not appear to have an intensity component, although the time spent in REM sleep can increase following sleep loss. In mammals that engage in extended periods of wakefulness, the time spent in SWS is the greatest and SWS-related SWA is the highest early in the subsequent sleep bout, whereas the proportion of time devoted to REM sleep increases toward the end of the sleep bout.

Monotremes

The study of sleep in monotremes, an egg-laying order of mammals that are the closest extant relatives to therian mammals (**Figure 1**), may provide insight into sleep in the most recent common ancestor to therian mammals. An early study of the short-beaked echidna (*Tachyglossus aculeatus*) found only EEG activity indicative of SWS during sleep, thereby suggesting that REM sleep evolved after the appearance of the therian lineage. The electrophysiological correlates of sleep in the echidna were reexamined using a combination of EEG and brain stem neuronal recordings. As in the earlier study, REM sleep with cortical activation was not observed. However, during sleep with cortical slow waves, brain stem reticular neurons fired in an irregular pattern similar to that observed in placental mammals during REM sleep. This suggests that aspects of REM sleep in the brain stem and SWS in the cortex occur concurrently. Moreover, an EEG-based study of the duck-billed platypus (*Ornithorhynchus anatinus*) suggests that this mixed sleep state is typical of monotremes (**Figure 1**). Although neuronal activity in the brain stem was not recorded, the platypus showed frequent rapid eye movements and twitching of the head and bill, similar to that associated with REM sleep in

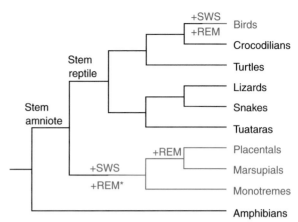

Figure 1 Cladogram for amniotes showing the evolutionary appearance of slow-wave sleep (SWS) and rapid eye movement (REM) sleep. Here, SWS and REM sleep arose independently in the most recent common ancestor to mammals and the most recent common ancestor to birds. However, in mammals, REM sleep first appeared in the common ancestor to all mammals as a heterogeneous state with neuronal activity in the brain stem indicative of REM sleep (+REM*) occurring concurrently with EEG activity indicative of SWS. In the most recent common ancestor to placental and marsupial mammals, REM sleep and SWS became segregated into two distinct states, with EEG activation occurring in conjunction with REM sleep-related brain stem activity. Evidence suggests that reptiles engage in SWS, but without the EEG slow-wave activity similar to that observed in mammals and birds.

therian mammals, whereas the cortex exhibited an EEG pattern indicative of SWS. Based on the incidence of twitches, the time spent in REM sleep was estimated at up to 8 h per day, the highest in any animal. A comparison of sleep times with therian mammals is difficult to interpret, however, given the heterogeneous nature of sleep and the absence of cortical correlates of REM sleep in monotremes.

Aquatic Mammals

Sleeping in the water poses a significant challenge for air-breathing mammals. Among Cetaceans, the electrophysiological correlates of sleep have been recorded only in the Odontocetes (dolphins and porpoises). Dolphins and porpoises exhibit unihemispheric SWS (USWS), a mixed state in which one cerebral hemisphere shows EEG activity characteristic of deep SWS while the other shows a pattern indistinguishable from wakefulness (**Figure 2**). Although the lighter stages of SWS can occur concurrently in both hemispheres, deep SWS occurs only unihemispherically. Evidence suggests that SWS is homeostatically regulated independently in the two hemispheres, which is an example of regional use-dependent SWS homeostasis, a phenomenon discovered only recently in terrestrial mammals including humans. Dolphins and porpoises swim and

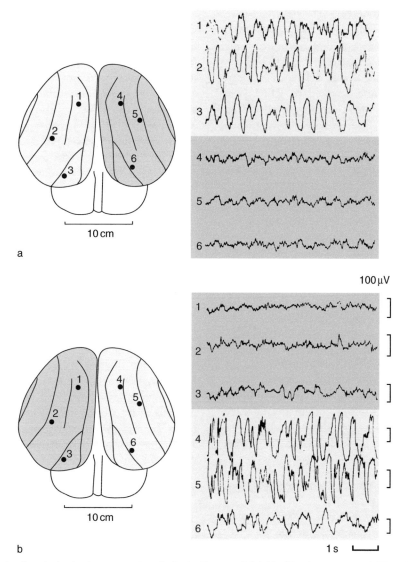

100 µV

1 s

Figure 2 Examples of unihemispheric slow-wave sleep in the bottle-nosed dolphin (*Tursiops truncatus*). The EEG was recorded from the anterior, medial, and posterior neocortex. Note the high-amplitude, low-frequency activity indicative of slow-wave sleep (blue) in only the left (a) or the right (b) hemisphere concurrent with low-amplitude, high-frequency activity indicative of wakefulness (red) in the other hemisphere. Reproduced from Mukhameotv LM, Supin AY, and Polyakova IG (1977) Interhemispheric asymmetry of electroencephalographic sleep patterns in dolphins. *Brain Research* 134: 581–584, with permission from Elsevier.

surface to breath during USWS, but they can also float at the surface or rest motionless underwater. Given that the eye opposite the awake hemisphere remains open, USWS may allow dolphins and porpoises to monitor the environment for conspecifics and predators during sleep. Although behavioral signs of REM sleep (e.g., twitching and rapid eye movements) have been observed infrequently in quiescent Cetaceans, unequivocal REM sleep has not been recorded electrophysiologically, suggesting that Cetaceans have lost REM sleep secondarily. Nevertheless, REM sleep could occur in very small amounts or in a modified form in Cetaceans. Interestingly, manatees (order: Sirenia) also exhibit USWS and

small amounts of REM sleep. Unlike Cetaceans and manatees, seals can sleep in the water and on the land. Seals in the family Phocidae hold their breath during periods of bilateral SWS and REM sleep underwater, whereas seals in the family Otariidae show interhemispheric asymmetries in the intensity of SWS while sleeping on the surface of the water.

Early Ontogeny of Mammalian Sleep

The time spent asleep changes greatly over early development. In mammals, newborns generally sleep the longest. This high amount of sleep declines over early ontogeny until it stabilizes at a species-specific level similar to that seen in adults. Sleep homeostasis has been demonstrated in postnatal day 5 rats, before SWS and REM sleep appear in the EEG as differentiated states. Evidence suggests that SWS and REM sleep actually differentiate before the EEG shows sleep-related changes in cortical activity. The proportion of total sleep time devoted to REM sleep (%REM sleep) and SWS also changes greatly during the first postnatal weeks. Interestingly, the magnitude of these changes appears to be influenced by the degree of precociality at birth (**Figure 3**). Altricial mammals (those that are relatively more dependent on their parents at birth for food, warmth, and protection; e.g., rats and cats) show a marked reduction in %REM sleep through the first postnatal weeks (**Figure 3**) and show more exaggerated behavioral characteristics of REM sleep, such as rapid eye movements and twitches. Conversely, newborn precocial mammals (e.g., guinea pigs) are more developed relative to altricial mammals, and they exhibit a %REM sleep comparable with that of adults (**Figure 3**). These observations led (in part) to the hypothesis that REM sleep provides endogenous stimulation necessary for the early development of the central nervous system, particularly that related to the visual system. Indeed, experimental evidence implicates REM sleep in maturational processes of the visual system during early ontogeny. As discussed later, phylogenetic comparative analyses also support the idea that REM sleep facilitates early brain development.

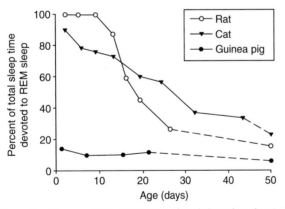

Figure 3 Changes in the percentage of total sleep time devoted to rapid eye movement (%REM) sleep across early development in the rat, cat, and guinea pig. In the altricial rat and cat, REM sleep constitutes most of total sleep time and decreases to levels close to that of adults approximately 1 month after birth. Conversely, %REM sleep in the precocial guinea pig is relatively stable over early ontogeny and into adulthood. Note, however, that even as adults, altricial species have higher %REM sleep than more precocial species. Reprinted Jouvet-Mounier D, Astic L, and Lacote D (1970) Ontogenesis of the states of sleep in rat, cat, and guinea pig during the first postnatal month. *Developmental Psychobiology* 2: 216–239, with permission from John Wiley & Sons, Inc.

Unlike newborn terrestrial mammals, newborn Cetacean calves are continuously active, swimming alongside their mothers during the first few weeks postpartum. Although electrophysiological recordings have not been obtained during this period of activity, the calves close one eye intermittently while swimming underwater and, therefore, might be engaging in USWS. However, calves are unlikely to be engaging in REM sleep if REM sleep is incompatible with swimming. The possible absence of REM sleep in newborn calves during their first weeks of life seemingly challenges theories for a functional role of REM sleep in brain development; however, given that Cetacean calves are extremely precocial, REM sleep could nonetheless play a role in brain development *in utero*.

Avian Sleep

Birds are a particularly interesting taxonomic group in which to study sleep. Birds have brains comparable in relative size to those of mammals and cognitive abilities that include vocal learning and tool making; however, much of the avian forebrain is organized in a manner markedly different from mammals. Although the dorsal two-thirds of the avian forebrain (a region formerly thought to be primarily striatal) is derived from the same embryonic neural tissue (the pallium) that gives rise to the mammalian neocortex, the avian pallium is arranged in a nuclear manner that lacks the true laminar organization of the neocortex (**Figure 4**). Interestingly, despite this difference in pallial organization, birds are the only nonmammalian taxonomic group to show unequivocal SWS and REM sleep (**Figure 1**).

Like mammalian SWS, avian SWS is characterized by an EEG of low-frequency, high-amplitude activity. Birds that sleep predominantly at night (e.g., domestic hens, *Gallus domesticus*; European blackbirds, *Turdus merula*; nonmigrating white-crowned sparrows, *Zonotrichia leucophrys gambelii*) show slow-wave activity during SWS that is greatest early in the night and gradually declines thereafter in a manner suggestive of mammalian-like SWS homeostasis. Indeed, we recently demonstrated that pigeons (*Columba livia*) show a compensatory increase in SWS-related low-frequency EEG power density following short-term sleep deprivation, suggesting that mammalian and avian sleep is regulated in a similar manner.

Avian REM sleep also shares several features in common with mammals. REM sleep is associated with closure of both eyes, rapid eye movements, occasional bill movements, and behavioral signs of reduced muscle tone such as dropping of the head. However, muscle atonia has been observed only in

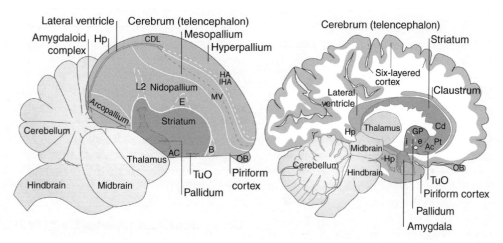

Figure 4 Modern consensus view of avian and mammalian brain relationships. Historically, much of the avian telencephalon was thought to be homologous to the mammalian striatum. However, converging lines of evidence demonstrate that the dorsal two-thirds of the avian telencephalon is actually derived from the pallium (green), the same neural tissue that gives rise to the mammalian neocortex. The new nomenclature reflecting this fundamental change in our understanding of brain evolution is depicted in the sagittal view of a zebra finch (left) and human (right) brain. Ac, accumbens; B, basorostralis; Cd, caudate nucleus; CDL, dorsal lateral corticoid area; E, entopallium; GP, globus pallidus (i, internal segment; e, external segment); HA, hyperpallium apicale; Hp, hippocampus; MV, mesopallium ventrale; IHA, interstitial hyperpallium apicale; L2, field L2; LPO, lobus parolfactorius; OB, olfactory bulb; Pt, putamen; TuO, olfactory tubercle. The several large, white regions are axon pathways (i.e., white matter) in the cerebrum. Lamina (cell-sparse zones separating brain subdivisions) are marked as solid white lines, primary sensory neuron populations are distinguished from neighboring regions by dashed white lines, and regions differing in cell density or size are demarcated by dashed gray lines. Reprinted by permission from Macmillan Publishers Ltd: *Nature Reviews Neuroscience* (Jarvis E, Gunturkun O, Bruce L, et al., 2005, 6: 151–159), Copyright 2005.

birds that can securely rest their head on the back. As in mammals, thermoregulatory responses are diminished during REM sleep. During REM sleep, the EEG reverts to a pattern similar to wakefulness, but often with lower amplitude. Unlike mammalian REM sleep, however, a hippocampal theta rhythm has not been found in birds. Episodes of REM sleep typically last 2–10 s and occur in clusters. The short duration of REM sleep episodes does not appear to be related to a need to maintain balance because REM sleep episodes are equally short when birds are sitting with their heads supported on their backs. In many birds, REM sleep increases throughout the night, in a manner similar to humans. In pigeons, this reflects an increase in the incidence and the duration of REM sleep episodes across the night. The proportion of total sleep time devoted to REM sleep appears to be lower in bird species (mean, 8%) than in mammalian species (mean, 17%). As in mammals, the time spent in REM sleep increases following sleep deprivation. Finally, since REM sleep and SWS have been recorded in every avian species investigated, both states were likely present in the most recent common dinosaur ancestor to birds (**Figure 1**).

Like aquatic mammals, birds often sleep with one eye open, a behavioral state associated with SWS in the hemisphere opposite the closed eye and EEG activity intermediate between wakefulness and SWS

in the hemisphere opposite the open eye. Birds have the ability to switch from sleeping with both eyes closed to sleeping with one eye open in response to a perceived increase in the risk of predation. Here, birds direct the open eye toward the potential threat and are able to respond to threatening stimuli presented to the open eye. In contrast to mammals, where unihemispheric SWS occurs only in aquatic mammals, such interhemispheric asymmetries in slow-wave activity are common in birds and may be an ancestral trait. As in Cetaceans that swim in a coordinated manner during unihemispheric SWS, sleeping with one eye open and half the brain awake may allow birds to sleep during flight. Although there is strong evidence showing that birds, such as common swifts (*Apus apus*) and frigatebirds (*Fregata* sp.), spend periods lasting days to weeks or longer in constant flight, sleep in flight has not been confirmed with electrophysiological recordings.

Early Ontogeny of Avian Sleep

The ontogenetic development of avian sleep has been studied in relatively few species. In precocial chickens (*Gallus gallus*), the EEG correlates of SWS and REM sleep can be distinguished 1 day before hatching, whereas in the altricial pigeon (*Columba livia*) EEG activity is absent at hatching and does not resemble that of adults until 14 days later. Although it is unclear whether chicken chicks show more or less REM sleep than adults shortly after hatching, the proportion of sleep allocated to REM sleep seems to reach adult levels within the first week post-hatch. Among altricial birds, the ontogeny of SWS and REM sleep has only been examined in magpies (*Pica pica*), in which juveniles sleep longer and have more REM sleep than adults. Additional studies on early sleep ontogenesis in species of greater taxonomic diversity are needed to determine whether the ontogenetic changes in sleep in altricial and precocial birds parallel those observed in mammals. Interestingly, chicken chicks that imprint on a moving object on the day of hatching exhibit more REM sleep during the first 3 h following imprinting than nonimprinted control chicks, suggesting that REM sleep may play a role in this acute form of learning. Sleep also appears to play a role in song learning in juvenile zebra finches (*Taeniopygia guttata*).

Sleep in Reptiles

The presence of SWS and REM sleep in all mammalian and avian species investigated suggests that either these states were present in the most recent common ancestor to extant mammals, birds, and reptiles or they evolved independently in mammals and birds. Several studies have attempted to distinguish between these alternatives by examining the electrophysiological correlates of sleep behavior in reptiles and, to a lesser extent, amphibians and fish. Unlike mammals and birds, however, where largely similar results have been obtained across species and laboratories, the results from reptiles have been less consistent and therefore subject to more diverse interpretations. Although some controversy persists, the EEG during reptilian sleep behavior typically shows intermittent high-voltage spikes arising from a background pattern similar to or slightly reduced in amplitude compared to quiet wakefulness. Because the incidence of spikes is correlated with arousal thresholds and increases following sleep deprivation, spikes appear to reflect sleep intensity. Studies have shown that these spikes originate in the reptilian hippocampus, thereby corroborating earlier pharmacological evidence indicating that reptilian spikes are comparable to similar spikes occurring in the mammalian hippocampus during SWS. Despite this similarity at the level of the hippocampus, however, the reptilian dorsal cortex does not generate concurrent high-amplitude slow waves typical of SWS in mammals and birds.

REM sleep has been reported in reptiles and fish based on the presence of eye, head, and limb movements during sleep. However, it remains unclear whether these behaviors truly reflect REM sleep-related twitching similar to that observed in mammals and birds or partial arousals from sleep. Although the presence of brain stem neural activity suggestive of REM sleep in sleeping echidnas raised the possibility that reptiles also exhibit REM sleep at the level of the brain stem, no sign of REM sleep was detected in the brain stem of sleeping turtles, despite the presence of neural structures involved in generating REM sleep in mammals. The presence of unequivocal SWS and REM sleep in mammals and birds, but not in reptiles, amphibians, or fish, suggests that these sleep states arose independently in the mammalian and avian lineages through convergent evolution (**Figure 1**).

Corticocortical Connectivity and SWS

Historically, the absence of slow waves in the EEG of sleeping reptiles has been attributed to the lack of a thick cortex similar to that which generates slow waves in mammals. However, the presence of slow waves in sleeping birds, despite the absence of a neocortex (**Figure 4**), demonstrates that the neocortex is not essential for the genesis of EEG slow waves. Instead, the extent of connections within the mammalian neocortex, avian hyperpallium, and reptilian dorsal cortex may explain why slow waves are present during sleep in mammals and birds but not reptiles (**Figure 5**). In mammals, corticocortical connections in layers II and III play an integral role in synchronizing the slow oscillation of neurons in a manner sufficient to generate slow waves in the EEG. In accord with the absence of slow waves in sleeping reptiles, the three-layered reptilian dorsal cortex lacks layers II and III, and it shows limited corticocortical connectivity (**Figure 5**). Furthermore, although birds lack a true neocortex, the hyperpallium shows extensive interconnectivity (**Figure 5**). Thus, the occurrence of sleep-related slow-wave activity in amniotes seems to be related to the extent of interconnectivity in the neocortex, hyperpallium, and dorsal cortex, although additional factors may also play a role.

A persistent question in sleep research is whether the EEG correlates of sleep are involved in the functions of sleep or simply reflect an epiphenomenon of the state. For example, the presence of slow waves in the EEG of sleeping mammals and birds may simply be an emergent property of a heavily interconnected neocortex and hyperpallium. Alternatively, as suggested by experimental work in mammals, the corticocortical connections that give rise to slow waves may also depend on slow waves to maintain the level of connectivity at an energetically and functionally adaptive level. Experimental evidence indicates that slow waves may also be involved in sleep-dependent memory processing and plasticity. Additional studies are needed, however, to determine whether slow

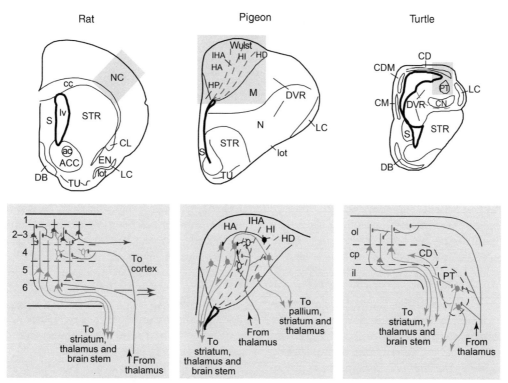

Figure 5 Comparison of the dorsal pallium in representative mammalian, avian, and reptilian species. The top row shows coronal cross sections through the brain. The area shaded in gray is expanded in the bottom row to show network connections. Note the comparatively high degree of interconnectivity in the rat neocortex and the pigeon Wulst (i.e., hyperpallium) compared to the turtle dorsal cortex. ac, anterior commissure; ACC, nucleus accumbens; cc, corpus callosum; CD, dorsal cortex; CDM, dorsomedial cortex; CL, claustrum; CM, medial cortex; CN, core nucleus of the DVR; cp, cell plate; DB, diagonal band of Broca; DVR, dorsal ventricular ridge; EN, endopiriform region; HA, hyperpallium apicale; HD, hyperpallium densocellulare; HI, hyperpallium intercalatum; HP, hippocampus; IHA, nucleus interstitialis hyperpallii apicalis; il, inner layer; LC, lateral cortex; lot, lateral olfactory tract; lv, lateral ventricle; M, mesopallium; N, nidopallium; NC, neocortex; ol, outer layer; PT, pallial thickening; S, septum; STR, striatum; TU, olfactory tubercle. Reproduced from Medina L and Reiner A (2000) Do birds possess homologues of mammalian primary visual, somatosensory and motor cortices? *Trends in Neurosciences* 23: 1–12, with permission from Elsevier.

waves evolved independently in mammals and birds (and not in reptiles) to maintain their heavily interconnected brains and associated cognitive abilities, or whether similar processes occur during reptilian sleep in a manner undetectable in the EEG. As indicated by the presence of hippocampal spikes during sleep, at least some of the neural correlates of SWS seem to be present in reptiles (**Figure 1**). A comparison of sleep-related gene expression among mammals, birds, and reptiles may clarify the evolutionary history of SWS and REM sleep.

Comparative Perspectives on the Functions of Sleep

More is known about sleep in mammals than in any other group of animals. Electrophysiological sleep data exist for just less than 100 mammalian species. An examination of this comparative data set reveals that the time spent in SWS and REM sleep varies greatly across Mammalia. Many studies have tried

to identify the evolutionary determinants responsible for maintaining such interspecific variation in the structure of sleep. The identification of significant predictors of sleep times may help shed light on the functions of SWS and REM sleep. For example, species with a higher mass-specific basal metabolic rate (i.e., basal metabolic rate per gram of body mass) were once thought to engage in more SWS, a relationship that could support an energy conservation role for SWS, because metabolic rates are lower during SWS. Additionally, it was once thought that species with greater encephalization allocated a lower proportion of time spent sleeping to REM sleep, seemingly refuting a neurophysiological role for REM sleep, such as memory processing and plasticity. However, excluding our own analysis, all comparative studies of sleep treated each species (or, in one case, family) as a statistically independent unit. Independence is a basic assumption of all statistical analyses, but species cannot be considered independent because they are related to one another through common ancestry. Phylogenetic comparative methods (e.g., independent contrasts) were developed by evolutionary biologists to deal with such interspecific relatedness in comparative studies; however, until recently, these procedures had not been used in comparative sleep research.

In our analysis, we controlled for shared evolutionary history among species using independent contrasts and analyzed our phylogenetically controlled data using a multivariate analysis that incorporated hypotheses for the functions of SWS and REM sleep. Many of our results were different from those in previous studies. Unlike all previous studies, we found that species with greater encephalization (i.e., brain mass controlling for body mass using regression) allocate a greater proportion of time asleep to REM sleep, thus providing the first comparative support for a neurophysiological role for REM sleep. Although no relationship was found between encephalization and the time spent in SWS, cumulative slow-wave activity may be the more accurate measure of SWS. Contrary to some comparative studies and expectations under the energy conservation hypothesis, we found that species with a higher residual basal metabolic rate (BMR) engage in less SWS and sleep less altogether. These relationships might reflect increased foraging demands associated with higher residual BMR and thus less time available for sleep. Nevertheless, mammals with higher residual BMRs may obtain functionally comparable amounts of SWS by engaging in more intense SWS, although such intensity data are largely unavailable. Alternatively, the restorative processes occurring during sleep could be achieved more quickly with higher residual BMR and thus take less time to accomplish.

Predation risk should be among the strongest selection pressures influencing how to structure sleep and how long to sleep because sleeping is dangerous. Interestingly, the vulnerability associated with sleep may depend on the sleep state. For example, due to high arousal thresholds, deep SWS and REM sleep may be particularly dangerous sleep states from an antipredator standpoint. Indeed, in the context of our multivariate models, we found that species sleeping at relatively exposed or risky locations in the wild engage in less REM sleep and allocate a lower proportion of sleep time to REM sleep in the laboratory. Although the time spent in SWS was largely independent of predation risk, once again, SWS intensity may respond more strongly to the risk of predation. As per previous studies, we found that species more precocial at birth engage in less REM sleep as adults, a relationship that was not mediated strongly by predation risk because precocial species did not generally sleep in more vulnerable locations than relatively altricial species. The degree of precociality as a

predictor of REM sleep time in adults may reflect an extension of the ontogenetic changes in REM sleep shown in **Figure 3**, which supports the idea that REM sleep is particularly important in early brain development. However, why high levels of REM sleep during early ontogeny should persist in adults remains enigmatic.

Correlates of Avian SWS and REM Sleep

As discussed previously, the electrophysiological correlates of avian sleep are remarkably similar to those observed in mammals. In addition to these similarities, there is also great variation in the time avian species spend in SWS and REM sleep. Thus, an obvious question is whether birds share the same evolutionary determinants of SWS and REM sleep as mammals? We addressed this question by conducting the first electrophysiologically based comparative analysis of avian sleep architecture using the same phylogenetically controlled variables as in our mammalian analysis. Overall, we found that birds that sleep at relatively exposed sites in the wild engage in less SWS in the laboratory, but this was the only significant relationship identified. Thus, if relationships identified in mammals reflect functional aspects of sleep architecture, then the same functions may not apply broadly to birds.

Sleep in Invertebrates

More than 97% of all animal life is invertebrates, of which 80% are arthropods, the taxonomic group which includes insects. Surprisingly, however, of the 30 or more animal phyla, sleep has been studied in only 2: Chordata (includes vertebrates) and Arthropoda (includes insects), with much work having been focused on the fruit fly (*Drosophila melanogaster*) and the honeybee (*Apis mellifera*). Sleep in honeybees is characterized as a sustained period of quiescence accompanied by increased arousal thresholds and specific postures (e.g., antennal immobility). Optomotor interneurons in the optic lobes of forager honeybees show lowered sensitivity during the subjective night, when forager bees are often quiescent, than during the subjective day. Sleep in honeybees was shown to be homeostatically regulated as antennal immobility increased following 12 h of sleep deprivation. Sleep has also been demonstrated in scorpions (*Heterometrus* and *Pandinus* spp.), cockroaches (*Blaberus giganteus* and *Leucophaea maderae*), and crayfish (*Procambarus clarkii*). Also, preliminary work suggests that sleep is present in three additional animal phyla: Nematoda (*Caenorhabditis elegans*), cephalopods (*Octopus vulgaris* and *Sepia pharaonis*) in the phylum Mollusca, and box jellyfish (*Chironex fleckeri*) in the phylum Cnidaria.

Conclusions

Our understanding of how and why we sleep has been enhanced by studies examining the evolution and ontogeny of sleep. The presence of SWS and REM sleep in all avian and mammalian species studied, and their apparent absence in reptiles, suggests that these states arose independently twice: once in the ancestor to birds and once in the ancestor to mammals (**Figure 1**). A comparison of neurocytoarchitecture among homeotherms suggests that the degree of corticocortical (or palliopallial) connectivity is responsible for EEG slow-wave activity in mammals and birds. Interestingly, REM sleep-related cortical activation evolved independently in the ancestor to therian mammals and the ancestor to birds (**Figure 1**). The similarity in the electrophysiological correlates of behavioral sleep in homeotherms suggests similarities in functions, possibly related to having heavily interconnected brains and associated cognitive abilities. However, in phylogenetically controlled comparative analyses, the evolutionary determinants of sleep times in birds were markedly different from those identified in mammals. Identifying the reasons for these differences is a promising area for future work.

Our understanding of the evolution, ontogeny, and functions of sleep would benefit greatly from studies in species representing broader phylogenetic diversity. Here, the study of species nearest the base of the Metazoan phylogenetic tree would be most revealing. A systematic approach to selecting taxa for study should be employed that takes into consideration neuroanatomy, neurophysiology, and phylogenetic position in the context of current hypotheses for the functions of sleep. Future work should also expand existing genetic work on fruit flies and rats to nonmammalian vertebrate taxa and to nonarthropod invertebrate taxa. The success of these endeavors will depend on the broad collaboration of animal behaviorists, evolutionary biologists, geneticists, and neurophysiologists, but will do much to aid our greater understanding of sleep.

See also: Gene Expression; Hippocampal–Neocortical Dialog; Sleep and Waking in Drosophila; Sleep Architecture; Sleep Oscillations and PGO Waves.

Further Reading

Amlaner CJ and Ball NJ (1994) Avian sleep. In: Kryger MH, Roth T, and Dement WC (eds.) *Principles and Practice of Sleep Medicine*, 2nd edn., pp. 81–94. Philadelphia: Saunders.

Eiland MM, Lyamin OI, and Siegel JM (2001) State-related discharge of neurons in the brainstem of freely moving box turtles, *Terrapene carolina major*. *Archives Italiennes de Biologie* 139: 23–36.

Hartse KM (1994) Sleep in insects and nonmammalian vertebrates. In: Kryger MH, Roth T, and Dement WC (eds.) *Principles and Practice of Sleep Medicine*, 2nd edn., pp. 95–104. Philadelphia: Saunders.

Jarvis E, Gunturkun O, Bruce L, et al. (2005) Avian brains and a new understanding of vertebrate brain evolution. *Nature Reviews Neuroscience* 6: 151–159.

Jouvet-Mounier D, Astic L, and Lacote D (1970) Ontogenesis of the states of sleep in rat, cat, and guinea pig during the first postnatal month. *Developmental Psychobiology* 2: 216–239.

Kaiser W and Steiner-Kaiser J (1983) Neuronal correlates of sleep, wakefulness and arousal in a diurnal insect. *Nature* 301: 707–709.

Lesku JA, Roth TC, Amlaner CJ, and Lima SL (2006) A phylogenetic analysis of sleep architecture in mammals: The integration of anatomy, physiology, and ecology. *American Naturalist* 168: 441–453.

Lesku JA, Roth TC, Rattenborg NC, Amlaner CJ, and Lima SL (2008) Phylogenetics and the correlates of mammalian sleep: A reappraisal. *Sleep Medicine Reviews* 12: 229–244.

Lima SL, Rattenborg NC, Lesku JA, and Amlaner CJ (2005) Sleeping under the risk of predation. *Animal Behavior* 70: 723–736.

Lyamin OI, Mukhametov LM, and Siegel JM (2004) Relationship between sleep and eye state in Cetaceans and Pinnipeds. *Archives Italiennes de Biologie* 142: 557–568.

Margoliash D (2005) Song learning and sleep. *Nature Neuroscience* 8: 546–548.

Martinez-Gonzalez D, Lesku JA, and Rattenborg NC (2008) Increased EEG spectral power density during sleep following short-term sleep deprivation in pigeons (*Columba livia*): Evidence for avian sleep homeostasis. *Journal of Sleep Research* 17: 140–153.

Medina L and Reiner A (2000) Do birds possess homologues of mammalian primary visual, somatosensory and motor cortices. *Trends in Neurosciences* 23: 1–12.

Mukhaneotv LM, Supin AY, and Polyokova IG (1977) Interhemispheric asymmetry of electroencephalographic sleep patterns in dolphins. *Brain Research* 134: 581–584.

Rattenborg NC (2006) Evolution of slow-wave sleep and palliopallial connectivity in mammals and birds: A hypothesis. *Brain Research Bulletin* 69: 20–29.

Rattenborg NC, Amlaner CJ, and Lima SL (2000) Behavioral, neurophysiological and evolutionary perspectives on unihemispheric sleep. *Neuroscience and Biobehavioral Reviews* 24: 817–842.

Roth TC, Lesku JA, Amlaner CJ, and Lima SL (2006) A phylogenetic analysis of the correlates of sleep in birds. *Journal of Sleep Research* 15: 395–402.

Siegel JM, Manger PR, Nienhuis R, Fahringer HM, and Pettigrew JD (1998) Monotremes and the evolution of rapid eye movement sleep. *Philosophical Transactions of the Royal Society Series B* 353: 1147–1157.

Szymczak JT, Kaiser W, Helb HW, and Beszczynska B (1996) A study of sleep in the European blackbird. *Physiology & Behavior* 60: 1115–1120.

Tobler I (1995) Is sleep fundamentally different between mammalian species? *Behavioural Brain Research* 69: 35–41.

Sleep in Adolescents

L Tarokh, Brown University, Providence, RI, USA
M A Carskadon, Brown University, and E. P. Bradley Hospital, Providence, RI, USA

Sleep behavior and sleep physiology manifest dramatic changes across adolescent development. Certain sleep pattern changes arise from psychological and societal sources, such as a growing sense of independence, increasing access to electronic apparatus in the bedroom, greater academic obligations, and more social opportunities. Some have noted that emergent changes in affect regulation and social affiliation, including 'first love,' can have striking influences on sleep patterns.

Certain hormonal changes that herald and accompany puberty are sleep related, such as the pulsatile sleep-dependent release of luteinizing hormone (LH), which leads to subsequent sleep-related release of sex steroids. Whether these hormonal changes have a direct effect on sleep is unknown; that sleep changes substantially in the pubertal transition is well established. As discussed later, maturation of the central nervous system (CNS) affects the phenomenology of sleep as well as the underlying intrinsic regulatory processes.

In addition to CNS changes, adolescent maturation is associated with new physical attributes and reproductive development. A common method to gauge pubertal development in humans is Tanner staging, which relies on identifying benchmarks in secondary sexual characteristics. Thus, in boys genital development and the growth and distribution of pubic hair are evaluated, and in girls breast development and pubic hair growth and distribution are evaluated. Tanner stage 1 is assigned if no signs of development are identified, Tanner stages 2–4 designate intermediate pubertal stages based on achieving developmental criteria, and Tanner stage 5 requires achievement of all the benchmarks identifying full maturation of secondary sexual characteristics. Note that central signs of puberty (e.g., nocturnal LH release and likely CNS changes) occur earlier than the external physical attributes of puberty.

Adolescent Development and the Brain

The behavioral, emotional, and cognitive changes of adolescence are accompanied by large-scale remodeling of the brain that provides a biological basis for many features of adolescent development. *In vivo* longitudinal studies of the developing human brain using magnetic resonance imaging (MRI) show that gray matter volume increases during preadolescence and declines postadolescence. This inverted U-shape change in gray matter volume corresponds on a cellular level with a marked proliferation of axons/ synapses in early puberty and synaptic pruning in later adolescence. Positron emission tomography studies have shown a similar inverted U-shape increase and then decrease in waking cerebral glucose metabolic rates. These cross-sectional studies demonstrate a peak in absolute cerebral glucose metabolism at the age of 9 or 10 years followed by a steady decline until the ages of 16–18 years, at which point levels reach adult values. Not surprisingly, there is a strong correspondence between the developmental curve for blood glucose metabolism, gray matter volume, and synaptogenesis. The combined results from these studies provide strong evidence that adolescence is a time of major neural reorganization.

Adolescent Development and the Waking Electroencephalogram

The changes in gray matter volume, glucose metabolism, and synaptogenesis parallel the decline of electroencephalographic (EEG) power across adolescent development. The EEG signal is thought to arise from synchronous synaptic activity at the cortex; therefore, a reduction in the number of cortical synapses would logically result in dampened EEG power. Although gray matter volume and EEG power decline across adolescence, white matter volume – an index of myelination – increases in a linear manner. Adolescent development is also accompanied by greater functional connectivity between distant cortical regions (i.e., connections between frontal and occipital lobes), as inferred from EEG coherence. Cross-sectional studies of waking EEG coherence show a pattern of increasing inter- and intrahemispheric coherence before and during adolescence.

Along with waking EEG power and coherence changes, the frequency of the dominant posterior rhythm (alpha rhythm) increases as the brain matures. This rhythm is associated with a relaxed waking state and increases in amplitude with eye closure. The dominant posterior rhythm is seen during eye closure in infants as young as a few months old and increases in frequency from approximately 4 Hz in infancy to approximately 8 Hz in mid-childhood, reaching adult frequencies (8–12 Hz) by adolescence.

Adolescent Development and the Sleeping EEG

The EEG is a central measure of sleep, and indeed, EEG patterns are a defining characteristic of sleep in mammals. Typically measured along with electrooculogram and electromyogram, the EEG in humans contributes to defining sleep states (nonrapid eye movement (NREM) sleep and rapid eye movement (REM) sleep) and four sleep stages within NREM sleep. As outlined previously, a significant amount is known about the maturation of the waking EEG. Less is known about adolescent development of the sleeping EEG, although a number of characteristic changes have been identified. As with waking, for example, the age-related decrease in EEG power is strikingly apparent during sleep, particularly in the lower frequencies.

Figure 1 illustrates a number of changes of nocturnal sleep that accompany adolescent maturation by displaying the sleep hypnogram of a representative 10-year-old Tanner stage 1 child and a 15-year-old Tanner stage 5 adolescent. When sleep quantity is held constant at approximately 9 or 10 h, adolescent development is accompanied by a profound decline of approximately 40–50% in the quantity of NREM

sleep stages 3 and 4, commonly referred to as slow-wave sleep (SWS). The dampening of EEG amplitude consequent to cortical synaptic pruning may account for the developmental reduction of SWS. At the same time, stage 2 NREM sleep, the definition of which involves EEG features unrelated to amplitude (i.e., sleep spindles and K complexes), increases. Another impressive developmental feature of sleep is the advance of REM sleep to an earlier point in sleep, from a latency of approximately 150–180 min to a latency of approximately 100 min in most studies.

Figure 2 illustrates the marked reduction of EEG amplitude for NREM sleep stages 2 and 4 and REM sleep across adolescence. The sleep EEG amplitudes are notably lower in the more mature adolescent. This EEG amplitude decline appears to be non-state-dependent because it is seen in wakefulness and NREM and REM sleep, although definitive longitudinal assessments have not been carried out. Figure 3 emphasizes this point by displaying the absolute EEG power spectra across the entire night for the pre- and postpubertal youngsters. Warmer colors (designating greater power) in the Tanner stage 1 child versus the Tanner stage 5 child are evident across frequencies and states. The differences are especially clear for low frequencies (1–4 Hz), particularly early in the night.

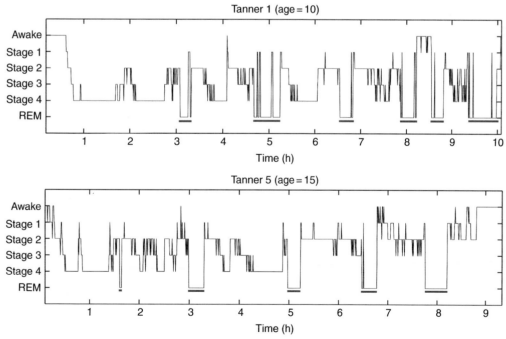

Figure 1 Sleep hypnograms of two boys, one at pubertal Tanner stage 1 (top) and the other at Tanner stage 5 (bottom). The y-axis designates stage/state, and the x-axis indicates elapsed time from lights out. REM episodes are underlined with a red bar. Stages were determined by visual scoring of 30 s epochs according to standard staging criteria. Significantly greater stage 4 sleep is typical of prepubertal adolescents compared to postpubertal adolescents. The decreased stage 4 sleep in more mature adolescents is accompanied by increased stage 2 sleep. Note also that REM sleep episodes increase in duration over the course of the night. Of note is the delayed onset of the first REM episode in the Tanner 1 preadolescent. Indeed, the first REM episode is often 'skipped' in Tanner 1 preadolescents.

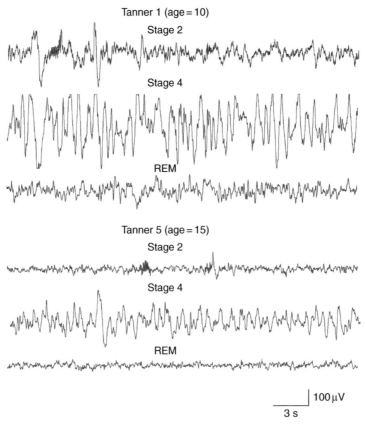

Figure 2 Comparison of Tanner stage 1 (top) and Tanner stage 5 (bottom) EEG tracings of NREM sleep stage 2, stage 4, and REM sleep from the same children and nights as in **Figure 1**. The calibrations designate the voltage and timescale of all tracings. The diminished EEG amplitude in the more mature adolescent is obvious from these tracings and is most evident in stage 4 sleep; however, this developmental change is not stage specific and occurs in all sleep stages and in the waking EEG. The reduction of EEG amplitude during early adolescence is believed to result from cortical synaptic pruning that occurs during this developmental stage.

This EEG balance of power toward the less mature brain is also clear for the sigma frequencies (approximately 14 Hz). Note that the sigma 'stripe' in these plots is evident only for NREM sleep and disappears during REM episodes.

Sleep Regulation and Adolescent Development

Homeostatic Process (Process S)

In the two-process model of sleep regulation, independent homeostatic (process S) and circadian (process C) processes operate interactively to regulate the amount and timing of sleep. Process S, which can be characterized colloquially as sleep pressure, has been modeled using the NREM sleep EEG slow-wave activity (0.5–4.5 Hz; SWA). As waking is extended, NREM EEG SWA power increases during subsequent sleep and is modeled using a saturating exponential function. With sleep, NREM EEG SWA spectral power manifests an exponential decay across the sleep episode.

The substantial decline in absolute SWA EEG power across adolescence raises the additional question of whether the homeostatic regulatory process is altered and perhaps secondarily accounts for certain expressed sleep behaviors in adolescents. The Brown University group examined this issue in several ways using cross-sectional data sets. First, they examined the dynamics of the decay of SWA across the sleep episode of Tanner stage 1 and Tanner stage 5 adolescents. As **Figure 4** illustrates, the exponential fits were overlapping functions, and the mean decay time constant for Tanner stage 1 subjects was 143.2 min (range, 87.0–195.5) and that for Tanner stage 5 subjects was 128.2 min (range, 80.4–176.9). This finding indicates that the nocturnal recovery process is similar for adolescents across this developmental phase, with an implication that neither sleep intensity nor sleep need change with adolescence.

On the other hand, when the accumulating function of process S is examined in the context of extended wakefulness, the data indicate the presence of a maturational change. One approach examined sleep pressure by assessing the speed of falling asleep

Tanner 1 (age = 10)

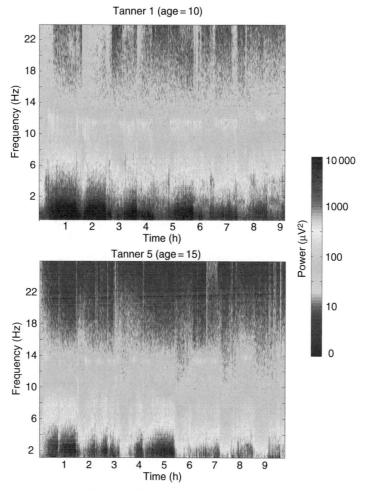

Tanner 5 (age = 15)

Figure 3 Time–frequency plot of the same nights and children depicted in the hypnograms displayed in **Figure 1**. The figure is derived from fast Fourier transform analysis of the C3/A2 EEG signal. The plots depict how the frequency content of the EEG signal changes over the course of the night. The *y*-axis is the frequency and the *x*-axis shows hours since lights out. EEG power is indicated by the color, in which warmer colors (reds and yellows) indicate higher power and cooler colors (aqua and blue) indicate lower power. During NREM sleep, the power of the signal at frequencies below 4 Hz is maximal, particularly early in the night, ebbing in strength across the sleep episode. Note also that the power in these slow frequencies is markedly greater for the younger boy. NREM episodes are also marked by high power in the spindle frequency range (12–16 Hz). In contrast, during REM episodes, the power in the lower frequency and spindle bins is diminished. The characteristic cycle-by-cycle nature of the sleep EEG is apparent in these plots. Overall EEG power differences between the Tanner 1 and Tanner 5 participants are readily apparent at all frequency bins.

in adolescents at Tanner stage 1 and Tanner stage 5. This study found that after approximately 14.5, 16.5, and 18.5 h awake, the less mature youngsters fell asleep significantly faster than the more mature group. A second approach modeled process S using SWA in a sleep deprivation study and found fitted functions for the exponential saturating function with a time constant of 15.4 ± 2.5 h for the mature Tanner stage 5 adolescents compared to 8.9 ± 1.2 h for a group of pre-/early pubertal (Tanner stages 1/2) children. These two findings indicate that the more mature adolescents, although they exhibit the same sleep recovery pattern, may find it easier to remain awake longer than the less mature adolescents. This

developmental alteration may contribute to the later timing of sleep that emerges across adolescence.

Circadian Timing (Process C)

Maturational changes in the circadian timing system (CTS) also appear to contribute to the behavioral changes in adolescent sleep patterns. The master 'clock' in mammals has been localized to a small paired nucleus in the hypothalamus, the suprachiasmatic nuclei (SCN). A direct pathway from the retina, particularly the light-sensitive melanopsin-containing retinal ganglion cells, carries light input directly to the SCN, where the light signals affect the molecular machinery of the CTS within SCN neurons.

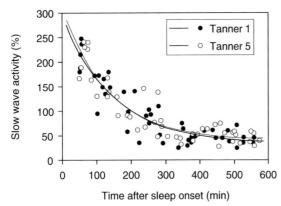

Figure 4 Dynamics of slow-wave activity (SWA; 0.6–4.6 Hz) across the sleep episode in eight prepubertal (solid circles) and eight mature (open circles) adolescents. SWA power for each NREM sleep cycle is shown as a percentage of the individual child's average overnight NREM sleep, and each symbol represents this value plotted at the midpoint of the time relative to the onset of sleep. The lines show exponential functions fitted to the Tanner 1 (thick line) and Tanner 5 (thin line) data. (Both fits achieved r^2 values of 0.83.) These data indicate that the dissipation of SWA, and inferentially of process S, does not change with pubertal development, although the power of the EEG SWA signal is markedly changed as shown in **Figure 3**. Reproduced from Jenni OG and Carskadon MA (2004) Spectral analysis of the sleep electroencephalogram during adolescence. *Sleep* 27: 774–783.

One of the first findings implicating the CTS in the adolescent delay of sleep timing was an association between self-reports of phase preference (e.g., morningness vs. eveningness) and puberty in sixth-grade girls: Girls who rated themselves more mature also rated themselves with greater evening-type phase preference. Subsequent studies examined these phenomena using better measures, such as Tanner staging and circadian phase measured using the dim light melatonin onset (DLMO) phase. DLMO is a reliable and noninvasive assessment of circadian phase acquired from serial plasma or saliva samples collected under dim light (less than approximately 40 lux) conditions. Because melatonin levels are low during the biological daytime and rise during the biological night, the circadian phase can be measured by assessing the onset phase of melatonin release. In adolescents, saliva is collected approximately every 30 min in dimly lit conditions during the participants' usual evening hours.

A major feature of the CTS is its sensitivity to environmental light. Effects of light on the CTS are phase dependent; to wit, light impinging on the clock in the early nighttime produces phase delays (later timing) and light occurring in the early morning produces phase advances (earlier timing). Because of this phase-dependent light sensitivity, mature adolescents

living in their normal circumstances will inevitably express later circadian phase positions because their delayed sleep pattern exposes them to light at a later time. Therefore, to determine whether the internal process undergoes developmental change, light exposure needs to be controlled, for example, by controlling sleep (dark) schedules. In such controlled circumstances, the DLMO phase is positively correlated with Tanner stage, whereby more mature adolescents have a later DLMO phase. This circadian phase delay across adolescence may affect the ability of older adolescents to fall asleep easily in the evening since the CTS favors wakefulness in the early circadian night, sometimes called the forbidden zone for sleep. In addition to humans, a pubertal delay of sleep behavior has been reported in rhesus monkeys and in the octodon degus, a small diurnal mammal with a relatively protracted developmental course from birth to reproductive maturity.

Attempts have been made to identify the underlying mechanism for this adolescent phase delay. One hypothesis holds that the intrinsic period of the circadian timing system may slow from prepubertal to postpubertal adolescence, thus delaying circadian phase. Studies that permit assessment of internal circadian period are both difficult and time-consuming and have not established the validity of this hypothesis. One study, however, has shown that the period is longer on average in adolescents than in adults. Another theory for the developmental change in circadian phase posits that the phase-dependent sensitivity to light's phase shifting effects may favor phase delays either through strengthening the delay response to evening light or through weakening the phase advance to morning light. Attempts to demonstrate this phenomenon have not been successful to date, although none has used the most circadian-sensitive shortwave (blue) light sources to examine the possibility of a maturational change in light sensitivity.

Thus, the circadian timing system does manifest a phase delay associated with pubertal/adolescent development; however, the mechanism for the delay is unknown. Unlike the sleep system in which the adolescent decline in EEG amplitude can be attributed directly to cortical synaptic pruning, no molecular, cellular, or systems basis has been identified to explain the maturation of the CTS across adolescence.

Interaction of Process S and Process C

As noted previously, the circadian timing system favors waking in the biological evening. Indeed, the strength of the arousing signal from the circadian timing system increases from the usual time of morning arousal until about the time of usual bedtime and decreases across the typical nighttime window. On the other hand, the homeostatic sleep–wake process builds as waking is extended, only to dissipate during sleep. Either process on its own is untenable for supporting our usual sleeping and waking patterns; however, the interaction provides an arousing circadian signal later in the day to support alertness when the homeostatic pressure would otherwise favor sleep. Conversely, the circadian arousing signal wanes during the night. As morning comes, with circadian arousal at its nadir, wakefulness is underwritten by the restorative function of sleep.

In the case of adolescent development, this interaction plays out in several important ways. When daytime alertness is evaluated in children keeping an optimizing sleep schedule (10 h per night from 10:00 p.m. to 8:00 a.m.), the interaction of processes S and C manifests as maximal alertness from rising until bedtime. With the onset of puberty, however, the pattern shifts to one that augments midday sleepiness. In the experimental paradigm that showed this phenomenon, young people were followed annually across several years with summertime assessments on the optimized sleep schedule and sleepiness measured with the Multiple Sleep Latency Test (MSLT), a standardized measure of the speed of falling asleep (sleep propensity) at 2 h intervals across the day. In prepuberty, the youngsters maintained alertness near the ceiling of this measure throughout the day; at midpuberty, the pattern changed to exhibit faster sleep onsets on tests given in the middle of the afternoon. Subsequent evaluation of these data indicated that the phase delay of pubertal development resulted in a reorganization of the phase relationship between sleep and the circadian timing system. With this reorganization, the circadian buffering of accumulating sleep pressure across the waking day did not occur until a later phase, and thus midday sleepiness emerged.

In normal circumstances, however, adolescents do not live with optimal sleep. As outlined below, sleep occurs later and is shorter in older adolescents. This pattern results in chronic insufficient sleep on school nights that is not recuperated fully by nighttime sleep. In these circumstances, morning alertness specifically is impaired since adolescents wake with elevated homeostatic pressure and also at a circadian phase that does not support alertness.

One study examined the consequences of this mismatch in students attending the 10th grade of high school in the United States. These students were required to attend school beginning at 7:20 a.m. When followed on their normal sleep schedule for 2 weeks, the students obtained approximately 7.5 h of sleep each night, going to bed on average at approximately 10:30 p.m. and waking at approximately

6:00 a.m. When their alertness was measured using the MSLT, these adolescents were sleepiest at 8:30 a.m. (average speed of falling asleep was 5 min), and alertness improved in the late morning and early afternoon, presumably due to the alerting effects of the circadian timing system. Thus, the speed of falling asleep at 2:30 p.m. was 11 min. Of greater note in this study, however, was the appearance of an abnormal pattern of sleep in nearly 50% of the young people on the MSLTs: They had at least one test (usually at 8:30 a.m.) in which they fell directly into REM sleep rather than NREM sleep (i.e., sleep-onset REM episodes). Furthermore, the speed of falling asleep in this group was even faster at 8:30 a.m. – 3.4 min on average. A distinguishing feature of this group was that the phase of DLMO was later, and the investigators concluded that they woke at a much earlier circadian phase than the other students, at a time favoring sleeping over waking and REM sleep over NREM sleep. The MSLT results in these young people looked more like those of patients with a serious sleep disorder – narcolepsy – than healthy teens.

Sleep Habits and Adolescent Development

As mentioned previously, sleep patterns undergo major alterations across adolescent development – changes that have been confirmed in many surveys of young people on every continent (except Antarctica). In each case, the timing of sleep is later overall in older adolescents. In societal circumstances requiring early morning awakening – for example, with early starting time for school as found in many communities in the United States – sleep quantity also is reduced, at least on school days. Indeed, the difference between school day and weekend schedules becomes greater during adolescence. At age 9 or 10 years, for example, children may delay weekend bedtime slightly but tend not to delay or extend sleep as much on weekend mornings. By age 14 or 15 years, on the other hand, bedtime is much later on school nights, delayed by 1–3 h on weekends, and weekend sleep can be several hours longer than on school days. These weekday–weekend differences express both the pressure for sleep to delay during adolescence and the compensatory need for sleep on weekends due to insufficient sleep on school nights in the older adolescents.

The 2006 Sleep in America poll of the National Sleep Foundation surveyed approximately 1600 adolescents and caregivers. Data from this poll provide a contemporary perspective of adolescent sleep. Although this article has emphasized the biological bases of sleep pattern alterations across adolescence, the influence of environmental factors is undeniable. Furthermore, the interactions among sleep regulatory changes and social/environmental influences and constraints can lead to troubling outcomes in susceptible youngsters. As the Sleep Foundation poll data show, school night sleep time reported by sixth-grade students in the United States is approximately 8.4 h and it is only 6.9 h in 12th-grade students, which by general consensus is significantly below the sleep need of teens.

The starting time for school plays an important role in setting the stage for short sleep in many adolescents. Thus, although teens' bedtimes delay significantly across adolescence, rising time on school days does not delay and indeed may actually become earlier (**Figure 5**). Later bedtimes fit well with the intrinsic maturational changes; earlier rising times, however, do not arise from a natural tendency on the part of adolescents but in direct response to waking in time for school. The result is truncated sleep time. In cases in which school begins later in the day, as shown by a number of ongoing studies of students in Brazil, Croatia, and the United States, teens sleep later and longer. One US study that examined sleep patterns of high school students before and after imposition of a later school schedule found that bedtimes did not change significantly; however, rising times were later and, consequently, youngsters slept more.

A newly emerging consensus also credits access to contemporary technologies as contributing to adolescent sleep patterns. For example, Belgian adolescents who report using electronic media to help them fall asleep in fact sleep less than other teens and report themselves as more tired. Furthermore, the National Sleep Foundation's 2006 Sleep in America Poll found

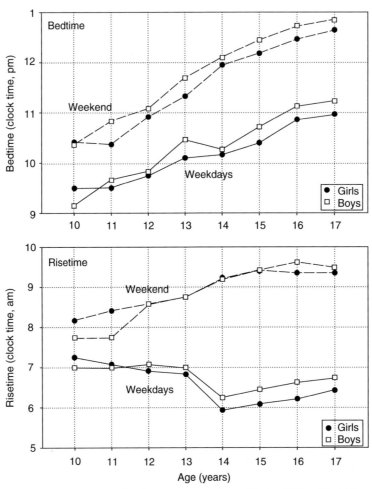

Figure 5 Adolescent self-reported sleep patterns taken from surveys performed in the 1990s with children living in the United States. Girls are represented by closed symbols and boys by solid symbols. Weekend bedtimes (top) and rise times (bottom) are shown with dotted lines; weekday (school day) schedules are indicated by solid lines. The sample included children ages 10–17 years who were enrolled in school at the time of the survey. Times on the *y*-axes are indicated using 24 h clock hours. Reproduced from Jenni OJ and Carskadon MA (2005) Infants to adolescents. In: *SRS Basics of Sleep Guide*, pp. 11–20. Westchester, IL: Sleep Research Society.

that compared to younger teens, older teens have increased numbers of technological devices (e.g., telephone, instant messaging, MP3 player, electronic games, television, computer, and Internet connection) accessible to them in their bedrooms. In addition, teens who reported having four or more such devices in the bedroom also reported sleeping 30 min less on average each night than those with fewer devices.

Consequences of Chronic Insufficient and Ill-Timed Sleep in Adolescents

The association of maturational changes of the intrinsic regulatory systems with the seductive nocturnal lifestyle options available to teenagers makes for a synergistic drive toward later timing of sleeping and waking for most teens. When confronted with the inevitability of an early school bell, a pattern emerges that comprises late, short, and irregular sleep. As noted previously, for some teens the combination results in extreme morning sleepiness and sleep-onset REM episodes. For others, the combination can be deadly. For example, in car crashes attributed to the driver having fallen asleep, the age of the driver is 16–25 years in more than 50% of cases.

Consequences of insufficient, irregular, and ill-timed sleep also include potential problems with mood regulation and depression, learning and memory, weight regulation, substance use or abuse, school tardiness and absenteeism, along with impulsivity and risk taking. Many of these behavioral tendencies are thought to be endemic problems of adolescence, yet the contributions of poor sleep may be involved. Serious consideration and scientific examination of these issues is needed. Trends point to continued deterioration of adolescent sleep patterns and at younger ages. Thus, whether adolescent sleep may affect the maturation of neurons, neural circuitry, or neurochemistry is in urgent need of study.

See also: Behavioral Change with Sleep Deprivation; Circadian Regulation by the Suprachiasmatic Nucleus; Circadian Rhythms in Sleepiness, Alertness, and Performance; Endocrine Function During Sleep and Sleep Deprivation; Narcolepsy; PET Activation Patterns; Phylogeny and Ontogeny of Sleep; Sleep Architecture; Sleep Deprivation and Brain Function; Sleep: Development and Circadian Control; Sleep-Dependent Memory Processing.

Further Reading

Carskadon MA (ed.) (2002) *Adolescent Sleep Patterns: Biological, Social, and Psychological Factors.* New York: Cambridge University Press.

Carskadon MA and Acebo C (2002) Regulation of sleepiness in adolescents: Update, insights, and speculation. *Sleep* 25: 606–614.

Carskadon MA, Harvey K, Duke P, Anders TF, Litt IF, and Dement WC (1980) Pubertal changes in daytime sleepiness. *Sleep* 2: 453–460.

Carskadon MA, Labyak SE, Acebo C, and Seifer R (1999) Intrinsic circadian period of adolescent humans measured in conditions of forced desynchrony. *Neuroscience Letters* 260: 129–132.

Carskadon MA, Vieira C, and Acebo C (1993) Association between puberty and delayed phase preference. *Sleep* 16: 258–262.

Carskadon MA, Wolfson AR, Acebo C, Tzischinsky O, and Seifer R (1998) Adolescent sleep patterns, circadian timing, and sleepiness at a transition to early school days. *Sleep* 21: 871–881.

Dahl RE and Spear LP (eds.) (2004) *Annals of the New York Academy of Sciences: Adolescent Brain Development: Vulnerabilities and Opportunities.*

Feinberg I (1982) Schizophrenia: Caused by a fault in programmed synaptic elimination during adolescence? *Journal of Psychiatric Research* 17: 319–334.

Grigg-Damberger M, Gozal D, Marcus CL, et al. (2007) The visual scoring of sleep and arousal in infants and children. *Journal of Clinical Sleep Medicine* 3: 201–240.

Jenni OG, Achermann P, and Carskadon MA (2005) Homeostatic sleep regulation in adolescents. *Sleep* 28: 1446–1454.

Jenni OG and Carskadon MA (2004) Spectral analysis of the sleep electroencephalogram during adolescence. *Sleep* 27: 774–783.

Lenroot RK and Giedd JN (2006) Brain development in children and adolescents: Insights from anatomical magnetic resonance imaging. *Neuroscience and Biobehavioral Reviews* 30: 718–729.

National Sleep Foundation (2006) *Sleep in America Poll Summary Findings.* Available at http://www.sleepfoundation.org.

Tanner JM (1962) *Growth at Adolescence.* Oxford: Blackwell.

Taylor DJ, Jenni OG, Acebo C, and Carskadon MA (2005) Sleep tendency during extended wakefulness: Insights into adolescent sleep regulation and behavior. *Journal of Sleep Research* 14: 239–244.

Sleep in Aging

M Rissling and S Ancoli-Israel, University of
California at San Diego, San Diego, CA, USA

Background

Sleep is one of many biological processes that invariably undergoes change with age. Older adults may experience an increase in sleep disturbance as the homeostatic and circadian rhythm processes regulating the sleep cycle become more fragile. Sleep is also affected by the many complications associated with physical and mental illness that often accompany aging. It is important to understand the distinction between expected changes in sleep associated with aging and those likely to arise from pathology unrelated to aging *per se*. This article reviews both.

Changes in Sleep with Age

It is generally well accepted that aging results in an overall decrease in slow-wave sleep (SWS). The amplitude of slow-wave delta waves is attenuated, and the percentage of SWS observed is markedly decreased. This process is gradual and begins in young adulthood, at around 20 years of age, but stabilizes by about age 60, with no further reduction in SWS. There is evidence of a gender difference, with men showing greater decreases in SWS than women. Latency to rapid eye movement (REM) sleep is also decreased in the older adult and is considered to be a direct consequence of the circadian phase advancement that occurs with aging.

The largest overall change observed in older adults is an increase in stage shifts, with more time spent awake in bed following sleep onset. The disruptions are often characterized by early morning awakenings and an inability to fall back to sleep and may be due to a number of sleep disturbances (such as insomnia or other specific sleep disorders) rather than to aging. The result is marked decreases in sleep efficiency (defined by the percentage of sleep that occurs while in bed) and consequently increases in daytime sleepiness. Stage 1 sleep is slightly increased in association with such arousals; stage 2 sleep appears to be more or less constant throughout the life span.

The need for sleep, or the sleep drive, however, does not diminish with age. Rather, it is the ability to satisfy the sleep drive that becomes impaired. Therefore, although the need for sleep remains the same, nighttime sleep is curtailed, with the remaining sleep drive redistributed to the daytime. The decreased ability to maintain nighttime sleep is most often attributed to changes in circadian rhythms, insomnia comorbid with medical or psychiatric illnesses and medication use, and the presence of other specific sleep disorders, such as periodic limb movements in sleep or restless leg syndrome (RLS), REM behavior disorder, and sleep-disordered breathing.

Changes in Circadian Rhythm

Circadian rhythms, such as hormone secretion, body temperature, and the sleep–wake cycle, fluctuate every 24 h. In humans, the sleep–wake cycle is regulated by the combined effect of an endogenous pacemaker and exogenous stimuli. The hypothalamic suprachiasmatic nucleus (SCN) is considered the endogenous clock of the brain and is believed to be the primary mediator of circadian rhythms. Circadian rhythms maintained by the SCN are naturally entrained by exogenous stimuli, or zeitgebers (time cues). The most significant zeitgeber is light, but other factors, such as physical activity, the timing of meals, and social interactions, have also been shown to be important regulators.

As people age, their circadian rhythms become weaker, desynchronize, and lose amplitude. It is hypothesized that the progressive deterioration of the SCN and its subsequent weakened functioning contribute to the disruption of circadian rhythms in older adults. The most common desynchronized pattern often seen in older adults is an advancement of the sleep phase.

Older adults with advancement of the sleep phase have increased numbers of early morning awakenings and difficulty maintaining sleep. They typically experience an increased drive for sleep between 7 p.m. and 9 p.m. and a drive to wake up between 3 a.m. and 5 a.m. Despite being sleepy earlier in the evening, they will often try to stay up later because of societal pressure, resulting in fewer hours spent in bed sleeping. The decreased total sleep time at night then results in daytime sleepiness and napping. Some older adults take inadvertent evening naps in front of the television or while reading early in the evening and then have difficulty falling asleep when they go to bed later in the evening; yet they still wake up early in the morning. This pattern gets interpreted as a complaint of insomnia, that is, difficulty falling asleep, difficulty staying asleep, and daytime sleepiness.

Treatment of Circadian Rhythm Disturbances

The most common treatment for circadian rhythm disturbances is bright light therapy. An advanced circadian rhythm can be delayed and the sleep–wake cycle strengthened with evening light exposure. As the sun is the strongest source of bright light, older adults with advanced rhythms should spend more time outdoors during the late afternoon or early evening and avoid bright light in the morning hours (as morning light would advance their rhythms even more). Studies have also shown that exposure to artificial light via a bright light box in the early evening has a similar effect and can improve sleep continuity in both healthy and institutionalized older adults. Adherence to a regular sleep schedule will promote a more robust sleep–wake cycle. Physical activity has also been shown to improve nighttime sleep quality. Such therapies are without adverse effects, and each has been shown to improve mood, performance, and quality of life and to reduce fatigue.

While melatonin has been shown to be safe, the efficacy for circadian rhythm disturbances has not been established. There is little consensus on the recommended dose or timing of administration, and melatonin purity and composition are currently not regulated in the United States by the Food and Drug Administration (FDA).

Insomnia

Insomnia is defined as difficulty falling asleep or staying asleep or nonrestorative sleep which results in daytime consequences. Insomnia is found in 30–50% of those more than 60 years old. Gender differences exist as well, with older women being more likely to complain of insomnia than older men. In older adults, sleep maintenance insomnia (an inability to maintain sleep throughout the night) and early morning or terminal insomnia (awakening early in the morning, with difficulty returning to sleep) are the most common complaints. Insomnia in the older adult is most often comorbid with medical and psychiatric illness or the medications used to treat those illnesses. Insomnia may also be a complaint in patients with specific sleep disorders, such as RLS or periodic limb movements in sleep.

Consequences of Insomnia

The consequences of disturbed sleep are well known and may include difficulty sustaining attention, slowed response time, difficulty with memory, and decreased performance. These consequences can be more pronounced in older adults and may even be misinterpreted as symptoms of dementia. Insomnia disrupts the sleep of 64% of bed partners of both healthy and demented older adults and 84% of caregivers (i.e., who may sleep in separate rooms). Nighttime restlessness secondary to insomnia is often cited as the primary reason for institutionalizing a demented relative.

Untreated insomnia is associated with a nearly 500% increase in the general risk for serious accidents and injuries, including car accidents due to daytime sleepiness and hip fractures due to falls. In the older adult, insomnia is most often comorbid with medical and psychiatric disorders and the medications used to treat them.

Medical and Psychiatric Illness and Insomnia

Older adults are the most susceptible to comorbid insomnia as they are more likely to have multiple medical conditions. Data from the National Sleep Foundation 2003 poll on sleep in older adults showed that those with three or four medical conditions, including cardiac disease, pulmonary disease, stroke, and depression, reported significantly more sleep complaints than did those with none, one, or two conditions.

Medical and psychiatric illnesses most commonly associated with insomnia include pain resulting from osteoarthritis or malignancies, chronic obstructive pulmonary disease, congestive heart failure, nocturia (e.g., due to enlarged prostate) and incontinence, neurological deficits (e.g., stroke, Parkinson's disease, and dementia), gastroesophageal reflux, depression, and anxiety disorder. Difficulty falling asleep has a prevalence rate of 31% for patients with osteoarthritis and 66% for those with chronic pain conditions. In addition, 81% of arthritis patients, 85% of chronic pain patients, and 33% of diabetes patients report difficulty staying asleep.

Two additional conditions that are associated with aging and contribute to poor sleep are nocturia and menopause. Nocturia or frequent voluntary voiding of urine during the night and nocturnal polyuria, the production of an abnormally large volume of urine during sleep, are common causes of awakenings and disruption of nighttime sleep. Although nocturia is often perceived as causing nocturnal awakenings, it has been hypothesized that the fragmented sleep experienced by the older adult leads to more awareness of the need to void. The circadian rhythm of urine excretion is also altered with age such that there is a fall of renal concentrating ability, sodium conservation, and secretion of renin–angiotensin–aldosterone, and arginine vasopressin secretion and an increase of atrial natriuretic hormone. The bladder's ability to store urine decreases with age, and age-related lower urinary tract problems, such as detrusor overactivity, can affect frequency and urgency.

Menopause is associated with hormonal, physiologic, and psychological changes that affect sleep and may play a critical role in modulating both the presence and the degree of sleep disruption. Menopausal vasomotor symptoms or hot flashes are known to lower sleep efficiency and cause multiple arousals during the night. Insomnia is a common symptom of menopause and may be partially related to psychological factors such as depression and anxiety.

In fact, insomnia can be prognostic or indicative of depression. While depressed mood may predict insomnia, unresolved insomnia also places adults at greater risk for developing depression, anxiety, and substance abuse. Studies in younger adults have suggested that treating the insomnia may improve depression.

Medication Use and Insomnia

Alerting or stimulating drugs, such as central nervous system stimulants, beta-blockers, bronchodilators, calcium channel blockers, corticosteroids, decongestants, stimulating antidepressants, stimulating antihistamines, and thyroid hormones are known to cause difficultly falling asleep when taken late in the day. Sedating drugs, such as sedative antihistamines, antihypertensives, and sedative antidepressants, may cause excessive daytime sleepiness when taken early in the day and contribute to daytime napping behavior and consequently delay nighttime sleep onset. Adjusting the dose or the time of day of administration often improves the insomnia.

Treatment of Insomnia

Effective treatments for insomnia in older adults include cognitive behavioral therapy (CBT) and pharmacological therapy.

CBT CBT is a combination of cognitive restructuring and behavioral modification for insomnia. The cognitive portion deals with misconceptions or unrealistic expectations about sleep (i.e., absolute requirement of 8 h or more of sleep), amplification of the consequences of not getting enough sleep, incorrect thinking about the cause of insomnia, and misunderstandings about healthy sleep practices. The behavioral component of CBT involves a combination of sleep restriction therapy, stimulus control therapy, and good sleep hygiene practices. **Tables 1–3** list the rules for each of these, respectively.

A recent National Institutes of Health (NIH) State-of-the-Science Conference on Insomnia concluded that CBT is as effective as prescription medications for the treatment of chronic insomnia and that there are indications that the beneficial effects of CBT may last well beyond the termination of treatment. If pharmacologic treatment is initiated, it is most effective when combined with CBT.

Table 1 Stimulus control therapy

Stimulus control is a behavioral therapy that seeks to remove the association of sleeplessness with the bed. The following steps reflect the principles of stimulus control:
Step 1. Go to bed only when sleepy.
Step 2. Use the bed only for sleeping – do not read, watch television, or eat in bed.
Step 3. If unable to fall asleep within about 20 min (don't watch the clock), move to another room. Stay up until really sleepy. The goal is to associate the bed with falling sleep quickly.
Step 4. If you are awake at night for more than about 20 min, get out of bed.
Step 5. Repeat Step 4 as often as necessary.
Step 6. Awaken at the same time every morning regardless of total sleep time.
Step 7. Avoid naps.

From Bootzin RR, Epstein D, and Wood JM (1991) Stimulus control instructions. In: Hauri PJ (ed.) *Case Studies in Insomnia*, pp. 19–28. New York: Plenum; Bootzin RR and Epstein D (2000) Stimulus control. In: Lichstein KL and Morin CM (eds.) *Treatment of Late-Life Insomnia*, pp. 167–184. Thousand Oaks, CA: Sage and Morin C (2005) Psychological and behavioral treatments for primary insomnia. In: Kryger MH, Roth T, and Dement WC (eds.) *Principles and Practice of Sleep Medicine*, 4th edn., pp. 726–748. Philadelphia: Elsevier/Saunders.

Table 2 Sleep restriction therapy

Sleep restriction therapy that seeks to keep patients from spending excessive time in bed. The following steps reflect the principles of sleep restriction therapy:
Step 1: Cut bedtime to actual amount patient reports sleeping but not less than 4 h per night.
Step 2: Prohibit sleep outside these hours.
Step 3: Get up at the same time each morning.
Step 4: Have patient report daily the amount of sleep obtained.
Step 5: Compute sleep efficiency (SE); based on moving average of five nights; when SE is >85%, increase bedtime by15 min.
Step 6: Compute sleep efficiency (SE); when SE is > 80%, increase time in bed by going to bed 15 min earlier. Continue to get up at same time each morning.

From Glovinsky PB and Spielman AJ (1991) Sleep restriction therapy. In: Hauri PJ (ed.) *Case Studies in Insomnia*, pp. 49–63. New York: Plenum; Morin C (2005) Psychological and behavioral treatments for primary insomnia. In: Kryger MH, Roth T, and Dement WC (eds.) *Principles and Practice of Sleep Medicine*, 4th edn., pp. 726–748. Philadelphia: Elsevier/Saunders; and Spielman AJ, Saskin P, and Thorpy MJ (1987) Treatment of chronic insomnia by restriction of time in bed. *Sleep* 10: 45–56.

Pharmacological therapy There are several classes of medications used to treat insomnia in older adults, in particular benzodiazepine receptor agonists (benzodiazepines and nonbenzodiazepines), melatonin receptor agonists, antidepressants, antipsychotics, and anticonvulsants. In addition, over-the-counter

Table 3 Sleep hygiene rules for older adults

1. Maintain a consistent bedtime and uptime.
2. Use the bed only for sleep and satisfying sex. Avoid watching television, eating, or discussing stressful issues while in bed.
3. Minimize noise and temperature extremes.
4. Keep the bedroom environment as dark as possible. Minimize the exposure to bright light if you get out of bed during the night. Use the dimmest night-light possible.
5. Avoid naps if you have difficulty falling asleep at night. If must nap, restrict napping to 30 min or less in the early afternoon.
6. Avoid nicotine near bedtime and during night awakenings.
7. Avoid all forms of caffeine for 4–6 h before bedtime.
8. Although alcohol may help you fall asleep, metabolism of alcohol will cause awakenings.
9. Avoid eating heavy meals and large amounts of protein close to bedtime. A light meal of carbohydrates or dairy products (particularly milk) is preferable.
10. Avoid vigorous exercise just before bedtime.
11. Avoid having pets in bed at nighttime.

From Morin C (2005) Psychological and behavioral treatments for primary insomnia. In: Kryger MH, Roth T, and Dement WC (eds.) *Principles and Practice of Sleep Medicine*, 4th edn., pp. 726–748. Philadelphia: Elsevier/Saunders; and Hauri PJ (1991) Sleep hygiene, relaxation therapy, and cognitive interventions. In: Hauri PJ (ed.) *Case Studies in Insomnia*, pp. 65–86. New York: Plenum.

medications such as antihistamines (diphrenhydramine) and nutritional or herbal supplements are often used. The NIH conference concluded that there is no systematic evidence for the effectiveness of antidepressants, antipsychotics, anticonvulsants, antihistamines, or herbal preparations in the treatment of insomnia. The panel also expressed significant concerns about the risks associated with the use of these medications, particularly in older adults.

The NIH panel also concluded that the benzodiazepine receptor agonists and in particular the newer nonbenzodiazepines have been shown to be the most effective and safest of the sedative hypnotics (at the time of the conference, the melatonin receptor agonist remelteon had not yet been approved by the FDA). The key to successful pharmacologic treatment of insomnia lies in choosing the sedative or hypnotic with the characteristics that best fit the patient's complaints. (It is important to note that only ten agents have been approved by the FDA for the treatment of insomnia. It is important to note that there are only 10 agents approved by the FDA for the treatment of insomnia, listed in **Table 4** with the recommended dose and half-life. **Table 5** lists the indications for the five newer nonbenzodiazepine receptor agonists and the melatonin receptor agonist.)

If sedatives or hypnotics are prescribed, the potential side effects must be taken into account. The administration of long-acting benzodiazepines can cause adverse daytime effects such as excessive daytime sleepiness and poor motor coordination, which

Table 4 Does and half-life of US Food and Drug Administration-approved agents for insomnia

Generic agent (trade name)	Dose (mg)[a]	Half-life (h)
Benzodiazepine receptor agonists: benzodiazepines		
Flurazepam HCL (Dalmane)	15, 30	47–100
Quazepam (Doral)	7.5, 15	39–73
Estazolam (ProSom)	0.5, 1, 2	10–24
Temazepam (Restoril)	7.5, 15, 30	3.5–18.4
Triazolam (Halcion)	0.125, 0.25	1.5–5.5
Benzodiazepine receptor agonists: nonbenzodiazepines		
Eszopiclone (Lunesta)	1, 2, 3	6
Zaleplon (Sonata)	5, 10, 20	1
Zolpidem (Ambien)	5, 10	1.4–4.5
Zolpidem MR (Ambien CR)	6.25, 12.5	2.8–2.9
Melatonin receptor agonist		
Ramelteon (Rozerem)	8	2–5

[a]The lowest dose is generally recommended as the starting dose for older people.
Physicians' Desk Reference (1991, 1999, and 2004).
FDA (2004) Lunesta (eszopiclone) Tablets: 1 mg, 2 mg, 3 mg. Retrieved January 24, 2008, from www.fda.gov/cder/foi/label/2004/021476lbl.pdf.
From Consensus conference (1984) *JAMA* 251: 2410–2414.

Table 5 Indications for the newer nonbenzodiazepine and melatonin receptor agonist agents

Agent	Indicated for sleep onset	Indicated for sleep maintenance	Required time in bed (h)
Eszopiclone[a]	X	X	8
Zaleplon[b]	X		4
Zolpidem[c]	X		7–8
Zolpidem MR[d]	X	X	7–8
Ramelteon[e]	X		No minimum time

[a]From Lunesta (eszopiclone) prescribing information. Sepracor.
[b]From SONATA (zaleplon) prescribing information. King Pharmaceuticals, Inc.
[c]From Ambien (zolpidem tartrate) prescribing information. Sanofi-Synthelabo Inc.
[d]From Ambien CR (zolpidem tartrate extended-release tablets) prescribing information. Sanofi-Synthelabo Inc.
[e]From Rozerem (ramelteon) prescribing information. Takeda Pharmaceutical Company Inc.

may lead to injuries. In older adults, the risk of falls, cognitive impairment, and respiratory depression are of particular concern, although recent data suggest that insomnia *per se*, and not hypnotics, might increase the risk of falls. Chronic use of long-acting benzodiazepines can lead to tolerance and withdrawal symptoms if they are abruptly discontinued. In addition, the potential for exacerbating coexisting medical conditions such as severe hepatic or renal disorders exists when some of these medications are used. The newer nonbenzodiazepine agents have been shown to be effective with a lower propensity for causing clinical residual effects, withdrawal, dependence, or tolerance.

Specific Sleep Disorders

There are also specific sleep disorders which are quite common in the older adult and which contribute to poor sleep. These would include sleep-disordered breathing (SDB), RLS and periodic limb movements in sleep, and REM behavior disorder.

Sleep-Disordered Breathing

SDB, also called sleep apnea, is a disorder in which the patient periodically stops breathing during sleep. Each apnea (complete cessation in respiration) or hypopnea (partial decrease in respiration) lasts a minimum of 10 s and generally ends with an arousal. This results in fragmented sleep and nocturnal hypoxemia. The total number of apneas plus hypopneas per hour of sleep is called the apnea–hypopnea index (AHI) or respiratory disturbance index (RDI). Depending on the laboratory, an AHI or RDI greater than or equal to 5–10 is required for the diagnosis of SDB.

The prevalence of SDB in older adults has been estimated to be between 30% and 60%, substantially higher than in younger adults. Cross-sectional studies have shown that there is a small increase in prevalence of SDB with increasing 10-year age groups. In the longest longitudinal study, older adults followed for 18 years showed changes in AHI only when associated with changes in body mass index.

There are two main types of SDB. The first, central sleep apnea, occurs when the respiratory centers in the brain shut down during sleep, and there is no attempt to breathe. Central sleep apnea is most common in congestive heart failure and is often associated with Cheyne–Stokes breathing. The second, obstructive sleep apnea (OSA), is the more common and occurs when the muscles of the airway collapse. The two main symptoms of OSA are loud snoring and excessive daytime sleepiness (EDS). EDS, which results from the recurrent nighttime arousals and sleep fragmentation, may manifest as unintentional napping or falling asleep at inappropriate times during the day. EDS can cause reduced vigilance and is associated with cognitive deficits, which may be particularly serious in older adults who may already have some cognitive impairment at baseline.

In addition to the cognitive deficits that may occur as a result of SDB, there is some evidence that many of the progressive dementias (e.g., Alzheimer's and Parkinson's disease) involve degeneration in areas of the brain stem responsible for the regulation of respiration and other autonomic functions relevant to sleep maintenance, thus placing the patient at increased risk for SDB.

Patients with SDB are at greater risk for hypertension, cardiac arrhythmias, congestive heart failure, myocardial infarction, and stroke. Most of the research, however, has not focused on older adults, so the relationship between SDB and some of these various morbidities remains unknown in the older population. Research that has targeted older adults has reported an association between SDB and the risk of developing cardiovascular disease, including coronary artery disease and stroke.

Treatment of SDB   Treatment of SDB in older adults should be guided not only by the severity of the SDB but by the significance of the patient's symptoms. Patients with more severe SDB (AHI > 20) deserve a trial of treatment. For those with milder SDB (AHI < 20), treatment should be considered if comorbid conditions such as hypertension, cognitive dysfunction, or EDS are present. Treatment should never be withheld because of age or because of assumed noncompliance.

The gold standard for treating SDB is continuous positive airway pressure (CPAP), a device which provides continuous positive pressure via the nasal passages or oral airway, thus creating a pneumatic splint to keep the airway open during inspiration. If used appropriately, CPAP has been shown to be safe and efficacious. Alternative treatments include oral appliances, weight loss (although most older adults with SDB are not obese), and surgery; however, no other treatments have been shown to be as effective as CPAP.

Significance of SDB in development   The presentation of sleep apnea in the older adult is not necessarily the same as in the younger adult. This has led to the idea that OSA in older adults might be a different disorder from OSA in younger adults. Two questions that arise are whether OSA is pathological in older adults and whether OSA in the older patient should be treated, and if so, when?

From a clinical standpoint, does it matter what is causing SDB in the older adult? Does it matter if the SDB in the older adult is the same as or different from the SDB seen in the younger adult? Probably not. What matters is whether SDB in the older adult has negative consequences. Most studies suggest that SDB does not increase the risk of mortality in the older adult, but that older adults with SDB are excessively sleepy and that SDB may contribute to decreased quality of life, increased cognitive impairment, and greater risk of nocturia, hypertension, and cardiovascular disease. In addition, older adults with both insomnia and SDB have more functional impairment than those with just insomnia.

Any patient, no matter the age, who presents with symptoms of impairment as well as with a history of traffic accidents and repeated falls, should be evaluated for a sleep disturbance. Treatment of a symptomatic older patient should not be withheld on the basis of age, as the consequence of SDB can be just as serious in older as in younger patients. The bottom line is that if the sleep apnea is associated with clinical symptoms (e.g., hypertension, cognitive dysfunction, nocturia, high levels of sleep-disordered breathing, cardiac disease), then it should be treated, regardless of the age of the patient.

Periodic Limb Movements and RLS

The condition known as periodic limb movements in sleep (PLMS) is characterized by repetitive leg movements during sleep, occurring every 20–40 s and recurring several hundred times over the course of the night, each causing a brief awakening, which results in sleep fragmentation. Patients with PLMS may complain of EDS and/or insomnia. The number of limb movements per hour of sleep is called the periodic limb movement index, with ten or more limb movements per hour needed for a clinical diagnosis. A diagnosis of PLMS therefore must be made with an overnight sleep study that records leg movements. The etiology of PLMS is unknown. PLMS can be seen in patients with fibromyalgia and in conjunction with other primary sleep disorders, including SDB and narcolepsy.

While the prevalence of PLMS in adults is estimated to be 5–6%, the rate increases dramatically with age, with reported prevalence rates of up to 45% in community-dwelling people over the age of 65.

RLS, a condition strongly linked to PLMS, is characterized by leg dysesthesia, often described as 'a creepy crawling' or 'restless' sensation, occurring in the relaxed awake state. The uncomfortable sensations can be relieved only by movement. The diagnosis of RLS is made on the basis of history alone, often with the question, When you try to relax in the evening or sleep at night, do you ever have unpleasant, restless feelings in your legs that can be relieved by walking or movement? Patients may have no knowledge that they kick, so interviewing the patient's bed partner may be helpful in elucidating the history of both RLS and PLMS. Patients with symptoms of RLS should also be assessed for anemia, uremia, and peripheral neuropathy as each of these conditions can cause or exacerbate RLS.

Treatment of PLMS and RLS   RLS and PLMS are treated primarily with dopamine agonists, which reduce the number of kicks and associated arousals. At this time, only ropinirole and pramipexole have been FDA-approved for the treatment of RLS.

REM Behavior Disorder

REM behavior disorder (RBD) is characterized by the intermittent absence of normal skeletal muscle atonia during REM sleep, resulting in patients' acting out their dreams. RBD generally occurs during the last third of the night, when REM sleep is more common. The movements during RBD may be violent and potentially harmful to patient and bed partner. Vivid dreams, consistent with the patient's aggressive and/ or violent behavior, may be recalled on waking.

The prevalence of RBD is unknown, but it has been shown to be most common in older men. The etiology of RBD is also unknown, but there is a strong association between idiopathic RBD and neurodegenerative diseases, including Parkinson's disease, multiple system atrophy, and Lewy bodies dementia, with the RBD often preceding the other disorders by years. Data suggest that close to half of RBD patients develop Parkinson's disease within several years of the RBD diagnosis. The onset of acute RBD has been associated with stress disorders and the use of monoamine oxidase inhibitors and fluoxetine as well as withdrawal of REM-suppressing agents (e.g., alcohol, tricyclic antidepressants, amphetamines, and cocaine).

Treatment of RBD   Treatment of RBD with clonazepam, a long-acting benzodiazepine, results in partial or complete cessation of abnormal nocturnal motor movements in 90% of patients. However, because of the long half-life, many patients report residual sleepiness in the morning and during the day. Recently, melatonin has also been shown to be an effective treatment. In addition to the pharmacologic treatment, other aspects of treatment should include good sleep hygiene and education for the patient and bed partner. The bedroom should be made safer by removing heavy or breakable or potentially injurious objects from the bed's vicinity. Heavy curtains should be placed on bedroom windows and doors, and windows should be locked at night. Finally, to avoid falling out of bed, patients may consider sleeping on a mattress on the floor.

Conclusion

The ability to have restful sleep decreases with age but is likely independent of age itself. Medical and psychiatric disorders, medication, circadian rhythm changes, and primary sleep disorders (such as SDB,

RLS or PLMS, and RBD) contribute to older adults' decreased ability to maintain a full night's sleep. Each of these conditions is treatable, and health care professionals should never withhold treatment because of age or because of assumed noncompliance.

See also: Acetylcholine; Adenosine; Cytokines and other Neuromodulators; Endocrine Function During Sleep and Sleep Deprivation; Gene Expression; Hippocampal–Neocortical Dialog; Histamine; Hypothalamic Regulation of Sleep; Locus Coeruleus and Raphe Nucleus; Network Reactivation; Phylogeny and Ontogeny of Sleep; Reticular Activating System; Sleep Apnea; Sleep Architecture; Sleep: Development and Circadian Control; Thalamic Regulation of Sleep.

Further Reading

Ancoli-Israel S (2005) Sleep and aging: Prevalence of disturbed sleep and treatment considerations in older adults. *Journal of Clinical Psychiatry* 66: 24–30.

Ancoli-Israel S (2007) Guest editorial: Sleep apnea in older adults: Is it real and should age be the determining factor in the treatment decision matrix? *Sleep Medicine Review* 11: 83–85.

Ancoli-Israel S, Kripke DF, Klauber MR, et al. (1991) Sleep disordered breathing in community-dwelling elderly. *Sleep* 14(6): 486–495.

Bliwise DL (1993) Sleep in normal aging and dementia. *Sleep* 16(1): 40–81.

Bootzin RR and Epstein D (2000) Stimulus control. In: Lichstein KL and Morin CM (eds.) *Treatment of Late-Life Insomnia*, pp. 167–184. Thousand Oaks, CA: Sage.

Bootzin RR, Epstein D, and Wood JM (1991) Stimulus control instructions. In: Hauri PJ (ed.) *Case Studies in Insomnia*, pp. 19–28. New York: Plenum Press.

Edinger JD, Heolscher TJ, Marsh GR, Lipper S, and Ionescu-Pioggia M (1992) A cognitive-behavioral therapy for sleep-maintenance insomnia in older adults. *Psychological Aging* 7: 282–289.

Foley DJ, Monjan A, Simonsick EM, Wallace RB, and Blazer DG (1999) Incidence and remission of insomnia among elderly adults: An epidemiologic study of 6,800 persons over three years. *Sleep* 22: S366–S372.

Glass J, Lanctot KL, Herrmann N, Sproule BA, and Busto UE (2005) Sedative hypnotics in older people with insomnia: Meta-analysis of risks and benefits. *British Medical Journal* 331: 1169.

Glovinsky PB and Spielman AJ (1991) Sleep restriction therapy. In: Hauri PJ (ed.) *Case Studies in Insomnia*, pp. 49–63. New York: Plenum.

Hauri PJ (1991) Sleep hygiene, relaxation therapy, and cognitive interventions. In: Hauri PJ (ed.) *Case Studies in Insomnia*, pp. 65–86. New York: Plenum.

Lichstein KL and Reidel BW (1994) Behavioral assessment and treatment of insomnia: A review with an emphasis on clinical application. *Behavior Therapy* 25: 659–688.

Martin JL and Ancoli-Israel S (2003) Insomnia in older adults. In: Szuba MP, Kloss JD, and Dinges DF (eds.) *Insomnia: Principles and Management*, pp. 136–154. Cambridge, UK: Cambridge University Press.

Morin C (2005) Psychological and behavioral treatments for primary insomnia. In: Kryger MH, Roth T, and Dement WC (eds.) *Principles and Practice of Sleep Medicine*, 4th edn., pp. 726–748. Philadelphia: Elsevier/Saunders.

Morin CM, Colecchi C, Stone J, Sood R, and Brink D (1999) Behavioral and pharmacological therapies for late life insomnia. *Journal of the American Medical Association* 281: 991–999.

National Institutes of Health (2005) State-of-the-Science conference statement on manifestations and management of chronic insomnia in adults. *Sleep* 28(9): 1049–1057.

Spielman AJ, Saskin P, and Thorpy MJ (1987) Treatment of chronic insomnia by restriction of time in bed. *Sleep* 10: 45–56.

Van Someren EJ (2000) Circadian rhythms and sleep in human aging. *Chronobiology International* 17(3): 233–243.

Vitiello MV (2000) Effective treatment of sleep disturbances in older adults. *Clinical Cornerstone* 2(5): 16–47.

Relevant Websites

http://www.aasmnet.org – American Academy of Sleep Medicine.
http://www.sleepfoundation.org – National Sleep Foundation.
http://www.sleepresearchsociety.org – Sleep Research Society.

CONTROL OF SLEEP AND SLEEP STATES

Thalamic Regulation of Sleep

A Destexhe, Centre National de la Recherche
Scientifique (CNRS), Gil-sur-Yvette, France
T J Sejnowski, Salk Institute for Biological Studies
and University of California at San Diego, La Jolla,
CA, USA

© 2009 Elsevier Inc. All rights reserved.

Introduction

During slow-wave sleep, brain activity is dominated by oscillations, such as delta oscillations and slower oscillations. The thalamus, together with the cortex, participates in generating and sustaining these oscillations. Delta waves were observed in the isolated cortex *in vivo* by Frost and colleagues in 1966, and delta-wave activity has been found in the isolated thalamus as well in thalamic slices, although these are more regular than delta waves *in vivo*. The latter studies demonstrated that this form of delta-wave activity is generated intrinsically by the interplay of I_T and I_h currents in thalamocortical (TC) neurons. Slow oscillations (<1 Hz) can be generated by the cortex; they were observed in cortical slices by Sanchez-Vives and McCormick in 2000. A similar type of oscillation has also been observed by the group of V Crunelli to be an intrinsic property of thalamic neurons when metabotropic receptors are stimulated.

Spindle oscillations, which are mostly visible in the early stages of sleep (stage II in humans), are the best-studied sleep oscillation and are generated by thalamic circuits under physiological conditions. In contrast to delta and slow oscillations, for which cortical generators also exist, spindles are generated exclusively in the thalamus. We focus here on sleep spindles, their genesis by thalamic networks, how these oscillations are distributed across the entire TC system, and how they are modulated.

In the human electroencephalogram (EEG), spindle oscillations are grouped in short 1–3 s periods of 7–14 Hz oscillations, organized within a waxing-and-waning envelope, that recur periodically every 10–20 s. In cats and rodents, spindle waves of similar characteristics appear during slow-wave sleep and are typically more prominent at sleep onset. They are enhanced by some anesthetics, such as barbiturates, which, when administered at an appropriate dose, generate an EEG dominated by spindles. In ferret thalamic slices, spindle oscillations occur spontaneously. This *in vitro* model of spindle waves has made it possible to precisely characterize the ionic mechanisms underlying spindle oscillations.

The first indication that spindles could originate outside the cerebral cortex was from Bishop in 1936, who showed that rhythmical activity was suppressed in the cerebral cortex following destruction of its connections with the thalamus. In 1938, Bremer showed that rhythmical activity is still present in the white matter following destruction of the cortical mantle. Later, Adrian in 1941 and Morison and Bassett in 1945 observed that spindle oscillations persist in the thalamus on removal of the cortex, providing firm experimental evidence that these oscillations originate in the thalamus. These experiments led to the development of the thalamic pacemaker hypothesis, according to which rhythmic activity is generated in the thalamus and communicated to the cortex, where it entrains cortical neurons and is responsible for the rhythmical activity observed in the EEG.

Thalamic Pacemakers for Sleep Spindle Oscillations

The first cellular mechanism for the genesis of spindle oscillations was proposed by Andersen and Eccles in 1962. From intracellular recordings from TC relay neurons during spindles, they reported that TC cells fired bursts of action potentials interleaved with inhibitory postsynaptic potentials (IPSPs). They suggested that TC cells fire in response to IPSPs (postinhibitory rebound), which was later demonstrated to be a characteristic electrophysiological feature of thalamic cells by Jahnsen and Llinas. Andersen and Eccles suggested that the oscillations arise from the reciprocal interactions between TC cells and inhibitory local-circuit interneurons. This mechanism was later incorporated into a computational model that provided a phenomenological description of the inhibitory rebound.

The mechanism proposed by Andersen and Eccles was seminal but not entirely correct because reciprocal connections between TC cells and thalamic interneurons have never been observed in anatomical studies. It was later found by Scheibel and Scheibel that the thalamic reticular (RE) nucleus, a sheetlike structure of inhibitory neurons surrounding the thalamus, could provide the inhibition of TC cells and that TC-RE loops could underlie the recruitment phenomena and spindle oscillations. They predicted that the output of the RE nucleus should be inhibitory and that the inhibitory feedback from RE cells onto TC cells should be critical for the genesis of thalamic rhythmicity. This hypothesis was supported by the observation that the pattern of firing of RE neurons was tightly correlated with IPSPs in TC neurons.

Several critical experiments by the group of Steriade firmly established the involvement of the RE nucleus in the generation of spindles in cats *in vivo*. The typical intracellular features of spindle oscillations in the two thalamic cell types is shown in **Figure 1(a)**. By using intracellular or extracellular experiments, it was shown that (1) cortically projecting thalamic nuclei lose their ability to generate spindle oscillations if deprived of input from the RE nucleus and (2) the isolated RE nucleus can itself generate rhythmicity in the spindle frequency range.

Thus, thalamic rhythmicity can be explained by three different mechanisms: the TC interneuron loops of Andersen and Eccles, the TC-RE loops of Scheibel and Scheibel, and the RE pacemaker hypothesis of Steriade (although Steriade's work also demonstrated the importance of the cortex; see the section titled 'Dialog between thalamus and cortex: emergence of large-scale synchrony'). The introduction of an *in vitro* model of spindle waves in ferrets by McCormick's group (**Figure 1(b)**) supported the second of these mechanisms. Slices of the visual thalamus that contain the dorsal (lateral geniculate nucleus

(LGN)) and reticular nuclei (perigeniculate nucleus (PGN)) as well as the interconnections between them generated spindles spontaneously, confirming earlier experimental evidence for the genesis of spindles in the thalamus.

The *in vitro* preparation allowed detailed pharmacological investigation of the ionic currents and synaptic receptors underlying the spindle oscillations. The spindle waves disappeared after the connections between TC and RE cells were severed, consistent with the mechanism based on intrathalamic TC-RE loops proposed by Scheibel and Scheibel. This *in vitro* experiment also confirmed the observation that the input from RE neurons is necessary to generate spindles, as found by Steriade's group in 1985. However, the RE nucleus maintained *in vitro* did not generate oscillations without connections with TC cells, in contrast with the observation of spindle rhythmicity in the isolated RE nucleus *in vivo* by Steriade's group in 1987.

Thus, *in vitro* experiments appear to support a mechanism by which oscillations are generated by the TC-RE loop, in contrast with the RE pacemaker hypothesis. Computational models suggested a way to

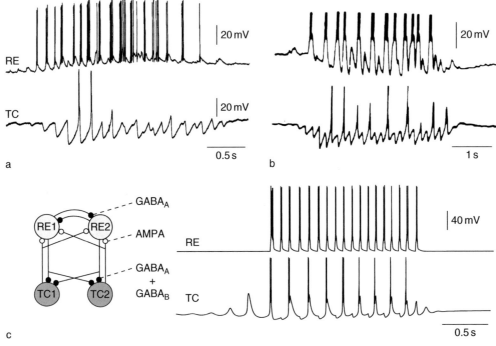

Figure 1   Spindle oscillations in thalamic circuits: (a) *in vivo* intracellular experiments in cats under barbiturate anesthesia, showing the activity of thalamocortical (TC) and thalamic reticular (RE) cells during spindle waves; (b) *in vitro* intracellular experiments in ferret visual thalamic slices, showing the activity of the same type of thalamic neurons during spindle waves; (c) computational model of spindle waves generated by TC-RE interactions. The left drawing in (c) shows a simple circuit, consisting of two TC and two RE cells interconnected with glutamatergic (AMPA) and GABAergic receptors (GABA$_A$ and GABA$_B$). The right drawing in (c) shows the activity of two model neurons during simulated spindle waves. AMPA, α-amino-3-hydroxy-5-methyl-4-isoxazole propionic acid; GABA, γ-aminobutyric acid. (a) Modified from Steriade M and Deschênes M (1984) The thalamus as a neuronal oscillator. *Brain Research Review* 8: 1–63. (b) Modified from von Krosigk M, Bal T, and McCormick DA (1993) Cellular mechanisms of a synchronized oscillation in the thalamus. *Science* 261: 361–364. (c) Modified from Destexhe A, Bal T, McCormick DA, and Sejnowski TJ (1996) Ionic mechanisms underlying synchronized oscillations and propagating waves in a model of ferret thalamic slices. *Journal of Neurophysiology* 76: 2049–2070.

reconcile these apparently contradictory experimental observations. Models showed that both types of rhythmicities are possible and suggested ways to reconcile the experiments. First, different modeling studies found that oscillations very similar to isolated RE preparation *in vivo* could be generated by networks of RE cells. The oscillations were generated by the interaction between the rebound properties of RE cells, as provided by the T-type Ca^{2+} current IT, and γ-aminobutyric acid (GABA)-mediated IPSPs, consistent with experiments. Second, several modeling studies found that spindle oscillations with characteristics identical to those in the *in vitro* preparation could be generated from TC-RE loops (**Figure 1(c)**). In this case, the oscillations were dependent on IT-mediated rebound properties in the TC cells and GABA-mediated IPSPs, together with excitatory

postsynaptic potential (EPSP)-evoked bursts in RE cells. Third, it has been proposed that the differences observed between spindles *in vitro* and *in vivo* could be explained by the limited connectivity between the RE neurons in the slice and/or insufficient levels of neuromodulation in the slice needed to maintain isolated RE oscillations.

Dialog between the Thalamus and Cortex: Emergence of Large-Scale Synchrony

The initiation and distribution of spindle oscillations in large circuits was investigated more recently by using multiple recordings *in vivo* and *in vitro* (**Figures 2(a)** and **2(c)**). Spindle oscillations *in vitro* show traveling wave patterns, with the oscillation typically starting on one side of the slice and propagating

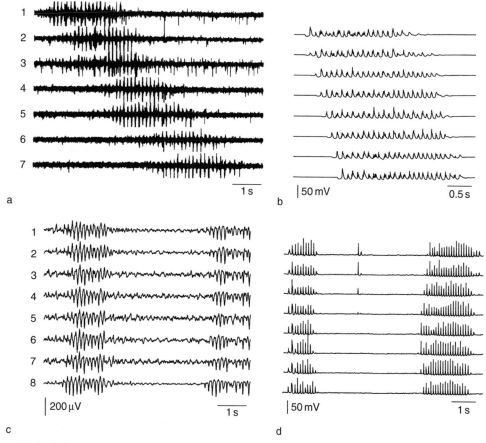

Figure 2 Control of spindle oscillations by the cerebral cortex: (a) multisite extracellular recordings *in vitro* showing the propagating activity of spindle waves in the visual thalamic slice; (b) model of propagating spindle wave activity in thalamic networks with topographic connectivity; (c) multisite extracellular recordings in the thalamus of cats *in vivo* showing the large-scale synchrony of spindle waves in the intact thalamocortical (TC) system; (d) model of the TC network showing large-scale synchrony. (a) Modified from Kim U, Bal T, and McCormick DA (1995) Spindle waves are propagating synchronized oscillations in the ferret LGNd *in vitro. Journal of Neurophysiology* 74: 1301–1323. (b) Modified from Destexhe A, Bal T, McCormick DA, and Sejnowski TJ (1996) Ionic mechanisms underlying synchronized oscillations and propagating waves in a model of ferret thalamic slices. *Journal of Neurophysiology* 76: 2049–2070. (c) Modified from Contreras D, Destexhe A, Sejnowski TJ, and Steriade M (1996) Control of spatiotemporal coherence of a thalamic oscillation by corticothalamic feedback. *Science* 274: 771–774. (d) Modified from Destexhe A, Contreras D, and Steriade M (1998) Mechanisms underlying the synchronizing action of corticothalamic feedback through inhibition of thalamic relay cells. *Journal of Neurophysiology* 79: 999–1016.

to the other side at a constant propagation velocity (**Figure 2(a)**). In contrast to thalamic slices, the intact TC system *in vivo* does not display such clear-cut propagation, but spindle oscillations are remarkably synchronized over extended thalamic regions and show very limited traveling-wave activity (**Figure 2(c)**), in agreement with early observations. Moreover, a study by Contreras and colleagues in 1996 showed that large-scale synchrony was lost when the cortex was removed, suggesting that, although the oscillation is generated by the thalamus, its synchrony depends on the cortex. However, cutting intracortical connections has no effect on large-scale synchrony, so cortical connections are not responsible for organizing the synchrony of sleep spindles.

The mechanisms for large-scale synchrony were investigated by computational models by first simulating the propagating properties found in slices (**Figure 2(b)**). These models assumed a topographic connectivity between TC and RE layers and could generate traveling spindle waves consistent with *in vitro* data. Second, a model by Destexhe and colleagues in 1998 simulated TC networks with bidirectional interactions between the cortex and thalamus and showed that all experiments could be reproduced under the assumption that the cortex recruited TC cells primarily through inhibition. This inhibitory-dominant cortical feedback to the thalamus is consistently observed experimentally and can explain large-scale synchrony by the mutual recruitment of thalamic and cortical networks (**Figure 2(d)**). The same mechanism can also explain the genesis of absence-type of epileptic seizures when the excitability of cortical neurons is too high. The concept of cortical control of thalamic spindle oscillations can be generalized to other oscillation types, and it was proposed by Steriade that the slow oscillation organizes spindle oscillations, delta oscillations, and more complex patterns such as K-complexes.

Control of Sleep Oscillations by Neuromodulators

Neuromodulators such as acetylcholine (ACh), norepinephrine, (NE), serotonin, (5-HT), histamine (HA), and glutamate abolish the low-frequency rhythms in TC systems during sleep. In the thalamus, where both the relay neurons and the reticular thalamic cells are hyperpolarized during sleep, the activation of the neuromodulatory systems depolarizes the thalamic cells by 5–20 mV, inactivating the low threshold Ca^{2+} current and inhibiting bursting. The transition to tonic firing enhances the participation of sensorimotor processing, which is blocked during sleep.

Conclusion

The thalamus shifts during sleep from a tonic mode, suitable for relaying information from the periphery to the cortex, to a rhythmic mode that generates activity and produces highly spatially and temporally coherent states through interactions with the cortex. During sleep, the feedback connections from the cortex to the thalamus become highly effective in globally coordinating activity in the thalamus. Thus, the thalamus becomes a mirror during sleep, linking distant parts of the cortex. This could be important for maintaining and adjusting the overall balance of activity in the cortex.

See also: Hypothalamic Regulation of Sleep; Phylogeny and Ontogeny of Sleep; Sleep Oscillations and PGO Waves.

Further Reading

Contreras D, Destexhe A, Sejnowski TJ, and Steriade M (1996) Control of spatiotemporal coherence of a thalamic oscillation by corticothalamic feedback. *Science* 274: 771–774.

Destexhe A, Bal T, McCormick DA, and Sejnowski TJ (1996) Ionic mechanisms underlying synchronized oscillations and propagating waves in a model of ferret thalamic slices. *Journal of Neurophysiology* 76: 2049–2070.

Destexhe A, Contreras D, and Steriade M (1998) Mechanisms underlying the synchronizing action of corticothalamic feedback through inhibition of thalamic relay cells. *Journal of Neurophysiology* 79: 999–1016.

Destexhe A and Sejnowski TJ (2001) *Thalamocortical Assemblies*. Oxford: Oxford University Press.

Destexhe A and Sejnowski TJ (2003) Interactions between membrane conductances underlying thalamocortical slow-wave oscillations. *Physiological Reviews* 83: 1401–1453.

Kim U, Bal T, and McCormick DA (1995) Spindle waves are propagating synchronized oscillations in the ferret LGNd *in vitro*. *Journal of Neurophysiology* 74: 1301–1323.

Sanchez-Vives MV and McCormick DA (2000) Cellular and network mechanisms of rhythmic recurrent activity in neocortex. *Nature Neuroscience* 10: 1027–1034.

Steriade M (2003) *Neuronal Substrates of Sleep and Epilepsy*. Cambridge, UK: Cambridge University Press.

Steriade M (2006) Grouping of brain rhythms in corticothalamic systems. *Neuroscience* 137: 1087–1106.

Steriade M and Deschênes M (1984) The thalamus as a neuronal oscillator. *Brain Research Review* 8: 1–63.

Steriade M, Jones EG, and McCormick DA (eds.) (1997) *Thalamus*, 2 vols. Amsterdam: Elsevier.

Steriade M, McCormick DA, and Sejnowski TJ (1993) Thalamocortical oscillations in the sleeping and aroused brain. *Science* 262: 679–685.

von Krosigk M, Bal T, and McCormick DA (1993) Cellular mechanisms of a synchronized oscillation in the thalamus. *Science* 261: 361–364.

Wang XJ, Golomb D, and Rinzel J (1995) Emergent spindle oscillations and intermittent burst firing in a thalamic model: Specific neuronal mechanisms. *Proceedings of the National Academy of Sciences of the United States of America* 92: 5577–5581.

Hypothalamic Regulation of Sleep

P M Fuller and J Lu, Harvard Medical School, Boston, MA, USA

Introduction

The fact that the hypothalamus plays a central role in the regulation of sleep–wake behavior was first suggested in 1916 when the Viennese neurologist and psychiatrist, Baron Constantine von Economo, began to study patients with a viral encephalitis that profoundly affected sleep–wake regulation (i.e., encephalitis lethargica or von Economo's sleeping sickness). The preponderance of patients with von Economo's encephalitis lethargica slept excessively (some more than 20 h per day). Postmortem brain analysis revealed that these individuals had lesions at the junction of the midbrain and posterior hypothalamus, suggesting that this area of the brain contained wake-promoting circuitry. Interestingly, a small percentage of individuals afflicted with encephalitis lethargica became insomniacs, often sleeping only a few hours each day. These insomniacs had lesions involving not the midbrain–diencephalon junction but, rather, the basal forebrain (BF) and anterior hypothalamus, suggesting that these areas of the brain contained sleep-promoting circuitry. von Economo's observations in the lethargica victims were later recapitulated experimentally by Ranson (1939) through his demonstration of hypersomnolence in rhesus monkeys sustaining hypothalamic lesions and by Nauta (1946) in his findings of profound changes in sleep–wake behavior in rats following surgical ablation of preoptic and BF tissue. Remarkably, however, the key role of the hypothalamus in the regulation of sleep–wakefulness was obscured for almost a half century by the discovery of a brain stem 'ascending arousal system' in the late 1940s.

Electrocortical and Behavioral Arousal: The Ascending Reticular Activating System

During cortical arousal (i.e., the neurologically defined vigilance state of wakefulness), the cortical electroencephalogram (EEG) is desynchronized (i.e., high frequency/low amplitude), presumably reflecting differences in the timing of processing of cognitive, motor, and perceptual functions. The EEG directly reflects the collective synaptic potentials of inputs largely to pyramidal cells within the neocortex and hippocampus. In 1935, Bremer uncovered evidence of an ascending arousal system necessary for forebrain and cortical arousal when he demonstrated that transection of the brain stem at the pontomesencephalic level (but not the spinomedullary junction) produced coma in anesthetized cats. More than a decade later, Morruzi and Magoun provided additional support for the concept of an ascending arousal system when they demonstrated that electrical stimulation of the rostral pontine reticular formation produced a desynchronized EEG (an electrophysiological correlate of the conscious state) in anesthetized cats. It was subsequently shown that transections, electrolytic lesions, and infarcts of this same region (i.e., the putative ascending arousal system) produced coma, although it remained unclear whether this was a result of damage to fibers of passage or specific cell groups. Because these ascending arousal pathways traversed (or appeared to originate in) the midbrain reticular formation, the term 'ascending reticular activating system' (ARAS) was coined and the functional integrity of the ARAS was henceforth considered the *sine qua non* for cortical arousal and consciousness. What remained unresolved, however, was the neuroanatomical pathways and population of neurons by which the ARAS might control cortical activity. In the 1970s and 1980s, experiments revealed that, contrary to popular conception, the origin of the ARAS was not a functionally homogeneous mass of reticular tissue in the brain stem tegmentum but, rather, composed of discrete neuronal populations (and neurotransmitters) which can be parsed into two distinct anatomical branches. The first branch of the ARAS consists of cholinergic neurons originating in the pedunculopontine (PPT) and laterodorsal (LDT) tegmental nuclei of the mesopontine tegmentum. The PPT and LDT send excitatory cholinergic projections to the midline and intralaminar nuclei of the thalamus and play a critical gating role in thalamocortical transmission by preventing relay neurons from being hyperpolarized and entering into burst mode, thus clearing the way for thalamocortical sensory transmission.

Regarding thalamic oscillations during wake and sleep, thalamocortical neurons fire in one of two stereotypic modes. When their membrane potential is near threshold (as during wake), they respond to incoming stimuli by firing single spikes. However, when they are deeply hyperpolarized, a low-threshold calcium channel is de-inactivated. The resultant excitatory postsynaptic potentials now produce calcium spikes which are prolonged and produce a depolarized plateau, from which the neuron fires a series or burst of action potentials (as during sleep). The increase in burst sequences inhibits thalamocortical sensory transmission, thus leading to deprivation of sensory signals from the environment. The state of these thalamocortical neurons is determined by inputs from the thalamic reticular nucleus, which surrounds the lateral surface of the thalamus. The reticular nucleus sends an inhibitory GABAergic projection to GABA$_B$ receptors on the thalamocortical neurons. These GABA$_B$ receptors are potassium channels that can deeply hyperpolarize the neuron sufficiently to de-inactivate the low-threshold calcium channels. Therefore, it is the state of the reticular nucleus that in turn determines the firing state of thalamocortical neurons.

Predictably then, neurons of the PPT and LDT fire most rapidly during wakefulness and rapid eye movement (REM) sleep, which is the stage of sleep associated with cortical activation, muscle atonia, and dreams. Historically, the thalamocortical system has been viewed as a critical component of the ARAS. In this view, the thalamus is thought to function as a major relay to the cortex for the ARAS, with the overall activity of the thalamocortical system forming the origin of the cortical EEG. Indeed, thalamic relay neurons fire in patterns that correlate with the cortical EEG. In turn, the overall activity in the thalamocortical system is thought to be regulated by the ARAS. Nevertheless, a critical role of the thalamocortical system in the ARAS has been challenged by studies demonstrating persistent cortical activation (i.e., desynchronized EEG) following thalamic lesions.

The second branch of the ARAS, which bypasses the thalamus and originates in the upper brain stem and caudal hypothalamus, consists of a series of wake-promoting monoaminergic cell groups that project to and activate neurons in the lateral hypothalamic area (LHA), BF, and cerebral cortex. The monoaminergic systems include the noradrenergic ventrolateral medulla and locus coeruleus, the dopaminergic neurons of the ventral periaqueductal gray matter, the serotoninergic dorsal and median raphe nuclei, and histaminergic neurons in the tuberomammillary nucleus. In general, neurons in all of these cell groups fire more during wakefulness than during non-rapid eye movement (NREM) sleep and show virtually no activity during REM sleep. Furthermore, of these monoaminergic cell groups, the dopaminergic neurons of the ventral periaqueductal gray appear to most potently influence arousal. More rostrally, the LHA, in addition to receiving input from these ascending monoaminergic systems, projects to the BF, cerebral cortex, and, in a reciprocal manner, components of the brain stem arousal systems. The LHA projection to the cortex is thought to augment

the arousal system as lesions of the LHA produce profound sleepiness, stupor, or even coma in monkeys and rats. In fact, the LHA contains at least two distinct populations of neurons that are thought to contribute to the regulation of wakefulness. LHA neurons that contain orexin (also called hypocretin) are active during wakefulness and contribute to cortical arousal through ascending projections to the cortex and descending projections to neurons in the tuberomammillary nucleus, locus coeruleus, and dorsal raphe. Mice lacking orexin demonstrate narcolepsy-like symptoms, including profound behavioral state instability (e.g., frequent state transitions and cataplexy). putative glutamatergic LHA neurons that contain melanin-concentrating hormone have similar projections to the orexin neurons but are mostly active during REM sleep, during which time they are thought to inhibit the ascending monoaminergic systems. A third group of putative glutamatergic LHA neurons is also thought to be involved in arousal; however, this group remains unresolved.

Finally, the BF (including the nucleus basalis and magnocellular preoptic nucleus in the substantia innominata, and the medial septal nucleus and nucleus of the diagonal band of Broca) consists of cholinergic neurons intermixed with noncholinergic (largely GABAergic) neurons which project to the cortex, hippocampus, and, to a lesser extent, the thalamus and are implicated in waking and EEG desynchronization.

New Perspective on the ARAS

As previously described, the traditional model for the ARAS posits an origin in the upper brain stem monoaminergic and cholinergic cell groups and reticular formation, and it emphasizes the role of the thalamus as an obligate relay in cortical activation. However, cell-specific lesions using excitotoxins (ibotenic acid and kainic acid) in these cell groups have caused relatively limited alteration of wakefulness in rats and cats. Hence, the origin of the arousing influence (i.e., the ARAS) has never been fully explained, and the source has been presumed to be the collective input from all of these structures so that lesions of any one of them will not cause loss of consciousness.

Experimental evidence has strongly suggested that the parabrachial nucleus (PB) and precoeruleus region (PC) of the dorsolateral pontine region, located caudal to the cholinergic PPT/LDT, are the likely neuroanatomical structures of origin for the ARAS, henceforth referred to as the 'ascending arousal system.' In general support of this concept are the findings that stimulation of the PB (in cats) produces marked electrophysiological activation of the cortex and cell-specific lesions of the PB (in rats) produce profound coma with persistent sub-delta EEG. It has also been shown that the PB sends excitatory projections to the BF that, in turn, project to the cortex, presumably to maintain cortical arousal. In parallel with the PB–BF projection, the PC projects to the medial septum that, in turn, projects to the hippocampus and is presumably important in ascending control of hippocampal function. This hypothesis is supported by the observation that lesions of the PC selectively eliminate hippocampal theta activity during REM sleep. Thus, in this more contemporary view, the PB/PC, like the cholinergic PPT/LDT, is active during all states of cortical activation, including both wakefulness and REM sleep. In the presence of concomitant activation of the monoaminergic arousal systems, the cortical response to PB/PC–BF activation would be wakefulness; in the absence of monoaminergic stimulation, a state of REM sleep would prevail. This ascending arousal model differs fundamentally from the prevailing one in which the wake–sleep state is determined by the level of activation only of the monoaminergic and cholinergic arousal systems. In this alternate view, monoaminergic and cholinergic systems may modulate the activity of the PB/PC and can determine (via their cortical and thalamic projections) whether the PB/PC induces wakefulness versus REM sleep, but they cannot sustain a waking (or REM) state without the PB/PC. In essence, it appears that the loss of wakefulness caused by electrolytic or mechanical lesions in the midbrain reticular formation may be due mainly to disruption of the ascending projections to the BF (but not thalamus) from the PB/PC. Although many other cell groups, including the monoaminergic and cholinergic components of the arousal system, may play specific roles in modulating arousal, particularly under specific physiological circumstances, no other area has this property of being absolutely necessary to maintain a waking state.

As indicated previously, the functional integrity of the thalamus has been widely considered to be necessary for cortical arousal and the maintenance of consciousness. A well-entrenched role for the thalamus in the ARAS is further indicated by the explicit representation in contemporary neuroscience textbooks of the thalamocortical system as an obligate system/relay for maintaining cortical awareness (i.e., consciousness). It is indeed likely that the thalamus, which transmits a wide range of sensory inputs to the cerebral cortex, is necessary for its normal cognitive function. However, experiments dating back to Ranson's 1939 work have repeatedly shown that rodents, nonhuman primates, and humans lacking a thalamic relay and/or thalamocortical system demonstrate relatively normal sleep–wake EEG patterns and cortical arousal. Work has provided additional data supporting a limited role for the thalamic relay as well as the (reticulo thalamocortical pathway (originating in the LDT/PPT) in the overall level of arousal or cortical EEG activation during wakefulness. Again, however, these data do not deny the thalamus a potentially critical role in defining the content of cortical function during wakefulness.

The reported evidence that the integrity of the thalamus is not necessary to maintain an aroused cortical state strongly indicates that activating influences from the brain stem may reach the neocortex via an extrathalamic route. Experimental data have indicated a critical role for the BF in cortical arousal, suggesting the BF might function as the primary structural relay in the ARAS. Indeed, lesions of the BF produce coma (i.e., persistent sub-delta <1 Hz EEG intrinsically generated by cortical neurons) in experimental animals. Work has demonstrated that BF cholinergic and noncholinergic (likely GABAergic) neurons jointly control cortical arousal because loss of either the cholinergic or the noncholinergic neuronal populations (i.e., each lesion sparing one population) has minimal effects on cortical arousal, but combined lesions produce coma.

In summary, work has suggested that (1) the reticulothalamocortical pathway may play a very limited role in the regulation of the overall level of behavioral arousal or EEG activation during wakefulness, (2) the PB/PC–BF-cortical axis is a critical system for maintaining cortical arousal, and (3) cholinergic and noncholinergic (possibly GABAergic) BF neurons jointly contribute to cortical arousal.

Sleep – Turning the Ascending Arousal System Off

As outlined previously, projections from cholinergic neurons, monoaminergic cell groups, and orexin neurons act in a coordinated manner to produce arousal. But what turns off this arousal system to produce sleep? As indicated previously, von Economo proposed a putative sleep-promoting area in the BF/anterior hypothalamic area based on postmortem brain analysis of his insomniac patients. In the 1990s, von Economo's prediction was borne out experimentally by a series of studies which demonstrated a critical role for the ventrolateral preoptic nucleus (VLPO) in inhibiting these arousal circuits during sleep. Neurons of the VLPO are sleep active, contain the inhibitory neurotransmitters galanin and GABA, and receive afferents from each of the major monoaminergic systems. Consistent with a sleep-promoting function, cell-specific lesion of the VLPO produces profound insomnia and sleep fragmentation. The VLPO contains two populations of neurons. The first is a cluster of neurons in the 'core' of the VLPO that projects most heavily to the tuberomammillary nucleus. A more diffusely distributed second population of VLPO neurons (i.e., the extended VLPO) project mainly to the locus coeruleus and the dorsal and median raphe nuclei and are thought to be important in gating REM sleep. The interaction between the VLPO and components of the arousal systems (e.g., tuberomammillary nucleus, dorsal raphe, and locus coeruleus) has been demonstrated to be mutually inhibitory and, as such, these pathways function analogously to an electronic 'flip-flop' switch/circuit (**Figure 1**). By virtue of the self-reinforcing nature of these switches (i.e., when each side is firing it reduces its own inhibitory feedback), the flip-flop switch is inherently stable in either end state but avoids intermediate states. The flip-flop design thus ensures stability of behavioral states and facilitates rapid switching between behavioral states. Within the framework of the flip-flop model, the VLPO represents the 'sleep side,' whereas the monoaminergic system represents the 'arousal side.' Importantly, it has not been determined if a mutually inhibitory

relationship exists between the VLPO and the PB/PC (putative origin of the arousal system).

Flip-Flop Switch Stability and the Lateral Hypothalamic Area

Flip-flop switches also possess the, at times, undesirable property of abruptly undergoing unwanted state transitions. The frequency of unwanted state transitions may increase if one side of the switch is 'weakened' as the weakened side becomes less able to inhibit the other side, thereby biasing the switch more toward a midpoint where smaller perturbations may trigger a state transition. For example, it has been suggested that cell loss in the VLPO during aging may weaken the switch, ultimately manifesting in sleep fragmentation and daytime napping, both of which are frequent complaints in the elderly. This phenomenon is seen in rats with VLPO lesions, which demonstrate a significant decrease in total sleep time that is largely attributable to shortened sleep bouts. That is, rats with VLPO lesions can initiate but not maintain sleep because the switch is no longer balanced but, rather, biased toward the arousal side.

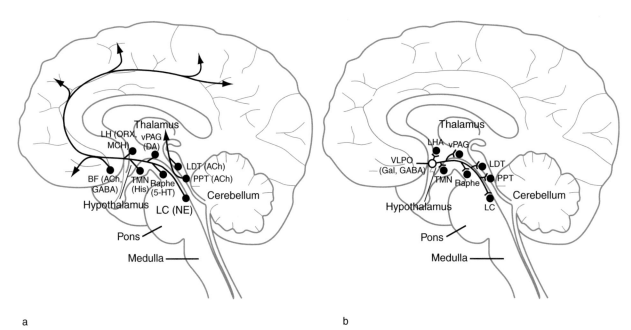

a   b

Figure 1   (a) The ascending arousal system consists of noradrenergic neurons of the ventrolateral medulla and locus coeruleus (LC), cholinergic neurons in the pedunculopontine (PPT) and laterodorsal tegmental (LDT) nuclei, serotoninergic neurons in the dorsal raphe nucleus, dopaminergic neurons of the ventral periaqueductal gray matter (vPAG), and histaminergic neurons of the tuberomammillary nucleus (TMN). These systems produce cortical arousal via two pathways: a dorsal route through the thalamus and a ventral route through the hypothalamus and basal forebrain (BF). The latter pathway receives contributions from the orexin and melanin-concentrating hormone (MCH) neurons of the lateral hypothalamic area (LHA) as well as from GABAergic or ACh neurons of the BF. (b) A schematic of the projections of the ventrolateral preoptic nucleus (VLPO; open circle) to the main components of the ascending arousal system. The VLPO neurons are primarily active during sleep and contain the inhibitory transmitters GABA and galanin.

Unwanted state transitions are a hallmark of narcolepsy. Patients with narcolepsy do not sleep more than normal people but experience profound fragmentation of sleep–wake behavior in the form of 'sleep attacks' during the day and frequent awakening during the night. In the late 1990s, two groups of investigators discovered that a lack of lateral hypothalamus (LH) orexin or their type 2 receptors can cause narcolepsy in experimental animals. It was subsequently demonstrated that humans with narcolepsy have few LH orexin neurons and low orexin levels in the cerebrospinal fluid. As indicated previously, orexin neurons are wake-active, and through their ascending projections to the cortex and descending projections to the monoaminergic and cholinergic system of the arousal system, they are thought to augment the ARAS in maintaining cortical arousal. These orexin neurons also project to the VLPO, although, curiously, the VLPO does not contain orexin receptors. This suggests that LH orexin neurons actively reinforce arousal tone as opposed to having a direct, mutually inhibitory relationship with the VLPO. Taken together, it has been hypothesized that LH orexin neurons and other, possibly glutamatergic neurons, are actually 'external' to the flip-flop switch but play an important stabilizing role for the switch. That is, the position of the orexin neurons outside the flip-flop switch permits them to stabilize behavioral state, thus reducing transitions during both sleep and wakefulness, whereas narcoleptic humans or animals lacking orexin have increased transitions in both states.

Homeostatic Regulation of Sleep

Although a teleological explanation for the function of sleep remains unresolved, it is clear that sleep has a restorative effect on the brain. This is perhaps best exemplified by the phenomenon of diminished cognitive and physical performance following sleep deprivation. Although the nature of 'sleep drive' is unknown, it has been conceptualized as a homeostatic pressure that builds during the waking period and is dissipated by sleep. This homeostatic process or 'sleep homeostat' thus represents the need for sleep (i.e., 'sleep propensity'). The cellular determinant of homeostatic sleep drive is unknown, although a putative endogenous somnogen, adenosine (AD), is thought to play a critical role. AD is a naturally occurring purine nucleoside that is hypothesized to accumulate during wake and, upon reaching sufficient concentrations, inhibits neural activity in wake-promoting circuitry of the BF and likely activates sleep-promoting VLPO neurons located adjacent to the BF. Consistent with this hypothesis are the observations that intracerebroventricular injections of AD promote sleep, extracellular concentrations of AD increase with prolonged waking and decline with sleep, administration of AD agonist near the VLPO increases sleep and induces Fos in VLPO neurons, and nonspecific AD antagonists (e.g., caffeine) potently increase waking and decrease sleep. However, data have demonstrated that accumulation of AD in the BF is not necessary for sleep drive, leaving the cellular basis of the sleep homeostat elusive.

Circadian Regulation of Sleep

Although a critical role of the circadian timing system (CTS) in providing temporal organization for behavioral states, including sleep–wakefulness, is well established, the role of the CTS in determining the duration, intensity, and propensity of sleep (and component sleep) is less clear. Nevertheless, regulation of sleep–wakefulness by the CTS is evident because the sleep–wake cycle continues on an approximately 24 h basis even when environmental conditions are constant (i.e., in the absence of environmental time cues). In addition, a clear circadian variation in sleep propensity and sleep structure has been demonstrated in humans by uncoupling the rest–activity cycle from the output of the circadian pacemaker (i.e., 'forced desynchrony protocol'). Circadian modulation of the cortical EEG during NREM sleep and REM sleep (i.e., independent of prior wake time) has also been demonstrated in humans. It has been hypothesized that this modulation of the cortical EEG, particularly REM sleep propensity, may occur through indirect projections from the suprachiasmatic nucleus (SCN) to the mesopontine tegmental nuclei involved in REM sleep generation. Rather strikingly, however, despite a demonstrated role for the circadian pacemaker in governing the timing of sleep, the SCN has only minimal monosynaptic outputs to sleep-regulatory centers, such as the VLPO and LHA orexin neurons, and none to brain stem arousal sites (**Figure 2**).

In 1982, Borbely proposed a two-process model of sleep regulation in which a homeostatic process (i.e., sleep drive) builds during the day and declines exponentially during sleep and interacts with a circadian process that is independent of sleep and waking. A further elaboration of Borbely's model for sleep–wake regulation, the 'opponent process' model, was proposed in 1993. Although conceptually similar to Borbely's two-process model, the opponent process model was the first to identify a role for the SCN in actively facilitating the initiation and maintenance of wakefulness and opposing homeostatic sleep tendency during the subjective day. Specifically, the model predicted a wake-, but not sleep-, promoting role for the SCN in sleep–wake regulation – that is, an 'alerting' signal produced by the SCN that enhanced wakefulness and actively opposed the sleep homeostat. Human sleep studies, however, have yielded data largely consistent with both a wake- and a sleep-promoting role for the SCN in sleep–wake regulation. Dijk and Czeisler provided data supporting the concept that the primary role of the SCN in sleep–wake regulation is the generation of an alerting signal necessary for generating a consolidated bout of wakefulness. This circadian drive for wakefulness (i.e., the alerting signal) becomes gradually stronger during the waking period so as to oppose increasing homeostatic sleep drive. Similarly, consolidated sleep is achieved by having maximal circadian drive for sleep near the end of the sleep cycle when homeostatic sleep drive is minimal. The concept of a circadian 'hypnotic' signal subserving the circadian drive for sleep has also been proposed; however, no endogenous correlate of a hypnotic signal has been uncovered.

Hypothalamic Circadian Integrator for Sleep–Wake Regulation

As indicated previously, despite the widespread role of the SCN in shaping daily rhythms of behavior,

physiology, and gene expression, SCN efferent projections to sleep systems (e.g., LHA orexin neurons and VLPO) are surprisingly limited (**Figure 2**). The densest projection from the SCN terminates dorsally and caudally in the adjacent subparaventricular zone (SPZ). Similar to the effects of SCN ablation, lesions that include the ventral SPZ abolish behavioral circadian rhythms, including rest–activity, sleep–wake cycles, and feeding. Surprisingly, lesions of the dorsal SPZ eliminate the circadian rhythms of body temperature but have little effect on the rhythms of locomotor activity or sleep. Thus, taken together, these observations suggest that neurons of the SPZ function as an obligate relay to maintain circadian rhythms of sleep and body temperature. Like the SCN, however, the SPZ has limited projections to the major components of the sleep–wake regulatory system. The SCN and ventral SPZ do however project densely to the dorsomedial hypothalamic nucleus (DMH). Lesions of the DMH also abolish circadian rhythms of locomotor activity, wake–sleep, feeding, and corticosteroid secretion, but not the rhythm of body temperature, which retains a normal circadian variation. These findings suggest that the principal neuronal output pathway that determines the timing of circadian behavior is mediated by a primary projection from the SCN to the SPZ, followed by a secondary projection to the DMH. Interestingly, the DMH sends

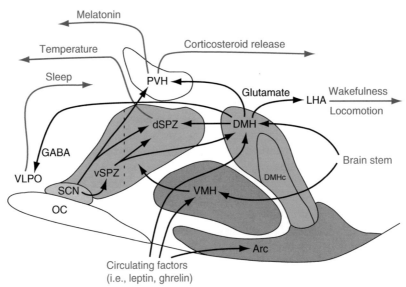

Figure 2 Hypothalamic integrator model for sleep–wake regulation. The suprachiasmatic nuclei (SCN) have relatively modest projections to sleep-regulatory networks, including the VLPO and orexin neurons. However, the bulk of its output goes to the ventral (vSPZ) and dorsal (dSPZ) subparaventricular zone and the dorsomedial hypothalamus (DMH). Lesions of the vSPZ disrupt the circadian rhythms of sleep–wakefulness and locomotor activity, whereas lesions of the dSPZ severely impair the circadian rhythms of body temperature. These findings suggest that vSPZ and dSPZ are obligate relays for the normal expression of the circadian rhythms of these functions. A major target of the SPZ is the DMH. Lesions of the DMH profoundly attenuate the circadian rhythms of sleep–wake, locomotor activity, corticosteroid secretion, and feeding. Interestingly, the DMH is one of the largest sources of input to the VLPO and lateral hypothalamic orexin neurons. This multistage regulation of circadian behavior in the hypothalamus allows for the integration of multiple time cues from the environment to shape daily patterns of sleep–wake.

a dense glutamatergic projection to the LHA orexin neurons as well as an intense GABAergic projection to the VLPO, suggesting a primary wake-promoting role for the DMH. Thus, these projections from the DMH neurons to sleep–wake centers of the hypothalamus provide a putative mechanism for the circadian regulation of sleep–wake cycles.

These findings raise the question of why circadian rhythms of behavior and physiology are regulated by a complex multisynaptic pathway in the hypothalamus rather than by a direct projection from the SCN to sleep–wake centers of the brain. In both diurnal and nocturnal species, the SCN is active during the light cycle, whereas the VLPO is always active during the sleep cycle (i.e., irrespective of circadian phenotype), suggesting an absolute requirement for additional (intervening) neural circuitry for organizing diurnal and nocturnal circadian programs despite identical clock input and sleep-control systems. In consideration of the fact that the timing of environmental pressures such as food availability and predation does not always track the solar day, and that the SCN normally remain phase-locked to the light–dark cycle, having flexibility in circadian organization may be extremely adaptive. Thus, for example, this series of multiple relays in the hypothalamus (SCN, SPZ, and DMH) could allow for the integration of light entrained circadian cues from the SCN with environmental time cues (e.g., feeding, temperature, and social cues) to sculpt patterns of rest–activity and sleep–wake cycles that are optimal for survival.

Corticolimbic Projections to the Hypothalamus – A New Frontier

In addition to the integration of external cues, these hypothalamic relays likely coordinate signals from visceral, emotional, and cognitive inputs which collectively influence arousal state. Although studies have demonstrated extensive corticolimbic (e.g., medial prefrontal cortex, infralimbic cortex, bed nucleus of the stria terminalis, and ventral subiculum) projections to the VLPO, SCN, and LHA, we still know very little about the mechanisms by which emotional or cognitive inputs interact with sleep–wake and arousal systems. Nevertheless, it has been hypothesized that during times of emergency these corticolimbic inputs (so-called 'allostatic load') may produce a state of hyperarousal that overrides the homeostatic and circadian drives for sleep, thereby enhancing cognitive and physical performance. Although the adaptive nature of this transient hyperarousal state is clear, chronic or inappropriate activity in corticolimbic sites secondary to psychological stress or depression may in fact be a primary contributing etiology to insomnia.

Summary

Approximately one century ago, von Economo proposed that sleep–wakefulness was governed by a waking influence arising from the midbrain and a sleep influence from the BF/anterior hypothalamic area. The remarkable accuracy of von Economo's predictions has been confirmed experimentally by the subsequent discovery of (1) a brain stem ascending arousal system that maintains consciousness and activates the forebrain both during wakefulness and during REM sleep and (2) the identification of a key hypothalamic switch that shuts off the arousal system to produce sleep. Work has further revealed the critical role of several additional hypothalamic nuclei in the circadian and homeostatic regulation of sleep–wakefulness. Despite this remarkable progress, significant gaps remain in our knowledge of how the arousal system and hypothalamic circuits interact to execute control of behavior state, including sleep–wake regulation.

See also: Acetylcholine; Circadian Rhythms in Sleepiness, Alertness, and Performance; Cytokines and other Neuromodulators; Histamine; Locus Coeruleus and Raphe Nucleus; Phylogeny and Ontogeny of Sleep; Sleep: Development and Circadian Control; Sleep Oscillations and PGO Waves; Thalamic Regulation of Sleep.

Further Reading

Blanco-Centurion C, Xu M, Murillo-Rodriguez E, et al. (2006) Adenosine and sleep homeostasis in the basal forebrain. *Journal of Neuroscience* 26: 8092–8100.

Borbely AA and Tobler I (1985) Homeostatic and circadian principles in sleep regulation in the rat. In: McGinty D (ed.) *Brain Mechanisms of Sleep*, pp. 35–44. New York: Raven Press.

Buzsaki G, Bickford RG, Ponomareff G, Thal LJ, Mandel R, and Gage FH (1988) Nucleus basalis and thalamic control of neocortical activity in the freely moving rat. *Journal of Neuroscience* 8: 4007–4026.

Chou TC, Scammell TE, Gooley JJ, Gaus SE, Saper CB, and Lu J (2003) Critical role of dorsomedial hypothalamic nucleus in a wide range of behavioral circadian rhythms. *Journal of Neuroscience* 23: 10691–10702.

Dijk DJ and Czeisler CA (1995) Contribution of the circadian pacemaker and the sleep homeostat to sleep propensity, sleep structure, electroencephalographic slow waves, and sleep spindle activity in humans. *Journal of Neuroscience* 15: 3526–3538.

Edgar DM, Dement WC, and Fuller CA (1993) Effect of SCN lesions on sleep in squirrel monkeys: Evidence for opponent processes in sleep–wake regulation. *Journal of Neuroscience* 13: 1065–1079.

Kinney HC, Korein J, Panigrahy A, Dikkes P, and Goode R (1994) Neuropathological findings in the brain of Karen Ann Quinlan.

The role of the thalamus in the persistent vegetative state. *New England Journal of Medicine* 330(21): 1469–1475.

Lu J, Greco MA, Shiromani P, and Saper CB (2000) Effect of lesions of the ventrolateral preoptic nucleus on NREM and REM sleep. *Journal of Neuroscience* 20: 3830–3842.

Lu J, Sherman D, Devor M, and Saper CB (2006) A putative flip-flop switch for control of REM sleep. *Nature* 441: 589–594.

Lu J, Zhang YH, Chou TC, et al. (2001) Contrasting effects of ibotenate lesions of the paraventricular nucleus and subparaventricular zone on sleep–wake cycle and temperature regulation. *Journal of Neuroscience* 21: 4864–4874.

Mochizuki T, Crocker A, McCormack S, Yanagisawa M, Sakurai T, and Scammell TE (2004) Behavioral state instability in orexin knock-out mice. *Journal of Neuroscience* 24: 6291–6300.

Moruzzi G and Magoun HW (1949) Brain stem reticular formation and activation of the EEG. *Electroencephalography and Clinical Neurophysiology* 1: 455–473.

Nauta WJH (1946) Hypothalamic regulation of sleep in rats. An experimental study. *Journal of Neurophysiology* 9: 285–314.

Porkka-Heiskanen T, Alanko L, Kalinchuk A, and Stenberg D (2002) Adenosine and sleep. *Sleep Medicine Reviews* 6: 321–332.

Ranson SW (1939) Somnolence caused by hypothalamic lesions in monkeys. *Archives of Neurology and Psychiatry* 41: 1–23.

Saper CB, Lu J, Chou TC, and Gooley J (2005) The hypothalamic integrator for circadian rhythms. *Trends in Neurosciences* 28: 152–157.

Saper CB, Scammell TE, and Lu J (2005) Hypothalamic regulation of sleep and circadian rhythms. *Nature* 437: 1257–1263.

Sherin JE, Shiromani PJ, McCarley RW, and Saper CB (1996) Activation of ventrolateral preoptic neurons during sleep. *Science* 271: 216–219.

Starzl TE, Taylor CW, and Magoun HW (1951) Ascending conduction in reticular activating system, with special reference to the diencephalon. *Journal of Neurophysiology* 14: 461–477.

Steriade M, McCormick DA, and Sejnowski TJ (1993) Thalamocortical oscillations in the sleeping and aroused brain. *Science* 262: 679–685.

Vanderwolf CH and Stewart DJ (1988) Thalamic control of neocortical activation: A critical re-evaluation. *Brain Research Bulletin* 20(4): 529–538.

Villablanca J and Salinas-Zeballos ME (1972) Sleep–wakefulness, EEG and behavioral studies of chronic cats without the thalamus: The 'athalamic' cat. *Archives Italiennes de Biologie (Pisa)* 110: 383–411.

von Economo C (1930) Sleep as a problem of localization. *Journal of Nervous Disorders* 71: 249–259.

Histamine

H L Haas, O Selbach, and O A Sergeeva, Heinrich-Heine-Universität, Düsseldorf, Germany

Overview

Histamine is a signaling molecule in the immune system, the skin, the stomach, and the brain of vertebrates. Histaminergic neurons are located exclusively in the posterior hypothalamus and extend their axons throughout the central nervous system (CNS) (**Figure 1**). They are active during the awake state but not during sleep. Three of the four known histamine receptors are widely expressed in the brain. Close mutual interactions between histaminergic and other aminergic (acetylcholine, norepinephrine, dopamine, serotonin) and peptidergic (orexins/hypocretins) systems form a network with basic homeostatic functions that are closely connected with the sleep–wake regulation. Because the fluorescence histochemistry techniques that have delivered complete pictures of the other aminergic systems for decades fail to make the histaminergic system visible, histamine has been a relatively neglected transmitter. Only after the discovery of antibodies against histamine and histidine decarboxylase did the existence of histamine gain general acceptance. Histaminergic neurons are now recognized as the major regulator of the waking center of the brain.

Histamine Metabolism

Histidine is taken up by histaminergic neurons through the L-amino acid transporter and is decarboxylated via the specific histidine decarboxylase to form histamine. This synthesis can be inhibited by α-fluoromethylhistidine (α-FMH) – a useful tool to study the histaminergic system functions. Histamine is carried by the vesicular monoamine transporter VMAT-2 into vesicles, from which it can be released upon arrival of action potentials. Inactivation in the extracellular space is achieved by methylation via histamine methyltransferase. Tele-methylhistamine undergoes oxidative deamination to tele-methylimidazoleacetic acid. Diamine oxidase leads directly to the production of imidazoleacetic acid, mainly in invertebrates; this effective γ-aminobutyric acid (GABA$_A$) receptor agonist does not seem to occur to a significant degree in mammalian tissues. The level of histamine in brain tissue is somewhat lower than that of other biogenic amines but its turnover is considerably faster (on the order of minutes) and varies with functional state. The synthesis of histamine is controlled by feedback through H3 autoreceptors.

Morphology and Location of Histaminergic Neurons

The histaminergic neurons, consisting of five subgroups, are located in the tuberomamillary nucleus. Many of the neurons lie close to the outer and inner (third ventricle) surfaces of the posterior hypothalamus. They are relatively large cells, with typically two or three thick major dendrites and an axon emanating from each cell at considerable distance from the soma. The dendrites contain vesicles with histamine, which is released to act as a feedback signal, reducing the firing of action potentials and the synthesis of histamine at the soma level. Histamine synthesis in and release from histaminergic endings, varicosit0ies on the multifold branching axons, is also reduced. These varicosities have been seen to form synaptic contacts with postsynaptic membranes, but a vast majority of the endings are located at some distance from the target membranes.

Several species have variable numbers of brain mast cells which contain and release histamine and can contribute substantially to the histamine content. A massive invasion of mast cells into the brain has been reported in doves during courting.

Physiology of Histaminergic Neurons

The histaminergic neurons display a rather regular firing pattern below 4 Hz and their membrane potential dwells at about -50 mV during waking (**Figure 2**). Their action potential displays a clear Ca^{2+} component, and two major opposing currents give the responses to current injection a very typical appearance, with a strong hyperpolarization-dependent inward rectification and marked outward rectification upon return to the membrane potential from more negative voltages. These conductances carry currents of the I_h- and the A-type, respectively. Ca^{2+}-dependent prepotentials in the dendrites and a tetrodotoxin-sensitive, noninactivating Na^+ current, together with a postspike K^+ current, create a pacemaker cycle. The actual firing rate is variable during waking and is dependent on behavioral state. A minimal polarization suffices to stop the firing as it occurs during sleep. This inhibition is attributed to the (GABA)ergic input mainly from the ventrolateral preoptic (VLPO) area, a known sleep active nucleus.

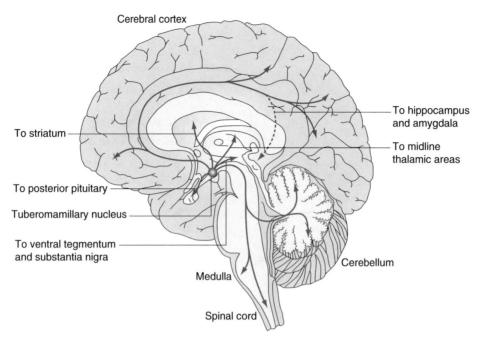

Figure 1 The histaminergic system in the human brain: origin and projections. Reproduced with permission from *Nature Reviews Neuroscience* 4: 121–130, Haas HL and Panula P, The role of histamine and the tuberomamillary nucleus in the nervous system, copyright 2003, Macmillan Magazines Ltd.

Local injections of GABA agonists in the tuberomamillary nucleus (TMN) cause sedation while GABA antagonists have the opposite effect.

The histaminergic neurons contain further transmitters with unknown specific functions. GABA and glutamic acid decarboxylase (GAD) are expressed in most TMN neurons, but so far no evidence for GABA release from the somata or the axon varicosities has been presented. Should GABA be released from a substantial fraction of such a projecting system the physiological impact would be expected to be significant and possibly opposite to the normal histamine release. Similarly, the inhibitory neuropeptide galanin is found in the TMN neurons and in the GABAergic afferents to them. Furthermore, enkephalins, thyrotropin-releasing hormone, and substance P are present in populations of the TMN neurons. The histamine neurons display a high level of expression of adenosine deaminase, which may be involved in a number of basic cellular functions.

Inputs to Histaminergic Neurons

Amino Acids

Sources for the inputs to histaminergic neurons are widespread: for example, glutamatergic fibers from the cortex and the hypothalamus are a source of glutamate, which excites histaminergic neurons, and the fibers carry both α-amino-3-hydroxy-5-methyl-4-isoxazole propionic acid (AMPA) and N-methyl-D-aspartate (NMDA) receptors. Glycine inhibits a subpopulation of histaminergic neurons, but the role of glycinergic fibers is uncertain. Taurine, an osmolyte that can reach relevant concentrations in the extracellular space, gates strychnine-sensitive glycine receptors and GABA$_A$ receptors.

GABAergic inputs come from several mostly hypothalamic locations; functionally most prominent with respect to sleep–waking regulation is the innervation from the VLPO area, which fires high during sleep and thus suppresses the firing of histamine neurons. GABA$_A$ receptors are quite heterogeneous among histamine neurons; three groups with different GABA sensitivity have been identified, depending on the expression of the γ subunit of the ionotropic GABA receptor. The sedative component of general anesthetic effects (e.g., propofol) is attributed to an action on the TMN, with a key to this action being the low expression of the GABA$_A$ receptor ε subunit. Cessation of histaminergic neuron firing is associated with the loss of consciousness. The GABAergic inputs to the TMN are under feedback control of GABA$_B$ receptors.

Amines

All aminergic nuclei give rise to projections and they are all functionally excitatory, but use a variety of mechanisms. A nicotinic fast-desensitizing action occurs through α7-type receptors, but muscarinic actions have not been detected. Noradrenaline does not affect histaminergic neurons directly but

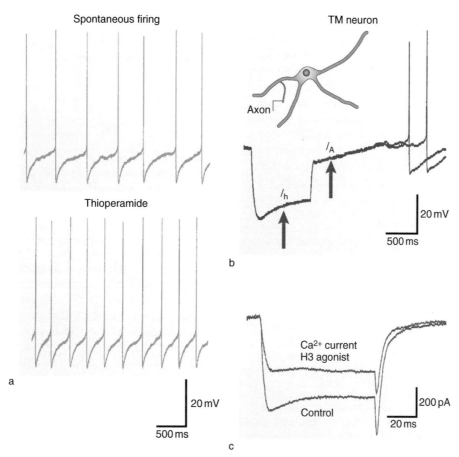

Spontaneous firing

TM neuron

Thioperamide

Axon

I_A

I_h

20 mV

500 ms

b

Ca²⁺ current
H3 agonist

Control

200 pA

20 ms

a

20 mV

500 ms

c

Figure 2 Properties of histaminergic neurons. (a) Spontaneous firing of a tuberomamillary neuron recorded *in vitro* at a membrane potential of −50 mV. Under superfusion with a histamine H3-receptor inverse agonist (thioperamide) the firing is markedly enhanced, indicating a tonic inhibition of the neuron through its autoreceptors. (b) Camera lucida drawing and recordings from a TM neuron. (Top) Response to negative current injection: a hyperpolarization-activated inward rectification (due to the current Ih, HCN1, and HCN3) and an outward rectification that delays the return to the membrane potential (due to A-type currents). (c) Voltage clamp recording (switch clamp) showing an inward Ca²⁺ current that is markedly reduced by a histamine H3-receptor agonist (R-alpha-methylhistamine). Reproduced with permission from *Nature Reviews Neuroscience* 4: 121–130, Haas HL and Panula P, The role of histamine and the tuberomamillary nucleus in the nervous system, copyright 2003, Macmillan Magazines Ltd.

effectively controls GABAergic input through α2 receptors, mediating an inhibition of inhibitory postsynaptic currents (IPSCs). Dopamine excites histamine neurons through a yet undefined indirect action, whereas serotonin excites the histaminergic neurons through activation of Na⁺−Ca²⁺ exchange (NCX). This electrogenic transporter has to let three Na⁺ ions enter the cell to expel one Ca²⁺ ion; resulting in a depolarization and excitation in the absence of any conductance change.

Peptides

Galanin is expressed in histaminergic neurons and in the GABAergic inputs to them and it inhibits their firing. It is thus possible that galanin participates in both the autogenic (feedback) inhibition and in the extrinsic inhibition from the VLPO area. Neurons containing orexins (also called hypocretins) are neighbors of the histamine neurons; the nuclei intermingle

partially. Both types of orexins (hypocretin-1 and -2) excite histamine neurons through the Hcrt2 receptor and largely via activation of NCX. This action is secondary to a rise in intracellular Ca²⁺ that probably depends on both extra- and intracellular sources of the Ca²⁺. Hypocretin neurons also express dynorphin, which can contribute to the excitation of histaminergic neurons, as it suppresses GABAergic input. Leptin, the hormone from fat that controls food intake, has no obvious effect, but the appetite-stimulating ghrelin inhibits a potassium channel (Kir3). Nociceptin strongly inhibits TMN neurons at the postsynaptic level while morphine (μ receptors) excites them, probably through disinhibition.

Purines

The tuberomamillary nucleus displays a very strong expression of adenosine deaminase, which has led to the suggestion that it may also use adenosine as a transmitter. So far such a role of adenosine is elusive, and there is no evidence for synaptic release of this nucleoside but it is sedative through A1 receptors. A2A adenosine receptors have also been implicated in sleep regulation, through enhancement of the inhibition of histamine neurons. Adenosine accumulates during wakefulness and is likely an endogenous sleep pressure substance. Interestingly, adenosine, which strongly inhibits many neurons in the brain, has no such effect on histamine neurons. In contrast, ATP excites histamine neurons through P_2Y_1 and P_2Y_4 receptors.

Histaminergic Pathways

Histaminergic fibers innervate virtually all parts of the CNS, from the spinal cord to the retina. A dense innervation is found in the hypothalamus, cortex, amygdala, and basal ganglia and more variably in the thalamus and hippocampus. The mutual innervation of aminergic systems forms a network that acts in concert to ensure vegetative, endocrine, metabolic, osmotic, and energetic homeostasis.

The waking action of histamine is attributed to projections to thalamus and cortex, as well as to other aminergic nuclei and their emanating axons. Relay cells in the thalamus are depolarized and thus are shifted to the regular firing mode, which represents an 'open door' for perception. In the absence of histaminergic tone these cells return to the burst firing mode, which closes the door to the cortex for sensory input. Most targets are depolarized and excited through a number of different mechanisms; it is only in the hippocampus that pyramidal cells are hyperpolarized and inhibited through intracellular Ca^{2+} increase, which activates a K^+ conductance. On the other hand, histamine can block Ca^{2+}-dependent K^+ conductance, and thus accommodate firing through another receptor (H2). Some hippocampal interneurons are rather strongly excited by H2 receptor activation, making the overall histamine action in this structure an inhibitory one.

Histamine Receptors

Histamine can activate Cl^- conductances in thalamus and hypothalamus: in the supraoptic nucleus this effect is blocked by picrotoxin and H2 receptor antagonists. In thalamic interneurons Cl^- conductance blocking is also H2 receptor mediated but is not picrotoxin sensitive. Histaminergic ionotropic receptors may play only a minor role, if any, in the vertebrate brain. In contrast, arthropod sensory neurons, the insect eye photoreceptor, and lobster olfactory neurons display synaptic transmission through histamine-gated Cl^- channels. Four metabotropic

histamine receptors have been cloned; H1, H2, and H3 occur in the brain and H_4 receptors are mainly expressed in peripheral tissues such as blood, spleen, lung, and liver.

H1 Receptors

Histamine H1 receptors occur throughout the central nervous system, with a particularly high density in regions involved in arousal and waking, such as the thalamus and cortex and the cholinergic, noradrenergic, dopaminergic, and serotonergic nuclei (**Figure 3**). H1 receptor activation causes excitation in most brain regions (brain stem, thalamus, hypothalamus, cortex, amygdala, striatum) through G_{q11} protein and direct block of leak K^+ conductance or phospholipase C, inositol trisphosphate (IP_3), and diacylglycerol (DAG) mediation. IP_3 releases Ca^{2+} from internal stores and activates a number of Ca^{2+}-dependent processes, including the opening of a cation channel of the transient receptor potential canonical (TRPC) type or the stimulation of a Na^+–Ca^{2+}-exchanger. Furthermore, the elevated intracellular Ca^{2+} can stimulate NO synthase and consequently guanylate cyclase. On the other hand, Ca^{2+}-dependent K^+ channels can be opened, leading to hyperpolarization and inhibition, for instance, in hippocampal pyramidal neurons (**Figures 4**).

H1 receptor antagonists are classic antihistaminics and are widely prescribed for allergies. Their well-known sedative actions have prompted early suggestions for an involvement of endogenous

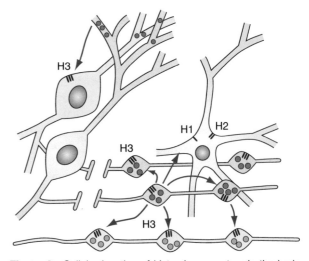

Figure 3 Cellular location of histamine receptors in the brain. TM neurons (left) with H3 receptors on their somata and their axons, as well as on the axons of other cells (containing glutamate or other amines). H1 and H2 receptors on a target cell body. Reproduced with permission from *Nature Reviews Neuroscience* 4: 121–130, Haas HL and Panula P, The role of histamine and the tuberomamillary nucleus in the nervous system, copyright 2003, Macmillan Magazines Ltd.

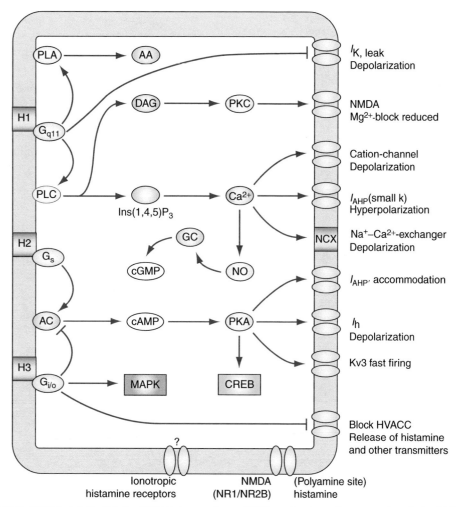

Figure 4 Signaling pathways activated by histamine receptors. cGMP, cyclic guanosine monophosphate; HVACC, high-voltage activated Ca^{2+} current; MAPK, mitogen-activated protein kinase; PKA, protein kinase A; PKC, protein kinase C. Reproduced with permission from *Nature Reviews Neuroscience* 4: 121–130, Haas HL and Panula P, The role of histamine and the tuberomamillary nucleus in the nervous system, copyright 2003, Macmillan Magazines Ltd.

histamine in sleep–waking regulation. There is strong evidence for H1 receptors being the main mediators of histamine waking action. In H1 receptor knockout (KO) mice, however, the sleep–waking pattern is hardly changed, but the waking response to H3 receptor antagonists, which relieve the autoinhibition of histamine release, is abolished. Most antipsychotic and antidepressive pharmaceuticals occupy H1 receptors, but this is not part of the therapeutic effect. Genetic and molecular studies have recently attributed an important role in peripheral and central immunity to histamine receptors. The H1 receptor is identical to an autoimmune locus receptor (*Bordetella pertussis*-induced histamine sensitization; Bphs), with relevance to inflammatory brain diseases (e.g., multiple sclerosis). Through an action on astrocytes, histamine regulates the permeability of the blood–brain barrier and the immigration of immune cells in the otherwise privileged CNS. Histidine decarboxylase

(HDC) KO mice, compared to wild type, produce more proinflammatory signals and display a particularly marked experimental autoimmune encephalitis.

H2 Receptors

Histamine from enterochromaffin-like cells in the stomach stimulates acid secretion by parietal cells, an action that is insensitive to the classic antihistamines (H1). Development of H2 receptor antagonists by Jim Black revolutionized the therapy of stomach ulcers. H2 receptors are also widespread in the brain, in particular in the basal ganglia, the hippocampus, and the amygdala. Activation of histamine (H2) as well as some β-adreno-, dopamine, serotonin, corticotropin-releasing hormone (CRH), and vasoactive intestinal peptide (VIP) receptors has similar actions on neuronal excitability: blockage of a Ca^{2+}-activated K^+ conductance (small K), which is responsible for the accommodation of firing and a long-lasting (seconds) afterhyperpolarization following action potentials (**Figure 5**). The accommodation of firing determines the number of action potentials in response to a longer (~100 ms) depolarizing stimulus – that is, the number of action potentials in a burst, and the amount of Ca^{2+} inflow through high-voltage activated Ca^{2+} channels (HVACCs) or NMDA channels. This sets up a neuron state of readiness to respond. These actions are mediated by cyclic AMP and protein kinase A (PKA) and lead to phosphorylation of the transcription factor cAMP response element-binding protein (CREB; **Figure 4**). Cyclic AMP can also interact directly with another current, the hyperpolarization-activated cation current (I_h), by shifting its activation curve to more positive levels, leading to depolarization in thalamic and hippocampal neurons. Brief exposure of hippocampal slices to histamine, noradrenaline, or cyclic AMP causes long-term enhancements of CA1 population spikes and pyramidal neuron firing in the absence of high-frequency stimulation. Adenosine A1 receptors are negatively coupled to adenylyl cyclase and have effects opposite to those of histamine H2 receptor activation (i.e., increasing the long-lasting afterhyperpolarization and accommodation of firing). Adenosine has a sedative action (in contrast to the waking action of histamine) and is a major factor in sleep induction.

H3 Receptors

The histamine autoreceptor was termed H3 by the Paris group of J-C Schwartz, who first described it in 1983. H3 receptors are present on TMN somata, dendrites, and axons (**Figure 3**). They are coupled directly through a $G_{i/o}$ protein to Ca^{2+} channels, causing inhibition of TMN firing, histamine release from varicosities, and histamine synthesis. Adenylyl cyclase is depressed by H3 receptor activation. Importantly H3 receptors are found on many nonhistaminergic varicosities, such as those containing glutamate, acetylcholine (ACh), dopamine, noradrenaline, serotonin, GABA, and various peptides. Excitatory transmission in the glutamatergic perforant pathway from the entorhinal cortex to the dentate area is depressed by about 30% by histamine in rats, but not in mice. The glutamatergic corticostriatal pathway and the dopaminergic nigrostriatal pathway are also suppressed. In the substantia nigra, GABAergic potentials are reduced.

In contrast to the H1 and H2 receptor genes the H3 receptor gene features two or three introns. Alternative splicing yields a large number of H3 receptor isoforms with differential distribution and constitutive activity. The H3 receptors are thus highly promising targets for the development of selective

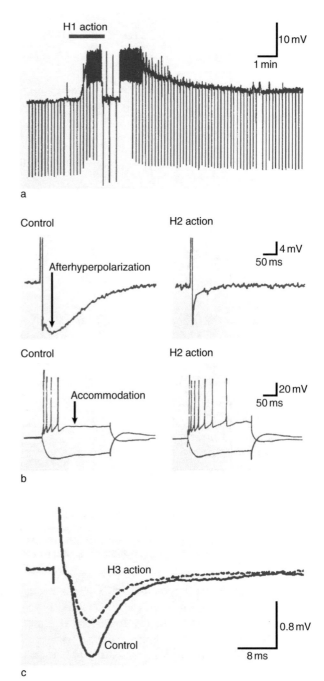

Figure 5 Electrophysiology of histamine actions. (a) H1-receptor: a depolarization of a pontine reticular formation neuron through block of a potassium conductance. (b) H2-receptor: block of the long-lasting afterhyperpolarization after Ca^{2+} inflow (Ca^{2+}-dependent potassium conductance, sK-channel) and block of the accommodation of action potential firing in a human hippocampal pyramidal cell *in vitro*. (c) H3-receptor: decrease of field excitatory postsynaptic potential (EPSP) in the dentate gyrus of the hippocampus evoked by perforant path stimulation. Histamine reduced release of the excitatory transmitter glutamate in these two pathways. Reproduced with permission from *Nature Reviews Neuroscience* 4: 121–130, Haas HL and Panula P, The role of histamine and the tuberomamillary nucleus in the nervous system, copyright 2003, Macmillan Magazines Ltd.

pharmaceuticals to treat neuropsychiatric disorders affecting sleep–waking regulation, feeding, mood, learning, and memory, as well as epilepsy, migraine, and inflammatory disorders. H3 receptor antagonists have, indeed, facilitatory actions on learning and memory, presumably in part by removal of a tonic inhibition of the release of other transmitters, such as glutamate and acetylcholine. In the light of the many possible actions on H3 heteroreceptors, the systemic effects of interference are difficult to predict.

Histamine and Synaptic Plasticity

Histamine (and hence the activity of the histaminergic neurons) can cause long-lasting changes in synaptic transmission in several structures by itself, and it can prime or modulate other forms of synaptic plasticity such as long-term potentiation (LTP) and long-term depression (LTD) (**Figure 6**). These phenomena represent coincidence detection and are associated with learning and memory. An important signal transduction pathway is the transient rise of intracellular Ca^{2+} and its binding to a Ca^{2+}/calmodulin kinase (CaMK).

Interference of the polyamine spermidine with early attempts to stain histamine neurons came to mind when direct interaction of the diamine histamine with the NMDA receptor on hippocampal pyramidal cells was first detected: this effect is similar to the potentiation by spermidine of the NMDA receptor (NR2B subunit) and opposite to the inhibitory action of protons. It is facilitated in slightly acidic conditions, such as occur in the CA3

region during synchronous discharges, with 'sharp waves' that evoke specifically distributed LTP in CA1 and that are likely a decisive element in memory trace formation. Histamine strongly promotes burst firing in the CA3 area, frequently leading to 200 Hz oscillations (ripples) during quiet waking or sleep. The replay of ripples during sleep has been proposed as critical for memory consolidation.

H1 receptor activation increases intracellular Ca^{2+} and potentiates the H2 effects that are proved proplasticity signals. H2 receptor activation stimulates adenylyl cyclase and controls a Ca^{2+}-dependent potassium current through CaM kinase II and the extent of firing in response to an excitatory stimulus. These mechanisms are deeply involved with triggering long-term potentiation and the cellular basis of learning and memory. Brief exposure of hippocampal slices to histamine causes LTP in the absence of any electrical stimulation. Adenosine A1 receptors are negatively coupled to adenylyl cyclase and have effects opposite to those of histamine on LTP. The histamine H3 receptor is an auto- and heteroreceptor with inhibitory action. The glutamatergic corticostriatal pathway undergoes long-term depression after H3 receptor stimulation.

Memory trace formation occurs in the hippocampus where a selected population of CA3 pyramidal cells produces a synchronous burst that is relayed to the CA1 area and acts as the natural trigger for long-term potentiation. Whether, and at which intensity, this burst is fired depends on the subcortical control by largely aminergic systems. Histamine interacts at all levels of this process.

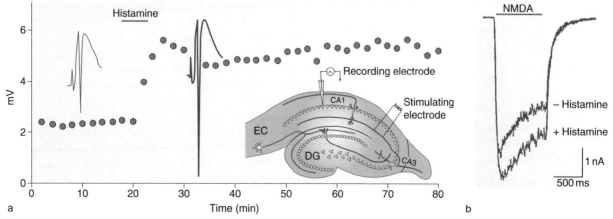

Figure 6 Histamine and synaptic plasticity. (a) Histamine (H2-receptor activation) induces LTP of population spike in CA1 (Schaffer-collateral stimulation) in hippocampal slice. Inserted are population spikes at the times where they are shown and a slice. EC, entorhinal cortex; CA, cornu ammonis; DG, dentate gyrus. (b) Whole cell recording from hippocampal pyramidal cell. Neuron exposed to NMDA for 1 s displays inward current, which is larger in the presence of histamine. Reproduced with permission from *Nature Reviews Neuroscience* 4: 121–130, Haas HL and Panula P, The role of histamine and the tuberomamillary nucleus in the nervous system, copyright 2003, Macmillan Magazines Ltd.

Histamine and Behavior

Histaminergic systems are phylogenetically old and are concerned with basic body functions of organisms in both the invertebrate and the vertebrate world. They are found in vertebrates in the hypothalamus, where feeding, drinking, osmoregulation, temperature control, energy administration, and the endocrinium are integrated with the sleep–waking and circadian rhythms. Histamine neurons fire exclusively during waking and particularly during attention. Excitation of the large cells in the supraoptic nucleus leads to antidiuresis. Injection of histamine into other hypothalamic sites induces drinking, and dehydration increases synthesis and release of histamine. Stimulation of oxytocin-containing cells and those releasing the prolactin-releasing hormone have far-reaching influences on mammalian reproduction. The suprachiasmatic nucleus, our circadian zeitgeber, receives strong histaminergic excitation. Locomotion is maintained by histaminergic activation and serves the basic function of feeding. $HDC^{-/-}$ mice display reduced exploratory activity, but normal habituation to novel environments, increased fear of heights, and superior motor coordination. H3 receptor$^{-/-}$ mice are lazy walkers, but H3 antagonists can cause hyperactivity. H1 receptor$^{-/-}$ mice display reduced ambulatory and exploratory activity and a reduced sensitivity to pain. Histamine activity on H2 receptors is an essential link in opioid receptor-mediated analgesia. In the periphery, histamine participates in the stimulation of nociceptors and itching, and the central histamine system is involved through both H1 and H2 receptors in pain perception.

Leptin-induced anorexia is mediated by the histaminergic system. Glucose-responsive neurons in the ventromedial nucleus are excited by histamine. In the anterior hypothalamus, H1 receptor activation lowers the set point of the hypothalamic thermostat; in the posterior hypothalamus, H2 receptors are involved with initiating heat loss mechanisms. The vegetative nervous system is under the control of the hypothalamus, with a significant contribution by histamine; this also includes cardiovascular control.

Control of Waking and Arousal

Sleep and awake states are closely correlated with essential body functions. Histaminergic neurons interact with the neighboring orexin/hypocretin-containing neurons to determine the right time for sleeping and waking. Loss-of-function of orexin/hypocretin neurons results in narcolepsy. Cataplexy, the major symptom of narcolepsy, is characterized by loss of muscle tone in the presence of consciousness. In contrast to other aminergic systems, histamine neurons are fully active during cataplexy. Thus histamine is mainly responsible for consciousness and noradrenaline is mainly responsible for muscle tone.

Destruction or inhibition of the tuberomamillary region causes somnolence, whereas activation evokes arousal. During the devastating influenza epidemic of 1918–1919, a significant number of patients suffered from sleep disturbances. Von Economo in Vienna investigated the brains of these patients and found posterior hypothalamic lesions in those who had suffered from hypersomnia (encephalitis lethargica) and anterior hypothalamic lesions in those who had suffered from insomnia. It is now clear that the first group must have lost their histamine neurons and probably the orexin/hypocretin neurons, whereas the insomnia patients displayed lesions in an area now known to be exceptionally active during slow-wave sleep (the VLPO area). The VLPO houses GABAergic neurons that project to wake-active aminergic neurons, specifically the histaminergic neurons, the firing of which is suppressed by the input. $GABA_A$ receptors display a large variability due to differential assembly of subunits in a receptor-channel complex. Depending on γ subunit expression, populations with different sensitivities to GABA have been identified – presumably reflecting an involvement in different aspects of the waking function.

The TMN is a major target of general anesthesia because it mediates the sedative component. $GABA_A$ receptor stimulation (local muscimol) is sedative whereas the local injection of a $GABA_A$ antagonist causes arousal. Systemic administration of general anesthetics that potentiate $GABA_A$ action increases activity of the sleep-active preoptic area and reduces activity of the TMN. Low-level expression of the $GABA_A$ receptor ε subunit in most TMN neurons (in contrast to the neighboring orexin/hypocretin neurons) may be key to the high sensitivity of the TMN to anesthetics.

Much of the more recent knowledge in this field comes from the rare, technically difficult recordings from identified histaminergic neurons over several hours in freely behaving animals, but results from experiments using local excitation, inhibition, or destruction all point in the same direction. In combination, measures of histamine release by microdialysis or c-fos activation and the selective, complete or reversible removal of the histamine system or its receptors have also given congruent evidence.

Hibernation, an adaptation to low external temperature and the lack of food, is a state of radically reduced physiological functions – brain activity, body temperature, heart rate, and metabolism – that highlights the link between feeding and sleeping. Interestingly, the turnover of histamine, the waking transmitter, is considerably enhanced during hibernation. It seems to play a major role in regulating maintenance of and arousal from the hibernation state.

Learning and Memory

Correlation of synaptic plasticity with memory mechanisms is often claimed but the evidence is rarely convincing. The effects of histamine H1 and H2 receptor activation on principal cells in hippocampus, cortex, amygdala, or striatum are quite compatible with a positive influence on synaptic plasticity and memory. The excitatory action and the intracellular increase of Ca^{2+} ions (H1) are important mechanisms for maintaining wakefulness and attention and are a prerequisite for effective learning. Block of the accommodation of firing (H2) shifts target neurons toward a state of quiet readiness, characterized by little difference in membrane potential and spontaneous action potential firing, but a dramatically changed response to longer lasting (<100 ms) depolarizing signals. This is a response to compartmentalized Ca^{2+} signals. Classical conditioning (eye-blink reflex in rabbits) can reduce the amplitude and duration of the Ca^{2+}-dependent afterhyperpolarization in hippocampal pyramidal cells. These mechanisms are also of interest with respect to pathophysiological events leading to exaggerated activity and epilepsy. Contrary to expectation, histamine H1 antagonists (the classical antihistamines) are slightly epileptogenic. Two clues can be offered to explain this: (1) H1 receptor activation leads to hyperpolarization and inhibition of principal neuron firing in the (rodent) hippocampus through activation of Ca^{2+}-dependent K^+ channel opening; (2) interneurons in the hippocampus are strongly excited by H2 receptor activation. The same interneurons suffer a cutoff at high-frequency firing. Thus, the final action on a whole structure and its output cannot be predicted by studying principal cells alone and *in vitro*. H2 and H3 receptor blocking has revealed ambiguous results on learning paradigms. The histaminergic system can affect memory through at least two independent mechanisms, directly on the synaptic plasticity in hippocampus, striatum, and amygdala, and by an indirect inhibitory effect through the brain's reinforcement system. H1 receptor antagonists can act as reinforcers, alone or together with opiates, cocaine, or amphetamine.

H1 antagonists improve water maze performance but reduce radial maze performance. Locomotor activity and exploratory behavior are reduced, but learning and memory seem unaffected in H1 receptor knockout mice. H3 receptor$^{-/-}$ mice display enhanced spatial learning and reduced anxiety. Lesions of the histaminergic nucleus can lead to facilitation of learning, and

pharmacological inactivation by posttrial injection of lidocaine leads to memory improvement. Such lesions or stimulations of the tuberomamillary nucleus are very difficult or impossible to perform without affecting neighboring structures or bypassing fibers.

Histidine decarboxylase KO mice cannot synthesize histamine. JS Lin has described an important defect in HDC$^{-/-}$ mice that is quite relevant for memory formation: their response to novelty is strongly impaired and they fall asleep rather than explore a new environment. JP Huston's group revealed a number of rather specific changes in HDC$^{-/-}$ mice: an improvement in negatively reinforced water maze performance but a defect in (nonreinforced) episodic object memory.

Histamine Pathophysiology

Hepatic encephalopathy (HE) develops as a result of high ammonia levels. Histidine and some other amino acids are significantly increased in the brains of HE patients. This leads to a higher synthesis of histamine, which does not cause alertness in these patients; instead, they suffer from sleepiness. Histamine turnover is not increased. Model rats with a portacaval anastomosis display defects in histamine-induced plasticity and are sleepy and uninterested in their environment; their histamine receptors may be desensitized or incorporated in the cells. In contrast to low levels of histamine, high levels inactivate NMDA receptors – in keeping with cognitive deficits.

Within the hypothalamus only the TMN neurons show tangles and degeneration in Alzheimer's disease and Down syndrome. In Parkinson's disease, histaminergic endings seem to take over the empty places left by degenerated dopaminergic axon varicosities. Histamine turnover is increased and H1 receptor binding is low in the frontal cortex of schizophrenic patients. H2 receptor antagonism can reduce negative symptoms in schizophrenia. Histamine receptor expression is altered in rats with ethanol preference. Finally, the histaminergic system has an antiepileptic influence; antihistamines (H1) can cause seizures and H3 receptor antagonists are anticonvulsive.

Conclusion

The brain histaminergic system is our major waking center and is deeply involved in many basic functions. It keeps the CNS alert and ready to react and is the gatekeeper of consciousness. H1 and H2 receptors mediate excitation and (long-term) potentiation of excitation, while the H3 autoreceptors provide a feedback control of synthesis, release, and electrical activity. As heteroreceptors, these receptors also control the varicosities of most other transmitter systems, making them a prime target for pharmaceutical research and development.

See also: Circadian Rhythms in Sleepiness, Alertness, and Performance; Sleep: Development and Circadian Control.

Further Reading

Black JW, Duncan WA, Durant CJ, et al. (1972) Definition and antagonism of histamine H2-receptors. *Nature* 236: 385–390.

Brown RE, Stevens DR, and Haas HL (2001) The physiology of brain histamine. *Progress in Neurobiology* 63: 637–672.

Haas H and Panula P (2003) The role of histamine and the tuberomamillary nucleus in the nervous system. *Nature Reviews Neuroscience* 4: 121–130.

Haas HL and Reiner PB (1988) Membrane properties of histaminergic tuberomamillary neurones of the rat hypothalamus *in vitro. Journal of Physiology (London)* 399: 633–646.

Hill SJ, Ganellin CR, Timmerman H, et al. (1997) International Union of Pharmacology. XIII. Classification of histamine receptors. *Pharmacological Reviews* 49(3): 253–278.

Hough LB, Nalwalk JW, Li BY, et al. (1997) Novel qualitative structure–activity relationships for the antinociceptive actions of H2 antagonists, H3 antagonists and derivatives. *Journal of Pharmacology & Experimental Therapeutics* 283(3): 1534–1543.

John J, Wu MF, Boehmer LN, et al. (2004) Cataplexy-active neurons in the hypothalamus: Implications for the role of histamine in sleep and waking behavior. *Neuron* 42: 619–634.

Leurs R, Bakker RA, Timmerman H, et al. (2005) The histamine H3 receptor: From gene cloning to H3 receptor drugs. *Nature Reviews Drug Discovery* 4: 107–120.

Lin JS, Sakai K, and Jouvet M (1988) Evidence for histaminergic arousal mechanisms in the hypothalamus of cat. *Neuropharmacology* 27(2): 111–122.

Nelson LE, Guo TZ, Lu J, et al. (2002) The sedative component of anesthesia is mediated by GABA(A) receptors in an endogenous sleep pathway. *Nature Neuroscience* 5: 979–984.

Panula P, Yang HY, and Costa E (1984) Histamine-containing neurons in the rat hypothalamus. *Proceedings of the National Academy of Sciences of the United States of America* 81(8): 2572–2576.

Passani MB, Lin JS, Hancock A, et al. (2004) The histamine H3 receptor as a novel therapeutic target for cognitive and sleep disorders. *Trends in Pharmacological Sciences* 25: 618–625.

Schwartz JC, Arrang JM, Garbarg M, et al. (1991) Histaminergic transmission in the mammalian brain. *Physiological Reviews* 71(1): 1–51.

Vorobjev VS, Sharonova IN, Walsh IB, et al. (1993) Histamine potentiates N-methyl-D-aspartate responses in acutely isolated hippocampal neurons. *Neuron* 11: 837–844.

Wada H, Inagaki N, Yamatodani A, et al. (1991) Is the histaminergic neuron system a regulatory center for whole-brain activity? *Trends in Neurosciences* 14(9): 415–418.

Cytokines and other Neuromodulators

J M Krueger, L Churchill, and D M Rector,
Washington State University, Pullman, WA, USA

Introduction

Sleep is of central importance to neurobiology, yet our understanding of the biochemical and neurophysiological mechanisms of sleep is rudimentary and sleep function has not been experimentally demonstrated. Our quality of life, performance, and mental well-being are all adversely affected by even a single night's loss of sleep. Chronic sleep loss is associated with pathologies such as the metabolic syndrome. Primary sleep disorders increase morbidity and mortality. Sleep plays a central role in many mental and physiological functions – for example, memory improves during sleep. It is thus important that we understand sleep regulatory mechanisms. This article focuses on the biochemical regulation of sleep and discusses how recent knowledge has led to new insights into how the brain is organized to produce sleep.

There are two somewhat independent literatures concerning the fundamental mechanisms of sleep regulation. One is based on neurophysiological methods, and this literature has led to the identification of circuits involved in non-rapid eye movement sleep (NREMS) regulation, such as corticothalamic projections and the hypothalamic ventrolateral preoptic and median preoptic circuits, and those involved in rapid eye movement sleep (REMS) regulation, such as the laterodorsal tegmental nucleus. Satisfactory explanations of how these circuits impose sleep on the brain and how they keep track of past sleep–wake activity likely will involve the biochemical mechanisms that interact with these circuits. Such interactions are particularly important for sleep homeostasis, the phenomenon exemplified by sleep rebound after prolonged wakefulness. The second sleep regulatory literature is based on biochemical methods. This work has its basis in the homeostatic nature of sleep and the nearly 100-year-old finding that has been replicated many times showing that the transfer of cerebrospinal fluid from sleep-deprived, but not control, animals enhances sleep in the recipients.

During the past 25 years, several sleep regulatory substances have been identified and extensively tested. However, only a handful are strongly implicated in sleep regulation. The list includes tumor necrosis factor-α (TNF-α), interleukin-1β, growth hormone releasing hormone (GHRH), prostaglandin D$_2$, and adenosine for NREMS regulation and vasoactive intestinal peptide, prolactin (PRL), and nitric oxide for REMS. When injected into animals, each of these substances promotes sleep; if they are inhibited, spontaneous sleep is reduced, their levels in brain vary with sleep propensity, and they act on sleep regulatory circuits to promote sleep. Here, we focus on TNF-α in NREMS regulation because it is the only substance identified whose circulating levels in humans correlate with sleepiness and sleep. We also discuss PRL because it has been implicated in REMS regulation by several laboratories and is a cytokine. In addition to these humoral agents, all neurotransmitters have been implicated in sleep–wake regulation; this literature is not discussed here. There is yet another literature demonstrating changes in gene expression, usually determined by changes in mRNA levels using gene chips, with sleep or sleep loss. That literature is distinguished from the sleep regulatory literature because every physiological function changes during sleep; thus, it is not possible from gene expression data alone to know if sleep *per se* is causative of the change. It is important to recognize that those humoral agents implicated in NREMS and REMS regulation affect each other's production and seem to act in concert with each other to affect sleep (**Figure 1**).

TNF-α in Sleep Regulation

The ability of TNF-α to promote NREMS was first described in 1987. TNF-α given directly into the brain or systemically enhances the duration of NREMS. For instance, mice receiving 3 µg TNF-α intraperitoneally spend approximately 90 min extra in NREMS during the first 9 h postinjection. In addition, NREMS after TNF-α treatment is associated with supranormal electroencephalogram (EEG) delta (0.5–4 Hz) waves, which are indicative of a greater intensity of NREMS. TNF-α is somnogenic in all species thus far tested: rabbits, mice, rats, and sheep. TNF-α has little effect on REMS if low NREMS-promoting doses are used; however, higher doses inhibit REMS. Sleep following TNF-α treatment appears to be normal in that sleep architecture remains normal, sleep remains easily reversible, postures remain normal, and animals remain responsive to handling. Changes in sleep-coupled autonomic functions, such as the decreases in brain temperature upon entry into NREMS, also persist after TNF-α treatment.

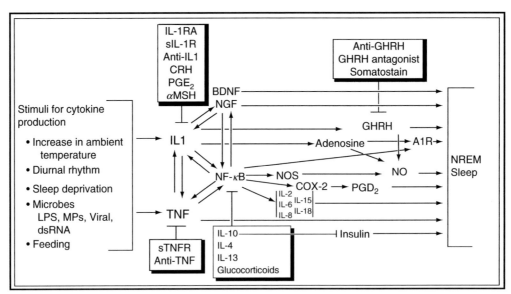

Figure 1   Molecular networks are involved in sleep regulation. Substances in boxes inhibit sleep and inhibit the production or actions of sleep-promoting substances illustrated via feedback mechanisms. Inhibition of one step does not completely block sleep since parallel sleep-promoting pathways exist. These redundant pathways provide stability to sleep regulation. Our knowledge of the biochemical events involved in sleep regulation is more extensive than that illustrated. The molecular network shown here possesses many of the characteristics of biological networks and engineered systems. Thus, the network is modular in that several proteins (cytokines) are working in 'overlapping coregulated groups' in this pathway. Second, the molecular network is robust in that removal of one of the components does not result in complete sleep loss. Third, the network operates as a recurring circuit element with multiple feedback loops affecting other pathways to the extent that similar networks involving many of the same substances and component network parts are used to regulate body temperature, inflammatory responses, microcirculation, memory, food intake, etc., and these systems, to a limited degree, coregulate. Specificity for any one physiological process, such as sleep, results from multiple interacting molecular and cellular circuits, each possessing different, but similar, reactivity. A1R, adenosine A1 receptor; αMSH, α-melanocyte-stimulating hormone; anti-IL1, anti-IL-1 antibodies; anti-TNF, anti-TNF antibodies; BDNF, brain-derived neurotropic factor; COX2, cyclooxygenase 2; CRH, corticotrophin-releasing hormone; GHRH, growth hormone releasing hormone; IGF1, insulin-like growth factor 1; IL-1RA, IL-1 receptor antagonist; LPS, lipopolysaccharide; MPs, muramylpeptides; NF-κB, nuclear factor-kappa B; NGF, nerve growth factor; NOS, nitric oxide synthase; PGD$_2$, prostaglandin D$_2$; sIL-1R, soluble IL-1 receptor; sTNFR, soluble TNF receptor; TGFβ, transforming growth factor-β.

Inhibition of TNF-α inhibits spontaneous NREMS. Thus, treatment with anti-TNF-α antibodies, the full-length soluble TNF receptor, or TNF soluble receptor fragments containing the TNF recognition site all inhibit spontaneous NREMS in rabbits and rats. Furthermore, pretreatment of animals with TNF inhibitors prior to sleep deprivation reduces the expected sleep rebound that normally occurs after sleep loss. Substances that inhibit TNF-α action or production also inhibit spontaneous sleep (e.g., interleukin-4, -10, and -13) (**Figure 1**). In addition, these substances also inhibit the production of certain other cytokines, such as interleukin-1β; therefore, their action on sleep may not be specific to TNF. However, they do form part of the negative feedback loops that help to regulate these nuclear factor-kappa B (NF-κB)-sensitive cytokines. Furthermore, inhibition of TNF also blocks the increases in NREMS observed in response to an acute mild increase in ambient temperature. Mice lacking the TNF 55-kDa receptor fail to exhibit NREMS responses if given TNF-α, thereby implicating this receptor in

TNF-α-enhanced sleep. These mice also sleep less than corresponding control strains; the reduced NREMS occurs mostly during daylight hours.

Brain levels of TNF-α and the TNF-α mRNA vary diurnally and are influenced by sleep deprivation. The highest levels in rats occur at daybreak. The amplitude of the day–night changes in TNF protein is approximately tenfold and that in mRNA approximately twofold; this reflects the predominate posttranscriptional regulation of TNF. After sleep loss, hypothalamic TNF-α mRNA also increases. Sleep deprivation also increases the expression in brain of the 55 kDa TNF receptor mRNA. TNF-α serum levels increase in mice after sleep deprivation but not after stress. In normal humans, blood levels of TNF-α correlate with EEG delta wave activity. After sleep loss, circulating levels of TNF-α and the 55 kDa soluble TNF receptor, but not the 75 kDa TNF soluble receptor, increase.

Clinical conditions associated with sleepiness correlate with higher blood levels of TNF. Thus, patients with sleep apnea, chronic fatigue, chronic insomnia, myocardial infarct, excessive daytime sleepiness, and preeclampsia all have elevated TNF plasma levels. AIDS patients have disrupted TNF and sleep rhythms. Postdialysis fatigue is associated with higher TNF levels, and cancer patients receiving TNF report fatigue. Furthermore, the 308A TNF-α polymorphic variant is associated with the metabolic syndrome and sleep apnea. Rheumatoid arthritis patients receiving the soluble TNF 75 kDa receptor report reduced fatigue. Sleep apnea patients treated with the soluble TNF receptor have reduced sleepiness. If obstructive sleep apnea patients are surgically treated, their elevated TNF-α plasma levels return to normal.

Systemic TNF, like several other cytokines, signals the brain via multiple mechanisms; one involves vagal afferents since vagotomy attenuates intraperitoneal TNF-α-induced NREMS responses. The effects of systemic bacterial products such as endotoxin, a component of gram-negative bacterial cell walls, may also involve TNF. For instance, in humans, endotoxin doses that induce transient increases in sleep also induce concomitant increases in circulating TNF-α. In addition, the soluble TNF receptor fragment attenuates bacterial cell wall peptidoglycan-enhanced NREMS in rabbits.

The site(s) of action of TNF-α-induced NREMS includes the preoptic area of the anterior hypothalamus, the locus coeruleus, and the somatosensory cortex. Thus, microinjection of TNF-α into the preoptic area enhances NREMS in rats. In contrast, injection of a soluble TNF receptor fragment into this site inhibits spontaneous NREMS. Microinjection of TNF-α into the locus coeruleus, after a brief period of excitation, induces prolonged increases in sleep and EEG synchronization. These effects are antagonized by anti-TNF-α antibodies. Microinfusion of TNF-α into the subarachnoid space just beneath the basal forebrain in rats enhances NREMS and reduces REMS. Finally, unilateral application of TNF-α onto the surface of the somatosensory cortex induces unilateral state-dependent increases in EEG delta wave power. Conversely, the soluble TNF receptor unilaterally reduces EEG power during the NREMS occurring after sleep deprivation. These latter two findings are germane to the brain organization of sleep and are discussed further later.

TNF Cell Biology

TNF signaling is an intense area of research due to its clinical importance and its inherent interest. TNF-α is synthesized as a 26 kDa membrane-associated protein. Soluble TNF-α, a 17 kDa protein, is cleaved from the 26 kDa membrane-associated protein by TNF-α converting enzyme. TNF-α production is tightly regulated in a tissue-specific manner, with transcription, translation, and secretion all controlled at multiple points. For instance, the 3′ UTR of TNF contains 24 miRNA complementary sites, suggesting translational regulatory sites. TNF-α induces its own expression, in part, via an NF-κB regulatory site (**Figure 1**). TNF-α can exist as a transmembrane protein that has biological activity. This membrane integrated form of TNF-α also receives signals and thus acts as a receptor, as well as a ligand.

There are two TNF cell surface receptors, the 55 kDa TNF receptor and the 75 kDa TNF receptor. The intracellular domains of these receptors lack intrinsic enzymatic activity. Rather, both receptors signal by recruitment of cytosolic proteins via protein–protein interaction domains. The diversity of these adaptor proteins and their ability to interact with other members of the TNF receptor superfamily help explain the pleiotropic actions of TNF-α. The TNF receptor family is divided into two large groups based on the adaptor proteins recruited in response to ligand binding. One major TNF-activated signaling pathway activates new gene transcription, whereas the other leads to cell death. Activation of the cell death pathway via the caspases 8 and 3 appears to be a rare physiological event because TNF-responsive gene products function to prevent cell death.

Both TNF receptors are cell surface receptors; they form trimeric complexes with TNF. The spatial distribution of the receptor and associated intracellular adaptor molecules is likely an important determinant of cellular specificity of action. Consequently, TNF activates signaling at multiple subcellular compartments, including the plasma membrane, mitochondria, and the nucleus. The ectodomains of both receptors are shed to form soluble receptors. The 55 kDa soluble TNF receptor is a constituent of normal cerebrospinal fluid, although its physiological role is unknown. However, given its effects on sleep and fatigue, it likely plays a role in sleep regulation.

TNF-α is expressed by microglia, astrocytes, and neurons and has a variety of biological actions in the brain, including a role in mediating both brain damage and neuroprotection. Whether TNF-α is protective or damaging may depend on the receptor type present – either the TNF 55 kDa or the TNF 75 kDa receptor – as well as the stimulus context and the presence or absence of substances that modify TNF-α activity. In addition to sleep, TNF-α has a role in thermoregulation, food intake, brain development, and neuronal connectivity. For instance, TNF-α potentiates AMPA-induced postsynaptic potentials,

AMPA-induced cytosolic Ca^{2+} increases, as well as several voltage-dependent calcium channels. TNF has a role in synaptic scaling, a process involved in maintaining synaptic diversity and plasticity.

Other NREMS Cytokine and Peptide Regulatory Substances

There is extensive evidence implicating both interleukin-1β (IL-1β) and GHRH in NREMS regulation that parallels that described previously for TNF-α. Thus, for example, inhibition of either GHRH or IL-1β inhibits spontaneous sleep. Mutant mice lacking functional GHRH receptors or mice lacking the IL-1β type I receptor have substantially less NREMS than respective strain controls. IL-1β induces growth hormone release via a hypothalamic mechanism that includes GHRH; anti-GHRH antibodies block IL-1β-induced growth hormone release. These antibodies also block IL-1β-induced NREMS. In fact, hypothalamic GABAergic neurons are receptive to both IL-1β and GHRH; both substances enhance intracellular calcium levels in these neurons. It is speculated that these neurons are the sleep-active hypothalamic neurons that increase their firing rates in response to IL-1β. In contrast, wake-active hypothalamic neurons decrease their firing rates in response to IL-1β. IL-1β and GHRH are related to TNF via the pathways outlined in **Figure 1**; collectively, it seems likely that these three substances form part of an NREMS regulatory network. Many other cytokines and members of the somatotropic axis can also affect sleep, but extensive knowledge concerning their roles in sleep regulation is lacking. They could in their own right be sleep regulatory substances or they may affect sleep via TNF-α, IL-1β, or GHRH.

Brain Organization of Sleep and TNF-α

Many investigations support the idea that sleep is regulated by specific brain circuits, such as the anterior hypothalamus. Perhaps the best evidence for this paradigm with regard to NREMS is that electrical, chemical, or thermal stimulation of the anterior hypothalamus induces NREMS. However, this paradigm inadequately addresses the issue of how the brain keeps track of its prior sleep–wake history over long periods of time. The transfer of cerebrospinal fluid from a sleep-deprived animal, but not a control animal, into a fully rested recipient enhances sleep in the recipient. This finding suggests that the homeostatic processes for sleep involve sleep regulatory substances.

There are also numerous results suggesting that sleep can be targeted to brain areas disproportionately activated during prior wakefulness. For instance, unilateral stimulation of the somatosensory cortex in awake humans produces asymmetry in EEG delta wave activity during subsequent NREMS. Similar results were obtained from rats and mice. Furthermore, dolphins only show EEG slow wave sleep on one side of the brain at a time. After a specific learning task, the subsequent EEG delta wave activity during sleep is enhanced in the area activated by the learning task. This finding reinforced previous work showing that either EEG activity or brain metabolism during sleep is regionally localized depending on prior use during wakefulness.

Our view of brain organization of sleep posits that sleep is initiated within neural assemblies as a function of prior cellular activity and this provides the mechanism for the localization and targeting of sleep, as well as for sleep homeostasis. In this view, for example, TNF-α is produced and released in response to neural and glial activity. These substances in turn are posited to be made and act locally in autocrine, paracrine, and juxtacrine ways to alter the input–output relationships of neural assemblies (e.g., cortical columns, barrels, or a cortical–thalamic–cortical loop) and thereby induce a functionally different state in these assemblies. Indeed, cortical TNF-α levels increase during prolonged wakefulness and during seizures – two conditions thought to be associated with enhanced cellular activity – and activity-dependent increases in TNF-α appear to be important for the developing nervous system. Furthermore, whisker stimulation upregulates TNF-α expression in neurons in the stimulated whisker's corresponding somatosensory barrel, but not in adjacent barrels, thereby directly demonstrating activity-dependent production of TNF-α.

Individual auditory and somatosensory cortical columns oscillate between functional states as defined by the amplitude of auditory- or whisker stimulation-induced surface evoked potentials. One of the functional states usually occurs simultaneously with whole animal sleep. This sleeplike state is dependent on its prior history in the sense that the probability of its occurrence is higher if, for the 15 min period prior to the measurement, it had been in the other functionally defined state. Furthermore, if two whiskers are stimulated, one at twice the rate of the other, the probability of the cortical column being in the sleep state is greater in the column receiving afferent input from the more rapidly stimulated whisker. If such changes in the functional state are dependent

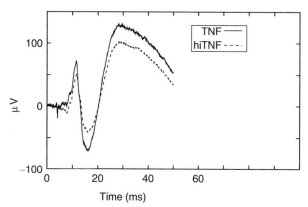

Figure 2 TNF-α enhances the amplitude of surface evoked potentials induced by whisker stimulation. A rat with two-bilateral 12 electrode arrays plus a guide cannula was trained to accept the restraint associated with the recording apparatus and slept during recordings. If the rat was pretreated with TNF-α applied to the surface of the somatosensory cortex, surface evoked potentials were enhanced during quiet sleep compared to those occurring after heat-inactivated TNF-α (hiTNF). We propose that cytokines are driving neurons within cortical columns into a down state and thereby amplify surface evoked potentials. We also propose that the sleep functional state of the column is promoted by cytokines produced in response to cellular activity and is a function of the number of neurons in the down state.

on activity-dependent sleep regulatory substances such as TNF-α, then one would predict that direct application of TNF-α onto the surface of the cortex would drive the affected neural assemblies to simultaneously express the same functional state. Indeed, when TNF-α is applied in this manner, EEG delta power is enhanced unilaterally. Furthermore, if TNF-α is applied to the surface of the cortex and individual cortical columns are probed by whisker twitches, the probability of the column being in the sleeplike state increases (**Figure 2**). Such experiments strongly support the notion that sleep begins as a local process, being induced by the production of activity-dependent sleep regulatory substances such as TNF-α. Such a view has profound implications for our understanding of sleep regulation, sleep function, and unconsciousness.

Regulation of REMS by Prolactin and Related Molecules

The first anecdotal account implicating PRL in REMS regulation was from an experiment in which REMS was enhanced after systemic administration of PRL in hypophysectomized pontine cats. Subsequently it was shown that systemic PRL causes dose-dependent enhancements of REMS in rabbits. Thereafter, several

lines of evidence implicated PRL in the regulation of REMS. The following were the major observations:

1. Injection of exogenous PRL stimulates REMS in cats, rabbits, and rats.
2. The increases in REMS develop slowly, over 2 or 3 h postinjection, and are maintained for several additional hours.
3. PRL-induced changes in sleep are selective for REMS.
4. PRL enhances REMS only during the light period in the rat.
5. In contrast, REMS is suppressed during the light period and is enhanced at night in mutant, PRL-deficient rats and PRL knockout mice.
6. Rats rendered chronically hyperprolactinemic by means of grafting pituitaries under the kidney capsule exhibit increases in REMS.
7. Both REMS and NREMS are enhanced in hyperprolactinemic pseudopregnant rats.
8. Restraint stress and ether stress are associated with enhanced circulating PRL and enhanced REMS. Restraint stress also enhances hypothalamic release of PRL, although stressor-induced release of pituitary PRL is thought to be responsible for the increases in REMS after stress.

Intracerebral administration of PRL or antibodies to PRL stimulates or inhibits REMS, respectively. The modulation of REMS is therefore a central action of PRL. PRL may come from two sources: it is produced by neurons in the hypothalamus and brain stem, and it is also released into the blood from the anterior pituitary. Circulating PRL is transported into the brain via specific receptor-mediated transport mechanisms residing in the choroid plexus. The REMS-promoting activity of blood-borne PRL is supported by the observation that REMS is enhanced in response to systemically injected PRL, excess PRL released from pituitary grafts, or stimulation of endogenous PRL secretion from the pituitary. Nevertheless, only slight decreases in REMS are observed when basal, nonstimulated blood PRL is suppressed by means of immunoneutralization. Although plasma PRL levels in humans are sleep related, being highest in the early morning hours, patients with PRL-producing adenoma fail to display changes in REMS; instead, their NREMS is increased. In summary, increases in blood PRL stimulate REMS in animal models, but circulating PRL at normal concentrations in the blood has only a minor effect on the regulation of REMS. The impact of the increased blood PRL concentrations on REMS is, in part, attributed to stimulation of the expression of PRL receptors in the choroid plexus, whereby PRL enhances its own transport into the brain. It is thus posited that intracerebral PRL modulates REMS under normal conditions, whereas pituitary PRL provides additional stimulating influence when PRL secretion is high (e.g., during stress). PRL entering the brain from the systemic circulation acts on the same structures normally under the influence of intracerebral prolactinergic neurons.

Intracerebroventricular injections of hypothalamic peptides that cause pituitary release of PRL – vasoactive intestinal peptide (VIP) and pituitary adenyl cyclase activating peptide, both of which promote REMS – elicit expression of PRL mRNA in the hypothalamus and anti-PRL antibodies can block VIP-enhanced REMS responses. Another hypothalamic releasing factor for pituitary PRL, prolactin releasing peptide, also enhances REMS in rats if injected centrally.

The mechanism by which PRL increases REMS is unclear. PRL may modulate the circadian regulation of REMS at the level of the hypothalamus, or it may act in the brain stem structures generating REMS. The long (2 or 3 h) latency of the REMS response to exogenous PRL suggests that some metabolic actions mediate the sleep effects. Finally, growth hormone, which is structurally related to PRL, also stimulates REMS. Human growth hormone has a high affinity to PRL receptors and promotes REMS in both humans and rats.

The neuronal circuits responsible for the generation of REMS reside in the brain stem. The brain stem control of REMS involves the interactions between REMS-on neurons, which promote REMS, and REMS-off neurons, which inhibit REMS. Acetylcholine (ACh) is the best characterized neurotransmitter that promotes REMS. Cholinergic neurons localized in the laterodorsal tegmental nucleus (LDT) and the pedunculopontine tegmental nucleus (PPT) are involved in the generation of REMS signs such as EEG desynchronization, ponto-geniculo-occipital waves, and muscle atonia. Cholinergic neurons increase their firing rates approximately 20 s before the EEG signs of REMS and maintain higher activity levels during REMS than during NREMS, suggesting that these neurons participate in the initiation and maintenance of REMS. Numerous microinjection and microinfusion studies show that REMS is enhanced by cholinergic agonists and inhibited by cholinergic antagonists. Furthermore, kainate lesions of the PPT result in prolonged decreases in REMS. Anatomical data indicate that single cholinergic neurons in the LDT or PPT project to multiple brain regions that are responsible for the generation of individual signs of REMS. Therefore, these cholinergic neurons might be involved in the orchestration of different REMS signs to generate REMS.

A Proposed Mechanism for PRL-Enhanced REMS

REMS is influenced by metabolism. The length of the REMS cycles correlates with brain size and general metabolic activity. The brain primarily utilizes glucose as an energy source. Glucose is transported from the blood and into astrocytes through an active transport system and metabolized into pyruvate, the main fuel for neurons. In neurons, pyruvate may be metabolized into lactate by the enzyme lactate dehydrogenase in the absence of oxygen, or it may be metabolized into CO_2 and H_2O through the oxidative pathway which involves many enzyme actions, including pyruvate dehydrogenase (PDH). REMS is selectively suppressed during hypoxic hypoxia, whereas NREMS or waking may be enhanced. In contrast, an increase in oxygen availability induces a significant increase in REMS in chronic pontine cats kept at a fixed temperature. The high glucose and oxygen consumption levels during REMS compared to those during NREMS suggest that PDH activity is enhanced during REMS. It is suggested that glucose may be metabolized through the anaerobic pathways during wakefulness, converted to glycogen during NREMS, and metabolized through the aerobic pathways during REMS.

Stimulation of PDH in the mesencephalic/pontine cholinergic cells may serve as a final common pathway for various experimental interventions that promote REMS, such as REMS deprivation, immobilization stress, or administration of PRL. The connection between PRL and PDH was suggested by observations that PRL stimulated PDH activity in the prostate and the mammary gland. Increases in brain stem PDH activity occur in response to systemic injection of PRL.

The synthesis of ACh is critically dependent on the supply of acetyl-CoA, which in turn relies on the oxidation of pyruvate by the PDH complex during high glucose and oxygen consumption. PDH is particularly enriched in brain cholinergic neurons. Enhanced PDH activity results in increases in the concentration of acetyl-CoA and thereby stimulates the synthesis of ACh. Furthermore, brain cholinergic neurons are vulnerable to the deficiency of PDH. ACh levels are decreased by the PDH inhibitor 3-bromopyruvate in brain homogenates or in the brain after local injection of this compound into the basal forebrain. Decreases in PDH activity elicited by experimental thiamine deficiency are also associated with the suppression of REMS.

Based on these observations, the proposed mechanism of the effects of PRL on sleep is as follows: PRL in the brain stimulates PDH and thereby energy production in neurons. This stimulation of PDH results in enhancement of ACh synthesis in cholinergic neurons because of an increased supply of acetyl-CoA. In our opinion, increased ACh production in itself does not trigger REMS. Due to the increased ACh pool, however, the quantity of ACh released increases whenever REMS spontaneously occurs. Therefore, PRL does not trigger REMS; instead, it facilitates spontaneous REMS. Furthermore, there is no evidence to suppose that PRL is specific to REMS-promoting cholinergic neurons. Cholinergic neurons also exist in the basal forebrain. These neurons are implicated in the mechanism of arousal, and their activities are associated with wakefulness. The importance of normal PDH activity has been demonstrated in the function of cholinergic neurons of the basal forebrain in experiments modeling the pathology of Alzheimer's disease. Stimulation of PDH in these neurons may explain the promotion of wakefulness by PRL during the dark period in rats. If further experiments substantiate that PRL acts via an increased synthesis of ACh due to enhanced activity of PDH, then PRL might be the first metabolic hormone whose effects on sleep can be linked to a distinct biochemical mechanism.

Conclusion

Our understanding of the biochemical regulation of sleep has progressed rapidly during the past 20 years. We now have a firm understanding of at least some of the humoral substances involved in sleep regulation. No doubt, other sleep regulatory substances will be described. Perhaps more important, these substances have already given insight into how the brain is organized to produce sleep. These mechanisms include a dynamic oscillation of states within neural assemblies coordinated into coherent whole animal states by sleep regulatory circuits and homeostatically driven by the accumulation of sleep regulatory substances.

See also: Acetylcholine; Dopamine; Histamine; Hypothalamic Regulation of Sleep; Locus Coeruleus and Raphe Nucleus; Thalamic Regulation of Sleep.

Further Reading

Eissner G, Kolch W, and Scheurich P (2004) Ligands working as receptors: Reverse signaling by members of the TNF superfamily enhance the plasticity of the immune system. *Cytokine & Growth Factor Reviews* 15: 353–366.

Krueger JM and Obal F Jr. (1993) A neuronal group theory of sleep function. *Journal of Sleep Research* 2: 63–69.

Krueger JM and Obal F Jr. (2003) Sleep function. *Frontiers in Bioscience* 8: 511–519.

Krueger JM, Obál F Jr., Opp M, Toth L, Johannsen L, and Cady AB (1990) Somnogenic cytokines and models concerning their effects on sleep. *Yale Journal of Biology and Medicine* 63: 157–172.

Majde JA and Krueger JM (2005) Links between the innate immune system and sleep. *Journal of Allergy and Clinical Immunology* 116: 1188–1198.

Maquet P, Peigneux P, Laureys S, et al. (2003) Memory processing during human sleep as assessed by functional neuroimaging. *Revue Neurologique (Paris)* 159: 6S27–6S29.

Marshall L and Born J (2002) Brain-immune interactions in sleep. *International Review of Neurobiology* 52: 93–131.

Obal F Jr. and Krueger JM (2003) Biochemical regulation of sleep. *Frontiers in Bioscience* 8: 520–550.

Opp MR (2005) Cytokines and sleep. *Sleep Medicine Reviews* 9: 355–364.

Pollmacher T, Schuld A, Kraus T, Haack M, Hinze-Selch D, and Mullington J (2000) Experimental immunomodulation, sleep, and sleepiness in humans. *Annals of the New York Academy of Sciences of the United States of America* 917: 488–499.

Rector DM, Topchiy IA, Carter KM, and Rojas MJ (2005) Local functional state differences between rat cortical columns. *Brain Research* 1047: 45–55.

Roky R, Obal F Jr., Valatx JL, et al. (1995) Prolactin and rapid eye movement sleep. *Sleep* 18: 536–542.

Shoham S, Davenne D, Cady AB, Dinarello CA, and Krueger JM (1987) Recombinant tumor necrosis factor and interleukin 1 enhance slow-wave sleep. *American Journal of Physiology* 253: R142–R149.

Steriade M (2003) The corticothalamic system in sleep. *Frontiers in Bioscience* 8: d878–d899.

Locus Coeruleus and Raphe Nucleus

B E Jones, McGill University, Montreal, QC, Canada

Background

The monoamines, noradrenaline (NA) and serotonin (5-HT), were found to have the capacity to influence sleep–wake states in the 1960s through early pharmacological and histochemical, combined with lesion, studies. Particularly through the work of Jouvet with his colleagues, evidence was presented for an important role of NA and the major NA-containing locus coeruleus (LC) nucleus in wakefulness and for a role of 5-HT and the 5-HT-containing raphe nuclei in sleep, particularly slow-wave sleep (SWS). Pharmacological depletion of NA (with inhibition of its synthesis by α-methyl-paratyrosine (AMPT)) and lesions of the ascending pathway from the LC led to decreased wakefulness. Conversely, depletion of 5-HT (with inhibition of its synthesis by para-chloro-phenylalanine (PCPA)) and lesions of the raphe, including the dorsal raphe (DR) nucleus, led to decreased SWS. These early studies and conclusions indicating opposite roles for the noradrenergic and serotonergic systems in sleep–wake state control have since been further scrutinized by finer techniques, including single-unit recording, and revised to reveal different – but not opposite – roles of these systems in the sleep–wake cycle.

Projections and Role of Noradrenergic Locus Coeruleus Neurons in Arousal

The LC is a collection of NA-containing neurons located in the periventricular region of the rostral pons, which can be likened to a sympathetic ganglion because from it emerge multiple major projection pathways of varicose and highly collateralized axons that reach out to innervate the entire central nervous system (**Figure 1**). Although the neurons are somewhat segregated according to their predominant projections (such that the most dorsal cluster projects primarily to the thalamus and lateral cortex; the more intermediate projects to the hypothalamus, basal forebrain, and medial cortex; and the most ventral cluster projects to the brain stem and spinal cord), a single neuron can project through multiple collaterals to all regions. This highly diffuse noradrenergic system is thus uniquely organized to simultaneously influence the entire central nervous system in a way that could alter its fundamental mode of operation, as occurs in a change of state from sleeping to waking. It was indeed thus proposed in early studies that the noradrenergic LC neurons formed the most important contingent of an ascending activating system of the brain stem. In support of this concept, one of the major stimulants, amphetamine, was found to act by release of NA and also dopamine (DA). The early lesion studies involving multiple systems did diminish wakefulness and cortical activation which were thus proposed to depend on the noradrenergic LC innervation of the cortex. However, subsequent more selective lesions or destruction by neurotoxins (6-hydroxydopamine (6-OHDA)) of the noradrenergic system did not result in the same loss of cortical activation and waking as did the earlier nonselective lesions. These results thus indicated that, although the noradrenergic neurons have the capacity to stimulate cortical activation and wakefulness, they are not essential for these functions. They represent, instead, one of multiple partially redundant, although differentiated, arousal systems in the brain, which include most notably the acetylcholine (ACh)-releasing cholinergic neurons of the brain stem and basal forebrain and the orexin (Orx, or hypocretin) neurons of the posterior hypothalamus (**Figure 1**).

The NA LC neurons discharge during active waking, gradually decrease their firing during SWS, and cease firing during rapid eye movement (REM) sleep, also called paradoxical sleep (PS) (**Figure 2(a)**). They accordingly discharge during active or attentive waking associated with behavioral arousal and the presence of postural muscle tone recorded on the electromyogram (EMG; **Figure 1**). They discharge during orientation and response to sensory stimuli. They also respond to physiological stimuli which activate the sympathetic nervous system and situations which are associated with stress. Given their diffuse projections and actions, the noradrenergic neurons can simultaneously stimulate fast cortical activity on the electroencephalogram (EEG) and motor activity with postural muscle tone on the EMG during waking in a tonic and phasic manner. Acting through α_1-adrenergic receptors (α_1-AR), NA can depolarize and excite many neurons through the central nervous system, including thalamo-cortical and basalo-cortical projection neurons, which respectively release glutamate (Glu) and ACh in the cortex, and excitatory (Glu) reticulo-spinal projection or motor neurons (**Figure 1**). These actions appear to be very important or at least potent for stimulating and maintaining an aroused and attentive state in that the condition of hypersomnia and narcolepsy is best treated by drugs which enhance NA release, including amphetamine-like compounds or

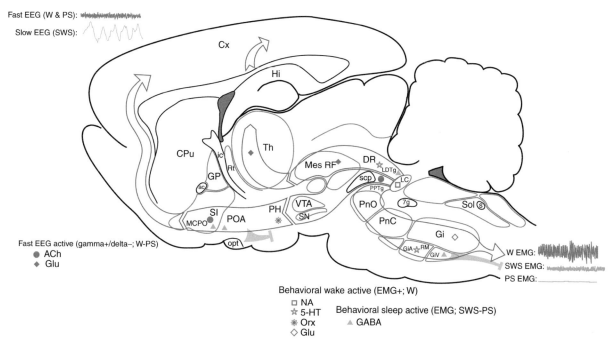

Fast EEG (W & PS):
Slow EEG (SWS):

Fast EEG active (gamma+/delta–; W-PS)
● ACh
◆ Glu

W EMG:
SWS EMG:
PS EMG:

Behavioral wake active (EMG+; W)
□ NA
☆ 5-HT   Behavioral sleep active (EMG; SWS-PS)
✳ Orx   ▲ GABA
◇ Glu

Figure 1 Noradrenergic (NA), locus coeruleus (LC), and serotonergic (5-HT) raphe nuclei along with other neurons involved in sleep–wake state control. Sagittal schematic view of the rat brain depicting neurons with their chemical neurotransmitters and pathways by which they influence cortical activity or behavior across the sleep–wake cycle. Wake (W) is characterized by fast gamma activity on the cortical EEG (upper left) and high postural muscle tone on the neck EMG (lower right); slow-wave sleep (SWS) is characterized by slow delta EEG and low tone on the EMG; and paradoxical sleep (PS) is characterized by fast gamma EEG and atonia on the EMG. Neurons that are active during waking (red symbols) include cells with ascending projections toward the cortex, which stimulate fast cortical activity, and cells with descending projections toward the spinal cord, which stimulate postural muscle tone and behavioral arousal. The NA LC neurons give rise to diffuse projections with highly collateralized and varicose axons that innervate multiple regions as way stations along their major pathways. One major projection pathway ascends dorsally through the reticular formation into the thalamus and beyond to the cortex. Another ascends ventrally through the ventral tegmentum into and through the hypothalamus via the medial forebrain bundle to reach the preoptic area and basal forebrain (MCPO and SI) and continue forward to the cortex, which is diffusely innervated. A third projection pathway passes ventrally into the pontine tegmentum to descend through the brain stem reticular formation and continues into the spinal cord, where multiple, including motor, neurons are innervated. 5-HT raphe nuclei also give rise to multiple and relatively diffuse projections, which, however, are segregated such that the dorsal raphe (DR) gives rise to prominent ascending projections and the medullary raphe (including the raphe magnus, RM) gives rise to prominent descending projections to the spinal cord. The NA and 5-HT neurons along with orexinergic (Orx) and some putative glutamatergic (Glu) neurons which commonly have relatively diffuse or descending projections discharge in association with behavioral arousal and EMG activity (EMG+) and cease firing with muscle atonia to be active during W and silent during PS (W, empty red symbols). The cholinergic (ACh) neurons and other Glu neurons which have predominantly ascending projections discharge in association with fast EEG activity (gamma+) and cease firing with delta activity (delta–) to be active during W and PS (W-PS, filled red symbols). Neurons that are active during sleep (aqua symbols) include cells with descending projections into the brain stem and spinal cord which diminish behavioral arousal and muscle tone. Important among such cells are GABAergic (and glycinergic) neurons in the ventral medulla which project to the brain stem and spinal cord, where they can inhibit motor neurons during sleep. The cholinergic neurons have the capacity to stimulate cortical activation during waking and PS and also to promote PS with muscle atonia by acting in part on such GABAergic neurons, when the monoaminergic and orexinergic neurons are silent and also inhibited by GABAergic neurons. 5-HT, serotonin; 7g, genu seventh nerve; ac, anterior commissure; ACh, acetylcholine; CPu, caudate putamen; Cx, cortex; DR, dorsal raphe nucleus; EEG, electroencephalogram; EMG, electromyogram; GABA, γ-aminobutyric acid; Glu, glutamate; Gi, gigantocellular RF; GiA, gigantocellular, alpha part RF; GiV, gigantocellular, ventral part RF; Glu, glutamate; GP, globus pallidus; Hi, hippocampus; ic, internal capsule; LC, locus coeruleus nucleus; LDTg, laterodorsal tegmental nucleus; Mes RF, mesencephalic RF; MCPO, magnocellular preoptic nucleus; NA, noradrenaline; opt, optic tract; Orx, orexin; PH, posterior hypothalamus; PnC, pontine, caudal part RF; PnO, pontine, oral part RF; POA, preoptic area; PPTg, pedunculopontine tegmental nucleus; PS, paradoxical sleep (also called REM sleep); RF, reticular formation; RM, raphe magnus; Rt, reticularis nucleus of the thalamus; s, solitary tract; scp, superior cerebellar peduncle; SI, substantia innominata; SN, substantia nigra; Sol, solitary tract nucleus; SWS, slow-wave sleep; Th, thalamus; VTA, ventral tegmental area; W, wake. Modified from Jones BE (2005) From waking to sleeping: Neuronal and chemical substrates. *Trends in Pharmacological Sciences* 26: 578–586.

catecholamine-uptake blockers. NA also excites and can thereby recruit Orx neurons, which in turn give rise, like the LC neurons, to highly diffuse projections and act by exciting multiple arousal systems together with cortical and spinal motor neurons. Orx, which also excites the LC neurons in return, is essential for the maintenance of waking because in its absence narcolepsy with cataplexy results.

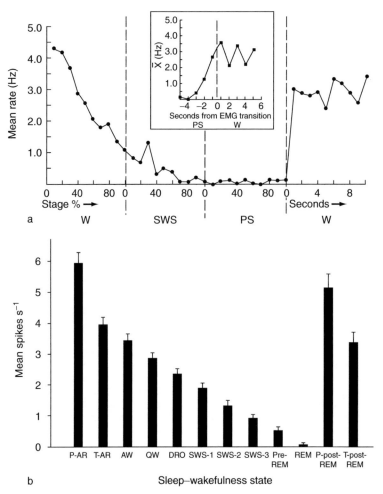

Figure 2 Discharge of (a) locus coeruleus (LC) and (b) dorsal raphe (DR) units across sleep–wake states: (a) mean firing rate (in hertz) per percentage of stage or seconds across a full sleep–wake cycle for multiple LC neurons (normalized according to completion of the entire cycle); (b) mean firing rate (in spikes s^{-1}) per state or stage across a full sleep–wake cycle for multiple DR neurons. As shown in (a), there is a progressive decrease in discharge with transition from waking into SWS and a cessation of firing during PS prior to an increase with the return of muscle tone (EMG) that actually anticipates the EMG change (inset). The graph in (b) also shows a progressive decrease with transition from aroused waking to late SWS and a virtual cessation of firing during PS prior to an increase with awakening. AW, active waking; DRO, drowsy; EMG, electromyogram; P-AR, phasic arousal; P-post-REM, phasic post-REM; Pre-REM, 60 s before REM onset; PS, paradoxical sleep; QW, quiet waking; REM, rapid eye movement sleep; SWS, slow wave sleep; SWS-1 early slow-wave sleep; SWS-2, middle slow-wave sleep; SWS-3, late slow-wave sleep; T-AR, tonic arousal; T-post-REM, tonic post-REM; W, waking. Vertical bars = SEM. (a) From Aston-Jones G and Bloom FE (1981) Activity of norepinephrine-containing locus coeruleus neurons in behaving rats anticipates fluctuations in the sleep-waking cycle. *Journal of Neuroscience* 1: 876–886. (b) From Trulson ME and Jacobs BL (1979) Raphe unit activity in freely moving cats: Correlation with level of behavioral arousal. *Brain Research* 163: 135–150.

NA also promotes waking by inhibiting sleep-promoting γ-aminobutyric acid (GABA)ergic neurons in the preoptic area (POA) and basal forebrain which bear α_2-AR (**Figure 1**). These GABAergic neurons discharge during SWS and PS in a reciprocal manner to the NA LC neurons and, accordingly, in negative correlation with muscle tone on the EMG. By descending projections, these sleep-promoting GABAergic neurons can inhibit Orx neurons in the posterior hypothalamus (PH) and also NA cells in the LC more caudally. The NA LC neurons thus promote arousal by synergistic activity with other arousal systems, including the Orx neurons, and prevent sleep by antagonistic activity

with sleep-promoting systems, including GABAergic preoptic and basal forebrain neurons.

Specifically with regard to PS, the NA LC neurons were found long ago to have a reciprocal relationship in their discharge with respect to presumed cholinergic neurons of the brain stem, as formalized in a model by McCarley and Hobson. Whereas LC neurons cease firing during PS, presumed cholinergic ponto-mesencephalic neurons discharged at maximal rates during PS. From pharmacological and electrophysiological studies, it appeared that the monoaminergic, including, significantly, noradrenergic neurons, had to be off for the cholinergic neurons to stimulate the constellation of events, including cortical activation with REMs and muscle atonia, of PS (**Figure 1**). Indeed, at least some cholinergic neurons are inhibited by NA through α_2-AR. On the other hand, ACh excites the NA LC neurons and thus must act to inhibit the NA cells through GABAergic interneurons in the region, which from c-Fos expression, as an indication of neural activity, are active together with the cholinergic neurons during PS (**Figure 1**). The NA LC neurons have thus been conceived as PS-permitting (by permitting PS to occur through their cessation of activity and resulting disinhibition of cholinergic cells along with disfacilitation of motor systems), whereas the cholinergic and local GABAergic neurons are PS-promoting (and act in part by inhibiting the NA LC neurons).

Projections and Role of Serotonergic Raphe Neurons in Waking and Sleeping

Following pharmacological and lesion studies which indicated that 5-HT and raphe neurons play a role in promoting sleep, single-unit recording studies were undertaken to determine when they discharged. These early studies by McGinty and Harper surprisingly revealed that presumed 5-HT neurons in the DR discharged during waking, decreased their firing during SWS, and ceased firing during PS, in a similar manner to the NA LC neurons (**Figures 1** and **2**). Subsequent studies examining the discharge of presumed 5-HT neurons in the other raphe nuclei through the brain stem similarly found the same profile of activity across the sleep–wake cycle, and studies examining release in the forebrain also indicated that 5-HT release was greatest during waking and minimal during sleep. So, it became clear through these results that 5-HT could not generate SWS, as was originally proposed.

However, many lines of pharmacological and physiological evidence indicated that 5-HT and raphe neurons did not act in the same way that the NA neurons did. First, as already stated, depletion of NA resulted in decreased waking, whereas depletion of 5-HT resulted in decreased sleep. Second, whereas NA LC neurons discharged during orientation to a novel sensory stimulus, presumed 5-HT DR neurons ceased firing. Moreover, whereas NA LC neurons discharged in response to physiological challenges which activated the sympathetic nervous system, 5-HT raphe neurons did not. So, perhaps 5-HT neurons promote a less attentive or responsive waking state from which sleep onset is facilitated.

The 5-HT DR neurons project along similar pathways as the NA LC neurons to reach the thalamus, hypothalamus, basal forebrain, and cortex while leaving collaterals along the way (**Figure 1**). Their ascending projections are complemented by those of the central superior nucleus. The major descending 5-HT projections originate from neurons in the medulla (raphe magnus, pallidus, and obscurus) and extend through the entire spinal cord. 5-HT acts on multiple, different receptors and thus has the capacity to excite or inhibit different cell populations through the central nervous system. Whereas, its actions on certain thalamo-cortical projection (i.e., Glu) neurons and basalo-cortical (i.e., ACh) neurons are often inhibitory through 5-HT1 receptors, they are excitatory on some cortical, including pyramidal, cells, through 5-HT2 or other receptors. In the brain stem and spinal cord, 5-HT depolarizes and excites, like NA, motor neurons and can thus enhance muscle tone and activity. It can inhibit sensory transmission.

With respect to PS, the presumed 5-HT neurons of the DR, like the NA LC neurons, cease firing in a reciprocal manner to the presumed cholinergic cells of the pontomesencephalic tegmentum. As for the NA LC neurons, the 5-HT DR neurons are inhibited by local GABAergic neurons during PS. 5-HT hyperpolarizes and inhibits cholinergic cells in both the brain stem and basal forebrain. It can thus prevent the occurrence of PS with muscle atonia promoted by brain stem cholinergic neurons and diminish cortical activation evoked by basal forebrain cholinergic neurons.

Given their discharge during waking behaviors involving rhythmic motor activities, such as locomotion or grooming, and their cessation of discharge during orientation to sensory stimuli, it has been proposed by Jacobs that 5-HT neurons are primarily involved in facilitating rhythmic motor patterns and attenuating responses to extraneous sensory stimuli during those activities. Their discharge thus facilitates muscle tone with motor activities, yet diminishes cortical activation in response to external sensory stimuli during those activities. In this way, they function in opposition to the cholinergic cells, which diminish muscle tone and motor activities and enhance cortical activation during attentive waking and PS. The discharge of the NA LC cells facilitates muscle tone and arousal, but also enhances cortical activation in response to external sensory stimuli during waking. NA, which excites and depolarizes the basal forebrain cholinergic cells and can act in synergy with certain cholinergic neurons during waking but operates in opposition to other pontomesencephalic cholinergic cells whose discharge occurs during PS. Collectively, both the NA and 5-HT systems prevent loss of muscle tone and the apparition of PS, and drugs which are used to treat narcolepsy include reuptake blockers, which act to enhance both 5-HT and NA transmission.

See also: Acetylcholine; Cytokines and other Neuromodulators; Dopamine; Histamine; Hypothalamic Regulation of Sleep; Thalamic Regulation of Sleep.

Further Reading

Aston-Jones G and Bloom FE (1981) Activity of norepinephrine-containing locus coeruleus neurons in behaving rats anticipates fluctuations in the sleep–waking cycle. *Journal of Neuroscience* 1: 876–886.

Jacobs BL and Fornal CA (1999) Activity of serotonergic neurons in behaving animals. *Neuropsychopharmacology* 21: 9S–15S.

Jones BE (2004) Paradoxical REM sleep promoting and permitting neuronal networks. *Archives of Italian Biology* 142: 379–396.

Jones BE (2005) Basic mechanisms of sleep–wake states. In: Kryger MH, Roth T, and Dement WC (eds.) *Principles and Practice of Sleep Medicine*, pp. 136–153. Philadelphia: Elsevier Saunders.

Jones BE (2005) From waking to sleeping: Neuronal and chemical substrates. *Trends in Pharmacological Sciences* 26: 578–586.

Jouvet M (1972) The role of monoamines and acetylcholine-containing neurons in the regulation of the sleep-waking cycle. *Ergebnisse der Physiologie* 64: 165–307.

McCormick DA (1992) Neurotransmitter actions in the thalamus and cerebral cortex and their role in neuromodulation of thalamocortical activity. *Progress in Neurobiology* 39: 337–388.

Nishino S and Mignot E (1997) Pharmacological aspects of human and canine narcolepsy. *Progress in Neurobiology* 52: 27–78.

Saper CB, Chou TC, and Scammell TE (2001) The sleep switch: Hypothalamic control of sleep and wakefulness. *Trends in Neuroscience* 24: 726–731.

Siegel JM (2004) The neurotransmitters of sleep. *Journal of Clinical Psychiatry* 65(supplement 16): 4–7.

Trulson ME and Jacobs BL (1979) Raphe unit activity in freely moving cats: Correlation with level of behavioral arousal. *Brain Research* 163: 135–150.

Zeitzer JM, Nishino S, and Mignot E (2006) The neurobiology of hypocretins (orexins), narcolepsy and related therapeutic interventions. *Trends in Pharmacological Sciences* 27: 368–374.

Acetylcholine

B E Jones, McGill University, Montreal, QC, Canada

Background

Since early histochemical and pharmacological studies in the 1960s, acetylcholine (ACh) has been known to comprise a key element of the activating systems in the brain. The catabolic enzyme acetylcholinesterase (AChE) was found to be contained in fibers which ascended from the brain stem reticular formation into the forebrain and from the basal forebrain into the cerebral cortex in what Shute and Lewis called the 'cholinergic reticular activating system.' This designation advanced the concept that the neurons which utilized ACh as a neurotransmitter formed the principal component of the 'ascending reticular activating system' previously delineated by Moruzzi and Magoun and colleagues in the 1940s and 1950s and shown to be critical for the cortical activation of waking. Enhancement of synaptic ACh through drugs such as physostigmine, which inhibits AChE, stimulated fast cortical activity on the electroencephalogram (EEG), which is indicative of cortical activation. The ACh receptor agonists, nicotine and muscarine, were long known as stimulants and also found to elicit cortical activation. Conversely, blocking nicotinic or muscarinic ACh receptors diminished fast cortical activity. Yet interestingly, atropine, which blocks muscarinic ACh receptors, produced slow wave activity on the cortex but did not prevent behavioral activity and arousal, and thus led to a disassociation of the normal EEG activity and behavior of waking. This dissociation revealed the selective role of ACh in stimulating cortical activation. Indeed, ACh was subsequently found to stimulate cortical activation during the behavioral quiescence and immobility that occurs during the state of rapid eye movement (REM) sleep (also called paradoxical sleep (PS); **Figure 1**). The cholinergic systems are thus well known to play an important role in cortical activation that occurs during both waking and REM sleep. Conditional upon the activity of other systems, they can also promote either waking or REM sleep states.

Cholinergic Systems in the Brain

The two major groups of cholinergic neurons with ascending projections are located in the pontomesencephalic tegmentum and the basal forebrain (**Figure 1**). They have been identified and localized definitively by employing immunohistochemical staining for choline acetyl transferase (ChAT), the synthetic enzyme for ACh. The enzyme is found in select neurons that synthesize ACh for use as a neurotransmitter and were thus presumed to be 'cholinergic.' As originally visualized by Shute and Lewis, AChE, the catabolic enzyme, is contained in many more neurons, including monoaminergic and glutamatergic reticular and thalamic neurons, which are likely target neurons of the cholinergic cells. Nonetheless, ChAT immunohistochemistry revealed the presence of cholinergic neurons within the brain stem reticular formation and basal forebrain as well as the presence of cholinergic fibers ascending through the brain and from the basal forebrain up to the cerebral cortex in parallel with other ascending systems to confirm the essential characterization of a cholinergic reticular activating system. The presence of the vesicular transporter protein for ACh (VAChT), which permits storage and synaptic release of ACh, in the brain stem and forebrain ChAT-immunostained cell bodies and terminals confirmed the use of ACh as a neurotransmitter by these accordingly established 'cholinergic' neurons.

The pontomesencephalic cholinergic neurons are located within the laterodorsal tegmental (LDTg) and pedunculopontine tegmental (PPTg) nuclei (**Figure 1**). Together with other neurons of the reticular formation (which predominantly utilize glutamate (Glu) as a neurotransmitter), these cholinergic cells send ascending projections dorsally up to the thalamus and ventrally to the hypothalamus and, to a lesser extent, the basal forebrain. In the thalamus, they innervate some of the specific thalamocortical relay nuclei (particularly the medial and lateral geniculate nuclei) but innervate most prominently and ubiquitously the neurons of the nonspecific thalamocortical projection system (the midline and intralaminar nuclei, which, like all thalamic projection neurons, utilize Glu), which in turn project in a widespread manner to the cerebral cortex to have the capacity to stimulate widespread cortical activation. Either through collaterals of the forebrain projecting cells or by unique projections, the pontomesencephalic cholinergic neurons also project into the brain stem reticular formation, where they can influence reticular neurons with ascending, local, or descending projections, which are important for REM sleep generation.

The basal forebrain cholinergic neurons are located within the medial septum and diagonal band of Broca (MS-DBB) rostrally and the magnocellular preoptic nucleus, substantia innominata, and globus pallidus (MCPO-SI-GP) more caudally (**Figure 1**). These two major collections of cells provide prominent innervation to the hippocampus and paleocortex (MS-DBB)

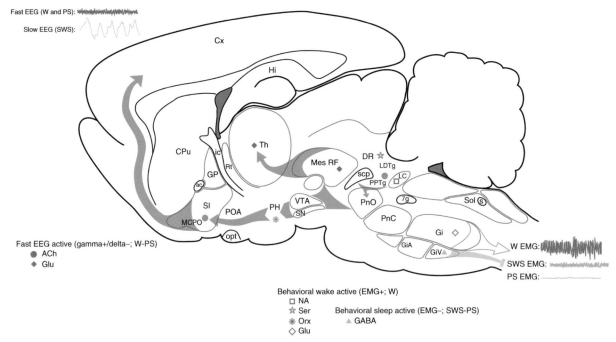

Fast EEG (W and PS):

Slow EEG (SWS):

Fast EEG active (gamma+/delta–; W-PS)
● ACh
◆ Glu

Behavioral wake active (EMG+; W)
□ NA
☆ Ser   Behavioral sleep active (EMG–; SWS-PS)
☀ Orx   ▲ GABA
◇ Glu

W EMG:

SWS EMG:

PS EMG:

Figure 1 Cholinergic and other neurons involved in sleep–wake state control. Sagittal schematic view of the rat brain depicting neurons with their chemical neurotransmitters and pathways by which they influence cortical activity or behavior across the sleep–wake cycle. Wake (W) is characterized by fast gamma activity on the cortical electroencephalogram (EEG; top left) and high postural muscle tone on the neck electromyogram (EMG; bottom right), slow wave sleep (SWS) is characterized by slow delta EEG and low tone on the EMG, and paradoxical sleep (PS) is characterized by fast gamma EEG and atonia on the EMG. Neurons that are active during waking (red symbols) include cells with ascending projections toward the cortex which stimulate fast cortical activity and cells with descending projections toward the spinal cord which stimulate postural muscle tone and behavioral arousal. The cholinergic (ACh) neurons and other glutamatergic (Glu) neurons which have predominantly ascending projections discharge in association with fast EEG activity (gamma+) and cease firing with delta activity (delta-) to be active during W and PS (W-PS; solid red symbols). The serotonergic (Ser), noradrenergic (NA), orexinergic (Orx), and some putative glutamatergic (Glu) neurons which have more diffuse or descending projections discharge in association with behavioral arousal and EMG activity (EMG+) and cease firing with muscle atonia to be active during W and silent during PS (W; open red symbols). Neurons that are active during sleep (aqua symbols) include cells with descending projections into the brain stem and spinal cord which diminish behavioral arousal and muscle tone. Important among such cells are GABAergic (and glycinergic) neurons in the ventral medulla which project to the brain stem and spinal cord, where they can inhibit motor neurons during sleep. The cholinergic neurons have the capacity to stimulate cortical activation during waking and PS and also to promote PS with muscle atonia by acting in part upon such GABAergic neurons, when the monoaminergic and orexinergic neurons are silent. 7g, genu seventh nerve; ac, anterior commissure; ACh, acetylcholine; CPu, caudate putamen; Cx, cortex; DR, dorsal raphe nucleus; Gi, gigantocellular RF; GiA, gigantocellular, alpha part RF; GiV, gigantocellular, ventral part RF; Glu, glutamate; GP, globus pallidus; Hi, hippocampus; ic, internal capsule; LC, locus coeruleus nucleus; LDTg, laterodorsal tegmental nucleus; Mes RF, mesencephalic RF; MCPO, magnocellular preoptic nucleus; NA, noradrenaline; opt, optic tract; Orx, orexin; PH, posterior hypothalamus; PnC, pontine, caudal part RF; PnO, pontine, oral part RF; POA, preoptic area; PPTg, pedunculopontine tegmental nucleus; PS, paradoxical sleep (also called REM sleep); RF, reticular formation; Rt, reticularis nucleus of the thalamus; s, solitary tract; scp, superior cerebellar peduncle; Ser, serotonin; SI, substantia innominata; SN, substantia nigra; Sol, solitary tract nucleus; Th, thalamus; VTA, ventral tegmental area. Modified from Jones BE (2005) From waking to sleeping: Neuronal and chemical substrates. *Trends in Pharmacological Sciences* 26: 578–586, with permission from Elsevier.

and to the neocortex (MCPO-SI-GP). Projecting in a topographically organized, though widespread manner, the basal forebrain cholinergic neurons thus have the capacity to stimulate widespread cortical activation through both paleo- and neocortex.

Cholinergic Systems and Cortical Activation of Waking

The pontomesencephalic cholinergic neurons send their major projections into the thalamus by which they can excite thalamocortical neurons to in turn stimulate cortical activation. Release of ACh in the

thalamus is maximal in association with fast cortical activity during waking and during REM sleep, indicating that the thalamic projecting cholinergic cells discharge during both waking and REM sleep. Acting on both nicotinic and muscarinic ACh receptors, ACh depolarizes and excites thalamocortical projection neurons. It thereby prevents their slow bursting discharge, which occurs during slow wave sleep (SWS), and stimulates their fast tonic discharge, which occurs during, while also stimulating, fast cortical activity of waking or REM sleep.

The basal forebrain cholinergic cells project directly to the hippocampus and neocortex, where they act on cortical neurons to alter their discharge. Release of ACh from hippocampus and neocortex is high in association with fast cortical activity during both waking and REM sleep. ACh prevents the slow bursting discharge of cortical neurons, which occurs during SWS, and stimulates the fast tonic discharge that occurs during waking and REM sleep. ACh's action in the cortex occurs largely through muscarinic receptors of different types on pyramidal cells and interneurons. Some interneurons are hyperpolarized through muscarinic ACh receptors, whereas others, such as pyramidal cells, are depolarized and excited through muscarinic and nicotinic ACh receptors. Overall, however, ACh and the basal forebrain cholinergic neurons stimulate fast cortical activity particularly in the gamma frequency range (30–60 Hz), which is maximal during attentive or active waking and REM sleep. ACh also stimulates theta activity (4–10 Hz), which is similarly maximal during attentive or active waking and REM sleep. Indeed, it may stimulate these in parallel since the high-frequency gamma activity rides on the slow rhythmic theta waves. Local injections of glutamate or other agonists acting on the cholinergic cells in the basal forebrain evoke theta with enhanced gamma activity. The basal forebrain cholinergic neurons can modulate this rhythmic cortical activity by their discharge, which, as established for identified ChAT-immunostained cells, occurs in rhythmic bursts with theta along with high gamma activity during active waking and REM sleep (**Figure 2(a)**).

Cholinergic Systems and REM Sleep

Following identification of the cholinergic neurons by ChAT immunohistochemistry, it became apparent that the major cell groups in the LDTg and PPTg were located in a region previously shown by Jouvet to be important in the generation of REM sleep, named by him 'paradoxical sleep.' Indeed, complete transections of the brain stem through the caudal midbrain which allowed the persistence of PS in the periphery (marked by postural muscle atonia on electromyogram) would have left intact the pontomesencephalic cholinergic cells innervating the brain stem (**Figure 1**). In later studies, neurotoxic lesions destroying the pontomesencephalic cholinergic neurons resulted in the loss of PS, including the phasic pontogeniculo-occipital (PGO) spikes accompanying REM sleep and tonic muscle atonia of that state. It had also been shown in both early and later studies that local injections of the cholinergic agonist carbachol into the pontine tegmentum could elicit PS, including PGO spikes with REM, theta activity, and muscle atonia.

On the other hand, the pharmacological elicitation of REM sleep through cholinergic agonists was not

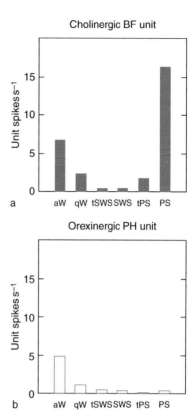

Cholinergic BF unit

Orexinergic PH unit

Figure 2 Discharge cholinergic and orexinergic units across sleep–wake states. Average discharge rate (spikes s^{-1}) across the sleep–wake states and transitions (t) of a cholinergic cell (a) in basal forebrain (BF; within MCPO-SI, identified by juxtacellular labeling with neurobiotin and immunohistochemical staining for ChAT) and an orexinergic neuron (b; in the posterior hypothalamus (PH) within the perifornical area, identified by juxtacellular labeling with neurobiotin and immunohistochemical staining for orexin). ACh neurons discharge in association with cortical activation (fast EEG active, gamma+/Delta−) during wake and paradoxical sleep (W-PS; solid red symbols in **Figure 1**), whereas orexinergic neurons discharge, like NA and Ser neurons, in association with behavioral activity and postural muscle tone (behavioral wake active, EMG+) during W (open red symbols in **Figure 1**). The discharge of cholinergic neurons when orexinergic and monoaminergic neurons are silent leads to PS. aW, active wake; qW, quiet wake; tSWS, transition to slow wave sleep; SWS, slow wave sleep; tPS, transition to paradoxical sleep; PS, paradoxical sleep (also called REM sleep). Data from Lee MG, Hassani OK, Alonso A, and Jones BE (2005) Cholinergic basal forebrain neurons burst with theta during waking and paradoxical sleep. *Journal of Neuroscience* 25: 4365–4369; and Lee MG, Hassani OK, and Jones BE (2005) Discharge of identified orexin/hypocretin neurons across the sleep–waking cycle. *Journal of Neuroscience* 25: 6716–6720.

always automatic. AChE inhibitors such as physostigmine commonly stimulated cortical activation with wakefulness in otherwise untreated normal animals or humans. However, in early studies it was found that if physostigmine was administered following depletion of the monoamines (with reserpine), it elicited REM sleep. By these pharmacological studies, it was evident that the occurrence of REM sleep depends on the activity of cholinergic neurons, but it also depends on the inactivity of monoaminergic neurons, a principle which was subsequently formalized in a model of reciprocal interaction by McCarley and Hobson.

Single unit recording studies indicated that whereas many neurons in the pontomesencephalic tegmentum discharged robustly during REM sleep, including those in the region of the cholinergic cells, the presumed serotonergic neurons in the raphe and presumed noradrenergic neurons in the locus coeruleus ceased firing during REM sleep (**Figure 1**). Although definite identification of recorded neurons awaits absolute confirmation of these principles, c-Fos expression, which reflects neural discharge, was found to be maximal in identified cholinergic pontomesencephalic neurons and minimal in serotonergic and noradrenergic neurons during REM sleep. Finally, ACh release is high during REM sleep in both the forebrain and the brain stem, whereas the release of the monoamines is minimal during REM sleep. Converging lines of evidence thus indicate that cholinergic neurons trigger REM sleep if monoaminergic neurons are silent. ACh can act in different ways on various brain stem and spinal neurons, including GABAergic (and glycinergic) neurons which can inhibit motor neurons, such as to provoke decreased activity and muscle tone. In contrast, serotonin and noradrenaline can stimulate behavioral arousal along with postural muscle tone through excitatory actions on brain stem and spinal neurons, including reticular glutamatergic and motor neurons.

It has also been revealed that hypothalamic neurons containing the peptide orexin (or hypocretin) normally stimulate behavioral arousal with postural muscle tone in a critical manner since in their absence narcolepsy with cataplexy results in humans and animals. Like the presumed monoaminergic neurons and in contrast to the cholinergic neurons, the orexin neurons fire selectively during waking and are silent during REM sleep (**Figures 1** and **2(b)**). Like monoaminergic neurons, orexin neurons would thus also oppose cholinergic neurons in the generation of REM sleep, although they can work together with cholinergic neurons to stimulate behavioral arousal together with cortical activation during waking.

Cholinergic Systems, Sleep, and Memory

The importance of ACh in memory processes has been known since early pharmacological studies showed serious decrements in memory consolidation following administration of the muscarinic ACh receptor antagonists atropine or scopolamine. Both short- and long-term memory processes have been shown to be enhanced by ACh. ACh enhances long-term

potentiation (LTP) and stimulates theta, which in turn enhances LTP and plasticity. Stimulation of the basal forebrain has been shown to promote learning and plasticity. It thus follows that states with heightened ACh release and maximal cholinergic unit discharge would be associated with enhanced learning and memory. These states are active or attentive waking and REM sleep. Although it has long been known that learning is maximal during aroused or attentive waking, it has also been shown in multiple studies that memory consolidation for implicit and explicit or particularly procedural, emotional, and spatiotemporal elements is enhanced during REM sleep.

See also: Cytokines and other Neuromodulators; Dopamine; Histamine; Hypothalamic Regulation of Sleep; Locus Coeruleus and Raphe Nucleus; Sleep-Dependent Memory Processing; Thalamic Regulation of Sleep.

Further Reading

Jones BE (2004) Paradoxical REM sleep promoting and permitting neuronal networks. *Archives Italiennes de Biologie* 142: 379–396.
Jones BE (2005) From waking to sleeping: Neuronal and chemical substrates. *Trends in Pharmacological Sciences* 26: 578–586.
Lee MG, Hassani OK, Alonso A, and Jones BE (2005) Cholinergic basal forebrain neurons burst with theta during waking and paradoxical sleep. *Journal of Neuroscience* 25: 4365–4369.
Lee MG, Hassani OK, and Jones BE (2005) Discharge of identified orexin/hypocretin neurons across the sleep–waking cycle. *Journal of Neuroscience* 25: 6716–6720.
McCormick DA (1992) Neurotransmitter actions in the thalamus and cerebral cortex and their role in neuromodulation of thalamocortical activity. *Progress in Neurobiology* 39: 337–388.
Nishino S and Mignot E (1997) Pharmacological aspects of human and canine narcolepsy. *Progress in Neurobiology* 52: 27–78.
Rauchs G, Desgranges B, Foret J, and Eustache F (2005) The relationships between memory systems and sleep stages. *Journal of Sleep Research* 14: 123–140.
Sarter M and Bruno JP (2000) Cortical cholinergic inputs mediating arousal, attentional processing and dreaming: Differential afferent regulation of the basal forebrain by telencephalic and brainstem afferents. *Neuroscience* 95: 933–952.
Siegel JM (2004) The neurotransmitters of sleep. *Journal of Clinical Psychiatry* 65(supplement 16): 4–7.
Steriade M (2004) Acetylcholine systems and rhythmic activities during the waking–sleep cycle. *Progress in Brain Research* 145: 179–196.
Stickgold R, Hobson JA, Fosse R, and Fosse M (2001) Sleep, learning, and dreams: Off-line memory reprocessing. *Science* 294: 1052–1057.
Weinberger NM (2003) The nucleus basalis and memory codes: Auditory cortical plasticity and the induction of specific, associative behavioral memory. *Neurobiology of Learning and Memory* 80: 268–284.
Zeitzer JM, Nishino S, and Mignot E (2006) The neurobiology of hypocretins (orexins), narcolepsy and related therapeutic interventions. *Trends in Pharmacological Sciences* 27: 368–374.

Dopamine

P M Fuller and J Lu, Harvard Medical School, Boston, MA, USA

Introduction

The traditional account of the central dopaminergic system includes the important role for dopamine (DA), a catecholamine neurotransmitter, in the regulation of a myriad of neurobiologic, physiologic, and pathophysiologic processes, including cognition, motivation, memory, salience detection, motor disturbances of Parkinson's disease (PD), depression, schizophrenia, and hypophyseal function. The central dopaminergic system comprises three major and well-described tracts, all of which originate in ventral mesencephalic neurons. The mesolimbic and mesocortical pathways both originate in DA neurons of the ventral tegmental area (VTA) and substantia nigra pars compacta (SNc) and project to structures in the ventral striatum, hypothalamus, nucleus accumbens and other limbic structures, and the prefrontal association cortex. These projections are implicated in mediating many motivated behaviors, addiction, and salience detection. The nigrostriatal pathway connects the substantia nigra with the striatum (caudate and putamen) and is implicated in movement and the pathogenesis of PD. In addition to the three major central dopaminergic tracts described here, several other dopaminergic systems exist, including the diencephalic A11–A15 cell groups, which include the tuberoinfundibular pathway that regulates the release of prolactin from the anterior pituitary; the retinal interplexiform cells (A16); the periglomerular cells of the olfactory bulb (A17); and the wake-active DA neurons located in the ventral periaqueductal grey (vPAG) (**Figure 1**).

In contrast to other monoaminergic and cholinergic systems, the central DA system has historically been ascribed only a limited role in the regulation of sleep–wakefulness and cortical arousal. Similarly, in the past, DA has been assigned only a marginal role in mediating the effect of wake-promoting therapeutics. Here, we review recent work, including our own, that has challenged this popular conception. In addition to firmly establishing an important role for DA in the regulation of both sleep–wake behavior and electrocortical arousal, this work has identified, for the first time, the neuroanatomical locus of wake-active DA neurons.

Sleep Regulation: Wake-Active Dopamine Neurons

The role of DA in the regulation of sleep–wakefulness as well as the neuroanatomical locus of wake-active DA neurons has only recently been established. As described previously, the DA neurons of origin for the mesocorticolimbic and nigrostriatal projections are distributed in several midbrain areas (designated A8–A10) that include the VTA (A10), the SN (A9, including the pars reticulata and pars compacta), and the retrorubral area (A8). Whereas the A8 and A9 cell groups form the origin of the projections to the striatum (predominantly the caudate and putamen), the A10 group projects most heavily to limbic forebrain areas, including the medial prefrontal cortex (PFC) and septal area. Because A10 (and to some extent A9) neurons respond to alerting stimuli, mesocorticolimbic DA transmission originating in these cell groups has been implicated in both behavioral and EEG arousal. Yet, surprisingly, neurotoxic lesions of the A10 cell group do not decrease behavioral wakefulness. Moreover, recording studies have revealed that the firing patterns of mesencephalic DA neurons do not correlate with overall levels of behavioral wakefulness. By contrast, all other monoaminergic cell groups exhibit robust state-dependent activation. Although these results seem to suggest a limited role for midbrain DA neurons and mesocorticolimbic DA transmission in sleep regulation, an important role for DA in the regulation of sleep–wakefulness is nevertheless indicated by several findings. First, it has been demonstrated that DA transporter (DAT) knockout mice have approximately 20% more wakefulness than control mice and are refractory to psychostimulant-induced (presumably DA-mediated) arousal. A similar phenotype (i.e., increased wakefulness) results from mutation of the *Drosophila* DAT gene. Second, administration of exogenous dopaminomimetics affects the sleep–wake state in a complex dose- and receptor-dependent manner. For example, in general, lower dopaminomimetic doses have a soporific effect, presumably mediated by presynaptic D2-like inhibitory autoreceptors, whereas higher doses (which typically promote an attendant increase in locomotor activity and suppress rapid eye movement (REM) sleep) enhance arousal, likely via postsynaptic D1-like receptors. Third, patients with PD who have extensive loss of DA neurons in the SN (A9), and less extensive loss of DA neurons in the VTA (A10), often demonstrate

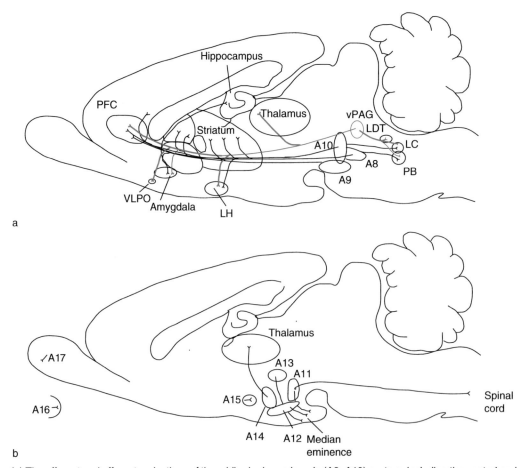

a

b

Figure 1 (a) The afferent and efferent projections of the midbrain dopaminergic (A8–A10) system, including the ventral periaqueductal area (vPAG) dopaminergic neurons. The wake-active vPAG DA neurons are reciprocally connected with several components of the sleep regulatory and arousal systems, including the ventrolateral preoptic nucleus (VLPO), the locus coeruleus (LC), the pontine laterodorsal tegmental nucleus (LDT), the lateral hypothalamic (LH) orexin neurons, the midline and intralaminar thalamus, the basal forebrain (BF) cholinergic cells, and the prefrontal cortex (PFC). (b) The projections of the diencephalic dopaminergic system, including the descending subparafascicular thalamus (A11) projection to the dorsal horn at all levels of the spinal cord.

excessive daytime sleepiness (EDS), which is made worse by D2 receptor agonist (which activates inhibitory presynaptic autoreceptors on DA neurons and therefore inhibits the firing of DA neurons). Finally, arousal and waking behaviors are associated with increased forebrain DA secretion.

Work has uncovered a previously unrecognized group of wake-active DA neurons in the vPAG that may provide the long-sought ascending dopaminergic waking influence (**Figure 2**). In this study, 6-hydroxydopamine-induced lesions of vPAG DA neurons (sparing intermingled dorsal raphe serotonergic neurons, the VTA, and SNc) resulted in an increase in total daily sleep (non-REM and REM) by approximately 20%. Importantly, the magnitude of this increase in sleep seen following vPAG DA lesions is significantly larger than that produced by lesions of other monoaminergic and cholinergic cell groups

thought to be important in arousal, including the locus coeruleus (LC), lateral dorsal tegmentum (LDT), dorsal raphe (DRN), lateral hypothalamic (LH) orexin neurons, basal forebrain cholinergic neurons, and the histaminergic tuberomammillary nucleus (TMN). Interestingly, DA neurons in the vPAG have been identified previously in humans and rats. Yet because the vPAG DA neurons share many efferent projections with A10 DA neurons, such as the ventral striatum and PFC, they have long been considered a rostral extension of the A10 DA group. Anatomical and physiological characteristics of the vPAG DA neurons suggest, however, that these cells constitute a functionally distinct neuronal population. For example, unlike VTA DA neurons, vPAG DA neurons project heavily to the central and extended nucleus of the amygdala as well as to the ventrolateral preoptic nucleus (VLPO), a critical

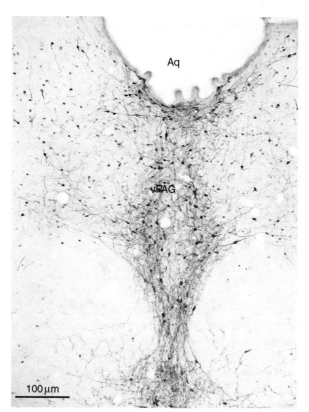

Figure 2 Photomicrograph showing the distribution of DA neurons in the ventral periaqueductal area (vPAG). Aq, aqueduct.

sleep-promoting center. These vPAG DA neurons also project heavily to (and receive reciprocal innervations from) most of the major components of the sleep–wake and arousal systems, including the VLPO, basal forebrain, LH orexin cells, LC, the LDT cholinergic cells, and the midline and intralaminar thalamus. Also, as mentioned previously, unlike the vPAG DA neurons, VTA DA neurons do not demonstrate state-dependent changes in firing. Taken together, these findings indicate that vPAG DA neurons contain a functionally distinct group of DA neurons which likely form the origin of a potent dopaminergic arousal system. Thus, for example, loss/dysfunction of vPAG DA neurons may underlie EDS in PD. Although it remains unclear how DA influences sleep regulation and arousal, it is likely through both inhibition of sleep active neurons in the VLPO and activation of the basal forebrain and monoaminergic systems (i.e., as part of the extrathalamic cortical arousal system).

A role for DA in REM sleep (also called 'paradoxical sleep' and 'active sleep') control has also been indicated by several studies. REM sleep is a behavioral state characterized by activation of the cortical and hippocampal EEG, REM, and muscle atonia, although the neurobiologic significance of this behavioral state remains unresolved. Nevertheless, electrophysiological work has revealed increased bursting activity of VTA (A10) DA neurons during REM sleep. DA levels are also elevated in the PFC during REM sleep. It remains unclear, however, if the A10 bursting is linked to the increase in DA in the PFC. Additional work suggests that DA may also directly regulate the activity of REM active neurons located in the peri-LC alpha. Specifically, Sakai and colleagues have shown that microdialysis of DA in the peri-LC alpha inhibits REM sleep in cats, likely via adrenergic α_2 receptors. Finally, numerous cholinergic receptors have been found in the midbrain DA neurons, indicating that PPT cholinergic neurons may influence the midbrain DA system, particularly during wake and REM sleep states. Collectively, these findings provide further evidence linking changes in the activity of DA neurons with changes in behavioral state.

Dopaminergic Disorders and Sleep: PD

Patients with PD suffer a progressive loss of DA neurons, largely in the SNc and VTA, leading to a marked reduction in DA content in the basal ganglia and ultimately manifesting in motor abnormalities that include akinesia, rigidity, resting tremor, and postural instability. In addition to motor disturbances, PD patients often complain of sleep disturbances ranging from sleep fragmentation to abnormal motor activity and EDS. These problems tend to worsen with disease progression. During the early stages of PD, dopaminomimetic drugs such as L-3,4-dihydroxyphenyalamine (L-DOPA) can treat these sleep disturbances; unfortunately, patients become refractory to L-DOPA treatment during the more advanced stages of PD.

Ascertaining the contribution of central DA dysfunction to the sleep disturbances of PD is complicated by several factors. First, patients with PD also exhibit neurochemical changes in cholinergic and monoaminergic systems, both of which are implicated in sleep regulation. Second, medications used to treat PD may alter sleep. Third, nocturnal tremors and inappropriate phasic motor bursting may produce fragmented sleep. Nevertheless, several compelling lines of evidence provide support for the concept that DA dysfunction significantly contributes to the pathological sleep disturbances of PD, the most common of which are EDS and nocturnal sleep disruption in the form of involuntary motor disturbance (e.g., REM sleep behavior disorder (RBD)).

Significant hypersomnia in PD, manifesting as EDS, is common and yet underrecognized with respect to diagnosis and treatment. Clinic-based objective measurements of sleepiness employing the standardized Multiple Sleep Latency Test has revealed a high rate of EDS (approximately 20–50%) in PD patients. As indicated previously, the determinant(s) of EDS in PD is difficult to resolve because EDS is likely secondary to severe sleep fragmentation in PD, which itself may be attributable to other motor (e.g., abnormal nocturnal movements) or respiratory (e.g., central/obstructive sleep apnea) disturbances that accompany the pathology of PD.

Perhaps a more insidious form of sleep disruption that has been tacitly linked to central DA dysregulation is RBD. RBD is a parasomnia that typically manifests as 'dream enactment' behavior – that is, involuntary nocturnal movements that include kicking, punching, shouting, and screaming during REM sleep. RBD may represent an early pathophysiologic manifestation of evolving PD and other Lewy body diseases (LBDs) (e.g., dementia with Lewy bodies, multiple systems atrophy, and pure autonomic failure). Indeed, RBD typically manifests a decade prior to the motor and cognitive sequela of PD, and thus the diagnosis of RBD may provide an early therapeutic window for delaying or preventing the full development of PD.

Of note, because Lewy bodies and Lewy neurites are composed of α-synucleinopathies, these disorders are termed 'synucleinopathies.' Although the brain stem is clearly implicated in RBD pathogenesis, the identity of the neural networks that become dysfunctional to manifest RBD is unknown. Because approximately 50% of people with RBD develop PD or dementia with Lewy bodies and the majority of PD patients have RBD, it has been suggested that the intrinsic pathology of PD (i.e., severe nigrostriatal DA neuron loss) may contribute to the development of excessive nocturnal movement, particularly RBD. Despite the intuitive appeal of this hypothesis, however, studies have linked RBD to dysfunction of brain stem REM sleep circuitry. These structures include the subcoeruleus region (SC; equivalent to the sublateral dorsal nucleus (SLD) in rats) and, possibly, the intermediate region of the ventromedial medulla; both structures appear critical for generating atonia during REM sleep. SC/SLD dysfunction, as the neuropathologic substrate for RBD, is seemingly consistent with the temporal pattern of neuronal degeneration in PD and other LBDs, which starts in the brain stem and includes the coeruleus–subcoeruleus complex in stages I and II of PD and progresses inexorably rostrally toward the forebrain. The temporal pattern of lesions (i.e., caudal-to-rostral progression) is also consistent with RBD (secondary to SC/SLD degeneration) as an early manifestation of these neurodegenerative conditions. Nevertheless, although SC/SLD dysfunction has been demonstrated in PD, it remains unclear if RBD in evolving PD is caused by SC/SLD degeneration or loss of critical inputs such as DA nigrostriatal projections.

Dopamine and Stimulant-Induced Arousal

Drugs such as amphetamines and amphetamine-like stimulants (e.g., cocaine, methylphenidate, and methamphetamines) have potent wake-promoting effects. These drugs are thought to produce their arousal effects by blocking DA reuptake/transport (i.e., acting on the cell membrane DAT and/or stimulating DA release), resulting in increased synaptic DA concentrations. Determining DA's contribution to behavioral arousal in this context has nevertheless been complicated by the fact that psychostimulant administration also promotes the synaptic accumulation of other monoamines, particularly norepinephrine (NE). This is particularly true at higher doses, where amphetamines interact with vesicular monoamine transporter-2 to increase the cytoplasmic pool of monoamines. In fact, for many years, the wake-promoting effect of amphetamine-like stimulants was attributed almost exclusively to NE mechanisms since adrenergic signaling is known to modulate arousal state. Further reinforcing this notion is the fact that LC neurons (the major source of brain NE) display robust state-dependent activity and discharge most rapidly during enhanced arousal. Although, taken together, these observations suggest NE signaling might form the molecular basis for the wake-promoting actions of psychostimulants, several studies have suggested a more dominant role for DA than NE in this regard. Centrally acting DA antagonists, for example, cause drowsiness, and drugs that selectively block the DAT are more effective in promoting wakefulness than drugs that selectively block the NE transporter (NET). Indeed, both desipramine and nisoxetine, two potent and selective NET blockers, have minimal effects on EEG arousal in narcoleptic canines (even at high doses which suppress REM sleep). The wake-promoting effect of amphetamine-like stimulants is also abolished in mice with deletion of the DAT gene. Conversely, the wake-promoting effect of amphetamines is preserved following lesions of the LC (and hence a dramatic reduction in central NE) in cats and following chemical ablation of NET-bearing NE forebrain projections from the LC in mice. The potency of most wake-promoting psychostimulants is also best predicted by binding affinity to DAT. Finally, elevated arousal levels correlate with increases in synaptic DA but not NE. Taken together, these findings support the concept that presynaptic modulation of DA transmission (at least at the level of the forebrain) is the key pharmacological mechanism by which amphetamines and their derivatives mediate cortical and behavioral arousal.

The anatomical substrates for DA's effects on arousal also remain unresolved, although several lines of experimental evidence suggest this may occur through inhibition of sleep-promoting systems. For instance, as previously indicated, a mutually inhibitory circuit between wake-active vPAG DA neurons and sleep-active neurons of the VLPO has been elucidated. Moreover, *in vitro* administration of DA inhibits neurons of the VLPO. Curiously, however, DA-mediated inhibition of the VLPO is blocked by α_2 receptor antagonists but not by D2 receptor antagonists. Although these findings appear difficult to reconcile, they likely indicate, at a minimum, cross-talk between these central nervous system catecholaminergic systems. Thus, for example, the drug modafinil (which shares the wake-promoting properties of traditional psychostimulants) hyperpolarizes VLPO neurons *in vitro*, presumably by blocking the NET. Yet DAT knockout mice are unresponsive to modafinil treatment, suggesting instead that DA neurotransmission (i.e., blockade of the DAT) underlies the wake-promoting effect of modafinil. Combined, these findings suggest a possible dual mode of action for modafinil (i.e., interference with both NE and DA uptake), and this may explain why modafinil can exert its wake-promoting effects without inducing dopaminergic side effects such as addiction. Indeed, data indicate that modafinil may produce its arousal effects through dopaminergic-dependent adrenergic signaling. For example, DA may function as a physiological ligand at adrenergic receptors. In support of this concept, DA stimulation of NE receptors has been documented previously in the pontine brain stem. However, the mechanism of action by which DA produces its effects on cortical and behavioral arousal remains unresolved.

In addition to inhibiting VLPO neurons, amphetamine-like stimulants may also promote wakefulness by activating, for example, the wake-promoting basal forebrain cholinergic neurons and LH orexin neurons. To this end, DA has been reported to stimulate cholinergic cells in the basal forebrain by D1-like receptors (i.e., D1 and D5) *in vitro*. Research has also uncovered an extended network of basal forebrain and preoptic sites that mediate amphetamine-induced increases in behavioral and electrocortical arousal. Remarkably, however, although the anatomical resolution of this mapping study was limited, close inspection of the data reveals that infusion of amphetamine into the region best approximating the location of the VLPO produced one of the largest increases in arousal, providing further support for the concept that the VLPO may be a critical structure for mediating the arousal-promoting effects of amphetamines and amphetamine-like stimulants.

Summary

Work during the past decade has firmly delineated an important role for DA in sleep–wake regulation (both normal and pathologic) and behavioral and electrocortical arousal. Wake-active DA neurons of the vPAG likely exert their potent arousal influence through a mutually inhibitory interaction with the VLPO as well as less well-defined interaction with components of the ascending arousal system (e.g., basal forebrain, LC, and lateral hypothalamus). Based on recent findings, it is tempting to speculate that alterations in DA neurotransmission may underlie the excessive sleepiness of evolving PD as well as the manifestation of abnormal nocturnal movements in the form of RBD. Finally, a critical role for DA in mediating the wake-promoting effects of psychostimulants has begun to emerge, although the indirect or direct mechanism by which this occurs remains to be clarified.

See also: Acetylcholine; Cytokines and other Neuromodulators; Histamine; Hypothalamic Regulation of Sleep; Locus Coeruleus and Raphe Nucleus; Thalamic Regulation of Sleep.

Further Reading

Aston-Jones G and Bloom FE (1981) Activity of norepinephrine-containing locus coeruleus neurons in behaving rats anticipates fluctuations in the sleep–waking cycle. *Journal of Neuroscience* 1: 876–886.

Berridge CW, O'Neil J, and Wifler K (1999) Amphetamine acts within the medial basal forebrain to initiate and maintain alert waking. *Neuroscience* 93: 885–896.

Crochet S, Onoe H, and Sakai K (2006) A potent non-monoaminergic paradoxical sleep inhibitory system: A reverse microdialysis and single-unit recording study. *European Journal of Neuroscience* 24(5): 1404–1412.

Dahan L, Astier B, Vautrelle N, Urbain N, Kocsis B, and Chouvet G (2007) Prominent burst firing of dopaminergic neurons in the ventral tegmental area during paradoxical sleep. *Neuropsychopharmacology* 32(6): 1232–1241.

Gallopin T, Luppi PH, Rambert FA, Frydman A, and Fort P (2004) Effect of the wake-promoting agent modafinil on sleep-promoting neurons from the ventrolateral preoptic nucleus: An *in vitro* pharmacologic study. *Sleep* 27: 19–25.

Jones BE, Harper ST, and Halaris AE (1977) Effects of locus coeruleus lesions upon cerebral monoamine content, sleep–wakefulness states and the response to amphetamine in the cat. *Brain Research* 124: 473–496.

Kume K, Kume S, Park SK, Hirsh J, and Jackson FR (2005) Dopamine is a regulator of arousal in the fruit fly. *Journal of Neuroscience* 25: 7377–7384.

Lu J, Chou TC, and Saper CB (2006) Identification of wake-active dopaminergic neurons in the ventral periaqueductal gray matter. *Journal of Neuroscience* 26(1): 193–202.

Miller JD, Farber J, Gatz P, Roffwarg H, and German DC (1983) Activity of mesencephalic dopamine and non-dopamine neurons across stages of sleep and waking in the rat. *Brain Research* 273: 133–141.

Wisor JP and Eriksson KS (2005) Dopaminergic–adrenergic interactions in the wake promoting mechanism of modafinil. *Neuroscience* 132: 1027–1034.

Wisor JP, Nishino S, Sora I, Uhl GH, Mignot E, and Edgar DH (2001) Dopaminergic role in stimulant-induced wakefulness. *Journal of Neuroscience* 21: 1787–1794.

SLEEP AND AROUSAL STATES

Reticular Activating System

E Garcia-Rill, University of Arkansas for Medical Sciences, Little Rock, AR, USA

Introduction

Sleep and arousal states are characterized in the electroencephalogram (EEG) by synchronization of slow rhythms during slow-wave sleep and synchronization of fast rhythms during waking and rapid eye movement (REM) sleep. These rhythms are generated by reverberating oscillations between the thalamus, hypothalamus, and basal forebrain and the cortex. The reticular activating system (RAS) serves to modulate these oscillations (ascending inhibition or decreased excitation will serve to slow these rhythms, whereas ascending excitation or disinhibition will speed up rhythms) depending on sensory inputs and ongoing activity. In addition, descending modulation of motor neurons serves to decrease muscle tone during REM sleep (atonia) or prime the system to fight or flee during waking by modulating the startle response.

Wiring Diagram

RAS is composed of three main cell groups: the pedunculopontine nucleus (PPN) (and its medial partner, the laterodorsal tegmental nucleus, both of which are mainly cholinergic), the locus coeruleus (LC; mainly noradrenergic), and the raphe nucleus (RN; mainly serotonergic). RAS receives input from all afferent sensory systems in parallel to primary afferent sensory projections. That is, the 'nonspecific' projection system through RAS relays 'arousal' information via the intralaminar thalamus (ILT) to the cortex. This system functions in parallel with the flow of 'specific' sensory input through the primary sensory thalamic nuclei that is then relayed to the cortex. It is the temporal summation of ILT nonspecific inputs with primary sensory specific inputs at the level of the cortex that is thought to participate in the conscious perception of a sensory event. Disturbances in RAS driving of the ILT and thalamocortical oscillations thus can be expected to lead to disturbances in perception.

RAS also sends descending projections to posture and locomotion systems that act as the 'fight-or-flight' system, simultaneously activating higher centers while priming motor systems to respond appropriately to sudden stimuli. These descending projections modulate the pontine inhibitory area (PIA) that controls the atonia of REM sleep, pontine neurons that generate the startle response (a flexor response that primes the motor system), and reticulospinal systems that drive locomotion. RAS, especially cholinergic, projections to the dorsal subcoeruleus region also trigger the generation of ponto-geniculo-occipital (PGO) waves during REM sleep. This region generates high-frequency bursts of activity (such as those required for long-term potentiation) that have been proposed to participate in sleep-dependent plasticity.

The hypothalamic sleep–wake modulatory system is composed mainly of the tuberomammillary nucleus (TMN; mainly excitatory histaminergic), the lateral hypothalamus (LH; mainly excitatory orexinergic), and the ventrolateral preoptic nucleus (VLPO; mostly inhibitory GABAergic). These hypothalamic systems are thought to help stabilize sleep–wake states. Excitatory LH orexin projections can be thought of as an 'on' switch promoting waking, particularly through activation of excitatory TMN neurons, which are tonically active during waking (and decrease firing during sleep), and of excitatory basal forebrain and RAS neurons. The basal forebrain cholinergic system is mainly active during waking and serves to raise the excitability of the cortex, although acetylcholine release from the basal forebrain is greater during REM sleep than during waking. Inhibitory VLPO GABAergic projections can be thought of as an 'off' switch promoting slow-wave sleep through inhibition of RAS, LH, basal forebrain, and cortex.

Figure 1 outlines the main projections described previously. The dorsal ascending cholinergic (PPN) and monoaminergic (LC and RN) projections from RAS to ILT serve to activate the cortex via thalamocortical projections. There is also a ventral projection system that bypasses the thalamus to terminate diffusely throughout the cortex that originates in RAS (noradrenergic LC and serotonergic RN) and ascending projections from TMN, LH, and VLPO (**Figure 1, VP**), as well as from the basal forebrain (not shown). In turn, TMN, LH, and VLPO send descending projections to RAS that may act reciprocally to stabilize sleep–wake states.

In addition, adenosine (**Figure 1, A**; diffusely localized) is a ubiquitous homeostatic factor believed to be involved in sleep–wake regulation. High metabolic activity or prolonged wakefulness leads to a buildup of adenosine, which decreases subsequent to sleep. Adenosine may also modulate sleep–wake states via its inhibitory actions on most cells. Adenosine injections into RAS are known to decrease waking, whereas adenosine levels in the basal forebrain (but not in the thalamus) progressively increase during

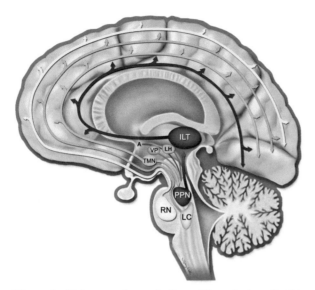

Figure 1   Main ascending projections of the reticular activating system. A, adenosine; ILT, intralaminar thalamus; LC, locus coeruleus; LH, lateral hypothalamus; PPN, pedunculopontine nucleus; RN, raphe nucleus; TMN, tuberomammillary nucleus; VP, ventrolateral preoptic nucleus.

prolonged wakefulness and decrease during subsequent recovery of sleep.

Figure 2 shows the connectivity of neurons within RAS and their firing patterns during sleep and waking. RN inhibits PPN and LC, and LC inhibits PPN, whereas PPN activates LC. During waking, all three cell groups are active, but in slow-wave sleep the cholinergic cells decrease firing while monoaminergic cells remain active. However, in REM sleep, the cholinergic cells are highly active while monoaminergic cells decrease their firing rates markedly, thereby disinhibiting cholinergic cell groups. Thus, cholinergic RAS neurons are active during waking and REM sleep (i.e., during synchronization of fast cortical rhythms) but slow their firing during synchronization of slow cortical rhythms in slow-wave sleep.

Intrinsic Properties

Neurons in the region of PPN have been reported to show 'wake-on,' 'REM-on,' and 'wake-REM-on' firing patterns whose properties are implied by their names. It is not clear how these patterns correspond to intrinsic membrane properties. There are three types of PPN neurons based on intrinsic membrane properties: type I have low-threshold spike (LTS) calcium-dependent inward currents, are noncholinergic, and tend to burst especially when released from hyperpolarization; type II have Ia transient outward potassium currents, are mostly cholinergic, and tend to fire slowly and tonically; and type III have both

LTS and Ia currents, are mainly noncholinergic, and fire tonically and in bursts. A large proportion of these cells also have hyperpolarization-activated Ih currents that tend to restore the membrane potential following hyperpolarization. Evidence suggests that at least some neurons in PPN, possibly GABAergic, are electrically coupled, promoting ensemble activity. Cholinergic RAS neurons are active tonically during waking depending on the level of depolarization. In a slightly hyperpolarized state, some of these cells show LTS bursts, perhaps corresponding to the bursting activity observed during REM sleep. Then, further hyperpolarization leads to cessation of firing, as in slow-wave sleep. During moderate inhibition, PPN neurons thus may undergo a change from tonic to bursting activity such as that observed in waking versus REM sleep.

LC and RN cells fire relatively slowly during slow-wave sleep, cease firing during REM sleep (disinhibiting PPN cells), and are most active and variable during waking, particularly in response to novel stimuli. LC neurons show electrotonic coupling, which allows them to act as a group. Therefore, a novel stimulus could elicit a synchronized response from this population of neurons. However, recurrent inhibition follows such activation so that the initial response is quickly followed by inhibition of further output. RN cells are similar to LC neurons in that many of the conductances of these neurons tend to limit the firing rate (e.g., long-duration action potential and high-amplitude afterhyperpolarizations). It has been postulated that during waking serotonin inhibits REM-on neurons without affecting wake-REM-on neurons, thereby allowing synchronization of fast rhythms; however, these slow down during REM sleep, thus disinhibiting REM-on, presumably bursting, neurons to promote REM sleep. In general, noradrenergic and serotonergic neurons have high input resistance so that relatively small currents (such as those elicited by marginal inputs) will elicit considerable changes in membrane potential. There are also both excitatory and inhibitory interneurons (GABAergic) and projection (glutamatergic) neurons intermingled within all RAS cell groups.

Descending cholinergic projections to the subcoeruleus region affect several types of neurons, including cells with LTS currents, Ia currents, and LTS and Ia currents – similar to the types found in PPN. Electrical coupling appears to be more common in this region, mainly among GABAergic neurons that are induced to fire rhythmically at theta rhythm by cholinergic agonists. Such rhythmic coherent activation may underlie the mechanism behind REM sleep and perhaps PGO wave generation.

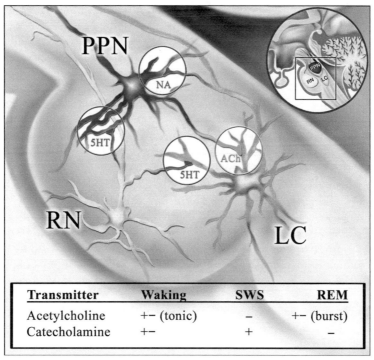

Transmitter	Waking	SWS	REM
Acetylcholine	+− (tonic)	−	+− (burst)
Catecholamine	+−	+	−

Figure 2 Wiring diagram within the main cell groups of the reticular activating system. LC, locus coeruleus; PPN, pedunculopontine nucleus; RN, raphe nucleus; SWS, slow-wave sleep.

Ascending Effects

Waking and the 40 Hz Rhythm

The waking state is characterized by fast EEG oscillations in the 15–60 Hz range. The waking EEG is of low amplitude and composed of synchronization of fast rhythms or of high frequencies >15 Hz. Faster rhythms appear upon arousal and occur mainly in the 35–45 Hz range during selective attention to sensory input, referred to as the '40 Hz rhythm.' This rhythm is thought to be generated by cortical neurons receiving inputs from thalamocortical specific and nonspecific nuclei, such that the specific inputs to cortex provide the content of cognition (via input through primary sensory pathways), whereas the nonspecific inputs provide the context (via ILT), with their coherence providing the temporal binding required for the unity of cognitive experience.

The mechanism behind the 40 Hz rhythm appears to include layer IV cortical inhibitory interneurons, which exhibit a sequential activation of a persistent low-threshold sodium current followed by a subsequent potassium conductance. These cells can transmit the 40 Hz rhythm onto thalamic cells, allowing the entrainment of thalamocortical resonance. This 'ringing' is thought to be the basis of conscious perception so that for a transient sensory input to be perceived

or consciously appreciated, it must generate a lasting reverberation or reentrant signaling in thalamocortical systems. RAS modulates these oscillations through its ascending projections (**Figure 1**).

The process of awakening entails two stages: a rapid (5 min) reestablishment of consciousness that is marked by increases in cerebral blood flow in RAS and thalamus, followed by a slower (15 min) increase in cerebral blood flow, primarily in anterior cortical regions. It should also be noted that cholinergic RAS neurons have some of the highest concentrations of nitric oxide synthase in the brain. Therefore, wherever PPN projects, a corollary effect may include vasodilation and other processes modulated by nitric oxide.

Slow-Wave Sleep

This state is characterized by high EEG with synchronization of low frequencies <15 Hz. These oscillations include spindles (7–14 Hz), delta waves (1–4 Hz), and slow oscillations (<1 Hz). This state is marked by decreases in cerebral blood flow throughout the brain, especially in RAS, thalamus, hypothalamus, and basal forebrain. It is characterized by decreased activity of cholinergic RAS, basal forebrain neurons, LH orexinergic cells, and TMN histaminergic cells.

The main excitatory projection systems decrease their outputs during slow-wave sleep.

Spindles Spindles are present early in sleep and the inhibition they promote basically deafferents the cerebral cortex (primary sensory pathways are gated), which may be a prerequisite for falling asleep. The reticular nucleus of the thalamus is thought to be the main cell group generating spindles, which result from bursting in a syncitium of electrically coupled reticular nucleus neurons that produce synchronized inhibitory postsynaptic potentials in thalamocortical neurons. These lead to postinhibitory rebound bursts that are relayed to cortex, producing slow, rhythmic, excitatory postsynaptic potentials. Spindles begin to occur as brain stem cell groups, particularly cholinergic RAS neurons, diminish their firing rates, disinhibiting reticular nucleus neurons.

Delta waves Delta waves increase gradually as slow-wave sleep deepens. These waves are generated by cortical and thalamocortical neurons that are synchronized by inhibitory reticular nucleus neurons and scattered dorsal thalamic cells. They represent waveforms elicited at a more hyperpolarized membrane potential than spindles so that as slow-wave sleep deepens, the occurrence of spindles decreases, whereas that of delta waves increases. These waves mark the virtual cessation of activity in mesopontine cholinergic cells.

Slow waves Slow waves appear to be generated spontaneously in cortical neurons, but this rhythm is also present in thalamic reticular nucleus neurons. All major classes of cortical cells display slow waves. The slow depolarization of these neurons corresponds to the depth-positive EEG wave, whereas their hyperpolarization corresponds to the depth-negative EEG wave. Because of the widespread synchronization of slow waves in cortical neurons, it is likely that thalamic neurons, specifically inhibitory reticular thalamic cells, help entrain these slow oscillations.

REM Sleep

Every 90 min during sleep, slow rhythms are interrupted by fast oscillations triggered by cholinergic mesopontine cell groups that induce REM sleep. The EEG displays low-amplitude synchronization of fast oscillations identical to those in the waking state. This 'paradoxical' state is accompanied by eye movements, contains most of our dreaming time, and leads to a decrease in skeletal muscle tone or atonia. The atonia is particularly strong on antigravity muscles, whereas the diaphragm and extraocular muscles

retain substantial tone. REM sleep is characterized by increased blood flow in RAS, thalamus, anterior cingulate, and other cortical areas, with decreases in blood flow in the dorsolateral prefrontal cortex. It has been proposed that the unregulated (by other RAS nuclei) activity of RAS cholinergic cells is responsible for REM sleep and, via unknown mechanisms, for decreased frontal lobe blood flow. The 'hypofrontality' of REM sleep may account for the lack of critical judgment during dreaming (and during hallucinations).

In advance of and during REM sleep, paroxysmal events called PGO waves can be detected in the brain stem, the lateral geniculate nucleus, and occipital cortex, and they appear to be due to bursting activity in brain stem nuclei, including the PPN and especially the subcoeruleus. Interestingly, the 40 Hz cortical oscillation evident in waking is also present during REM sleep; however, it does not appear to be reset by sensory input, despite the fact that specific primary sensory thalamic systems are not gated. This suggests that the intrinsic activity of the brain does not place sensory input in the context of the functional state being generated by the brain. That is, REM sleep may be a state of intrinsic activity in which external sensory input cannot access thalamocortical reverberation or conscious experience.

It should be noted that transitions in state from waking to slow-wave sleep, or from slow-wave sleep to REM sleep, are gradual. During these transitions, waveforms characteristic of more than one state appear intermingled in the EEG. The frequency of these events increases until the new state is fully expressed. It appears that new states are gradually recruited rather than elicited instantaneously.

Muscle Tone and Locomotion

RAS is a phylogenetically conserved system that modulates fight-or-flight responses. During waking, our ability to detect predator or prey is essential to survival. RAS is linked to the motor system in order to optimize attack or escape. During REM sleep, atonia keeps us from acting out our dreams. Only our diaphragm and eye muscles appear to be acting out dream content. During waking, RAS can modulate muscle tone and locomotion via reticulospinal systems. For example, in a standing individual, there is tonic activation of antigravity, mainly extensor, muscles. Before the first step can be taken, there must be flexion (i.e., a release from extensor bias). The startle response, a rapid response to a supramaximal stimulus, is basically an induced flexor response, placing the body in a 'ready' position. The startle response is composed of a short latency activation of muscle activity (the ready condition), followed by a brief inhibition (the 'reset' state) and then a long latency activation (the 'go' condition). The intermediate latency inhibition is thought to be part of the modulation of the startle response by cholinergic RAS neurons, and it may represent a 'resetting' of motor programs which allow the subsequent selection of response strategies – the triggering of attack or escape movements.

There appear to be multiple descending pathways by which RAS modulates muscle tone and locomotion. PIA can be activated by cholinergic agonists to induce REM sleep with atonia and PGO waves. Lesions of this region produce an animal exhibiting REM sleep without atonia, although such lesions may damage passing axons. Presumably, outputs from this region activate reticulospinal systems that lead to profound hyperpolarization of motor neurons, which is the mechanism responsible for the atonia of REM sleep. Cholinergic projections to the medioventral medulla appear to elicit increases in muscle tone and controlled locomotion. Outputs from this region activate reticulospinal systems that trigger spinal pattern generators to induce stepping. Electrical stimulation of the pontine and medullary reticular formation is known to induce decreased muscle tone at some sites while producing stepping movements at other sites. This suggests the presence of a heterogeneous, distributed system of muscle tone and locomotor control.

Neuropharmacology of Sleep and Arousal

Alcohol

Direct effects of alcohol on sleep–wake control regions appear to involve potentiation of GABAergic transmission. Significant impairment in motor performance, such as driving, occurs at very low blood alcohol concentrations, an effect potentiated by sleep deprivation or sleep loss. Most alcoholic patients suffer from insomnia, which is clinically important since alcoholism can exacerbate the adverse consequences of insomnia, such as mood changes and anxiety, and because insomnia has been associated with alcohol relapse.

Anesthetics and Sedatives

The proposed mechanisms of action of anesthetics have typically involved a number of cellular effects at different sites. Evidence suggests that the primary site of action of most anesthetics may be the sleep–wake control system. The parallel manifestations between sleep and anesthesia suggest that anesthetics basically 'hijack' the sleep–wake control system to induce anesthesia.

Most anesthetics, including barbiturates, etomide, propofol, neuroactive steroids, and volatile anesthetics, act on GABA$_A$ receptors, among others. Sedation and natural sleep occur mostly as a result of enhanced GABAergic transmission, which in turn affects the release of a number of excitatory transmitters, such as acetylcholine, excitatory amino acids, and histamine. These actions may take place specifically in such regions as RAS, TMN, and basal forebrain (all of which have local circuit GABAergic neurons and receive GABAergic input from VLPO), thereby regulating the level of arousal.

Interestingly, halothane and propofol appear to block gap junctions, and oleamide, which promotes sleep, is also known to block electrical coupling. Anandamide, which enhances adenosine levels to promote sleep, can block gap junctions. Carbenoxolone, a putative gap junction blocker, is known to decrease gamma band oscillations (high-frequency EEG), whereas connexin 36 (the neuronal gap junction protein) knockout mice exhibit low gamma band power. These findings all support a role for electrical coupling via gap junctions in the control of sleep–wake states and their modulation by certain anesthetic agents.

The benzodiazepines act by binding to a site that modulates GABA receptors, especially GABA$_A$ receptors. These agents produce sedative, hypnotic, anxiolytic, and anticonvulsant activities. They act generally by amplifying GABAergic transmission, such that short-acting agents have been used to promote sleep in insomnia patients.

Antihistamines

Pathology and lesions of TMN cause hypersomnia (these neurons are highly active during waking). Histaminergic inputs to RAS suppress slow-wave sleep and promote waking, although they do not affect REM sleep significantly. Administration of antihistamines (histamine receptor blockers) results in sedation due to blockade of histaminergic inputs to RAS, basal forebrain, and/or LH.

Stimulants

The popularity of caffeine is related to its stimulant properties, which are mediated by its ability to reduce adenosine release in the brain. Caffeine appears to block adenosine A1 and A2a receptors, producing a psychomotor stimulant effect.

Amphetamine induces the release of monoamines, especially dopamine, but also blocks reuptake and may have neurotoxic effects on nigral neurons. Methylphenidate does not appear to have such neurotoxic effects. Amphetamine psychosis occurs in two forms – acute intoxication after a single large dose (characterized by confusion and disorientation) and chronic abuse that produces a schizophrenia-like syndrome. The increased release of dopamine by amphetamine is considered a model of schizophrenia, and it led to the 'dopamine theory' of schizophrenia.

Modafinil is a newer stimulant that does not appear to act through dopaminergic mechanisms, and it has been found to be effective in the treatment of daytime sleepiness in patients with narcolepsy. Modafinil has also been reported to alleviate spatial neglect and to increase vigilance in stroke patients. This suggests that sensory neglect arising from stroke can be treated successfully by 'waking up' the involved cortex, implying that the consequences of stroke may be to decrease activity, blood flow, and/or metabolism, one or more of which modafinil may counteract.

Development

The human newborn sleeps 16 h per day – 8 h in REM sleep and 8 h in slow-wave sleep. Between birth and the end of puberty, there is a developmental decrease in REM sleep such that the adult has 1 h of REM sleep and 6 or 7 h of slow-wave sleep per day. That is, the decrease in REM sleep is replaced by a postpubertal increase in waking. During development, REM sleep may serve to direct brain maturation, providing endogenous stimulation at a time when the brain has little or no exogenous input. However, REM sleep is not essential for survival in the adult. REM sleep suppression does not lead to death or disease. Evidence suggests that a number of transmitter systems change during the developmental decrease in REM sleep and may modulate that decrease. In addition, levels of the gap junction protein connexin 36 are high before the developmental decrease (10 days in the rat) and are considerably lower at the end of the decrement (30 days in the rat). It thus appears that there is a large syncitium of coupled neurons which segregates into distinct functional networks with age; however, the capacity to revert to a developmental condition may be present in the adult, leading once more to increased REM sleep drive.

Arousal and Sleep Disorders

The connectivity of RAS shown in **Figure 2** can be used to deduce the potential sites of dysregulation related to hypervigilance and increased REM sleep drive that are evident in depression, anxiety disorder, and schizophrenia.

The treatment of depression previously included tricyclic antidepressants that mainly blocked reuptake of noradrenaline and serotonin and blocked histamine and acetylcholine release, thus accounting for increased sleepiness. The selective serotonin reuptake inhibitors more selectively affect RAS by increasing the inhibition of PPN and LC neurons (**Figure 2**). It is not clear if the etiology of depression is related to disinhibition of PPN and LC by a decrement in serotonergic tone, although this would seem a likely origin for the sleep–wake disturbances seen in depression.

The treatment of anxiety disorder has involved the use of benzodiazepine amplification of GABAergic inhibition. In addition, the use of the α_2 receptor agonist clonidine produces anxiolytic effects, probably by inhibiting autoreceptors in LC and postsynaptic receptors in PPN, thus downregulating vigilance. The etiology of anxiety disorder has been proposed to include downregulation or degeneration of LC outputs (possibly induced by stress hormones), which would act to disinhibit PPN neurons.

The treatment of schizophrenia previously involved the use of the dopaminergic receptor blocker haloperidol, which induced tardive dyskinesia, among other side effects. Newer antipsychotics appear to block dopaminergic, noradrenergic, and serotonergic receptors. More striking antipsychotic effects resulted from the use of clozapine, which was designed as a muscarinic cholinergic blocker for the treatment of Parkinson's disease. A later generation agent that has maintained anticholinergic activity without this side effect is olanzapine. These agents appear to have partial penetrance at serotonergic, cholinergic, and dopaminergic receptors, basically reducing muscarinic cholinergic activation of the substantia nigra (SN), as well as partially blocking dopaminergic actions in the striatum. The etiology of schizophrenia has been suggested to include increased PPN output, accounting for marked hypervigilance and hallucinations (REM sleep intrusion into waking). Excessive PPN output would overactivate the SN and, in turn, increase striatal release of dopamine, complying with the dopamine theory of schizophrenia.

Disturbances in RAS function are evident in a number of parasomnias that are discussed in other articles in this encyclopedia. Suffice it to say that RAS dysregulation is part of the psychopathology in a host of disorders and may even presage the onset of some of these disorders.

See also: Adenosine; Dopamine; Histamine; Hypothalamic Regulation of Sleep; Modafini, Amphetamines, and Caffeine; Narcolepsy; Sleep and Consciousness; Sleep in Adolescents; Sleep in Aging; Sleep Oscillations and PGO Waves; Thalamic Regulation of Sleep.

Further Reading

Chase MH and Morales FR (1994) The control of motoneurons during sleep. In: Kryger MH, Roth T, and Dement WC (eds.) *Principles and Practice of Sleep Medicine*, pp. 163–176. London: Saunders.

Datta S and Patterson EH (2003) Activation of phasic pontine wave (p-wave): A mechanism of learning and memory processing. In: Maquet J, Stickgold R, and Smith C (eds.) *Sleep and Brain Plasticity*, pp. 135–156. Oxford: Oxford University Press.

Garcia-Rill E, Kobayashi T, and Good C (2003) The developmental decrease in REM sleep. *Thalamus & Related Systems* 2: 115–131.

Greene R and Siegel J (2004) Sleep. A functional enigma. *Neuro-Molecular Medicine* 5: 59–68.

Heister DS, Hayar A, Charlesworth A, Yates C, Zhou Y, and Garcia-Rill E (2007) Evidence for electrical coupling in the SubCoeruleus (SubC) nucleus. *Journal of Neurophysiology* 97: 3142–3147.

Jouvet M (1999) *The Paradox of Sleep. The Story of Dreaming.* Cambridge, MA: MIT Press.

Kilduff TS and Peyron C (2000) The hypocretin/orexin ligand-receptor system: Implications for sleep and sleep disorders. *Trends in Neurosciences* 23: 359–365.

Kleitman N (1963) *Sleep and Wakefulness.* Chicago: University of Chicago Press.

Llinas R, Ribary U, Joliot M, and Wang XJ (1994) Content and context in temporal thalamocortical binding. In: Buzsaki G (ed.) *Temporal Coding in the Brain*, pp. 251–272. Berlin: Springer.

Lydic R, Baghdoyanm HA, and McGinley J (2003) Opioids, sedation and sleep. Different states, similar traits, and the search for common mechanisms. In: Malviya S, Naughton N, and Tremper KK (eds.) *Contemporary Clinical Neuroscience: Sedation and Analgesia for Diagnostic and Therapeutic Procedures*, pp. 1–31. Totowa, NJ: Humana Press.

Moruzzi G and Magoun HW (1949) Brain stem reticular formation and activation of the EEG. *Electroencephalography and Clinical Neurophysiology* 1: 455–473.

Siegel JM (1994) Brainstem mechanisms generating REM sleep. In: Kryger MH, Roth T, and Dement WC (eds.) *Principles and Practice of Sleep Medicine*, pp. 125–144. London: Saunders.

Steriade M and McCarley RW (2005) *Brain Control of Wakefulness and Sleep,* 2nd edn., New York: Springer.

Nightmares

M Schredl, Central Institute of Mental Health,
Mannheim, Germany

Introduction

Nightmares are very common; 70–90% of young adults have reported that they experienced a nightmare at least once in their lifetime. In this article, the definition and etiology of nightmares are reviewed, and a brief and simple treatment technique is presented.

Definition

Nightmares are usually defined as frightening dreams that awaken the dreamer. However, there are two problems with this restricted definition. First, several researchers have demonstrated that very negatively toned dreams that do not wake the dreamer can be as disturbing as nightmares and have coined the term 'bad dreams' for this class of dreams. On average, the emotional intensity of nightmares is higher than that of bad dreams, but there is a large overlap. Whether or not the sleeper is able to judge if the dream emotion was really the cause for waking is still an open question. Some characteristic dreams, such as being chased and waking up just before the attacker grabs you or falling dreams with waking up just before hitting the ground, are obvious examples of dreams that awaken the dreamer, but other dream themes may not be that clear. Second, detailed studies revealed that up to 30% of the nightmares are not dominated by fear but, rather, by other emotions such as extreme sadness, anger, disgust, and confusion.

Nightmares are rapid eye movement (REM) sleep phenomena and often occur in the second half of the night (**Table 1**). Upon awakening, the person is oriented about her or his surrounding and can give a detailed description of the dream action. Nightmares should be differentiated from night terrors and post-traumatic reenactments. Night terrors are associated with slow-wave sleep and occur predominantly in the first part of the night (usually 1 h after sleep onset). The attack often begins with a piercing cry and – although the eyes are open – the person is not fully awake, experiences intense anxiety (with scarce or no mental content), and does not recall the event in the morning. Sleepwalking often follows a night terror attack. Posttraumatic reenactments also differ from nightmares in that they can occur during REM sleep as well as non-REM sleep (and also during the day – known as 'flashbacks'). Regarding content, there is a continuum from exact replays of the trauma to the nightmares without any obvious reference to the trauma.

Typical nightmare contents are being chased (50%), death or injury (20%), death or injury of close persons (15%), and falling (10%).

Prevalence

The prevalence of children experiencing nightmares once a week or more often is estimated to be approximately 5%. Although nightmare frequency decreases with age (peak prevalence rates occur at ages 6–10 years), large-scale representative studies indicate that approximately 5% of the adult population suffers from nightmares (presumably nightmare frequency is less than once a week in many people). Nightmare frequency is higher in girls and women (with medium effect sizes for the gender difference).

Etiology

Throughout the centuries, many hypotheses regarding the causes of nightmares have been formulated, ranging from devils and demons having sexual intercourse with the dreamer ('succubus' for men and 'incubus' for women) to modern disposition-stress models (**Table 2**). For the factors depicted in **Table 2**, empirical evidence of their role in nightmare etiology has been compiled. A large-scale Finnish twin study clearly demonstrated that genetic factors play a role in the etiology of nightmares. Whereas trait anxiety was often not related to nightmare frequency in adults, studies in children and adolescents often found an association between the two. Ernest Hartmann and others have shown that the personality dimension called 'thin boundaries' is associated with nightmares. People who are creative, sensitive, and have intense but problematic relationships, unusual sensory perceptions, and extraordinary occupations suffer from nightmares more often. With respect to the Big Five Factor model of personality, the neuroticism factor is strongly related to nightmare frequency.

Regarding state factors, the presence of a current stressor increases nightmare frequency in children, in nonclinical student samples, and also in nightmare sufferers. Interestingly, one study was able to demonstrate that the effect of personality (neuroticism and thin boundaries) is mediated by the occurrence of stress; that is, people with high neuroticism scores

Table 1 Nocturnal awakenings associated with anxiety

	Nightmare	Night terror (Pavor nocturnus)	Posttraumatic reenactment
Time of the night	Predominantly in the second part of the night	Predominantly in the first part of the night	Both parts of the night
Sleep stage	REM sleep	Slow-wave sleep	REM sleep or sleep stage 2
Physiological anxiety measures	Moderate	Very strong	Strong to very strong
Dream content	Detailed dream	Rarely a threatening image	Replay of the trauma
Orientation upon awakening	Often fully oriented, although the dream anxiety persists	Not oriented, does not respond properly if addressed	Often fully oriented, strong fear persists
Recall in the morning	Good recall of the dream story	No recall	Good recall
Highest prevalence	6–10 years old	3–7 years old	Event related

Table 2 Etiology of nightmares

Disposition-stress model
  Genetic factors
  Trait factors ('thin boundaries,' trait anxiety, neuroticism)
  State factors (stress)
  Trauma
  Drugs
Maintaining factors (cognitive avoidance)

experience more stress in their waking life and, therefore, experience nightmares more often. Experiencing a trauma (war experiences, sexual abuse, natural disasters, and severe accidents) can result in a full-blown posttraumatic stress disorder (PTSD) with nightmares as one of the core symptoms, but often nightmare frequency is elevated even if no other PTSD symptoms are present. Within a detailed nightmare history the question about current medication should always be included since quite a few compounds, such as antidepressants (e.g., selective serotonin reuptake inhibitors), acetylcholinesterase inhibitors, or antihypertensive medication, can cause nightmares.

In addition to disposition and triggering factors, one has to keep in mind that nightmares are anxiety phenomena like phobias which persist if the person avoids an active confrontation with the fear. In the case of nightmares, the largest percentage of an adolescent sample reported that trying to forget the nightmare was their primary coping strategy (**Table 3**). Statements such as "This was only a dream" can be considered as cognitive strategy to avoid confronting the anxiety experienced within the dream. Therefore, the avoidance behavior maintains the nightmares even if the original trigger is no longer active.

Nightmares and Sleep Quality

Several researchers conceptualize nightmares as a primary sleep disorder, although they are categorized into the group of parasomnias and not into the group of dyssomnias such as primary insomnia. Empirical evidence has clearly demonstrated low subjective sleep quality in nightmare sufferers, presumably due to direct effects (waking up from the nightmare and prolonged periods of wakefulness after the nightmare before falling asleep again) and indirect effects (fear of falling asleep in the evening because of the reoccurrence of nightmares). Since stress is associated with poor sleep quality and nightmares, partial correlational analyses were carried out to determine whether nightmares are associated with poor sleep quality even if the stress level is statistically controlled. The results clearly indicate that nightmare frequency and severity of insomnia complaints were associated. The therapy studies performed by Barry Krakow and co-workers also showed that a reduction of nightmare frequency is paralleled by an increase in sleep quality.

Nightmare Frequency versus Nightmare Distress

The hypothesis has been put forward that poorly adjusted people (i.e., those with high neuroticism scores) are more distressed by their nightmares – regardless of frequency – than well-adapted people. This would explain the inhomogeneous findings regarding the relationship between waking-life psychopathology and nightmare frequency and the higher correlation coefficients for the relationship between nightmare distress and psychopathology compared to the correlation coefficients between nightmare frequency and psychopathology. However, most studies have relied on the Belicki scale for measuring nightmare frequency, but this measure is strongly confounded with nightmare frequency (correlation coefficients ranging from $r = 0.24$ to 0.47). To avoid this methodological issue, we developed a scale that measures the distressing effect of a single nightmare and correlated this measure with neuroticism. As expected, neuroticism was correlated with

Table 3 Coping strategies used for bad dreams and nightmares

Category	Total (N = 606) (%)	Boys (n = 329) (%)	Girls (n = 277) (%)
No bad dreams	13.5	19.5	6.5
Tell the dream	22.9	14.3	33.2
Draw a picture	2.0	2.4	1.4
Record (diary)	2.5	0.9	4.3
Try to forget the dream	39.9	41.0	38.6
Image a positive ending	10.1	11.6	8.3
Other	9.1	10.3	7.6

nightmare frequency $(r = 0.39)$ and with the sum score of the Belicki scale $(r = 0.42)$ but not with the single nightmare distress scale $(r = 0.19)$ or nightmare intensity $(r = 0.04)$. The hypothesis that people with high neuroticism scores overestimate nightmare distress can be rejected; nightmare frequency is the major factor in determining distress attributed to nightmares.

Nightmares and Sleep Physiology

In the nineteenth century, several scientists held the view that nightmares are caused by a shortage of oxygen. It was sometimes said that a living creature was sitting on the chest ('Nachtmahr' or 'Alpdrücken' in German) or choking the dreamer. Based on experience with his own nightmares, Boerner concluded that blocking the mouth and nose by the bedspread or by the pillow (sleeping in the prone position) caused nightmares. Using a cloth to block the mouth and nose, Boerner was able to induce nightmares in three different sleepers. It seems obvious that studying patients with nocturnal breathing disorders, such as the sleep apnea syndrome, is a promising approach to test the oxygen hypothesis. The sleep apnea syndrome is characterized by apneas (minimal duration of 10 s) which can be accompanied by severe oxygen desaturation. An alarm reaction with an arousal terminates the apnea by increasing muscle tone and reopening the upper airways. In pilot studies, a few examples of breathing-related nightmares have been collected. A 39-year-old male patient with sleep apnea (respiratory disturbance index, 68.1 apneas per hour; maximal decrease in blood oxygen saturation, 43%) stated the following:

> During the dream I felt tied up or chained. I saw thick ropes around my arms and was not able to move. I experienced the fear of suffocation without being able to cope with the situation. Powerlessness and also resignation came up.

One patient reported having recurrent nightmares of the following type: He was lying in a coffin – dead.

He cried, "My god, please don't let me die, I am still young and I have small children," and woke up, gasping for breath.

Despite the illustrative examples, a large-scale study with 323 apnea patients did not find any correlation between nightmare frequency measured retrospectively via questionnaire and respiratory parameters (respiratory disturbance index and nadir of oxygen desaturation). Overall, nightmare frequency was not elevated in comparison to that of healthy controls. That is, the oxygen hypothesis is not supported by modern research findings.

Anne Germain and Tore Nielsen carried out a polysomnographic study in 11 nightmare sufferers. They reported a nightmare frequency of one or more nightmares per week. Although they reported considerable nightmare distress, their sleep parameters recorded in the sleep laboratory did not differ from those of healthy controls (e.g., sleep efficiency (92.1 vs. 92.7%), sleep onset latency (14 vs. 11 min), and REM sleep percentage (21.2 vs. 21.5%)). However, they found an elevated index of periodic limb movements during sleep which might be an effect of intense negative dreaming. The sleep laboratory findings (very few studies with small numbers have been carried out) contrast with the subjective reports of the patients. This might be an effect of the laboratory surroundings: many sleep researchers have reported that nightmares occur less often in the lab than at home. This effect might be attributed to the 'caring' environment – the presence of a sleep technician, someone interested in their history, listening to them empathically and taking them seriously. Similar effects (discrepancy between subjective sleep quality and polysomnographic findings) have been reported for other patient groups, such as patients with borderline personality disorder, patients with panic disorders, and adult patients with attention deficit/hyperactivity syndrome.

Charles Fisher and co-workers carried out an extensive laboratory study to investigate the physiology of nightmares. They included 38 patients suffering from three or more nightmares per week and studied their sleep over 162 nights. Although they recorded a fair number of night terror attacks ($n = 50$), only 22 nightmares occurred in the laboratory. From the patients' evaluations, the attacks in the lab were much less severe than those at home – again the lab effect. Only in three very intense nightmares were marked increases in heart rate (an increase of 20 beats per minute), respiration rate (an increase from 18 to 30 breaths per minute), and the number of eye movements measured in the 8 min prior to the awakening. This is in contrast with the night terror findings: documented increases in heart rate from 64 to approximately 120–150 beats per minute during the attacks. Despite the intense anxiety experienced during a nightmare, the autonomic reaction is moderate (**Table 1**).

Therapy of Nightmares

The question arises as to when nightmares should be treated. Similar to other mental disorders, it is important to evaluate whether the person suffers from her or his nightmares in a considerable way. Treatment should be offered to patients reporting negative effects on daytime mood and functioning, worries at sleep onset (fear that a nightmare might occur), and insomnia complaints. If a cutoff is needed, for example, for evaluating nightmares in children, a frequency of one or more nightmares per week is reasonable.

Whereas drugs such as REM sleep-suppressing tricyclic antidepressants are often ineffective in the treatment of nightmares, several different psychological interventions seem promising. Although several case report series indicate that lucid dreaming is an effective tool in coping with nightmares, the major problem with this technique is the fact that induction of lucid dreams (learning to have dreams when you know you are dreaming) is a highly expensive and protracted process. Systematic desensitization was shown to be effective in the treatment of nightmares in two carefully performed studies. The most effective and simplest treatment strategy was developed by Barry Krakow and co-workers and is called imagery rehearsal treatment (IRT). Based on my clinical experience, a reduced form of IRT including the basic steps necessary for the effective treatment of nightmares is depicted in **Table 4**. For most patients, it is helpful to point out the analogy of coping with fear and anxiety in the waking life.

The first step consists of confrontation, which is simply writing down the dream or, for children, the drawing of the most important dream scene. In the second step, the person is asked to imagine a new ending to the dream. Children were asked to draw

Table 4 Therapeutic principles

Step 1: Confrontation
  Record or draw the dream
Step 2: Coping with the situation in the nightmare
  Construct a new dream ending or add something to the drawing that reduces the fear
Step 3: Training of the coping strategy
  Approximately 5–10 min per day over a 2-week period

something in the picture which reduced anxiety. Ideally, the therapist should encourage the person to develop his or her own personal coping strategy for the nightmare situation and refrain from giving suggestions. The coping strategy should include active behavior. Flying away or hiding is often not helpful (because these reflect avoidance), so one simply asks whether other strategies might be applicable as well. The 'solution' of the dream must not be realistic; most psychologists advise not to increase violence (e.g., killing the opponent) while coping with the dream situation. The 'new dream' should be written down. The last step consists of internalizing the new strategy by imagining the new ending 5–10 min every day during a 2-week period. The repetition during waking life affects subsequent dreams. Interestingly, the effect of the intervention often generalizes; that is, nightmares with other topics are affected and include more adequate coping strategies.

For IRT, several randomized controlled trials have been carried out including life-long nightmare sufferers (idiopathic nightmares) and sexually assaulted women with PTSD. In one of the studies by Barry Krakow and co-workers, a single $2\frac{1}{2}$ h group session was effective in reducing nightmare frequency in 39 patients with idiopathic nightmares from six nightmares per week to two. In addition, sleep quality and daytime mood improved. To illustrate the procedure, two case reports from my clinical practice are presented here.

Case Report – Adult

A 22-year-old woman consulted our sleep center due to nightmares that started several months ago. The anamnesis also indicated that several night terror attacks also occurred. Two polysomnographic recordings (including breathing parameters and tibialis EMGs) were normal, with the exception of low sleep efficiency (first night, 67.8%; second night, 76.7%). Nightmares and night terrors did not occur in the sleep laboratory. The psychometric measurement of sleep behavior and sleep quality revealed that sleep quality and the feeling of being refreshed in the morning were not reduced but daytime sleepiness score was elevated (1.6 standard deviations above the mean of a comparable student sample).

In the first session, the patient reported current stressors. A severe conflict in the core family resulted in a complete cessation of her relationship with her parents. The therapeutic principles, including confronting, coping, and training, were explained to the patient. It was easy for her to understand this approach because it was similar to the treatment of anxiety. In the nightmare of the previous night, the patient was in the dressing room of a fitness club. At first, her complete family was present. Then, her mother was in the center and verbally attacked and criticized the dreamer. She felt very helpless with respect to the accusations. Answering the question as to how the dream could be altered, she came up with several suggestions: ignoring the mother and active acting, such as saying something to her mother and expressing her needs. The 'new dream' included the active confronting of the mother and the sentence, "I can manage my life myself."

In the second session 2 weeks later, the patient reported that she trained the new dream ending regularly during the first week and that no new nightmares including her mother occurred. However, several other negatively toned dreams occurred, one of which was a recurring nightmare. In it, the dreamer stays at the house of her grandmother and is now in the kitchen. At first, the overall emotional tone is positive. Then she senses that there is something threatening outside the house that might come inside and threaten her. The scene changes: the dreamer finds herself in a white room with several fantasy characters (pleasant characters). Two of the characters encourage her to step outside the house and confront the threatening something alone. At this point, she usually wakes up. The awakening from the last dream was at 5 a.m., and the anxiety she felt in the dream was still present after waking up. After talking about the different dream elements (e.g., the relationship to her grandmother, to whom she has a positive connection), the patient was again asked to imagine a new ending to the dream. The patient visualized how she goes outside and confronts the unknown threat with the help of the friendly fantasy characters.

In the follow-up session 2 weeks later, the patient reported a complete cessation of nightmares and that the dreams she recalled incorporate new and active behavior patterns. The dreamer felt more confident in dealing with other dream characters and was able to express her needs. She also reported a marked lessening of daytime sleepiness. Although the stressors did not change over the treatment period, the intervention did reduce nightmare frequency by strengthening the dream ego – that is, the negative effect of the stressor on dreaming and sleep was significantly reduced.

Case Report – Child

A mother consulted our sleep center because her son (5 years old) suffered from nightmares every night during the past month. Before that, nightmares rarely occurred. Common topics are ghosts, shadows, and horrid monsters. At night, he calls out for his mother and needs comfort for up to 2 h because after the nightmare he cannot fall asleep easily. The mother was not aware of any incidents triggering the period of intense nightmares. The boy was somewhat shy but free from any psychopathology or somatic problems.

After taking the nightmare history, the boy was asked to draw a recent nightmare (**Figure 1**). The dreamer is walking within a castle (**Figure 1**, lower left) and carrying a book. He is very frightened by the sight of two enormous ghosts (the differences in size between the figures and himself clearly reflect the fear involved). After he was asked what might help reduce his anxiety, the boy drew a large spider between him and the ghosts, facing the ghosts and protecting him.

After the first session, the boy continued this work with the help of his mother by drawing several pictures of other nightmares and concentrating on one of them each day for approximately 10–15 min. Within 2 weeks, nightmare frequency was drastically reduced. Although the topics of the dreams did not change much, the fear associated with these pictures was almost completely gone. Sleep behavior and quality were back to normal. A follow-up session 1 year later revealed that nightmares were still rare (approximately once a month) and the boy did not feel distressed by them. For several nightmares, he invented new solutions (one involving a magic spray). He has developed a positive attitude toward his dreams which can be seen in the fact that he occasionally asks his mother to record a dream in the booklet which she had kept during the study period.

Conclusion and Future Directions

Nightmares are common and approximately 5% of the population report that they suffer from nightmares. Fortunately, nightmares can be treated effectively with a brief and simple intervention method which can be classified as a cognitive technique. The beneficial effect has been demonstrated for idiopathic nightmares as well as posttraumatic nightmares in children and in adults. The basic principle of confronting and coping with the nightmare situation is easily understandable to patients and can be done without professional assistance, thus encouraging self-help.

Figure 1 Drawing of a nightmare with a coping strategy included.

In addition to the clinical aspect, nightmares are also an interesting topic for basic sleep and dream researchers. The body–mind interaction can be studied – for example, the correlation between emotional intensity and brain activation related to emotions in waking life (limbic system). Techniques such as magnetic resonance imaging will be applicable to sleep if the scanner noise can be markedly reduced. Since nightmares occur quite rarely in the sleep laboratory, large-scale studies with ambulatory measurement units should be conducted to shed light on the physiological parameters during REM sleep before awakening from a nightmare. The lack of consistent results from sleep laboratory studies may be attributable solely to the 'lab effect' (reducing emotional dream intensity due to the 'caring' environment). Since nightmares reflect stressors and PTSD-related nightmares often replay the trauma experienced by a person, they may help in elucidating the role of dreams in general in sleep-related memory consolidation. Research has shown that sleep facilitates consolidation of declarative and procedural memory tasks. Whether dreams play a crucial role in that relationship remains to be determined.

See also: Dreams and Nightmares in PTSD; REM/NREM Differences in Dream Content; Sleep Apnea; Sleep in Adolescents; Sleep in Aging; Theories and Correlates of Nightmares; Theories of Dream Function.

Further Reading

Germain A and Nielsen TA (2003) Sleep pathophysiology in posttraumatic stress disorder and idiopathic nightmare sufferers. *Biological Psychiatry* 54: 1092–1098.

Hublin C, Kaprio J, Partinen M, and Koskenvuo M (1999) Nightmares: Familial aggregation and association with psychiatric disorders in a nationwide twin cohort. *American Journal of Medical Genetics (Neuropsychiatric Genetics)* 88: 329–336.

Krakow B and Zadra A (2006) Clinical management of chronic nightmares: Imagery rehearsal therapy. *Behavioral Sleep Medicine* 4: 45–70.

Schredl M (2003) Effects of state and trait factors on nightmare frequency. *European Archives of Psychiatry and Clinical Neuroscience* 253: 241–247.

Schredl M, Blomeyer D, and Görlinger M (2000) Nightmares in children: Influencing factors. *Somnologie* 4: 145–149.

Schredl M, Landgraf C, and Zeiler O (2003) Nightmare frequency, nightmare distress and neuroticism. *North American Journal of Psychology* 5: 345–350.

Spoormaker VI, Schredl M, and Van den Bout J (2006) Nightmares: From anxiety symptom to sleep disorder. *Sleep Medicine Reviews* 10: 19–31.

Zadra A and Donderi DC (2000) Nightmares and bad dreams: Their prevalence and relationship to well-being. *Journal of Abnormal Psychology* 109: 273–281.

Coma

S Laureys, M Boly, G Moonen, and P Maquet,
University of Liege and Centre Hospitalier Universitaire,
Liege, Belgium

Two Dimensions of Consciousness: Arousal and Awareness

Consciousness is a multifaceted concept that has two dimensions: arousal, or wakefulness (i.e., level of consciousness), and awareness (i.e., content of consciousness). One needs to be awake in order to be aware (rapid eye movement (REM) sleep and lucid dreaming being notorious exceptions). **Figure 1** shows that in normal physiological states (green), level and content are positively correlated (with the exception of the oneiric activity during REM sleep). Patients in pathological or pharmacological coma (i.e., general anesthesia) are unconscious because they cannot be awakened (pink). The vegetative state (VS; blue) is a unique dissociated state of consciousness (i.e., patients are seemingly awake but lack any behavioral evidence of 'voluntary' or 'willed' behavior). There is, of course, an irreducible philosophical limitation in knowing for certain whether any other being possesses a conscious life. Awareness is a subjective first-person experience, and its clinical evaluation is limited to evaluating patients' motor responsiveness. In addition to its clinical and ethical importance, studying these disorders offers a still largely underestimated means of studying human consciousness. In contrast to unconscious states such as general anesthesia and deep sleep (in which impairment in arousal cannot be disentangled from impairment in awareness), VS is characterized by a unique dissociation of arousal and awareness, offering a lesional approach to identifying the neural correlates of awareness. A major challenge is to unequivocally disentangle patients residing in the clinical 'gray zone' between the vegetative and the minimally conscious state (MCS). The locked-in syndrome (LIS) is a rare but horrifying situation in which patients awaken from their coma fully aware but remain mute and immobile; it is called a 'pseudodisorder of consciousness' because patients superficially look unconscious but in reality are fully aware only unable to show it due to severe paralysis.

Arousal is supported by several brain stem neuronal populations (i.e., the reticular activating system) that directly or via nonspecific thalamic nuclei project to cortical neurons. Hence, focal damage of the brain stem or diffuse damage of the cerebral hemispheres may cause reduced arousal. The evaluation of eye-opening and brain stem reflexes is a key to the clinical assessment of the functional integrity of the arousal systems. Awareness depends on the functional integrity of the cerebral cortex and its reciprocal subcortical connections (each of its aspects partly residing in spatially defined brain areas), but its underlying neural code remains to be elucidated. Therefore there is, at present, no validated objective 'consciousness meter.' The estimation of the multiple dimensions of consciousness requires the interpretation of several clinical signs, mainly based on the observation of 'voluntary' interaction with the examiner or the environment. Brain death, coma, VS, MCS, and LIS (see **Figure 2**) are all solely defined by clinical criteria. Many scoring systems have been developed for a standardized assessment of consciousness in severely brain-damaged patients.

Nosology of Disorders of Consciousness

Brain Death

Brain death means human death determined by neurological criteria. The current definition of death is the permanent cessation of the critical functions of the organism as a whole (i.e., neuroendocrine and homeostatic regulation, circulation, respiration, and consciousness). Most countries, including the US, require death of the whole brain including the brain stem, but some (e.g., UK and India) rely on the death of the brain stem only, arguing that the brain stem is at once the through-station for nearly all hemispheric input and output, the center generating arousal (an essential condition for conscious awareness), and the center of respiration. Clinical assessments for brain death, however, are uniform and are based on the loss of all brain stem reflexes and the demonstration of continuing cessation of respiration (by performing a standardized apnea test) in a persistently comatose patient (**Table 1**). There should be an evident cause of coma, and confounding factors such as hypothermia, drugs, and electrolyte and endocrine disturbances should be ruled out.

Brain death is classically caused by a massive brain lesion (e.g., trauma, intracranial hemorrhage, anoxia) which increases intracranial pressure to values superior to mean arterial blood pressure and hence causes intracranial circulation to cease and damages the brain stem due to herniation. Using the brain stem formulation of death, however, unusual but existing cases of catastrophic brain stem lesion (often of hemorrhagic origin) sparing the thalami and cerebral cortex can be declared brain death in the absence of clinical brain stem function, despite intact

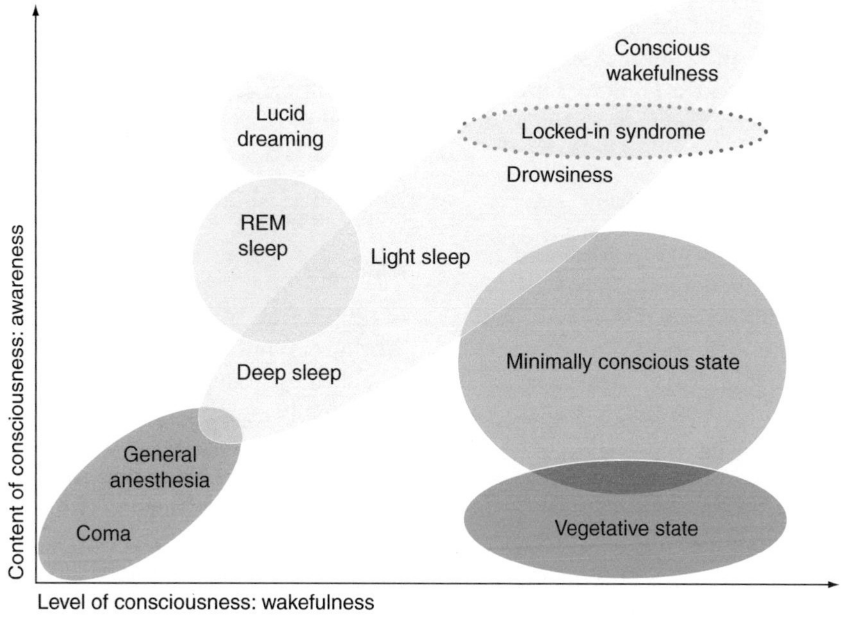

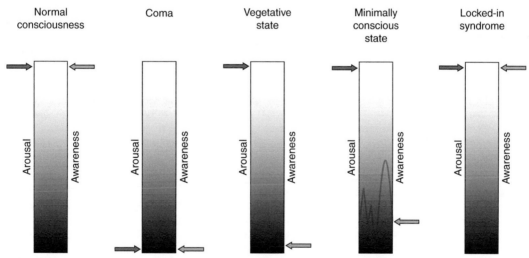

Figure 1 Oversimplified illustration of the two major dimensions of consciousness: the level of consciousness (i.e., arousal or wakefulness) and the content of consciousness (i.e., awareness or experience). Adapted from Laureys S (2005) The neural correlate of (un)awareness: Lessons from the vegetative state. *Trends in Cognitive Sciences* 9: 556–559.

Figure 2 Graphical representation of the two dimensions of consciousness: arousal (red arrow) and awareness (green arrow) and their alterations in coma, the vegetative state, the minimally conscious state, and the locked-in syndrome. Adapted from Laureys S, Owen AM, and Schiff ND (2004) Brain function in coma, vegetative state, and related disorders. *Lancet Neurology* 3: 537–546.

intracranial circulation. Hence a patient with a primary brain stem lesion (who did not develop raised intracranial pressure) might theoretically be declared dead by the UK doctrine but not by the US doctrine.

Coma

Coma is a state of unarousable unresponsiveness characterized by the deficiency of the arousal systems (clinically assessed as the absence of stimulation-induced eye opening after having ruled out bilateral ptosis) and thus also by absence of awareness. The comatose patient lacks the sleep–wake cycles that can be observed in the VS. To be clearly distinguished from syncope, concussion, or other states of transient unconsciousness, coma must persist for at least 1 h. In general, comatose patients who survive begin to awaken and recover gradually within 2–4 weeks. This recovery may go no further than VS or MCS, or these may be stages (brief or prolonged) on the way to more complete recovery of consciousness.

There are two main causes of coma: (1) bihemispheric diffuse cortical or white matter damage secondary to neuronal or axonal injury and (2) brain stem lesions bilaterally affecting the subcortical reticular arousing systems (i.e., pontomesencephalic tegmentum and/or paramedian thalami).

VS

VS was defined in 1972 by Bryan Jennett and Fred Plum to describe those patients who 'awaken' from their coma (meaning they open their eyes spontaneously or on stimulation) but remain unaware of self or environment (meaning they show only reflex motor responses; **Table 2**). According to the *Oxford English Dictionary*, the term 'vegetative' describes "an organic body capable of growth and development but devoid of sensation and thought." It is very important to stress the difference between persistent and permanent VS, which are, unfortunately, too often abbreviated identically as PVS, causing unnecessary confusion. 'Persistent VS' has been arbitrarily defined as a VS still present 1 month after the acute brain damage, but the term does not imply irreversibility. In 1994, the US Multi-Society Task Force on PVS concluded that 3 months following nontraumatic brain damage and 12 months after traumatic

injury, the condition of VS patients may be regarded as 'permanent.' Only in cases of 'permanent VS' do the ethical and legal issues surrounding withdrawal of treatment arise. It is essential that experienced examiners employing adapted standardized clinical assessment scales repeatedly establish the behavioral absence of any sign of conscious perception or deliberate action before making the diagnosis of VS.

MCS

In 2002, the Aspen Neurobehavioral Conference Workgroup published the diagnostic criteria for MCS to subcategorize patients above VS but unable to functionally communicate their thoughts and feelings. On a reproducible or sustained basis, MCS patients show limited but clearly discernible evidence of awareness of self or environment (**Table 3**). The emergence of MCS is characterized by the recovery of interactive communication or functional use of objects. Further improvement is more likely than in VS patients. However, some patients may remain permanently in MCS. At present, no time intervals for 'permanent MCS' have been agreed on.

'Akinetic mutism' (a condition characterized by severe poverty of movement, speech, and thought without associated arousal disorder or descending

Table 1 Criteria for brain death (American Academy of Neurology guidelines)

- Demonstration of coma
- Evidence for the cause of coma
- Absence of confounding factors, including hypothermia, drugs, electrolyte, and endocrine disturbances
- Absence of brain stem reflexes
- Absent motor responses
- Apnea
- A repeat evaluation in 6 h is advised, but the amount of time is considered arbitrary
- Confirmatory laboratory tests are required only when specific components of the clinical testing cannot be reliably evaluated

Table 2 Criteria for the vegetative state (US Multi-Society Task Force on Persistent Vegetative State guidelines)

- No evidence of awareness of self or environment and an inability to interact with others
- No evidence of sustained, reproducible, purposeful, or voluntary behavioral responses to visual, auditory, tactile, or noxious stimuli
- No evidence of language comprehension or expression
- Intermittent wakefulness manifested by the presence of sleep–wake cycles
- Sufficiently preserved hypothalamic and brain stem autonomic functions to permit survival with medical and nursing care
- Bowel and bladder incontinence
- Variably preserved cranial nerve and spinal reflexes

Table 3 Criteria for the minimally conscious state (Aspen Neurobehavioral Conference Workgroup)

Clearly discernible evidence of awareness of self or environment, on a reproducible or sustained basis, by at least one of the following behaviors:
- Purposeful behavior (including movements or affective behavior that occurs in contingent relation to relevant environment stimuli and are not due to reflexive activity), such as:
  - Pursuit eye movement or sustained fixation occurring in direct response to moving or salient stimuli
  - Smiling or crying in response to verbal or visual emotional (but not neutral) stimuli
  - Reaching for objects demonstrating a relationship between object location and direction of reach
  - Touching or holding objects in a manner that accommodates the size and shape of the object
  - Vocalizations or gestures occurring in direct response to the linguistic content of questions
- Following simple commands
- Gestural or verbal yes/no response (regardless of accuracy)
- Intelligible verbalization

Emergence from the minimally conscious state requires reliable and consistent demonstration of at least one of the following behaviors:
- Functional interactive communication: accurate yes/no responses to six of six basic situational orientation questions (including items as, "Are you sitting down?" and "Am I pointing to the ceiling?") on two consecutive evaluations
- Functional use of two different objects (such as bringing a comb to the head or a pencil to a sheet of paper) on two consecutive evaluations

motor tract impairment) is an outdated term that should be avoided and is now considered to be a subcategory of the minimally conscious syndrome.

LIS

The category of LIS (pseudocoma) was introduced by Fred Plum and Jerome Posner in 1966 to refer to the constellation of quadriplegia and anarthria brought about by the disruption of, respectively, the brain stem's corticospinal and corticobulbar pathways. The syndrome describes patients who are aroused and awake but selectively deefferented (i.e., have no means of producing speech or limb or face movements). Usually the anatomy of the responsible lesion in the brain stem is such that locked-in patients are left with the capacity to use vertical eye movements and blinking to communicate their awareness of internal and external stimuli. Acute vascular bilateral ventral pontine lesions are its most common cause. Patients with such lesions often remain comatose for some days or weeks, needing artificial respiration, and then gradually wake up, remaining paralyzed and voiceless, superficially resembling someone in a coma or a VS. **Table 4** lists the clinical criteria of LIS. The syndrome is subdivided, on the basis of the extent of motor impairment, into (1) classical LIS (characterized by total immobility except for vertical eye movements or blinking); (2) incomplete LIS (permitting remnants of voluntary motion); and (3) total LIS (consisting of complete immobility including all eye movements, combined with preserved consciousness).

Clinimetric Evaluation, Diagnosis, and Prognosis

Consciousness Scales

In 1974, Graham Teasdale and Bryan Jennett published the Glasgow Coma Scale (GCS). This standardized bedside tool to quantify arousal and awareness now is the gold standard in coma research. The GCS has tree components assessing eye opening and verbal and motor responses. Some authors have disagreed that evaluation of spontaneous or stimulation-induced opening of the eyes is sufficiently indicative of brain stem arousal system activity and have proposed coma scales that include brain stem reflexes. However, only the GCS has known a widespread use because the others generally are more complex. A simpler system, the Glasgow-Liège Scale (GLS), combines the GCS with five brain stem reflexes (i.e., fronto-orbicular, oculocephalic, pupillary, and oculocardiac reflexes). The increasing use of intubation and mechanical ventilation has rendered the verbal component of the GCS unassessable in many coma patients. Recently, the Full Outline of Un-Responsiveness (FOUR; also the number of components tested: eye, motor, brain stem reflexes, and respiratory function) scale proposed a hand-position test (i.e., asking the patient to make a fist or a 'thumbs-up', or 'V-for-victory' sign) as an alternative to the verbal component of the GCS.

The GCS, GLS, and FOUR scales lack reliability in assessing chronic disorders of consciousness. For these patients, more-sensitive scales are the Coma Recovery Scale-Revised (CRS-R), the Sensory Modality Assessment and Rehabilitation Technique (SMART), and the Wessex Head Injury Matrix (WHIM). The CRS-R is a recent scale specifically developed to differentiate VS from MCS and explicitly incorporating current diagnostic criteria. The basic structure of the CRS-R is similar to the GCS, but its subscales are much more detailed, targeting more-subtle signs of recovery of awareness.

Brain Death

Because many of the areas of the supratentorial brain (including the neocortex, thalami, and basal ganglia) cannot be tested for clinical functions accurately in a comatose patient, most of the bedside tests for brain death (such as cranial nerve reflexes and apnea testing) directly measure functions of only the brain stem. Since the first clinical definition of brain death nearly 50 years ago, no patient in apneic coma properly declared dead on the basis of neurological criteria has ever regained consciousness. Complementary neurophysiological tests such as electroencephalography (EEG), event-related potentials (ERPs), angiography, Doppler sonography, and scintigraphy reliably and objectively confirm the clinical diagnosis.

Coma

The management and prognosis of coma depend on many factors, such as etiology, the patient's general medical condition, age, clinical signs, and complementary examinations. After 3 days of observation,

Table 4 Criteria for the locked-in syndrome (American Congress of Rehabilitation Medicine)

- Sustained eye opening (bilateral ptosis should be ruled out as a complicating factor)
- Quadriplegia or quadriparesis
- Aphonia or hypophonia
- A primary mode of communication that uses vertical or lateral eye movement or blinking of the upper eyelid to signal yes/no responses
- Preserved awareness of the environment

absence of pupillary or corneal reflexes, stereotyped or absent motor response to noxious stimulation, isoelectrical or burst suppression pattern EEG, bilateral absent cortical responses on somatosensory evoked potentials, and (for anoxic coma) biochemical markers such as high levels of serum neuron-specific enolase are known to herald bad outcome. Visual and brain stem auditory evoked potentials are of limited prognostic use. Prognosis in traumatic coma survivors is better than in anoxic cases. Predicting outcome in toxic, metabolic, and infectious comatose states is challenging, and many unexpected cases of recovery have been reported.

VS

Whereas coma and brain death characteristically are acute conditions (lasting no more than days to weeks), the VS and MCS may become chronic entities. Vegetative patients, unlike brain dead or comatose patients, can move extensively, and studies have shown how difficult it is to differentiate 'automatic' from 'willed' movements in these patients. This results in an underestimation of behavioral signs of awareness and hence a misdiagnosis. It is known that when the diagnosis is made with insufficient care, up to one in three 'vegetative' patients actually are at least minimally conscious. Physicians also frequently tend to erroneously diagnose VS in older demented nursing home residents. Clinical testing for absence of awareness is much more problematic and slippery than testing for absence of arousal, brain stem reflexes, and apnea in irreversible coma. Given appropriate medical treatment, meaning artificial nutrition and hydration, VS patients may survive for many years.

Over the past decade, investigators have struggled to find an objective test that could reliably predict outcome for vegetative individuals. In contrast to coma and brain death, there are no validated diagnostic or prognostic markers for VS patients. The chances of recovery depend on the patient's age, etiology (worse for anoxic causes), and time spent in the VS. Recent data suggest that damage to the corpus callosum and brain stem indicate bad outcome in traumatic VS.

MCS

Because criteria for MCS have only recently been introduced, there are few clinical studies of patients in this condition. As stated above, it remains very challenging to behaviorally differentiate minimally conscious from vegetative patients because both are, by definition, noncommunicative. VS is one end of a spectrum of awareness, and the subtle differential diagnosis with MCS necessitates repeated evaluations by experienced examiners. As with VS, traumatic etiology has a better prognosis than does nontraumatic (anoxic) MCS. Preliminary data show that overall outcome for MCS is better than for VS.

LIS

In acute LIS, eye-coded communication may be difficult because of fluctuating arousal and limited control of voluntary eye movements. More than half the time, it is family members, not the physician, who first realize that the patient is aware. Distressingly, the diagnosis takes on average more than 2.5 months. In some cases it has taken 4–6 years before aware and sensitive patients, locked in an immobile body, have been recognized as being conscious. Some memoirs written by locked-in patients well illustrate the clinical challenge of recognizing the syndrome. Striking examples are *Look Up for Yes*, by Julia Tavalaro, and *Only the Eyes Say Yes*, by Phillippe Vigand.

While motor recovery remains very limited in LIS, life expectancy (with adequate medical care) may be several decades. Sensory evoked potentials are not reliable predictors of prognosis, but motor evoked potentials might evaluate the potential motor recovery. Eye-controlled computer-based communication technology now allows patients to control their environment, use a word processor coupled to a speech synthesizer, and access the Internet. Outsiders often assume that the quality of life with LIS is so poor that it is not worth living. Recent surveys, however, have revealed that chronic LIS patients self-report meaningful quality of life and that the demand for euthanasia is infrequent.

Residual Cerebral Function

Brain Death

The EEG in brain death shows absent electrocortical activity (i.e., isoelectric recording) with a sensitivity and specificity of around 90%. This makes the EEG the most validated and, because of its wide availability, preferred confirmatory test for brain death. Somatosensory evoked potentials typically indicate arrest of conduction at the cervicomedullary level, and brain stem auditory evoked potentials usually show only a delayed wave I (originating in the cochlear nerve). Cerebral angiography and transcranial Doppler sonography document with very high sensitivity and 100% specificity the absence of cerebral blood flow in brain death. Similarly, radionuclide cerebral imaging such as single-photon emission computed tomography (CT) and positron-emission tomography (PET) demonstrate the hollow-skull sign confirming

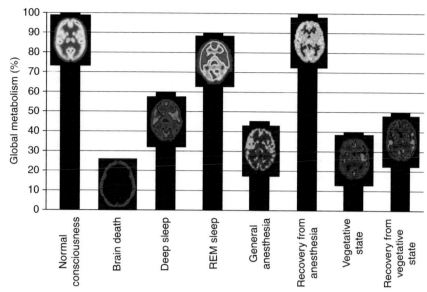

Figure 3 Cerebral metabolism in conscious wakefulness, in brain death (hollow skull sign confirming the absence of neuronal function in the whole brain in irreversible coma), physiological (slow wave sleep) and pharmacological (general anesthesia) modulation of arousal reflecting massive global decreases in cortical metabolism (in REM sleep, metabolic activity is paradoxically prominent); and in wakefulness without awareness (i.e., the vegetative state). Note that recovery from the vegetative state may occur without substantial increase in overall cortical metabolism, emphasizing that some areas in the brain are more important than others for the emergence of awareness. Images of halothane-induced loss of consciousness from Alkire M, Pomfrett CJ, Haier RJ, et al. (1999) Functional brain imaging during anesthesia in humans: Effects of halothane on global and regional cerebral glucose metabolism. *Anesthesiology* 90: 701–709.

the absence of neuronal function in the whole brain (**Figure 3**).

Anatomopathology in brain death patients receiving maximal artificial means of support will inevitably end up showing the so-called respirator brain, and after about a week, an autolyzed liquefied brain will pour from the opened skull.

Coma

The electrical activity of the brain as measured by the EEG tends to become nonreactive and slower as the depth of coma increases, regardless of the underlying cause. As stated above, the bilateral absence of cortical potentials (called N20 waves because they occur after about 20 ms) on somatosensory evoked potentials herald bad outcome. In cases in which cortical potentials are present, so-called endogenous ERPs might be useful. The presence of 'mismatch negativity,' that is, a negative component elicited after 100–200 ms by any change or 'mismatch' in a sequence of monotonous auditory stimuli (i.e., an 'oddball paradigm') indexes some persistent automatic information processing and is correlated with recovery of at least minimal consciousness.

Cortical metabolism in coma survivors is on average 50–70% of normal reference values. Cerebral metabolism correlates poorly with the level of consciousness, as measured by the GCS, in severely head-injured

patients. A global depression of cerebral metabolism is not unique to coma. When anesthetic drugs are titrated to the point of unresponsiveness, the resulting reduction in brain metabolism is similar to that observed in pathological coma. Another example of transient metabolic depression can be observed during slow wave sleep. In this daily physiological condition, cortical cerebral metabolism can drop to nearly 40% of normal values, although in REM sleep, metabolism returns to normal waking values (**Figure 3**).

VS

In VS the EEG most often shows a diffuse slowing (i.e., generalized polymorphic delta or theta rhythm); only sporadically is it of very low voltage or isoelectric. Somatosensory evoked potentials may show preserved primary somatosensory cortical potentials, and brain stem auditory evoked potentials often show preserved brain stem potentials in vegetative patients. Endogenous evoked potentials measuring, for example, the brain's response to complex auditory stimuli such as the patient's own name (compared with other names) record a so-called P300 response (i.e., a positive wave elicited around 300 ms poststimulus when patients detect an unpredictable target in the regular train of stimuli). In brain-damaged patients, the use of emotionally meaningful stimuli such as the patient's own name increases the chances of obtaining a P300 response. Recent data show that the P300 is not a reliable marker of awareness but rather indicates automatic processing, because it could be recorded in well-documented VS patients who never recovered (**Figure 4**).

In contrast to the functional decapitation observed in irreversible coma or brain death, vegetative patients show substantially reduced (40–50% of normal values) but not absent overall cortical metabolism. In some vegetative patients who subsequently recovered, global metabolic rates for glucose metabolism did not show substantial changes (**Figure 3**). Hence, the relationship between global levels of brain function and the presence or absence of awareness is not absolute. It rather seems that some areas in the brain are more important than others for its emergence.

Anatomopathologic features in anoxic VS encompass multifocal laminar cortical necrosis, diffuse leucoencephalopathy, and bilateral thalamic necrosis; VS following blunt head injury classically shows

diffuse white matter damage with neuronal loss in thalami and hippocampi. However, these postmortem studies have not permitted a detailed regional topography of cerebral damage characteristic of VS. Voxel-based statistical analyses of metabolic PET data have identified a metabolic dysfunction in a wide frontoparietal network encompassing the polymodal associative cortices: bilateral lateral frontal regions; parietotemporal and posterior parietal areas; and mesiofrontal, posterior cingulate, and precuneal cortices (**Figure 5**), known to be the most active 'by default' in resting nonstimulated conditions.

In some other conditions, patients also show merely reflex or automatic motor activity while behaviorally seeming 'awake.' In absence seizures (characterized by brief episodes of unresponsiveness and staring, frequently accompanied by purposeless eye blinking and smacking), functional magnetic resonance imaging (MRI) studies have shown decreases in blood oxygen level-dependent signals in a wide

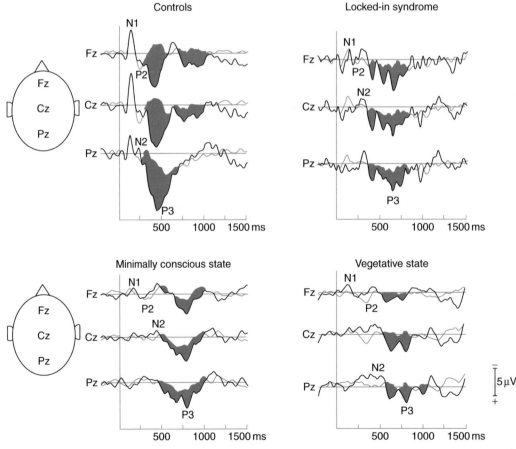

Figure 4 Endogenous event-related potentials to the patient's own name (thick traces) and to other first names (thin traces) in five healthy controls, four patients with locked-in syndrome, six minimally conscious state patients, and five vegetative state patients. Note that the differential P300 response (in pink) can also be observed in some well-documented VS patients (who never recovered) and hence is not a reliable indicator of awareness. Adapted from Perrin F, Schnakers C, Schabus M, et al. (2006) Brain response to one's own name in vegetative state, minimally conscious state, and locked-in syndrome. *Archives of Neurology* 63: 562–569.

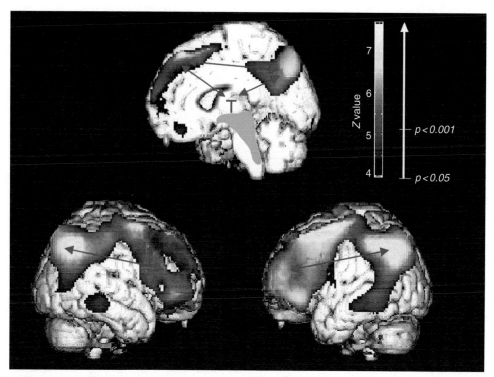

Figure 5 The common hallmark of the vegetative state is a metabolic dysfunctioning of a widespread cortical network encompassing medial and lateral prefrontal and parietal multimodal associative areas. This might be due to either direct cortical damage or corticocortical or thalamocortical disconnections (schematized by blue arrows). Also characteristic of the vegetative state is the relative sparing of metabolism in the brain stem (encompassing the pedunculopontine reticular formation, hypothalamus, and basal forebrain, represented in green), allowing for the patients' preserved arousal and autonomic functions. Adapted from Laureys S (2005) The neural correlate of (un)awareness: Lessons from the vegetative state. *Trends in Cognitive Sciences* 9: 556–559.

frontoparietal network, similar to that identified in VS patients. In some temporal lobe seizures that impair consciousness (coined 'complex partial') while preserving automatic activities such as picking, fumbling, or even cycling, single-photon emission CT has shown similar widespread deactivation in frontal and parietal association cortices. In contrast, temporal lobe epilepsy with sparing of consciousness (termed 'simple partial') is not accompanied by such frontoparietal dysfunction. Another example of transient nonresponsiveness with preserved automatic behavior such as ambulation can be observed in sleepwalkers. Again, deactivations in large areas of frontal and parietal association cortices were demonstrated by means of SPECT imaging during sleepwalking. Taken together, these observations point to the critical role of the frontoparietal associative cortices in the generation of conscious awareness.

However, awareness seems not exclusively related to the activity in this 'global workspace' cortical network but also, it is important to note, to the functional connectivity within this system and with the thalami. Long-range frontoparietal and thalamocortical (with nonspecific intralaminar thalamic nuclei)

'functional disconnections' have been identified in VS. Moreover, recovery is paralleled by a functional restoration of this frontoparietal network and part of its thalamocortical connections.

The most relevant question regards possible residual perception and cognition in vegetative patients. In well-documented VS patients, noxious high-intensity electrical stimulation (experienced as painful in controls) resulted in activation in brain stem, thalamus, and primary somatosensory cortex, but hierarchically higher-order areas of the pain matrix encompassing the anterior cingulate cortex failed to activate (**Figure 6**). It is important to note that the activated cortex was isolated and functionally disconnected from the frontoparietal network, considered critical for conscious perception. Similarly, auditory stimulation in VS was found to activate primary auditory cortices but not higher-order multimodal areas from which they were disconnected. The activation in primary cortices in awake but unaware patients confirms Crick and Koch's early hypothesis (based on visual perception and monkey histological connectivity) that neural activity in primary cortices is necessary but not sufficient for awareness.

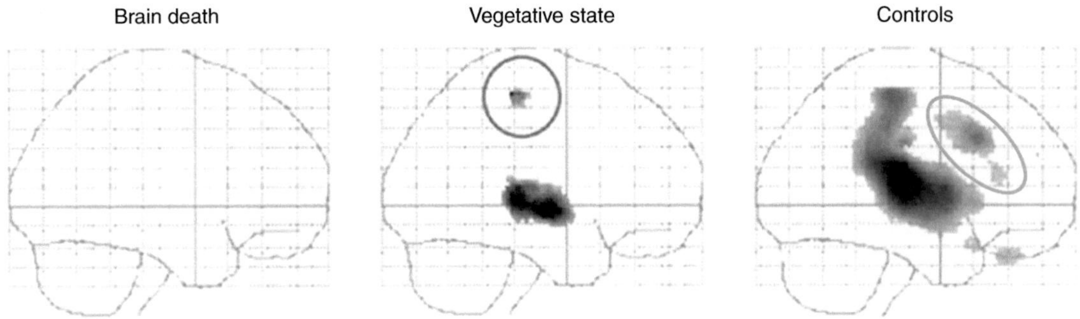

Figure 6 High-intensity somatosensory stimuli fail to induce any subcortical or cortical neural activation in irreversible coma with clinical absence of brain stem reflexes (i.e., brain death). In the vegetative state, subcortical (upper brain stem and thalami) and cortical (primary somatosensory cortex; red circle) activation can be observed. However, this preserved cortical activation is limited to the primary cortex and fails to reach higher-order associative cortices from which it is functionally disconnected. In healthy volunteers, reporting the stimuli as painful, stimulation resulted in a wide neural network activation (the so-called pain matrix) including the anterior cingulate cortex (green ellipse). Data from Laureys S, Faymonville ME, Peigneux P, et al. (2002) Cortical processing of noxious somatosensory stimuli in the persistent vegetative state. *Neuroimage* 17: 732–741 and shown on 'glass brains.'

MCS

PET studies measuring cerebral metabolism at rest are unable to reliably differentiate vegetative from minimally conscious individuals. Functional imaging can here be of utmost value in objectively differentiating activation patterns measured during external stimulation characteristic of either clinical entity. Complex auditory stimuli with emotional valence, such as personalized narratives or the patient's own name, activate language-related cortical areas not observed during presentation of meaningless stimuli in MCS. Such context-dependent higher-order auditory processing in MCS patients (often is not assessable at the patient's bedside) indicate that content does matter when talking to MCS patients.

However, given the absence of a thorough understanding of the neural correlates of consciousness, functional neuroimaging results must be used with caution as proof or disproof of awareness in severely brain-damaged patients. Recently, Adrian Owen, from Cambridge, proposed a more powerful approach for identifying 'volition without action' in noncommunicative brain-damaged patients (rather than the use of passive external stimulation paradigms): scanning patients when they are asked to perform a mental imagery task. Reproducible and anatomically specific activation in individual patients during tasks that unequivocally require 'willed action' or intentionality for their completion unambiguously reflect awareness. Of course, negative findings in the same circumstances cannot be used as evidence for lack of awareness. In one exceptional VS patient, task-specific activation was observed, unequivocally demonstrating conscious awareness in the absence of reliable behavioral motor signs of voluntary interaction with the environment. It is interesting that the patient subsequently recovered. Other studies also showed that VS patients with

atypical brain activation patterns on functional neuroimaging afterwards showed clinical signs of recovery of consciousness, albeit sometimes many months later.

Investigations using MRI diffusion tensor imaging, which gauges the integrity of the white matter, are increasing scientific understanding of the brain mechanisms underlying recovery from VS. A team led by Nicholas Schiff of Cornell University, for instance, recently used diffusion tensor imaging to show the regrowth of axons in the brain of Terry Wallis, an Arkansas man in posttraumatic MCS who started talking in 2003 after 19 years of silence.

LIS

Classically, the EEG is relatively normal (or minimally slow) and reactive to external stimuli in LIS, but unreactive alpha rhythms (i.e., 'alpha coma' patterns) may also be observed. Cognitive ERPs and brain–computer interfaces may document awareness and permit communication in the extremely challenging cases of complete LIS. Residual cognition in LIS has long remained *terra incognita*. Recently, standard neuropsychological testing batteries have been adapted and validated for eye response communication. In classical LIS caused by a brain stem lesion, these studies have shown preserved attention, memory, executive functioning, and phonological and lexicosemantic performance.

PET scanning has shown significantly higher metabolic levels in the brains of patients in LIS compared with patients in VS. Voxel-based comparisons with healthy controls showed that no supratentorial cortical area has a significantly impaired metabolism in classical LIS. Conversely, hyperactivity was observed in bilateral amygdala of acute, but not chronic, LIS patients. The amygdala is known to be involved in emotions, especially negative emotions such as fear and anxiety. The absence of metabolic signs of reduced function in any area of the gray matter reemphasizes the fact that LIS patients suffer from a pure motor de-efferentation and recover an entirely intact intellectual capacity. The increased activity in the amygdala might relate to the terrifying situation of an intact awareness in a mute but sensitive being. Healthcare workers should be aware of this condition, adapt their bedside behavior, and consider pharmacological anxiolytic therapy.

Conclusion

Brain death, coma, VS, and MCS represent different pathological alterations of both dimensions of consciousness (involving arousal and awareness) or, for LIS, of the motor signs of consciousness. At the bedside, the evaluation of conscious perception and cognition in these conditions is difficult and sometimes erroneous. Electrophysiological and functional neuroimaging studies can objectively describe the regional distribution of cerebral activity at rest, under various conditions of passive stimulation, and during 'active' mental imagery tasks. These studies are increasing our understanding of the neural correlates of arousal and awareness and will improve the diagnosis, prognosis, and management of disorders of consciousness. But at present, much more data and methodological validation are awaited before functional neuroimaging studies can be proposed to the medical community as a tool for disentangling the clinical gray zone that separates conscious from unconscious survivors of acute brain damage.

See also: Reticular Activating System.

Further Reading

Alkire M, Pomfrett CJ, Haier RJ, et al. (1999) Functional brain imaging during anesthesia in humans: Effects of halothane on global and regional cerebral glucose metabolism. *Anesthesiology* 90: 701–709.

American Congress of Rehabilitation Medicine (1995) Recommendations for use of uniform nomenclature pertinent to patients with severe alterations of consciousness. *Archives of Physical Medicine and Rehabilitation* 76: 205–209.

Bernat JL (2006) Chronic disorders of consciousness. *Lancet* 367: 1181–1192.

Giacino JT, Ashwal S, Childs N, et al. (2002) The minimally conscious state: Definition and diagnostic criteria. *Neurology* 58: 349–353.

Giacino JT, Hirsch J, Schiff N, and Laureys S (2006) Functional neuroimaging applications for assessment and rehabilitation planning in patients with disorders of consciousness. *Archives of Physical Medicine and Rehabilitation* 87: 67–76.

Laureys S, Faymonville ME, Peigneux P, et al. (2002) Cortical processing of noxious somatosensory stimuli in the persistent vegetative state. *Neuroimage* 17: 732–741.

Laureys S (2005) The neural correlate of (un)awareness: Lessons from the vegetative state. *Trends in Cognitive Sciences* 9: 556–559.

Laureys S (2005) Science and society: Death, unconsciousness and the brain. *Nature Reviews Neuroscience* 6: 899–909.

Laureys S, Owen AM, and Schiff ND (2004) Brain function in coma, vegetative state, and related disorders. *Lancet Neurology* 3: 537–546.

Laureys S, Pellas F, VanEeckhout P, et al. (2005) The locked-in syndrome: What is it like to be conscious but paralyzed and voiceless? *Progress in Brain Research* 150: 495–511.

Majerus S, Gill-Thwaites H, Andrews K, and Laureys S (2005) Behavioral evaluation of consciousness in severe brain damage. *Progress in Brain Research* 150: 397–413.

Multi-Society Task Force on PVS (1994) Medical aspects of the persistent vegetative state (1). *New England Journal of Medicine* 330: 1499–1508.

Owen AM, Coleman MR, Boly M, Davis MH, Laureys S, and Pickard JD (2006) Detecting awareness in the vegetative state. *Science* 313: 1402.

Perrin F, Schnakers C, Schabus M, et al. (2006) Brain response to one's own name in vegetative state, minimally conscious state, and locked-in syndrome. *Archives of Neurology* 63: 562–569.

Quality Standards Subcommittee of the American Academy of Neurology (1995) Practice parameters for determining brain death in adults (summary statement). *Neurology* 45: 1012–1014.

Schiff ND, Rodriguez-Moreno D, Kamal A, et al. (2005) fMRI reveals large-scale network activation in minimally conscious patients. *Neurology* 64: 514–523.

Tavalaro J and Tayson R (1997) *Look Up for Yes.* New York: Kodansha America.

Vigand P and Vigand S (2000) *Only the Eyes Say Yes.* New York: Arcade.

Voss HU, Uluc AM, Dyke JP, et al. (2006) Possible axonal regrowth in late recovery from the minimally conscious state. *Journal of Clinical Investigation* 116: 2005–2011.

CIRCADIAN

Circadian Regulation by the Suprachiasmatic Nucleus

D J Earnest, Texas A&M University Health Science Center, College Station, TX, USA

Functional Properties of the Mammalian Circadian Clock in the Suprachiasmatic Nucleus

With the prevailing evidence for the endogenous rather than exogenous regulation of circadian or 24 h rhythms in multicellular organisms, identification of the internal biological clock(s) responsible for the generation of these oscillations became the primary focus of many investigations. Through functional studies determining whether the removal of neuroendocrine structures and destruction of nervous tissue alter the expression of overt rhythmicity, the circadian clock in mammals was eventually localized to the hypothalamic suprachiasmatic nucleus (SCN). Ablation of the SCN eliminates circadian rhythms in many behavioral, endocrine, and biochemical processes throughout the body. Transplantation studies have provided unequivocal evidence for SCN function in the central neural control of mammalian circadian rhythms. Experiments using donors and hosts with different circadian genotypes demonstrate that transplantation of fetal SCN tissue restores behavioral rhythmicity in SCN-lesioned hosts and that a specific circadian property, namely period, is conveyed by transplanted pacemaker cells, not the surviving SCN cells in the host, to the reestablished rhythms. SCN-independent clocks play a role, albeit limited, in the generation of circadian rhythms because rhythmic behavioral activity persists in SCN-lesioned rodents treated with methamphetamine or maintained on restricted feeding schedules. Neural substrates for these extra-SCN clocks are unknown, but recent studies implicated the dorsomedial hypothalamic nucleus (DMH) in the circadian regulation of food-anticipatory activity.

In addition to orchestrating circadian rhythms in other brain regions and peripheral systems, the SCN generates self-sustained oscillations in its cellular activities independent of external input. Rhythms in neuropeptide and neurotransmitter levels, cellular metabolism, and neuronal activity are indigenous to the SCN *in vivo* and persist for multiple cycles following the isolation of SCN cells *in vitro*. Circadian oscillations in gene expression are another fundamental property of SCN cells. Rhythmic expression profiles are especially prevalent in the SCN among genes that

form the core of the circadian clock mechanism. The molecular clockworks consist of interlocked transcriptional/translational feedback loops in which the expression of core components, *Bmal1*, *Period1* (*Per1*), *Per2*, *Cryptochrome1* (*Cry1*), *Cry2*, and *Rev-erbα*, is rhythmically regulated through mutual interactions between their protein products.

Hierarchy and Cell-Autonomous Nature of Circadian Clocks in Mammals

The molecular dissection of the circadian timekeeping mechanism also brought about the discovery that clock properties are exhibited not only by the SCN but also by other tissues *in vivo* and fibroblast cell lines *in vitro*. The expression and rhythmic regulation of clock genes has been observed in many peripheral cells and tissues, ranging from the liver to endocrine tissues to the heart and skeletal muscles. Clock gene oscillations in peripheral tissues are comparable to those found in the SCN but phase delayed by 4–8 h. Similar to peripheral tissues, Rat-1 and NIH/3T3 fibroblasts exhibit induced oscillations of clock gene expression for several cycles in response to serum shock, forskolin, or corticosteroid treatment. The common molecular organization of the circadian clocks in the SCN, peripheral tissues, and fibroblast cell lines challenged existing dogma on SCN function in the mammalian circadian system by demonstrating that the SCN clock was one among many others throughout the body. These findings also raised many questions about the hierarchical organization of mammalian circadian clocks and functional implications of this canonical clock machinery in the SCN, peripheral tissues, and fibroblast cell lines: Is the SCN a master clock that regulates all or just some of these peripheral clocks? Beyond these molecular oscillations, what other properties are similar in SCN and peripheral clocks? Is clock gene rhythmicity sufficient for peripheral clocks to pace rhythms in downstream cells or tissues, and if not, what distinguishes the SCN from peripheral clocks? What SCN output signals are important in coordinating circadian oscillations in peripheral tissues?

Within the mammalian circadian hierarchy, the SCN appears to function as a master clock that coordinates rhythmicity in other brain regions and many, but not all, peripheral tissues (**Figure 1**). Consistent with this hierarchical organization, circadian oscillations of gene expression *in vitro* are more robust within the SCN than the peripheral tissues. Circadian rhythms of luciferase-reported *Per1* expression

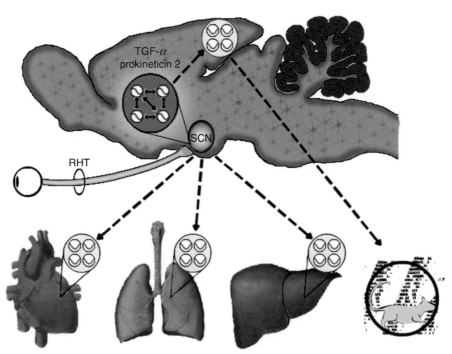

Figure 1 Hierarchy of cell-autonomous clocks in the mammalian circadian system. Entraining light–dark cues are communicated from the retina directly to the master circadian clock located in the suprachiasmatic nucleus (SCN) via the retinohypothalamic tract (RHT). The SCN clock consists of numerous cell-autonomous clocks that, in turn, synchronize the circadian oscillations of clock cells in other brain regions and peripheral tissues such as the heart, lung, and liver. Signals responsible for the local coordination of circadian oscillations among individual clock cells may play a role in the distinctive functional properties of the SCN. SCN control of circadian clocks in other tissues and of behavioral rhythmicity is mediated by diffusible or humoral signals. Many of the signals necessary for SCN pacemaker function are unknown, but transforming growth factor (TGF)-α and prokineticin 2 (PK2) are SCN output signals involved in the circadian regulation of locomotor activity.

persist for many weeks in explant cultures of the SCN but damp after several cycles in cultured lung, liver, kidney, and skeletal muscle tissue. Further evidence for the hierarchical relationship between the SCN and peripheral clocks is based on *in vitro* studies using *Per2-Luc* knockin mice to demonstrate that lesions of the SCN disrupt the normal phase coordination of clock gene rhythmicity among explant cultures of different peripheral tissues. However, some peripheral tissues, such as the liver, are not slaves to the SCN clock but instead can function independently in response to nonphotic cues. Restricted feeding cycles entrain the liver rhythm in *Per1-Luc* expression without influencing the circadian phase of this oscillation in the SCN. The prevalence of this common timekeeping program throughout the body and its functionality even in the absence of the SCN indicate that peripheral clocks provide for the local organization of tissue- or cell-specific processes in time.

Similarities between the SCN and peripheral tissues extend beyond core molecular components to even the basic cellular organization of the circadian clockworks. In the SCN, circadian oscillations are not only an ensemble property of the entire nucleus but are also expressed by individual neurons. Like the composite SCN *in vivo* and in brain-slice preparations,

many individual SCN neurons in dissociated cultures maintained on microelectrode arrays express independent oscillations in firing rate. The circadian phase and period of these SCN rhythms in firing rate differ widely among individual neurons despite the presence of synaptic connections. SCN cells in brain-slice preparations also exhibit ensemble rhythmicity in *Per-1*-driven green fluorescent protein (GFP) expression, and the waveform of this rhythm is derived from the multiphasic organization of autonomous oscillations in individual neurons. Recent findings suggest that peripheral tissues may contain cell-autonomous clocks similar to those found in the SCN. Individual NIH/3T3 fibroblasts exhibit circadian rhythms of gene expression that are largely synchronous following serum shock but are marked by broad phase differences in the absence of this treatment.

Functional Differences between Circadian Clocks in the Suprachiasmatic Nucleus, Peripheral Tissues, and Fibroblasts

Despite their apparent parallels at the molecular and cellular levels of organization, the SCN and peripheral clocks are fundamentally different. The foremost distinction is that the SCN contains the only known clock cells that can be directly entrained by light. Photoentrainment signals from the retina are processed through the retinohypothalamic tract (RHT), which terminates in the SCN (**Figure 1**). Evidence for the essential role of this retinofugal pathway in circadian photoentrainment is based on the finding that rodent activity rhythms persist but fail to entrain to light–dark cycles following localized transection of the RHT. In contrast, no data are currently available to unequivocally demonstrate that peripheral tissues in mammals also receive direct photic input. In the circadian hierarchy, the SCN clock thus provides a vital link between the external environment and the internal circadian timekeeping system. Further evidence for the preeminent role of the SCN as the gateway for the coordination of local processes with the light–dark environment is provided by the observation that clock gene rhythms are more rapidly entrained by light in the SCN than in peripheral tissues. Moreover, the 4–12 h phase delay of peripheral tissue oscillations relative to SCN rhythms is presumably indicative of the time required for entrainment signals to reach the periphery.

Distinctions between the SCN and peripheral clocks are especially evident in their functional properties.

SCN cells function as pacemakers and confer rhythmicity to SCN-lesioned rodents *in vivo* and to other cells *in vitro*, whereas peripheral tissues and cultured fibroblasts do not. Incontrovertible evidence for the unique pacemaker properties of the SCN is founded on transplantation studies. Fetal SCN tissue and immortalized rat SCN cells (SCN2.2), but not other types of tissue (e.g., cerebellum or cerebral cortex) or fibroblasts, reinstate circadian behavioral rhythmicity when transplanted into arrhythmic SCN-lesioned hosts. Similar to transplantation studies, the development of a coculture model has been instrumental in distinguishing the circadian pacemaker properties of SCN cells from the oscillatory behavior of fibroblasts. In this model, SCN2.2 cells are cocultured on companion plates with NIH/3T3 fibroblasts on cell-impermeable inserts. Consistent with SCN function *in vivo*, SCN2.2 cells endogenously generate circadian oscillations of glucose utilization and clock gene expression and confer this metabolic and molecular rhythmicity to cocultured NIH/3T3 fibroblasts (**Figure 2**). Serum-shock treatment produces rhythms of clock gene expression, but not glucose metabolism, in these fibroblasts. Despite the induction of clock gene oscillations, serum-shocked NIH/3T3 fibroblasts lack circadian pacemaker properties and fail to confer even this molecular rhythmicity to cocultured cells.

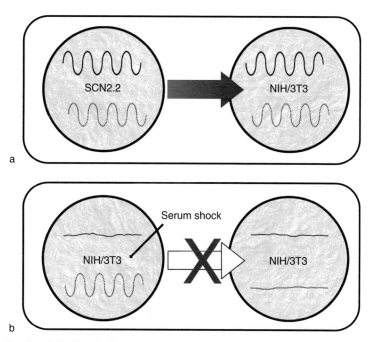

Figure 2 Diagram illustrating the distinctive rhythm-generating and pacemaker properties of SCN2.2 cells in coculture studies: (a) SCN2.2–NIH/3T3 cocultures; (b) serum-shocked (SS) NIH/3T3–untreated NIH/3T3 cocultures. For each cell type the cocultures, the upper (solid) waveform represents the temporal pattern of metabolic activity (2-deoxyglucose uptake) and the lower (dashed) waveform denotes the profiles of clock gene expression (*Per1* and *Per2*). The thick arrows between culture types indicate the presence or absence of signals that coordinate oscillations in cocultured cells. SCN2.2 cells, but not serum-shocked NIH/3T3 fibroblasts, are capable of pacing the metabolic and clock gene rhythms in cocultures of untreated NIH/3T3 cells.

Why peripheral tissues and fibroblasts cannot function as circadian pacemakers is presently unclear. One possible explanation is that all molecular elements of the clock mechanism may not be present or functionally configured. Experiments examining the effects of antisense inhibition of *Clock* reveal important distinctions between SCN2.2 cells and serum-shocked fibroblasts in the regulation of molecular oscillations. Treatment of SCN2.2 cells with *Clock* antisense oligonucleotides disrupted the circadian rhythm of *Per2* expression within these cells and cocultures of untreated NIH/3T3 fibroblasts. However, the circadian oscillations of *Per2*, *Bmal1*, and *Cry1* expression in serum-shocked fibroblasts showed some alterations in rhythm amplitude but otherwise persisted during antisense inhibition of *Clock*. These differences in *Clock* interactions within the molecular feedback loop could account for distinctions in the circadian pacemaker properties of SCN cells and fibroblasts.

The Search for Critical Output Signals of the Suprachiasmatic Nucleus

An alternative explanation for the distinctive pacemaker function of the SCN is that specific output signals necessary for coordinating rhythmicity among individual clock cells within the SCN and in the periphery may not be expressed or circadian-regulated in other tissues or stimulated fibroblasts. SCN pacemaker regulation of circadian rhythms in downstream cells and tissues is thought to involve diffusible or humoral factors rather than neurotransmission via synaptic connections. Functional implications for diffusible factors in the SCN pacemaker regulation of circadian rhythms were first provided by the observation that transplantation of fetal hypothalamic tissue into the third, or even lateral, ventricle of SCN-lesioned rodents restores circadian rhythms with few, if any, neural connections between the graft and host brain. However, the role of humoral outputs in the circadian pacemaker function of the SCN is directly supported by transplantation experiments determining whether the recovery of circadian function occurs in SCN-lesioned animals when grafted cells are encapsulated so as to prevent the establishment of neural connections with the host brain. Despite their isolation in polymeric capsules that allow graft–host communication only via diffusion of humoral signals, transplants of SCN cells still promote the recovery of circadian behavioral rhythmicity with the donor period in SCN-lesioned hamsters. Coculture studies similarly demonstrate that the rhythmic release of some paracrine signal(s) by the SCN is responsible for its function in pacing circadian oscillations in downstream cells. SCN2.2 cells impose rhythms of

metabolic activity and *Per* gene expression on cocultured NIH/3T3 fibroblasts, and these conferred oscillations are mediated by diffusible signals because the two cell types are separated in the coculture environment by a semipermeable membrane preventing physical contact (**Figure 2**). Furthermore, sustained rhythmicity in NIH/3T3 cells is dependent on continued exposure to this SCN2.2-communicated signal. Following their coculture and subsequent isolation from one another, SCN2.2 cells continue to express circadian rhythms of glucose utilization, whereas NIH/3T3 fibroblasts fail to exhibit self-sustained metabolic oscillations in the absence of SCN2.2-specific outputs. Therefore, identification of specific diffusible or humoral factors that coordinate rhythmicity among individual clock cells in the SCN and among circadian clocks in peripheral tissues is critical in determining which cellular processes are responsible for the distinctive function of the SCN as a circadian pacemaker.

Glucocorticoids were among the first signaling candidates implicated in the coordination of circadian oscillations within peripheral clocks. The SCN regulates the cyclical secretion of glucocorticoid hormones, and this signaling pathway is thought to mediate SCN control of peripheral clocks *in vivo* because adrenalectomy alters the light-induced phase resetting of circadian gene expression in the periphery. Glucocorticoid receptor agonists also induce clock gene oscillations in cultured fibroblasts. Transforming growth factor α (TGF-α) and prokineticin 2 (PK2) were subsequently identified as SCN output signals that play a role in its pacemaker function, at least with regard to the circadian regulation of locomotor activity *in vivo* (**Figure 2**). TGF-α is rhythmically expressed in the SCN, with peak mRNA levels occurring during the mid-subjective day, and chronic infusion of TGF-α into the third ventricle completely inhibits wheel-running activity and disrupts the circadian organization of the sleep–wake cycle. These effects of TGF-α are mediated through its primary receptor, the epidermal growth factor receptor (EGFR) because mice with a partial loss-of-function mutation for EGFR exhibit increased daytime activity and decreased inhibition of wheel-running activity in response to light during the subjective night. In a similar fashion, PK2 mRNA levels rhythmically fluctuate in the SCN with peak expression during the subjective day, and intraventricular injections of recombinant PK2 during the subjective night, when endogenous levels are low, inhibit nocturnal increases in rat locomotor activity. Significantly, the receptors for TGF-α and PK2 (EGFR and PKR2) are distributed in primary targets of SCN efferent pathways, including the subparaventricular zone and paraventricular nucleus. This observation suggests that TGF-α and PK2 are not involved in the synchronization of individual cell-autonomous clocks within the SCN but, instead, are critical SCN outputs necessary for pacemaker coordination of downstream clocks in other brain regions controlling locomotor behavior. The neuropeptide vasoactive intestinal polypeptide (VIP) is another output signal that mediates SCN pacemaker function in the circadian regulation of locomotor behavior. Mice with a null mutation of the VIP receptor, VPAC$_2$, show weak or low-amplitude rhythms of wheel-running behavior in conjunction with the disruption of circadian oscillations in SCN expression of core clock genes. VPAC$_2$ receptors are more prevalent within the SCN than in primary targets of its efferent pathways. Coupled with evidence for the essential role of VPAC$_2$ receptor signaling in the generation of the circadian rhythm in SCN neuronal activity, the local distribution of these receptors suggests that VIP may be involved in regulating cell-to-cell communication in the SCN. Thus, VPAC$_2$ receptor signaling may coordinate rhythmicity among populations of cell-autonomous clocks so as to provide for the generation of high-amplitude circadian outputs necessary for SCN pacemaker function.

It is unlikely that glucocorticoid hormones, TGF-α, PK2, and VIP are solely responsible for SCN pacemaker function in coordinating the entire spectrum of circadian processes in downstream cells and tissues. Global analysis of gene expression profiles has provided a valuable screening tool for other output signals that may distinguish the SCN from peripheral and fibroblast clocks. Comparisons of microarray and bioinformatics data indicate that the overall extent of circadian gene expression in the SCN and peripheral tissues *in vivo*, as well as cultured SCN2.2 cells and fibroblasts, was similarly low, ranging from approximately 2% to 10%. However, specific genes with circadian profiles are highly disparate among different tissue and cell types. The canonical clock genes constitute the principal portion of the overlap in circadian gene expression between SCN cells and peripheral tissues or fibroblasts. Other rhythmic genes in common to different mammalian tissues predominantly function as nuclear factors involved in the regulation of DNA-binding and activation of transcription. This intriguing observation suggests that the cyclical regulation of nuclear factors may have important implications in linking the common clockworks to tissue-specific cellular and physiological pathways so as to provide local control over circadian timing. In addition, functional analyses indicate that various mammalian cells and tissues are similar with regard to the diversity of rhythmically expressed transcripts. Within the SCN *in vivo*, SCN2.2 cells, fibroblasts, and peripheral tissues such as the liver, heart, and kidney, transcripts with

cyclical profiles control a wide range of biological processes. Yet circadian gene expression is typically more prevalent within certain functional categories, and there is some degree of tissue and cell specificity in this differential distribution of rhythmic genes. In SCN2.2 cells and the SCN *in vivo*, functionally annotated genes with circadian expression profiles are predominantly involved in the regulation of cellular communication, protein dynamics, and metabolic pathways mediating fatty acid recycling, steroid synthesis, glucose metabolism, and mitochondrial energy transduction. The prevalence of rhythmicity among genes mediating intercellular communication may have some significance with regard to SCN-specific signals necessary for the endogenous generation of self-sustained oscillations, local synchronization of clock cell populations, or pacemaker coordination of circadian oscillations in other tissues and cells.

Among rhythmic genes in intercellular communication pathways, inducible nitric oxide synthase (*iNos* or *Nos2*), an isozyme involved in nitric oxide (NO) production, is noteworthy as a circadian signal that is unique to the SCN. The role of *Nos2* in the SCN and the regulation of circadian rhythms has not been examined, presumably because this isozyme is thought to be expressed by macrophages, not SCN cells, and because functional studies suggest that the other isoforms of nitric oxide synthase in the central nervous system are not necessary for the generation or entrainment of circadian rhythms. The activity rhythms of mutant mice lacking neuronal *Nos* (*nNos* or *Nos3*) and endothelial *Nos* (*eNos* or *Nos1*) show no distinct abnormalities with regard to their free-running patterns, entrainment to light–dark cycles, and phase-shifting responses to light. Initial evidence for *Nos2* involvement in the endogenous oscillatory and pacemaker properties of the SCN is derived from the recent observation that antisense inhibition of NOS2 expression abolishes the metabolic rhythmicity of both cell types in cocultures containing antisense-treated SCN2.2 cells and untreated NIH/3T3 fibroblasts. Moreover, *Nos2* function in SCN regulation of circadian rhythms is compatible with the effects of a general NOS inhibitor on rhythmic glucose uptake in SCN2.2–NIH/3T3 cocultures. Similar to *Nos2* antisense treatment, N^G-nitro-L-arginine methyl ester (L-NAME), a reversible inhibitor of all three isoforms of NOS, disrupts the circadian pattern of glucose utilization in SCN2.2 cells. Following treatment, glucose uptake fluctuated rhythmically in both L-NAME-treated SCN2.2 cells and untreated NIH/3T3 fibroblasts, but the peak-to-peak interval for this rhythmicity was increased from 24 to 32 h in both cell types. It is unclear how L-NAME and *Nos2* antisense treatment specifically produce this loss of rhythmicity in SCN2.2 cultures. One possibility is that the treatment-induced inhibition of NOS2 may disrupt the clock mechanism by altering the biological effects of NO signaling through the intracellular second-messenger cyclic guanosine monophosphate (cGMP) and/or on posttranslational modification of cellular proteins. Another potential explanation for the abolition of metabolic rhythmicity in SCN2.2 cells is that L-NAME and *Nos2* antisense treatment may inhibit NO function as a diffusible messenger in the intercellular coupling or local synchronization of clock cell populations. As coupling strength among a multitude of autonomous clocks decreases from the population mean, changes in rhythm amplitude or period length are expected to occur and eventually result in the loss of rhythmicity from the ensemble. Thus, the disruption of metabolic rhythms in SCN2.2 cultures may reflect treatment-mediated loss of temporal synchrony among populations of clock cells. By affecting the generation of high-amplitude circadian output signals, this desynchrony between clock cells could also account for the effects of *Nos2* antisense treatment on the pacemaker function of SCN2.2 cells in regulating oscillations in cocultured fibroblasts. Together with evidence for the role of NO in resetting the SCN clock, these findings suggest that *Nos2* function in the production of this diffusible messenger may be essential for the endogenous rhythm-generating and pacemaker properties of the SCN.

Future Developments

The localization of cell-autonomous circadian clocks in the SCN, other brain regions, and peripheral tissues throughout the body has yielded a new understanding of the mammalian circadian hierarchy. It is not yet clear whether the canonical clock machinery in peripheral tissues is directly linked to the circadian regulation of any physiological processes. However, this discovery has kindled the search for SCN output signals responsible for the coordination of circadian oscillations in peripheral clocks. Although some candidates for SCN pacemaking signals have been identified, determining their role in the local coupling between individual clock cells and the coordination of circadian oscillations in peripheral clocks remains a major challenge.

Further Reading

Aton SJ and Herzog ED (2005) Come together, right . . . now: Synchronization of rhythms in a mammalian circadian clock. *Neuron* 48: 531–534.

Balsalobre A (2002) Clock genes in mammalian peripheral tissues. *Cell & Tissue Research* 309: 193–199.

Cermakian N and Sassone-Corsi P (2000) Multilevel regulation of the circadian clock. *Nature Reviews Molecular Cell Biology* 1: 59–67.

Herzog ED and Schwartz WJ (2002) A neural clockwork for encoding circadian time. *Journal of Applied Physiology* 92: 401–408.

Kriegsfeld LJ and Silver R (2006) The regulation of neuroendocrine function: Timing is everything. *Hormones & Behavior* 49: 557–574.

Okamura H (2004) Clock genes in cell clocks: Roles, actions, and mysteries. *Journal of Biological Rhythms* 19: 388–399.

Panda S and Hogenesch JB (2004) It's all in the timing: Many clocks, many outputs. *Journal of Biological Rhythms* 19: 374–387.

Reppert SM and Weaver DR (2002) Coordination of circadian timing in mammals. *Nature* 418: 935–941.

Schibler U (2006) Circadian time keeping: The daily ups and downs of genes, cells, and organisms. *Progress in Brain Research* 153: 271–282.

Sleep: Development and Circadian Control

H C Heller, Stanford University, Stanford, CA, USA
M G Frank, University of Pennsylvania, Philadelphia, PA, USA

The states of arousal, wake, rapid eye movement (REM) sleep, and nonrapid eye movement (NREM) sleep, are physiologically and behaviorally well defined in adult mammals, and their temporal expression is clearly organized by the circadian system. The situation is not clear, however, in developing mammals. Comparing studies of the development of sleep/wake and circadian rhythms in different species of mammals is complicated because of the major differences in maturity at the time of birth. However, understanding the development of these systems can contribute to understanding their basic biology and their possible contributions to certain postnatal pathologies, such as sudden infant death syndrome.

Three Stages of Sleep Development

Altricial mammals are less fully developed at birth than are precocial mammals, and this difference is clearly reflected in the development of sleep/wake and circadian rhythmicity. Whereas precocial species have brain activity at birth that shows characteristic differences associated with adult arousal states, the electrical activity recorded from brains of altricial species is undifferentiated. This difference is not surprising given the fact that in altricial brains at the time of birth most axons have not yet reached their targets and much synaptogenesis has yet to occur. However, in both altricial and precocial species, the development of sleep/wake can be divided into three stages, some of which take place *in utero* in the precocial species. These stages are dissociation, concordance, and maturation.

Dissociation is the stage of development when the defining electroencephalography (EEG) characteristics of arousal states are absent but the other correlates of arousal states, such as changes in heart rate and respiratory rate and phasic or tonic motor activity, are present. However, these correlates are not coordinated with each other. In studies of premature human neonates, Hamburger applied the term 'presleep' to refer to the period prior to the emergence of the EEG components of sleep states. Similar situations have been observed *in utero* in studies of precocial species such as lambs and baboons. In altricial species such as the kitten, the absence of sleep EEG patterns and dissociation of the other hallmarks of sleep states are seen postnatally.

Concordance occurs with the emergence of differentiated EEG patterns characteristic of adult arousal states and the alignment of the other correlates of wake and sleep states with those EEG parameters. In precocial species, concordance occurs in late gestation and EEG definable arousal states are observable in the newborn. Concordance in humans begins between gestational weeks 28 and 32 and is completed sometime in the last month of fetal life in full-term infants. Throughout the period of concordance in precocial species, there is no evidence of transitional or precursor sleep states that are homologous with adult sleep states. Rather, the EEG characteristics of REM and NREM sleep emerge at approximately the same time and the other sleep parameters come into alignment with those EEG patterns.

Concordance in altricial species is not as clear and has generated some controversy. EEG patterns are undifferentiated in the newborn altricial animal, but motor patterns that appear similar to those associated with arousal states are evident. Specifically, these motor patterns are coordinated movements, jerky movements (twitches), or quiescence. Periods of tonic, coordinated motor activity are considered wake, and the rest of the time is considered to be sleep. Sleep is then divided into behavioral quiet sleep (BQS), characterized by quiescence, or behavioral active sleep (BAS), characterized by phasic twitches. Because the undifferentiated EEG pattern is relatively fast and low amplitude, it resembles the activated EEG that in adults is associated with wake and REM sleep. This association of an apparent activated EEG with phasic muscle activity (BAS) has been assumed to be a precursor of mature REM sleep, whereas quiescent periods (BQS) have been assumed to be the precursor of mature NREM sleep. However, the problem of equating an undifferentiated EEG with an activated EEG casts doubt on what appears to be an obvious continuity between these apparent precursor sleep states and adult sleep states defined by differentiated EEG patterns. We examine more critically the assumption that BAS is the precursor of REM sleep and that BQS is the precursor of NREM sleep later in this article. Note, however, that the terms 'active sleep' and 'quiet sleep' used to describe sleep states in premature and term human infants refer to different suites of variables than are described by these same terms in altricial species. The difference is that the EEG patterns of active and quiet sleep are clearly differentiated in human preterm and full-term infants. In contrast, BAS and BQS in altricial species are not distinguished by any differential EEG activity; rather, these states are distinguished primarily by the presence or absence of myoclonic twitches.

Following concordance of parameters into distinct sleep states, there are rapid and sometimes dramatic maturational changes in the amount and timing of those states and in some of the EEG characteristics that define them. Early in the maturation of sleep, the amount of REM sleep is much higher than adult levels and declines, whereas the amount of NREM sleep decreases to a lesser degree and wake increases. The characteristic brain activities that define the sleep states in the adult become increasingly well defined and in some cases altered. In altricial species, for example, the activated EEG of REM sleep precedes the development of the slower, highly regular theta rhythm, which has a frequency in the 4 or 5 Hz range early on and gradually increases to the characteristic adult frequency of 7 or 8 Hz. Pontine–geniculate–occipital waves do not appear until a week or more after the emergence of REM sleep, and the characteristic muscle atonia of REM sleep is even slower to develop. The two major EEG correlates of NREM sleep, slow waves (delta band, 0.5–4.5 Hz) and sleep spindles (7–14 Hz), develop at different rates. First to appear is slow wave activity, which in the rat steadily increases in amplitude from approximately postnatal day (P) 9 to P21, reaching higher levels than are seen in the adult. Sleep spindles appear in the third postnatal week. Relatively late to mature are the regulatory mechanisms governing the expression of sleep – the ultradian, circadian, and homeostatic controls.

BAS and BQS May Not Be Homologous with REM and NREM Sleep

In altricial species such as the rat, the undifferentiated EEG is the same in the behaviorally defined states of BAS and BQS. Since the undifferentiated EEG lacks high-amplitude wave patterns, it somewhat resembles the activated EEG of REM sleep. This resemblance, however, is superficial, as might be expected based on the fact that most axons traveling to the cortex have not yet reached their targets and most synaptogenesis has not yet occurred. Since REM sleep is a cholinergically mediated state, it is especially important to note that the cholinergic system is very immature at the time BAS is maximally expressed. There is an inverse relationship between BAS expression and (1) expression of muscarinic receptors and their signal transduction mechanisms, (2) the development of cholinergic nuclei and their efferents, and (3) neuronal responsiveness to cholinergic drugs. In fact, cholinergic blockade which eliminates adult REM sleep has no effect on BAS in the rat until late in the second postnatal week after differentiated EEG patterns have developed. Lesions of brain stem structures essential for mature REM sleep have no effect on BAS

in neonatal rats or kittens. Even transection of the neonatal rat's spinal cord below the brain stem does not eliminate the phasic twitches that define BAS. In adult mammals, such transections eliminate all episodes of myoclonia that normally accompany REM sleep. These results pose the obvious question of what is the nature of the myoclonia of BAS. It is likely that the twitches of BAS are 'nonspecific fetal activity' that is associated with the development of spinal reflex circuits prior to the imposition of descending inhibitory control.

There is also no compelling evidence for assuming that BQS is a precursor state to NREM sleep. The comparison of different studies of BQS is complicated by the lack of a clear set of criteria that define BQS. If the definition of BQS is quiescent periods with regular respiration and no twitches, this amounts to only 2% of total recording time in 1-week-old rats. However, in other studies in which small movements or 'startles' are allowed in BQS, it can be more than 30% of total recording time. Similar variances in the designation of BQS in kittens exist, all depending on the behavioral criteria used by the investigators.

The first EEG hallmark of sleep states to appear in altricial species is slow waves, and when they appear, they are not restricted to episodes of BQS. In fact, when slow waves first appear in the neonatal rat, they are as likely, if not more likely, to occur during episodes of BAS than during episodes of BQS. By the third week of postnatal life, slow wave activity is mostly restricted to periods of motor quiescence and regular respiration, and the amplitude of the slow waves steadily grows until they reach maximum values at approximately the beginning of the fourth week. EEG theta activity, which marks REM sleep, appears in the neonatal rat at approximately the same time as slow waves, but it matures to adult levels more slowly than do slow waves.

In summary, the view most consistent with existing data is that both REM and NREM sleep emerge from an undifferentiated brain state, and they do so at approximately the same time even though NREM sleep patterns seem to mature more rapidly than REM sleep patterns. Thus, in both precocial and altricial species of mammals, there are not precursor sleep states that precede EEG differentiation. Rather, the development of the EEG hallmarks of sleep reflects the development of the underlying neuronal mechanisms, and presumably therefore the yet-to-be-discovered functions of the sleep states.

Sleep Regulation

There are three components of sleep regulation that develop after the concordance of EEG defined states: ultradian, circadian, and homeostatic. Ultradian refers to the cyclicity of sleep states, or the sleep cycle. In most mammals, the distribution of total sleep time is approximately 80% NREM sleep and 20% REM sleep, but these sleep times are distributed into bouts that alternate with each other. The sleep cycle in the human adult is approximately 90 min and the sleep cycle in a rat is approximately 10 min. This cyclicity of sleep states is poorly developed at the time the states first emerge, and it develops gradually, with the cycles becoming more regular and longer.

Circadian refers to the daily cyclicity of sleep and wake which is dependent on the circadian pacemaker that resides in the suprachiasmatic nucleus (SCN) of the hypothalamus. Although the pacemaker is probably oscillating in the fetus and is probably entrained by the periodicity of the mother, its coupling to the systems controlling sleep states occurs considerably after the emergence of sleep states. In rats, the circadian organization of sleep begins to appear on approximately P15. In humans, circadian organization of sleep emerges in the second or third month after birth, but it takes longer to become entrained to the environmental cycle of light and dark.

Homeostatic regulation refers to the relationship between sleep and the duration of prior wake. In adult mammals, sleep homeostasis is robust. An increase in wake time results in an increase in sleep time, but most important, it results in an increase in the EEG slow wave activity in NREM sleep. Thus, whereas the effect of sleep deprivation on REM sleep is to increase its duration, the main effect of sleep deprivation on NREM sleep is to enhance the expression of slow wave activity, also called delta power. One striking developmental change is an increased tolerance of sleep pressure. The amount of wake is very low in neonatal humans and altricial species, and they are unable to sustain long episodes of wake. Short periods of sleep deprivation that would be insignificant in the adult result in clear sleep rebound in the neonate. However, the nature of the sleep rebound changes with age. Early on when there are robust rebounds in NREM sleep, there are none in REM sleep. In the rat, responses in REM sleep time to prior sleep deprivation first appear toward the end of the third postnatal week. Thus, as with the development of the EEG characteristics of sleep, the homeostatic characteristics of REM sleep seem to emerge later and develop more slowly than those of NREM sleep. A most striking change in the homeostatic regulation of NREM sleep occurs in the fourth postnatal week. Prior to this time, the response to sleep deprivation is increased NREM sleep time with no changes in NREM sleep slow wave activity. In the fourth postnatal week, the response of the rat pup to sleep deprivation is no longer an increase in NREM sleep

time but, rather, an increase in slow wave activity or delta power. The significance of this sudden shift is not known, but it occurs after the time when normal delta power levels have reached their maximum and are declining and at the time when the ability of the animal to sustain longer episodes of wake is rapidly increasing. These observations indicate that the generation of EEG slow waves is a critical aspect of the restorative function of sleep.

Development of the Circadian System

The circadian system of mammals is practically synonymous with the SCN of the hypothalamus, but it also involves the pineal gland, which produces the hormone melatonin. In contrast to the complexity of the neural structures involved in sleep, the SCN and the pineal are discreet structures that can be studied anatomically, neurochemically, and functionally. Therefore, we can ask questions about the development of the circadian system at these levels.

The SCN is a compact but heterogeneous cluster of cells with a definite structure, yet the relationship between the organization of the SCN and its function as the circadian pacemaker is not understood. SCN cells arise from the germinal epithelium of the anterior, ventral diencephalon. In the rat, this begins on approximately embryonic day 14 (E14) and is complete by several days before birth. The fate determination of putative SCN cells is quite robust in that deafferentation of the area does not alter their differentiation, and even if they are excised and transplanted into the chamber of the eye, they still develop into SCN cells. Also, when SCN tissue is extracted in early postnatal life and placed in culture, the cells continue to express the properties of SCN cells, including circadian rhythmicity.

There are three major afferent pathways to the SCN, and their development has been described in the rodent. Information from photoreceptors reaches the SCN by a direct and an indirect route. The direct route is via the retinohypothalamic tract, in which the neurotransmitters are glutamate and pituitary adenylate cyclase-activating polypeptide (PCAP), and these connections develop between P1 and P10. The indirect pathway for photic information through the intergeniculate leaflet uses the neurotransmitter neuropeptide Y and develops between P4 and P11. The third set of projections are serotonergic from the midbrain raphe nuclei, and they develop between the day of birth and P10. These pathways are the major mechanisms of entrainment in the adult mammal, but as discussed later, entrainment of the fetal SCN occurs considerably before birth. Thus, additional mechanisms must be responsible for the initial entrainment of the fetal SCN. Strong evidence shows that melatonin and dopamine may be two such factors that mediate the maternal entrainment of the circadian systems of her pups, and fetal SCN cells do express the receptors for these two neurochemicals.

Functional development of the circadian system involves: (1) pacemaker function, or the generation of rhythms; (2) entrainment of rhythms; and (3) the coupling of SCN rhythmicity to drive other variables, such as body temperature and sleep/wake. The first of these properties to develop, of course, is pacemaker function. Circadian cycles are observable in the SCN within days after the cells become postmitotic. The first evidence of rhythmicity in the fetal SCN was documented in rats by the differences in uptake of 2-deoxyglucose (2DG) by SCN cells when the 2DG was injected at two different times of day. The earliest these oscillations were observed in rats was approximately 3 days before birth. Of course, these changes in fetal SCN metabolism could have been solely imposed by a maternal influence. However, subsequent studies revealed daily changes in 2DG uptake and in electrical activity of the SCN in brain slices taken from fetuses. Such rhythms have also been observed in fetuses or in brain slices from fetuses after the dam received SCN lesions early in gestation and was herself arrhythmic. The evidence supports the view that pacemaking function develops in the SCN prior to birth.

The ability of the SCN oscillations to be entrained also develops before birth in rodents, but the entraining stimuli come from the mother. In normal circumstances, the newborn pups are in synchrony with each other and with the maternal circadian rhythm. If the dam has received an SCN lesion early in gestation and is therefore arrhythmic, the newborn pups express circadian rhythms, but they are not in synchrony with each other. Most telling are cross-fostering experiments in which the newborn pups are presented to mothers that are 180 degrees out of phase from their natural mothers. In these litters, when rhythmicity is expressed in the pups, it is in phase with the rhythm of their birth mothers and not their foster mothers. The fetal SCN seems to be highly sensitive to entraining signals, including melatonin, dopamine, and probably other compounds such as the cholinergic drug nicotine. Whether or not disruption of fetal rhythms by exposure to such compounds *in utero* has effects on postnatal health is not known. However, when pregnant ewes were kept in constant light, their circadian rhythms were disrupted, as evidenced by vasopressin levels in the cerebrospinal fluid, and their fetuses did not survive. When kept on a light/dark cycle, their circadian rhythms were normal and their fetuses survived. In the postnatal period, the

sensitivity of the SCN to the entraining influence of light develops as the afferent pathways bringing photic information to the SCN develop.

The coupling of SCN oscillations to various physiological and behavioral properties of the newborn emerges at different postnatal ages. In the rat, a rhythm of pineal N-acetyltranferase (the rate-limiting enzyme for melatonin synthesis) activity develops in the first postnatal week, a rhythm of body temperature develops in the second postnatal week, and a rhythm of sleep/wake appears in the third postnatal week. In the human newborn, a clear day/night organization of sleep/wake develops by the third month of life.

Sleep and Circadian Development Continues throughout Life

Focusing on the human, it is known that many characteristics of sleep and circadian rhythmicity change from birth through childhood, adolescence, adult, middle age, and old age. Total sleep time (TST) is maximal in the neonate and may reach 17 or 18 h a day, with almost half of that being REM sleep. Circadian organization of sleep is absent in the neonate but gradually develops between the second and third months of life. When circadian organization appears, it is in phase with the environment, so entrainment of the pacemaker occurs before the coupling of the circadian system to sleep control. By age 5 years, TST declines to approximately 10–12 h per day, with only approximately 20% of that being REM sleep. Thus, the major change in sleep amounts over early childhood is the decrease in REM sleep.

Adolescence is a time of significant changes in characteristics of sleep and circadian rhythms. The spectral properties of the NREM sleep EEG show considerable change at this time. The power density in the slow wave band (delta power) is high during childhood and at a maximum at approximately 9 or 10 years. Delta power then declines significantly (approximately 25%) between ages 12 and 14 years. After that, delta power continues to fall slowly and steadily with age. Since delta power is the definitive characteristic of stages 3 and 4 NREM sleep, these stages decrease linearly with age and are practically nonexistent in the elderly. REM sleep amounts, in contrast, show no major change with adolescence and follow a modest, linear decline from an average of approximately 22% in childhood to approximately 18% in old age.

The major circadian changes that occur after infancy are a phase advance of the timing of sleep onset at adolescence and a gradual dampening of the amplitude of circadian rhythms with advanced age.

To a certain extent, this leveling of circadian organization of sleep/wake may be due to declining levels of physical activity during the day. Sleep during the day leads to a lessening of sleep pressure at night, and hence reduced quality of nocturnal sleep.

See also: Circadian Rhythms in Sleepiness, Alertness, and Performance.

Further Reading

Blumberg MS, Karlsson KA, Seelke AMH, and Mohns EJ (2005) The ontogeny of mammalian sleep: A response to Frank and Heller (2003). *Journal of Sleep Research* 14: 91–98.

Carskadon MA and Dement WC (2005) Normal human sleep: An overview. In: Kryger MH, Roth T, and Dement WC (eds.) *Principles and Practice of Sleep Medicine*, pp. 13–23. New York: Elsevier.

Davis FC, Frank MG, and Heller HC (1999) Ontogeny of sleep and circadian rhythms. In: Turek FW and Zee PC (eds.) *Regulation of Sleep and Circadian Rhythms*, pp. 19–79. New York: Dekker.

Frank MG and Heller HC (2003) The ontogeny of mammalian sleep: A reappraisal of alternative hypotheses. *Journal of Sleep Research* 12: 25–34.

Frank MG and Heller HC (2005) Unresolved issues in sleep ontogeny: A response to Blumberg et al. *Journal of Sleep Research* 14: 98–101.

Sleep and Waking in Drosophila

C Cirelli and **G Tononi**, University of
Wisconsin–Madison, Madison, WI, USA

The Molecular Correlates of Sleep May Offer Clues to Its Functions

All animal species studied so far sleep, and sleep may serve multiple functions, which may differ in different species. It is possible, however, that sleep also has a core function that is conserved from invertebrates to mammals. If this is the case, that function is most likely a cellular one, because flies and humans share most pathways for intercellular and intracellular signaling, from membrane receptors and ion channels to nuclear transcription factors, but differ significantly in the number, anatomy, and complexity of brain circuits. The idea that the functions of sleep may ultimately relate to cellular and molecular aspects of neural function is not new. Giuseppe Moruzzi argued that "... sleep concerns primarily not the whole cerebrum, nor even the entire neocortex, but only those neurons or synapses, and possibly glial cells, which during wakefulness are responsible for the brain functions concerned with conscious behavior...." (Moruzzi, 1972). Moruzzi suggested that neural cells or synapses supporting waking conscious activity undergo plastic changes that make sleep necessary, although he did not speculate about the mechanisms. Others have suggested that sleep may help in restoring brain energy metabolism, may be needed to maintain the synaptic efficacy of the neural circuits not frequently used during wakefulness, or may favor a generalized synaptic downscaling following waking-induced synaptic potentiation. It has also been suggested that the rich spontaneous activity of neocortical neurons during slow-wave sleep may consolidate memory traces acquired during wakefulness. Several of these hypotheses are not mutually exclusive, and sleep may actually favor different cellular processes.

The analysis of the molecular correlates of sleep and wakefulness may help to understand the benefits that sleep brings at the cellular level. More specifically, the identification of the genes whose expression changes in the brain between sleep and wakefulness may clarify if, and why, brain cells need sleep and why their functions are impaired if they are prevented from doing so during sleep deprivation. In the last 15 years, subtractive hybridization, mRNA differential display, and microarrays have emerged as new powerful methods in transcriptomics, the study of the complete set of RNA transcripts produced by the genome at any one time. Microarrays are currently the most comprehensive and sensitive method to detect genome-wide expression changes, and have allowed the identification of hundreds of transcripts whose expression is modulated by behavioral state. In addition, transcriptomics analysis has also identified functional categories of genes whose expression varies during sleep, wakefulness, and after different periods of sleep deprivation. At least some of these gene categories are conserved from invertebrates to mammals.

Table 1 lists the studies that used transcriptomics approaches to identify genes whose transcript (mRNA) levels vary as a function of sleep, wakefulness, and/or sleep deprivation. It is clear from the table that the main focus has been so far on the cerebral cortex, although other brain regions have recently been studied, including cerebellum, hypothalamus, and brain stem. The cerebral cortex is of major interest because it is responsible for the cognitive defects observed after sleep deprivation, and it is at the center of most hypotheses concerning the functions of sleep.

Several general conclusions can be drawn from the studies listed in **Table 1**. Each of them is further discussed below.

1. Hundreds of genes in the brain change their expression due to a change in behavioral state.
2. Some genes increase their expression during sleep, others during waking and sleep deprivation.
3. Sleep and waking genes belong to different functional categories.
4. Three functional categories of waking genes are conserved from flies to mammals.
5. The noradrenergic system plays a role in the induction of some waking genes.
6. The expression of several genes changes after chronic, but not acute sleep loss.

Hundreds of Genes in the Brain Change Their Expression between Sleep and Waking

A first important finding of the recent transcriptomics studies is that gene expression in the brain changes extensively as a function of behavioral states, sleep, spontaneous wakefulness, short-term (a few hours) or long-term (several days) sleep deprivation. Five percent or more of the transcripts tested in the brain show changes in gene expression according to behavioral state. Interestingly, a similar percentage of genes (from 1% to 10%) changes because of differences in

Table 1 Published studies that used transcriptomics approaches to identify molecular correlates of sleep (S), wakefulness (W), and sleep deprivation (SD)

Reference	Method used/N of tested transcripts	Experimental details	Major findings
Rhyner et al. *Eur J Neurosci* 2: 1063–73, 1990	Subtractive hybridization (~4000 transcripts)	24 h SD (forced locomotion) vs. time-matched controls; rat forebrain	four transcripts ↓ by SD, six transcripts ↑ by SD, all unknown
Cirelli and Tononi *Mol Brain Res* 56: 293–305, 1998	mRNA differential display (~9000 transcripts)	3 h S, W, SD (novel objects); rat cerebral cortex	one unknown transcript ↑ in S; 11 transcripts ↑ in W-SD, including *c-fos, NGFI-A,* and several mitochondrial genes
Cirelli and Tononi *Brain Res* 885: 303–321, 2000	mRNA differential display (~9000 transcripts) + cDNA arrays (Clontech, 1176 transcripts)	8 h S, W, SD (novel objects); rat cerebral cortex	12 transcripts ↑ in S (all unknown except membrane protein E25); 44 transcripts ↑ in W-SD, including IEGs and heat shock proteins
Terao et al. *Neurosci* 116: 187–200, 2003; 120: 1115–1124, 2003	cDNA arrays (Clontech, 1176 + 588 + 140 transcripts)	6 h SD (gentle handling), 6 h SD + 4 h recovery S vs. time-matched controls; mouse cerebral cortex (changes must be >100%); four array replicas	IEGs and heat shock proteins ↑ during SD Grp94 ↑ in recovery sleep
Cirelli et al. *Neuron* 41: 35–43, 2004	Affymetrix GeneChips, >24 000 probe sets, >15 000 reliably detected	8 h S, W, SD (novel objects); rat cerebral cortex (changes must be >20%); 3–5 array replicas; two statistical methods; qPCR; estimated false positives ~20%, based on qPCR confirmation of 52 transcripts (using independent groups of animals)	106 known transcripts and 155 ESTs ↑ in S; 95 known transcripts and 395 ESTs ↑ in W-SD
Cirelli and Tononi *J Neurosci* 24: 5410–5419, 2004	mRNA differential display (~5000 transcripts) + Affymetrix GeneChips, ~8000 probe sets, ~5000 reliably detected	8 h SD with or without prior LC lesion (DSP-4); rat cerebral cortex (changes must be >20%); three array replicas; two statistical methods; estimated false positives <20%, based on qPCR confirmation of 32 transcripts (using independent groups of animals)	20% (16/ 95) of the known transcripts ↑ in W-SD in Cirelli et al. (2004) are downregulated after cortical depletion of norepinephrine
Cirelli et al. *J Neurochem* 94: 1411–1419, 2005	Affymetrix GeneChips, 18 955 probe sets, ~9400 reliably detected	9 h S, W, SD (forced locomotion) fly head (changes must be >50% in two fly lines); four array replicas; one statistical method; estimated false positives <20%, based on qPCR confirmation of 19 transcripts (using independent groups of animals)	12 transcripts ↑ in S and 121 transcripts ↑ in W-SD
Terao et al. *Neurosci* 137: 593–605, 2006	Affymetrix GeneChips, 1322 probe sets, ~600 reliably detected	6 h SD (gentle handling), 6 h SD + 2 h recovery S vs. time-matched controls; rat cerebral cortex, hypothalamus, basal forebrain (changes must be >50%); two array replicas	
Cirelli et al. *J Neurochem* 98: 1632–1645, 2006	Affymetrix GeneChips, 26 261 probe sets, ~17 000 reliably detected	One week SD (disk-over-water method) compared to yoked controls, 9 h W, and 9 h SD; rat cerebral cortex (changes must be >20%); 3–8 array replicas; one statistical method; estimated false positives <5%, based on qPCR confirmation of 15 transcripts (using independent groups of animals)	75 transcripts ↑ and 28 transcripts ↓ after long-term SD

IEGs, immediate early genes; ESTs, expressed sequence tags.

circadian time. In the cerebral cortex of rats, out of the 15 459 transcripts we tested, 808 (5.2%) were affected by time of day, independently of behavioral state, and 752 (4.9%) were affected by sleep and wakefulness independently of time of day. These data indicate that, at least in the rat, day/night time and sleep/wakefulness influence gene expression in the cerebral cortex to a similar extent. Peripheral tissues such as liver and heart also show significant changes in gene expression as a function of circadian time. Whether this is also true for sleep-dependent and wakefulness-dependent genes is currently unknown, because none of the array studies published so far has focused on peripheral tissues.

Some Genes Increase Their Expression during Sleep, Others during Waking and Sleep Deprivation

We found that in the rat cerebral cortex 95 known genes and 395 expressed sequence tags (ESTs) (i.e., short transcribed sequences whose function is still unknown) show higher expression after 8 h of spontaneous wakefulness or sleep deprivation, while 106 known genes and 155 ESTs show increased expression after 8 h of sleep. In the fly head, in which we used the same experimental protocol previously applied to rats (three experimental groups, including 8 h of sleep, 8 h of spontaneous wakefulness, and 8 h of total sleep deprivation), we identified 121 waking transcripts, including both known genes and ESTs, and 12 sleep transcripts, mostly unknown. Finally, we recently compared gene expression in the cerebral cortex of long-term sleep deprived rats and sleeping controls, and found that 226 transcripts are upregulated after long-term sleep deprivation, and 113 after sleep. Thus, it seems that if all transcribed sequences are considered (including ESTs), there are more genes whose expression is upregulated during waking than during sleep. However, hundreds of genes are also upregulated during sleep, confirming the results of intracellular recording studies that show that sleep is far from being a quiescent state of global inactivity. In general, the largest fold changes are observed for waking genes. However, this is mainly due to the fact that waking increases the expression of transcription factors (e.g., c-fos) and effector genes (e.g., Arc) that behave as 'immediate early genes,' whose mRNA and protein levels may increase several folds relative to sleep. Most waking and sleep genes, instead, change their expression by 20–50%, a fold change similar to that seen for most genes whose expression varies because of circadian time.

Sleep and Waking Genes Belong to Different Functional Categories

Sleep and waking transcripts belong to different functional categories, and thus may favor different cellular processes. Most genes whose expression increases during wakefulness (or short-term sleep deprivation) code for proteins involved in energy metabolism, synaptic excitatory transmission, transcriptional activity, acquisition of new information, and the response to cellular stress. Sleep genes, instead, code for proteins involved in protein synthesis, synaptic downscaling and/or memory consolidation, lipid metabolism, and in membrane trafficking and maintenance.

Three Functional Categories of Waking Genes Are Conserved from Flies to Mammals

Three functional categories of genes are consistently induced in several animal species (fruit flies, mice, rats, hamsters) during waking and short-term sleep deprivation relative to sleep. The first one includes genes involved in energy metabolism, including mitochondrial genes such as cytochrome oxidase (CO) I and IV. One study in rats also found an increase in CO enzymatic activity after 3 h of wakefulness. In mice, 6 h of sleep deprivation also affect the mRNA levels of genes involved in glycogen metabolism (glycogen synthase and glycogen phosphorylase), and increase the activity of glycogen synthase, possibly to counterbalance the rapid depletion, at the transition from sleep to wakefulness, of astrocytic glycogen stores. Eight hours of spontaneous waking or sleep deprivation also induce the expression in the rat cerebral cortex of *Glut1*, the gene coding for one of the more widely expressed glucose transporter in the CNS. The upregulation during waking of genes involved in energy metabolism is perhaps not surprising, because glucose metabolism in many brain regions is higher in wakefulness than in non-rapid eye movement (NREM) sleep, and the brain relies almost exclusively on glucose as its energy substrate. These changes, however, do not seem to persist when sleep deprivation is protracted for 24 h or several days. This is consistent with imaging studies in animals, normal human subjects, and fatal familial insomnia patients, showing no change or a decrease in cerebral metabolic rate after prolonged sleep loss relative to normal wakefulness. Thus, the rapid upregulation of mitochondrial genes, *Glut1*, and glycogen-related genes may represent a mechanism by which the brain responds to the immediate increase in energy requirement during spontaneous waking or after a few hours of sleep deprivation, while prolonged sleep loss is associated with reduced energy demands.

A second group of genes consistently upregulated during wakefulness and downregulated during sleep includes heat shock proteins, chaperones, and, in general, genes involved in the response to cellular stress. Cellular stress is the cell's reaction to any adverse environmental condition that perturbs cellular homeostasis, with potential macromolecular damage, that is, damage to proteins, DNA, RNA, and lipids. Wakefulness and short-term deprivation are associated with higher mRNA levels of small heat shock proteins with actin-stabilizing properties, such as HSP27 and alpha-B crystallin, molecular chaperones (HSP70, HSP60), and chaperones in the endoplasmic reticulum, such as BiP and Grp94. BiP (immunoglobulin heavy chain-binding protein) shows the largest and most consistent upregulation in the brain of awake animals, including mice, rats, hamsters and sparrows (Cirelli et al., unpublished observations), as well as in the head of fruit flies. BiP assists in the folding and assembly of newly synthesized glycoproteins and secretory polypeptides and increases in *BiP* expression have been classically described after heat shock and ischemia, when unfolded proteins accumulate in the endoplasmic reticulum. This unfolded protein response is also induced by other conditions that adversely affect the function of the endoplasmic reticulum, such as anoxia, glucose deprivation (BiP is also called glucose-regulated protein 78), and depletion of Ca^{2+} stores in the endoplasmic reticulum. Importantly, the unfolded protein response also triggers a slowing down of protein synthesis. In the rat cerebellum, 8 h of spontaneous wakefulness or sleep deprivation are associated with an increase in the mRNA levels of *PEK*. *PEK* codes for the EIF2a kinase that mediates the inhibition of protein synthesis during the unfolded protein response. Moreover, in the mouse cerebral cortex, 6 h of sleep deprivation are also associated with signs of slowing down of protein synthesis, as indicated by the phosphorylation of the key component of the translational machinery and by changes in the ribosomal profile.

A third group of genes consistently upregulated during wakefulness and downregulated during sleep include transcripts sensitive to membrane depolarization and synaptic activity. In the rat cerebral cortex, they include *RGS2*, *Homer/Vesl*, *tissue-type plasminogen activator*, *casein kinase 2*, *cyclooxygenase 2*, *cpg2*, and *connexin 30*. Moreover, several genes involved in activity-dependent neural plasticity and long term potentiation, such as *Arc*, *BDNF*, *Homer/Vesl*, and *NGFI-A*, have been consistently found to be upregulated in the cerebral cortex of awake or

short-term sleep deprived rats. At least some of the same genes are induced during wakefulness and/or short-term sleep deprivation in mice, hamsters, and sparrows (Cirelli et al., unpublished results). In the fly head, many of the depolarization-sensitive genes involved in synaptic plasticity as identified in Guan et al. are also waking-related genes, including *Astray*, *Kek2*, *stripe*, and *Hr38*. The fly wakefulness-related gene *stripe* is a member of the Egr family of transcription factors. In the rat brain several members of this family, including *NGFI-A* (*Egr-1*), *krox 20* (*Egr-2*), and *NGFI-C* (*Egr-4*), are induced during wakefulness. *Hr38*, another fly wakefulness-related gene, is the *Drosophila* ortholog of the mammalian NGFI-B family of orphan nuclear receptors. *NGFI-B* is a wakefulness-related gene in the rat cerebral cortex. Thus, it seems that in both mammals and flies several plasticity-related genes involved in synaptic growth and potentiation are specifically induced during wakefulness.

The observation that genes involved in long-term potentiation and the acquisition of new memories are specifically induced during wakefulness should not be too surprising. After all, we learn when we are awake and able to adapt to the environment, and not when we are asleep. But if sleep is not involved in learning, is there a role for it in other aspects of synaptic plasticity? Well-controlled new experiments show that sleep may favor the performance of some, but not all, tasks. The strongest and more consistent data relate to procedural tasks, and suggest an important role for NREM sleep or a combination of NREM and REM sleep. What remains unknown, however, are the mechanisms by which sleep may enhance performance. Animal and human studies have shown that brain circuits activated during the execution or the acquisition of certain tasks are 'reactivated' during the subsequent sleep period. The presence of this memory trace, as shown by electrophysiological recordings and imaging studies, could reflect the fact that sleep replays the waking experience, and by doing so somehow allows the specific consolidation of the learned tasks. However, it is also possible that this trace is the inevitable consequence of the synaptic potentiation occurred during learning, and that sleep enhances performance by actually allowing a generalized downscaling of synapses. The first and most important effect of this downscaling is thought to be the decrease in the metabolic cost of synaptic activity, which increases during waking due to synaptic potentiation. A second effect of the generalized but proportional decrease in synaptic weight is an increase in the signal to noise ratio, which could improve the performance of some learned tasks.

If sleep is associated with a generalized depression of synaptic strength, molecular markers of synaptic depression should be increased during sleep relative to waking. Recent studies have identified and validated several specific and direct markers of synaptic strength, the best established among them being the density, composition, and phosphorylation state of the GluR1 subunit of the glutamatergic AMPA receptors. We are currently measuring the number and phosphorylation state of synaptic GluR1 in the rat cerebral cortex after 6 h of sleep or sleep deprivation. Preliminary results show that in several cortical areas, sleep is associated with a relative dephosphorylation of GluR1 at serine 845, a finding consistent with the hypothesis that sleep is associated with synaptic downscaling.

The Noradrenergic System Plays a Role in the Induction of Some Waking Genes

One of the mechanisms that underlie the widespread changes in cortical gene expression between sleep and wakefulness is the activity of the noradrenergic system of the locus coerulues (LC), whose neurons project diffusely over the entire brain. LC cells are tonically active during wakefulness, reduce their firing rate during NREM sleep, and cease firing during REM sleep. Moreover, LC activity increases phasically in response to salient events and in relation to the decision to act. In a series of experiments rats were subjected to unilateral or bilateral LC lesions to deplete one or both sides of the brain of norepinephrine. In these animals the raw EEG and its power density spectrum were not significantly affected by the lesion. However, all cortical areas depleted of norepinephrine showed a marked decrease, during wakefulness, of the expression of plasticity-related genes such as *Arc, BDNF, NGFI-A,* and *P-CREB,* as well as of stress response genes such as heat shock proteins and *BiP.* By contrast, the transcript for the translation elongation factor 2 was the only known sleep-related transcript whose expression increased after LC lesion (**Figure 1**). In a complementary experiment in mice the activity of the LC of one side was increased using a conditional transgenic approach. This manipulation resulted in an increased ipsilateral expression of *NGFI-A* in cortical and subcortical target areas. Thus, LC activity during wakefulness modulates neuronal transcription to favor synaptic potentiation and memory acquisition and to counteract cellular stress, while LC inactivity during sleep may play a permissive role to enhance brain protein synthesis.

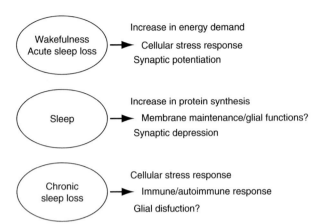

Figure 1 Schematic representation of the major functional categories of genes upregulated in the rat cerebral cortex after 3–8 h of spontaneous waking or sleep deprivation (acute sleep loss), 1 week of total sleep deprivation (chronic sleep loss), and 8 h of spontaneous sleep.

The Expression of Several Genes Changes after Chronic, but Not Acute Sleep Loss

In a recent array study we analyzed the expression of >26 000 transcripts in the cerebral cortex of long-term sleep deprived rats. Animals were kept awake for 7 days using the disk-over-water method (DOW). This method uses minimal stimulation to enforce chronic sleep deprivation in the sleep-deprived rat, while it simultaneously applies to the control rat the same stimulation, but without severely limiting its sleep. The sleep-deprived and the control rat are housed each on one side of a divided horizontal disk suspended over a shallow tray of 2–3 cm deep water. EEG and EMG are continuously recorded to detect sleep states. When the experimental rat falls asleep, the disk is automatically rotated at low speed, awakening the rat and forcing it to walk opposite to disk rotation to avoid being carried into the water. The yoked control rat receives the same physical stimulation because it is on the same disk. However, while sleep is severely reduced (by 70–80%) in the sleep deprivation rat, the control rat can sleep whenever the sleep deprived rat is spontaneously awake and eating, drinking, or grooming and thus its sleep is only reduced by 25–40%.

We compared long-term sleep deprived rats with rats spontaneously awake or sleep deprived for only 8 h, and identified 75 transcripts upregulated, and 28 transcripts downregulated after chronic sleep loss. Most transcripts specifically increased or decreased their expression due to the long-term sleep deprivation and not because of the disk stimulation, because they were not affected in the yoked controls. A large group of genes upregulated after prolonged sleep loss codes for immunoglobulins, including two autoantibodies. Another group includes stress response genes such as the macrophage inhibitor factor-related-protein MRP14, expressed in microglia. Aryl-sulfotransferase, an enzyme involved in the catabolism of catecholamine, is also upregulated after chronic sleep deprivation. These genes are either not induced (antibodies, MRP14), or induced to a lesser extent (aryl-sulfotransferase) after acute sleep loss, suggesting that short-term and long-term sleep deprivation differ to some extent at the molecular level. We also compared long-term and short-term sleep deprivation to sleep, and found that only a minority of genes were similarly up- or downregulated in the two sleep deprivation conditions. Among the genes whose expression was most significantly downregulated in long-term sleep deprived animals relative to sleeping animals were myelin constituents, including the most abundant structural protein component of myelin, the proteolipid protein. These findings suggest that sustained sleep loss may trigger a generalized inflammatory and stress response in the brain. Some glial cells (e.g., microglia) may help to protect neurons against this cellular insult, but others (e.g., oligodendrocytes) may also suffer some of its negative consequences. It will be important in the future to test whether sleep loss may be detrimental to the maintenance of cellular membranes and more specifically to myelin.

Genes Upregulated during Sleep

Figure 1 shows the sleep-related transcripts identified so far in the rat cerebral cortex . Unfortunately, so far only one study in mammals was designed to identify genes upregulated after several (8) hours of sleep relative to both spontaneous waking and sleep deprivation, to rule out effects of stress and of circadian time. Another study in flies used the same experimental conditions used in rats, but for technical reasons the entire fly head, rather than the brain only, was used to hybridize the arrays. This most likely decreased the ability to detect state-dependent genes. Only 12 genes were identified as sleep related in flies, and the function of most of them is currently unknown. Thus, the extent to which sleep genes are conserved across species is yet to be determined.

Two sleep genes in the rat cerebral cortex code for the translation elongation factor 2 and the initiation factor 4AII, and are involved in protein synthesis. The mRNA levels of the elongation factor EF2 are also increased during sleep in the brain of Djungarian hamsters (Cirelli et al., unpublished results). A positive correlation between sleep and protein synthesis has been suggested by previous studies, and recent evidence

suggests that waking and sleep deprivation are less conducive to protein synthesis than sleep. Whether sleep favors protein synthesis globally, or whether it enhances the synthesis of specific classes of proteins is still unclear.

Other sleep genes identified in the rat cerebral cortex are involved in the consolidation of long-term memory, as well as in synaptic depression and depotentiation. Among them are *calmodulin-dependent protein kinase IV*, *calcineurin*, *FK506 binding protein 12*, *inositol 1,4,5-trisphosphate receptor*, and *amphiphysin II*. As mentioned above, the mechanism by which sleep enhances memory and performance is still debated, but these molecular data are compatible with the idea that sleep benefits the brain by producing a global synaptic downscaling.

Another large group of sleep-related transcripts in the rat cerebral cortex is involved in membrane trafficking and maintenance. Some of these transcripts are important for the synthesis/maintenance of membranes in general and of myelin in particular, including oligodendrocytic genes coding for myelin structural proteins (e.g., *MOBP*, *MAG*, *plasmolipin*, *CD9*), myelin-related receptors and enzymes. Transcripts with higher expression in sleep also code for enzymes involved in the synthesis and transport of cholesterol, a major constituent of myelin and other membranes and an important factor in regulating synaptic efficacy. As mentioned above, long-term sleep deprivation decreases the expression of the gene coding for proteolipid protein, a major myelin component. Intriguingly, among the few known sleep genes in *Drosophila* is the gene *anachronism* (*ana*), as well as three genes involved in lipid metabolism (*CG8756*, *CG9009*, *CG11407*). Ana is a glycoprotein secreted by glial cells, and controls the timing of neuroblast development. Whether *ana* plays any role in membrane trafficking and synaptic plasticity remains to be determined.

Conclusions

As it is apparent from the small number of studies in **Table 1**, transcriptomics analysis is still a work in progress, and it would be inappropriate to draw firm conclusions from what has been published so far. Continuous improvement in the array technology will certainly favor more progress in this area. For instance, the current arrays can assess the expression of 20 000–30 000 transcripts, but this is still far from been a true 'whole-genome' analysis. Low-expression genes are often difficult to detect, especially in large and heterogeneous tissue samples such as most brain areas. Moreover, current arrays focus on the coding portion of the genome, but new 'tiling' arrays, in which probe sets are laid down at regular intervals across the entire genome, are becoming commercially available. This will permit the identification of new transcriptionally active regions of the genome that have currently not been tested in more conventional arrays. Moreover, proteomics, the study of the complete repertoire of proteins that contribute to a cell's physiological phenotype, is a natural next step in the study of molecular correlates of behavioral states.

Further Reading

Benington JH and Heller HC (1995) Restoration of brain energy metabolism as the function of sleep. *Progress in Neurobiology* 45: 347–360.

Cirelli C and Tononi G (1998) Differences in gene expression between sleep and waking as revealed by mRNA differential display. *Brain Research. Molecular Brain Research* 56: 293–305.

Cirelli C and Tononi G (2000) Differential expression of plasticity-related genes in waking and sleep and their regulation by the noradrenergic system. *Journal of Neuroscience* 20: 9187–9194.

Cirelli C and Tononi G (2000) Gene expression in the brain across the sleep-waking cycle. *Brain Research* 885: 303–321.

Cirelli C and Tononi G (2004) Locus ceruleus control of state-dependent gene expression. *Journal of Neuroscience* 24: 5410–5419.

Cirelli C, Pompeiano M, and Tononi G (1996) Neuronal gene expression in the waking state: A role for the locus coeruleus. *Science* 274: 1211–1215.

Cirelli C, Gutierrez CM, and Tononi G (2004) Extensive and divergent effects of sleep and wakefulness on brain gene expression. *Neuron* 41: 35–43.

Cirelli C, LaVaute TM, and Tononi G (2005) Sleep and wakefulness modulate gene expression in *Drosophila*. *Journal of Neurochemistry* 94: 1411–1419.

Cirelli C, Faraguna U, and Tononi G (2006) Changes in brain gene expression after long-term sleep deprivation. *Journal of Neurochemistry* 98: 1632–1645.

Everson CA, Smith CB, and Sokoloff L (1994) Effects of prolonged sleep deprivation on local rates of cerebral energy metabolism in freely moving rats. *Journal of Neuroscience* 14: 6769–6778.

Guan Z, Saraswati S, Adolfsen B, and Littleton JT (2005) Genome-wide transcriptional changes associated with enhanced activity in the *Drosophila* nervous system. *Neuron* 48: 91–107.

Hairston IS, Peyron C, Denning DP, et al. (2004) Sleep deprivation effects on growth factor expression in neonatal rats: A potential role for BDNF in the mediation of delta power. *Journal of Neurophysiology* 91: 1586–1595.

Kavanau JL (1997) Memory, sleep and the evolution of mechanisms of synaptic efficacy maintenance. *Neuroscience* 79: 7–44.

Krueger JM, Obal F Jr., Kapas L, and Fang J (1995) Brain organization and sleep function. *Behavioural Brain Research* 69: 177–185.

Maquet P (1995) Sleep function(s) and cerebral metabolism. *Behavioural Brain Research* 69: 75–83.

Moruzzi G (1972) The sleep–waking cycle. *Ergebnisse der Physiologie* 64: 1–165.

Naidoo N, Giang W, Galante RJ, and Pack AI (2005) Sleep deprivation induces the unfolded protein response in mouse cerebral cortex. *Journal of Neurochemistry* 92: 1150–1157.

Nelson SE, Duricka DL, Campbell K, Churchill L, and Krueger JM (2004) Homer1a and 1bc levels in the rat somatosensory cortex vary with the time of day and sleep loss. *Neuroscience Letters* 367: 105–108.

Nikonova EV, Vijayasarathy C, Zhang L, et al. (2005) Differences in activity of cytochrome C oxidase in brain between sleep and wakefulness. *Sleep* 28: 21–27.

Petit JM, Tobler I, Allaman I, Borbely AA, and Magistretti PJ (2002) Sleep deprivation modulates brain mRNAs encoding genes of glycogen metabolism. *European Journal of Neuroscience* 16: 1163–1167.

Reich P, Driver JK, and Karnovsky ML (1967) Sleep: Effects on incorporation of inorganic phosphate into brain fractions. *Science* 157: 336–338.

Rhyner TA, Borbely AA, and Mallet J (1990) Molecular cloning of forebrain mRNAs which are modulated by sleep deprivation. *European Journal Neuroscience* 2: 1063–1073.

Salbaum JM, Cirelli C, Walcott E, Krushel LA, Edelman GM, and Tononi G (2004) Chlorotoxin-mediated disinhibition of noradrenergic locus coeruleus neurons using a conditional transgenic approach. *Brain Research* 1016: 20–32.

Stickgold R (2005) Sleep-dependent memory consolidation. *Nature* 437: 1272–1278.

Taishi P, Sanchez C, Wang Y, Fang J, Harding JW, and Krueger JM (2001) Conditions that affect sleep alter the expression of molecules associated with synaptic plasticity. *American Journal of Physiology. Regulatory, Integrative and Comparative Physiology* 281: R839–845.

Terao A, Greco MA, Davis RW, Heller HC, and Kilduff TS (2003a) Region-specific changes in immediate early gene expression in response to sleep deprivation and recovery sleep in the mouse brain. *Neuroscience* 120: 1115–1124.

Terao A, Wisor JP, Peyron C, et al. (2005) Gene expression in the rat brain during sleep deprivation and recovery sleep: An Affymetrix geneChip((R)) study. *Neuroscience* 26: 26.

Terao A, Steininger TL, Hyder K, et al. (2003b) Differential increase in the expression of heat shock protein family members during sleep deprivation and during sleep. *Neuroscience* 116: 187–200.

Tononi G and Cirelli C (2006) Sleep function and synaptic homeostasis. *Sleep Medicine Reviews* 10: 49–62.

Voronka G, Demin NN, and Pevzner LZ (1971) Total protein content and quantity of basic proteins in neurons and neuroglia of rat brain supraoptic and red nuclei during natural sleep and deprivation of paradoxical sleep. *Dokland Akademii Nauk SSSR* 198: 974–977.

Genetics of Circadian Disorders in Humans

A-M Chang, Harvard Medical School, Boston, MA, USA

P C Zee, Northwestern University Feinberg School of Medicine, Chicago, IL, USA

Circadian Rhythms in Mammals

As terrestrial beings living on a planet with an axial rotation of about 24 h, giving rise to rhythms in light and darkness, almost all living organisms display circadian rhythms in their behavior. It is this near-24 h, period that gives the name 'circadian' or 'about a day' to the system. In mammals the circadian timing system, or biological clock, is located in the suprachiasmatic nuclei (SCN) of the hypothalamus. This circadian pacemaker synchronizes many physiologic processes to cyclic changes in the external environment and also maintains the temporal organization of these processes to each other. The daily light and dark cycles entrain the circadian system which signals the appropriate time for activity and rest. The circadian activity/rest rhythm as well as other physiological, biochemical, and molecular processes are synchronized to the external environment in response to changes across the 24 h day. From an evolutionary viewpoint, this kind of circadian organization creates specific temporal groups of animals sharing times of the day in which to do certain activities, for example, diurnal or nocturnal. This may have developed in organisms in response to several needs: food, search for prey, protection from predators, and reduced competition for resources in a physical environment.

Although humans are diurnal animals, preference for certain times of the day, or chronotype, varies among individuals. Some people prefer activity in the early part of the day; others prefer night-time activity, while still others have no preference. Early types have been described as larks and late types as owls. Subjective measures of diurnal preference by questionnaire have been used to assess the morningness or eveningness of individuals for several decades, driven in large part by the desire to identify workers' abilities to adjust to changing work schedules such as shift work. A number of studies have shown that diurnal preference has, in part, a genetic basis yet there are certainly environmental, social, and behavioral influences on the preference for sleep/wake schedules. Different subjective chronotypes have been associated with objective measures of circadian rhythms, including body temperature and hormonal rhythms. Diurnal preference has also been correlated to a sleep/wake pattern and performance measures.

The expression of circadian behavior is also, in part, genetically determined. The first circadian mutant mammal identified, the *tau* mutant hamster, has a shortened endogenous period compared to the wild-type hamster. This altered circadian period of locomotor activity was seen in the offspring of mutant animals demonstrating the heritability of circadian behavior and was eventually associated with a specific genetic mechanism. The identification of other mammalian circadian mutants that show heritability of their altered circadian rhythms has led to the identification of circadian genes that comprise the molecular mechanism for the generation of circadian rhythms. Homologs of these circadian genes have been identified in numerous species, including humans, showing a remarkable conservation of the molecular mechanisms of the circadian pacemaker. Evidence in different model systems shows that circadian timing is regulated transcriptionally, translationally, and posttranslationally by genes that make up the clock mechanism. They form autoregulatory feedback loops that drive the essential cycling patterns of daily rhythms and mutations in these genes result in alterations or abolition of properties of circadian rhythms.

Circadian Rhythm Sleep Disorders

Sleep is a physiological and behavioral state that is necessary for life and highly conserved in numerous species. There are several theories about the function of sleep and it appears likely that it fulfills multiple functions. It is also possible that sleep served one primordial function necessary for survival and that multiple purposes evolved. Moreover, the lower threshold for arousal that occurs during sleep, putting an organism at great risk, reflects its essential function. Sleep loss impacts general health, mood, and cognitive function. Results from acute and chronic sleep deprivation studies have shown that the consequences of sleep loss on health, determined by a number of physiological and endocrine measures, and cognitive function, assessed on several performance tasks, are severe and detrimental. In fact, the result of near-total sleep deprivation in rat models is death in about 3 weeks.

For optimal sleep, the desired sleep time should coincide with the timing of the circadian rhythm of sleep propensity. A misalignment between sleep and the 24 h social and physical environment results in circadian sleep/wake cycle disruption and circadian

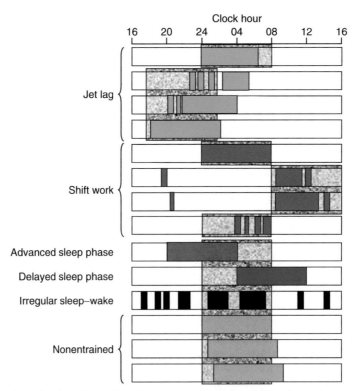

Figure 1 Circadian rhythm sleep disorders. The colored bars represent the sleep pattern for each type and the gray areas represent typical sleep times. For the first two types the environmental schedule is shifted, relative to the internal circadian clock. Examples are shown for jet lag (green) where the typical sleep time is advanced by 7 h (trans-Atlantic flight) and for shift work (purple) where a day shift (sleep at night) is followed by two consecutive night shifts (sleep during the day) followed again by a day shift (rotating shift schedule). For the last four types the circadian clock is altered relative to the environmental schedule.

rhythm sleep disorders, shown in **Figure 1**. Two types of circadian rhythm sleep disorders, jet lag and shift work, occur when the physical environment is altered relative to the internal circadian clock. Four types of circadian rhythm sleep disorders may result when the endogenous circadian rhythm is altered relative to the external environment: advanced sleep phase (ASPS), delayed sleep phase (DSPS), irregular sleep/wake, and nonentrained types. These types of disorders are thought to occur because of chronic alterations in the circadian pacemaker that result in the inability of the individual to achieve or maintain a conventional phase relationship with the external world. Of these four types, the most common are DSPS and ASPS, in which the timing of the major consolidated sleep period is altered, delayed, or advanced, in relation to local clock time. Despite being apparent disorders of circadian timing, little is known about how the circadian system and the diverse rhythms under its control are affected. As increasing numbers of patients with ASPS and DSPS are diagnosed and studied, more can be discovered about the pathophysiology of these disorders and rational strategies for treatments developed. In this article, the focus is on

the genetic basis for the circadian rhythm sleep phase disorders, ASPS and DSPS, because little is known regarding the genetic contributions to jet lag, shift work sleep disorder, irregular sleep/wake cycle, or nonentrained sleep/wake disorder.

Diagnosis

The diagnoses of both DSPS and ASPS are based on published clinical criteria. Their diagnosis requires a complaint of inability to fall asleep or to remain asleep at the desired bedtime. Diagnostic criteria also include an advanced or delayed sleep pattern that is stable, and normal sleep quality and duration when patients are allowed to choose their preferred sleep schedule. A stable advance or delay of the habitual sleep period must be demonstrated by either sleep log or actigraphic recording in combination with sleep log for a minimum of 1 week. These criteria exclude the existence of any other sleep disorder, medical or neurological disorder, mental disorder, medication use, or substance use that can better explain the altered sleep pattern.

DSPS, first described by Weitzman and colleagues, is characterized by bedtimes and wake times that are delayed 3–6 h, relative to desired sleep/wake times. The patient typically finds it difficult to fall asleep before 2–6 a.m. and wake up earlier than 10 a.m.–1 p.m. Classic symptoms include chronic sleep onset insomnia and difficulty waking in the morning. Patients also report daytime sleepiness, primarily during the earlier part of the day. Patients with DSPS often report feeling most alert and active in the late evening and most sleepy in the morning. They typically score as 'evening' types on the Horne and Östberg questionnaire of diurnal preference and are described as night people, or owls. The delay of the sleep period, particularly of the wake time, results in the patient's complaint of difficulty maintaining early morning work times, school schedule, or social obligations. Enforced conventional wake times frequently result in chronic sleep deprivation as the patient forces him- or herself to awaken early but still cannot fall asleep until later than desired. Often sleep is extended on days when the individual has no obligations that require an early wake time, such as on weekends, days off, and vacations.

In contrast to DSPS, ASPS is characterized by habitual and involuntary sleep (6–9 p.m.) and wake (2–5 a.m.) times that are several hours earlier than desired. ASPS patients complain of early morning awakening insomnia or sleep maintenance insomnia and also of sleepiness in the late afternoon or early evening. They typically consider themselves larks and score as morning types on subjective questionnaires. They report feeling best and most alert in the morning; however, some patients who delay their bedtime due to familial or social pressures still experience early morning awakenings and as a result are chronically sleep-deprived. An advance in the phase of the sleep/wake cycle is commonly seen with aging; however, sleep is more fragmented in this population as well, and alterations of the sleep homeostat could account for this advanced phase. In the few reported cases of young ASPS patients, sleep duration and architecture appeared normal without the fragmentation or sleep disturbances commonly seen in aging.

Prevalence

The prevalence of DSPS and ASPS is estimated to be low in the general population. The prevalence of DSPS is about 0.13–0.17%. The number appears to be much higher among adolescents and young adults, with a reported prevalence of 7–16%. This is supported by a number of studies showing that the sleep/wake rhythm is delayed in this group. The prevalence of ASPS is estimated to be much lower than DSPS. One reason for this disparity could be due to underreporting of ASPS, since an early

schedule may have advantages in fitting in with societal norms and schedules and therefore 'morning people' do not have complaints about their sleep habits or feel a need to seek treatment. Another possible explanation for the low prevalence of ASPS is the fact that many of the symptoms are commonly seen with aging; therefore, complaints of early morning awakening may be attributed to normal senescence by patients and clinicians alike.

Pathophysiology

Based on the fundamental properties of the circadian timing system and sleep, several mechanisms could account for a persistently altered phase relationship between the sleep/wake cycle and the environment. One possible explanation is that the endogenous period of the circadian pacemaker is unusually short in individuals with ASPS and long in individuals with DSPS. With respect to DSPS, there is evidence showing a greater frequency of longer periods in adolescents, the same demographic group to show a higher prevalence of the DSPS. A long period that lies outside the range of entrainment or is too long to make the necessary adjustments to maintain a 24 h day may cause non-24 h sleep/wake syndrome. A long endogenous cycle that is still entrained to a 24 h day may have an altered phase angle of entrainment, delaying the sleep period in relation to the light/dark cycle. Likewise, in the case of ASPS, a short period could have an effect on the phase angle of entrainment causing a subsequent advance in the sleep and wake times relative to the light/dark cycle. The report that one member with familial ASPS had a short circadian period supports this explanation and suggests that an alteration in the period of the pacemaker may be involved in the development of ASPS.

Another possible explanation underlying the phase advance or delay seen in ASPS and DSPS is the masking of the phase advance or delay regions of the phase response curve (PRC) to light due to the sleep schedule (**Figure 2**). ASPS patients who are consistently awakening early are likely to be exposed to light at the region of their PRC causing phase advances. If they are also going to bed earlier in the evening, they are less likely to be receiving light at the delay region, thereby reinforcing the advanced schedule. Similarly, DSPS patients who awaken later in the morning are presumably sleeping through the phase advancing portion of their PRC and are more likely to be receiving phase delaying light in the late evening hours. A longer interval between body temperature nadir and sleep offset has been described in a few patients with DSPS, suggesting that the sleep period extended into the morning may be masking the advance

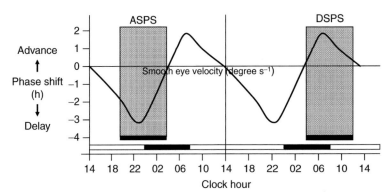

Figure 2 Representative human phase response curve to light, double-plotted. The light dark bar shows a 'typical' sleep/wake schedule (midnight–8 a.m.). The black bars at the bottom of the colored areas, represent the sleep times for ASPS and DSPS subjects, respectively. The colored bars show the proposed 'masking' of the PRC by the sleep schedule.

portion of the PRC in DSPS patients and therefore preventing them from phase advancing their circadian rhythms. This behavior acts as a vicious cycle and perpetuates the already advanced or delayed schedule in ASPS and DSPS, respectively.

An alteration in the entraining effects of light is yet another explanation for the altered phase relationship between the sleep/wake rhythm and the light/dark cycle. This has been described as a weak or low-amplitude advanced or delayed portion of the light PRC. For example, DSPS patients may have abnormally small advance portions of the light PRC explaining the inability to advance the phase of their sleep. Decreased sensitivity to the photic cues may also have an effect on the ability to synchronize an individual's sleep/wake rhythm to the environment. An increased sensitivity to night-time light has been shown in DSPS patients and a reduced response to morning light was demonstrated in subjects with a long free-running period, suggesting the possibility of an altered responsiveness to light cues in these conditions.

Finally, an alteration in the phase relationship between the circadian clock and the sleep/wake rhythm may play a role in these disorders. There have been reports of changes in the phase relationship between sleep times and measures of the circadian clock: specifically, an altered phase relationship between the melatonin rhythm and that of the sleep/wake cycle. The most consistent finding was a prolonged interval between melatonin acrophase and sleep offset. A change in the phase relationship of melatonin rhythm and core body temperature with an individual's diurnal preference has also been described with an earlier circadian phase of melatonin and core temperature in morning types and a later phase in evening types. Furthermore, the interval between circadian phase and habitual wake time was longer in morning types than in evening types. Because DSPS and ASPS patients segregate into the two diurnal groups, evening

types or morning types, this supports the explanation that an alteration in the phase relationship between the clock and the sleep/wake rhythm may affect the phase of the sleep episode. Interestingly, when the morning types were divided by age, there was a difference in this phase relationship. Older morning types who also had an earlier circadian phase showed a shorter interval between core body temperature nadir and sleep offset, opposite to that seen in the young morning types. This suggests that different mechanisms may be responsible for the earlier phase seen in association with aging and ASPS. Therefore, it is likely that not only changes in the free-running period but also entrainment mechanisms of the circadian system contribute to the alteration in the timing of sleep in DSPS and ASPS.

Though it is commonly accepted that both DSPS and ASPS are predominantly a result of alterations of circadian timing, there is recent evidence that alterations in the homeostatic regulation of sleep may play an important role in the pathophysiology of circadian rhythm sleep disorders and in diurnal preference. Polysomnographic recordings of sleep in DSPS patients showed that sleep architecture was not disrupted after the initiation of sleep when the subject was allowed to sleep until their desired wake time. However, following 24 h of sleep deprivation, DSPS patients, when compared to controls, showed a decreased ability to compensate for sleep loss during the subjective day and the first hours of subjective night. In addition, the phase angle between melatonin onset and sleep propensity was larger in DSPS patients. It is possible that poor sleep recovery perpetuates the inability of patients with DSPS to reset their sleep phase. In a recent study of morning and evening types, those with similar intermediate circadian phases showed significant differences in dissipation of sleep pressure. In contrast, the morning and evening types with extremely early and late circadian phases did not show this difference, suggesting that both the circadian system and the sleep homeostat contribute to diurnal preference in addition to sleep regulation and timing.

Genetic Basis of Human Circadian Rhythm Sleep Disorders

Our current understanding of circadian rhythm sleep disorders indicates that alterations in the interaction between circadian and homeostatic processes that regulate sleep and wakefulness play an essential role in the pathophysiology of DSPS and ASPS. Animal work examining the genetic regulation of both circadian and homeostatic mechanisms has revealed a complicated interaction between these two processes in determining sleep duration and quality. Sleep studies of *Clock* mutant mice showed that they not only sleep less but also appear to need less sleep, suggesting that the *Clock* mutation alters the homeostatic drive in these mutants. Similarly, sleep studies of *Cry1* and *2* double knockouts showed an increased sleep drive in the mutants and suggest that the *Cry* genes play an important role in both circadian behavior and the homeostatic regulation of sleep. In contrast, sleep analysis of m*Per1* and m*Per2* mutants showed no significant difference in homeostatic sleep regulation relative to wild-type mice. The only difference between the genotypes of each mutation was seen in the distribution of sleep throughout the day expressing the circadian regulation of sleep timing with the m*Per2* mutants in particular, showing an advance of the activity rhythm. This negative result suggests an independence of the circadian and homeostatic processes regulating sleep. However, it has been proposed that changes in the sleep homeostasis have been demonstrated in mutants having no functional circadian rhythms, such as arrhythmicity in constant darkness (e.g., *Clock*, *Bmal1*, *Cry* double-knockouts, and m*Per1* and m*Per2* double knockouts). The evidence in rodents that mutations in circadian genes can alter both circadian rhythmicity and sleep homeostasis suggests the possibility that a similar genetic predisposition could be involved in the etiology of the circadian and sleep homeostatic changes of sleep regulation seen in circadian rhythm sleep disorders.

In the last few years, a genetic basis for ASPS and DSPS has been established. Familial cases of ASPS and one published family with DSPS demonstrate the heritability of these disorders. The ASPS and DSPS phenotypes appear to segregate with an autosomal dominant mode of transmission. Furthermore, a family history was reported in 11 out of 53 patients with DSPS from a clinic in Japan. Genetic analysis of a family with ASPS led to the identification of a mutation in the human *PER2* gene causing the disorder. However, other published cases of fASPS had not been found to carry this mutation and showed heterogeneity for fASPS. More recently, a mutation in the *CKIδ* gene was linked to another fASPS kindred.

Association studies in idiopathic cases of circadian rhythm sleep disorders have also identified mutations in several circadian genes associated with altered circadian properties. Structural polymorphisms in the human *PER3* gene have been associated with the affected status in DSPS patients and different polymorphisms in the same *PER3* gene were also identified and associated with DSPS and eveningness. A missense mutation in the *CKIε* gene showed an inverse association with circadian rhythm sleep disorders DSPS and non-24 h sleep/wake syndrome. Genetic analysis of diurnal preference to identify polymorphisms in circadian genes has yielded similar results. A polymorphism in the *CLOCK* gene was correlated with evening type although two other groups did not find this association with eveningness in their study populations. More recently, association studies of diurnal preference have identified polymorphisms in the human *PER2* and *PER3* genes associated with extreme morningness and eveningness, respectively.

Summary

The study of disorders of the circadian timing of the sleep/wake rhythm in humans provides valuable insight into the interaction and regulation of various neurobiological processes. Further phenotypic characterization of aspects to the circadian regulation of sleep in people with an altered circadian phase of sleep (i.e., ASPS and DSPS) may serve to elucidate important pathways and functions of photic regulation and entrainment. Thorough evaluation of the sleep in these disorders may also yield information about the underlying processes affecting and causing the expressed changes of sleep and sleep timing. Genetic analysis of ASPS and DSPS familial and idiopathic cases is likely to reveal the molecular and cellular processes generating and modifying the phenotype and behavior. Together, these endeavors will undoubtedly provide greater understanding of the mechanisms underlying circadian rhythm and sleep regulation.

See also: Circadian Regulation by the Suprachiasmatic Nucleus; Circadian Rhythms in Sleepiness, Alertness, and Performance.

Further Reading

American Sleep Disorders Association, Diagnostic Classification Steering Committee (2005) *International Classification of Sleep Disorders: Diagnostic and Coding Manual*, ICSD-2, 2nd edn., Westchester, IL: American Academy of Sleep Medicine.

Aoki H, Ozeki Y, and Yamada N (2001) Hypersensitivity of melatonin suppression in response to light in patients with delayed sleep phase syndrome. *Chronobiology International* 18: 263–271.

Archer SN, Robilliard DL, Skene, et al. (2003) A length polymorphism in the circadian clock gene Per3 is linked to delayed sleep phase syndrome and extreme diurnal preference. *Sleep* 26: 413–415.

Ebisawa T, Uchiyama M, Kajimura N, et al. (2001) Association of structural polymorphisms in the human *Period3* gene with delayed sleep phase syndrome. *EMBO Reports* 2: 342–346.

Horne JA and Östberg O (1976) A self-assessment questionnaire to determine morningness–eveningness in human circadian rhythms. *International Journal of Chronobiology* 4: 97–110.

Jones CR, Campbell SS, Zone SE, et al. (1999) Familial advanced sleep-phase syndrome: A short-period circadian rhythm variant in humans. *Nature Medicine* 5: 1062–1065.

Ozaki S, Uchiyama M, Shirakawa S, and Okawa M (1996) Prolonged interval from body temperature nadir to sleep offset in patients with delayed sleep phase syndrome. *Sleep* 19: 36–40.

Regestein QR and Monk TH (1995) Delayed sleep phase syndrome: A review of its clinical aspects. *American Journal of Psychiatry* 152: 602–608.

Reid KJ, Chang AM, Dubocovich ML, et al. (2001) Familial advanced sleep phase syndrome. *Archives of Neurology* 58: 1089–1094.

Schrader H, Bovim G, and Sand T (1993) The prevalence of delayed and advanced sleep phase syndromes. *Journal of Sleep Research* 2: 51–55.

Toh KL, Jones CR, He Y, et al. (2001) An hPer2 phosphorylation site mutation in familial advanced sleep phase syndrome. *Science* 291: 1040–1043.

Uchiyama M, Okawa M, Shibui K, et al. (1999) Poor recovery sleep after sleep deprivation in delayed sleep phase syndrome. *Psychiatry and Clinical Neurosciences* 53: 195–197.

Weitzman ED, Czeisler CA, Coleman RM, et al. (1981) Delayed sleep phase syndrome. A chronobiological disorder with sleep-onset insomnia. *Archives of General Psychiatry* 38: 737–746.

Xu Y, Padiath QS, Shapiro RE, et al. (2005) Functional consequences of a CKIdelta mutation causing familial advanced sleep phase syndrome. *Nature* 434: 640–644.

Circadian Rhythms in Sleepiness, Alertness, and Performance

J D Minkel and D F Dinges, University of
Pennsylvania School of Medicine, Philadelphia, PA, USA

Introduction

Circadian rhythms (i.e., biological processes that repeat about every 24 h) reflect the incorporation of Earth's daily rotation into living systems. Internal circadian clocks are found in all eukaryotic and even some prokaryotic organisms indicating that natural selection has highly favored circadian rhythms from the beginnings of life for virtually all species.

The suprachiasmatic nucleus (SCN) in the hypothalamus serves as the brain's master clock, in humans and many other animals as well. Neurons in the SCN fire in a 24 h cycle that is driven by a transcriptional–translational feedback loop. Loss of the SCN abolishes the circadian rhythms of a range of behaviors and physiological processes, including sleep, if the animals are not given other external timing cues. Under normal circumstances, the SCN is reset on a daily basis by light inputs from the retina during the day and by melatonin secretion from the pineal gland at night.

The SCN temporally organizes physiological and behavioral tendencies, but the circadian rhythms it promotes are not fixed. This plasticity is believed to reflect adaptation to dynamic demands of the environment such as seasonal changes in daylight. Humans in industrialized societies, however, routinely push the limits of the adaptive capacity of circadian rhythms. Social demands and opportunities (e.g., night shift work, travel across time zones) often lead to abnormal sleep and wake times that are out of phase with biological requirements. The circadian clock has been shaped by natural selection to regulate physiological processes to occur at optimal times relative to each other or relative to the external environment. With the introduction of artificial lighting and 24 h services, however, many humans are awake when their biological clocks are telling them it is time to sleep and trying to sleep when the circadian clock is sending signals throughout the brain that it is time to be awake. The combination of restricting sleep and attaining it at biologically misaligned times is problematic. Approximately 6 million people in the US alone work during the biological night without adapting their internal physiology to these conditions.

In this article, we review what is currently known about circadian modulation of neurobehavioral and affective systems in humans.

Identifying the Circadian Rhythm in Sleep and Wakefulness

Subjective Measures of Fatigue and Alertness

Circadian rhythmicity in neurobehavioral variables has been successfully identified using a wide array of subjective measures of alertness and fatigue. Visual analog scales, Likert-type rating scales, the Karolinska sleepiness scale, the activation–deactivation adjective check list, and the profile of mood states have all demonstrated sensitivity to circadian fluctuations in sleepiness and mood.

Subjective measures may be particularly vulnerable to numerous confounding influences, however. Masking (i.e., the effects of noncircadian factors on the measurement of circadian rhythmicity) frequently prevents underlying circadian signals from being separated from noise. The context in which such measurements are taken (regardless of whether the setting is experimental or environmental) is a major source of masking effects. Demand characteristics, levels of motivation or boredom, distractions, stress, food intake, posture, background noise, ambient temperature, lighting conditions, and pharmacological factors (e.g., caffeine use) are some of the influences that have been identified to mask underlying circadian rhythmicity in subjective measures. These same variables can also create the false appearance of a circadian rhythm, especially in uncontrolled studies where daily schedules may be influenced by sociocultural factors as much as underlying biology.

Circadian rhythms in mood states and sleepiness can be masked by physical, mental, and social activity as well, even in highly controlled experiments. For example, subjects have reported feeling less alert after being challenged to perform. Thus, if not properly controlled, the drop in subjective alertness due to performance effort can mask circadian effects. Similarly, social interactions with staff and/or other subjects in a study can influence subjective variables. In experimental studies, this potential confound can be reduced through careful training of staff and attention to interpersonal interactions among study participants. However, experiments that fail to control social variables risk contaminating true endogenous circadian variance in self-report measures of alertness and fatigue.

These outcomes can also be masked by prior sleep–wake schedules. Sleep and sleep loss have large effects on alertness and performance that can interact with and alter subtler circadian fluctuations. Despite their sensitivity to masking factors, however, subjective scales have been used to index circadian rhythmicity by repeated administration across the day under carefully controlled conditions.

Cognitive Performance

Objective performance measures also demonstrate circadian rhythmicity and are generally less susceptible to masking influences than subjective reports of mood and sleepiness. Tasks that assess psychomotor and/or cognitive speed are particularly sensitive to circadian fluctuations in arousal and somewhat more resistant to masking effects than many other measures. For example, performance on the well-validated Psychomotor Vigilance Test, which shows almost no learning curve and little variance within and between healthy subjects who are fully rested, has been demonstrated to vary with other markers of circadian phase such as melatonin and core body temperature (see **Figure 1**).

Circadian variation has also been demonstrated in a wide range of cognitive tests including search-and-detection tasks, simple sorting, logical reasoning, memory access, and even school performance and meter-reading accuracy. A number of studies have concluded that different tasks and task parameters may yield different peak phases of circadian rhythmicity, but under strictly controlled laboratory conditions, most of these intertask differences disappear. Thus, it can generally be stated that circadian rhythms of cognitive and psychomotor performance covary. Furthermore, these rhythms are synchronized with physiological markers of circadian phase, such as body temperature.

Electroencephalographic and Ocular Measures

The circadian rhythm in task performance and subjective states reflects functional changes in the brain. Neural activity, as measured by evoked potentials or event-related potentials (ERPs), has been used to measure alertness. ERPs are relatively subtle signals that must be measured over many consecutive stimulus probes in order to separate them from background electroencephalographic (EEG) noise. Therefore, ERPs are usually recorded during repetitive tasks, such as reaction time tasks and search-and-detection tasks. Circadian rhythms in the changes to amplitude and location of ERP waves have been identified and interpreted to reflect fluctuations in alertness. Measurement of EEG during wakefulness has demonstrated circadian rhythmicity, especially in theta and alpha frequencies. The amounts of theta and alpha activity in the resting EEG are associated with levels of alertness.

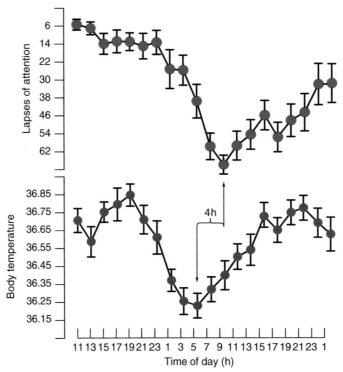

Figure 1 The mean (SEM) circadian minimum (nadir) in body temperature occurred approximately 4 h before the maximum of lapses of attention (note downward deflection on the upper graph indicates worse performance).

Sleep latency measured by EEG and electrooculographic (EOG) activity has been found to exhibit marked circadian variation. Since sleep latency is an indication of propensity of the brain to fall asleep, the circadian system is actively regulating sleepiness and sleep–wake cycles. In addition to shifts toward lower EEG frequencies at night, there is a circadian-mediated increase in slow eyelid closure and slow rolling movements, both of which herald the onset of sleep. Finally, autonomic tone has also been shown to covary with sleep pressure as measured by pupillometry (pupils dilate with sympathetic activation).

Midafternoon Dip

In addition to circadian fluctuations in performance that are tightly linked to physiological markers of phase, some individuals appear to demonstrate a short-term dip in performance in the afternoon (sometimes called the midafternoon dip, siesta, or postprandial dip). This dip has been observed in both field and laboratory studies, but is not consistently demonstrated, suggesting that it may be a relatively weak variable.

The best evidence for the existence of such a midafternoon dip comes from studies of sleep propensity and on the timing of daytime naps. There is much less evidence that the midafternoon dip has deleterious effects on performance measures, which makes the phenomenon's relationship to the biological clock somewhat unclear. Although field studies have reported a decline in performance at this time of day the uncontrolled nature of such experiments prevents strong conclusions about the underlying biology of the phenomenon. As mentioned earlier, external variables can produce the illusion of circadian rhythmicity as well as mask it.

Problems in Detecting Circadian Rhythmicity in Performance

Practice Effects and Other Artifacts

Although repeated administration of the same task is important for identifying circadian fluctuations in performance, increasing familiarity with a task also produces a practice effect (i.e., improved performance with practice on a given task). Practice effects can be difficult to distinguish from the circadian rhythm, but the problem can be circumvented by testing subjects in different orders across times of day, thus balancing out improvements in performance due to learning. This solution assumes however that the practice effect and circadian rhythm are merely additive influences on performance that have

the same relationship in every subject. For most tasks, neither assumption can be made with certainty. A better way to deal with the practice effect is simply to train subjects to their full capacity (i.e., until performance levels reach an asymptote) on tasks before attempting to assess circadian rhythms. When possible, the best way to deal with this problem is to use a measure that has no practice effects like simple psychomotor vigilance tasks.

In addition to the practice effect, many of the same variables that serve to mask circadian rhythmicity in subjective estimates of fatigue and alertness can also conceal, accentuate, or otherwise distort such rhythmicity on objective performance measures. The effects of masking can vary from changes in the shape of the circadian curve to fully obscuring the circadian influence. It is therefore difficult to extract meaningful information about the mathematical characteristics of the circadian phase without an understanding of the masking effects that influence the variables under investigation.

Not only is performance affected by learning, aptitude, and other masking factors, but also by neurocognitive processes and strategies that can be conscious or unconscious. For example, subvocalization (a strategy that can improve performance on certain tasks) has been reported to fluctuate with a circadian rhythmicity. Similarly, compensatory effort in the face of drowsiness can maintain objective performance at steady levels despite underlying circadian fluctuations. Finally, neuroimaging studies have begun to demonstrate that a given task can be accomplished using multiple neural pathways. Sleep-deprived subjects have been shown to compensate for their performance impairment by increasing activation in the parietal lobes. Such compensatory plasticity in neural function illustrates that the effects of sleepiness and arousal are not always expressed through differences in performance. For all of these reasons, successful demonstration of circadian rhythmicity in performance measures requires that tasks be structured in ways that minimize or eliminate masking effects.

Interindividual and Intraindividual Variability

A large body of literature supports relatively stable interindividual differences in many parameters related to the human circadian system. An important exception however is the period of the human circadian pacemaker. Although published reports spanning several decades asserted that the period could range from 13 h to 65 h, highly controlled data demonstrated that the period is very close to 24 h with little variation between or within subjects. The free running period of the human pacemaker is also not heavily influenced by age in healthy subjects. The large variance reported in less-controlled studies therefore probably represents individual differences in the sensitivity of the circadian system's response to masking factors such as physical activity, knowledge of time of day, and exposure to artificial light.

In contrast to the human circadian period, individual differences in circadian amplitude, phase, and mean performance level have been demonstrated. Differences in phase are particularly important because they determine differences in circadian variables relative to time of day. For example, personal preference in timing of work hours is related to individual differences in circadian phase.

The tendency for people to prefer to be active and alert earlier or later (often referred to as 'morningness' or 'eveningness,' respectively) is probably the most substantial source of interindividual variation in circadian rhythmicity. Individuals who prefer different times of day differ endogenously in the circadian phase of their biological clocks. Objective performance measures and physiological variables (e.g., core body temperature) support the morningness/eveningness distinction. This traitlike difference in performance may be seen as a phenotypic aspect of circadian rhythmicity in humans.

Intraindividual differences in circadian rhythms refer to changes within one individual over time, such as differences in morning alertness during the summer compared to the winter. The circadian trough is associated with increased intraindividual variability in performance, reflecting increased wake-state instability similar to what is seen with extended wakefulness. Age is also a source of interindividual variability in the circadian rhythm of alertness and performance, but no consistent relevance of sex has been found.

Circadian Rhythmicity and Sleep–Wake Cycles

Sleep Deprivation

Unmasking the circadian rhythm has been a priority in sleep research for many years. An important method in this domain has been the constant routine procedure (also called the unmasking procedure) which involves keeping subjects awake with a fixed posture in a constant laboratory environment for at least 24 h. By maintaining such a static environment, many of the potential confounding factors (changes in lighting and noise, social interaction, cognitive fatigue, etc.) are eliminated or minimized. Relatively subtle circadian fluctuations can then be readily separated from background noise.

Sleep deprivation, however, can still exert a masking influence on neurobehavioral variables because sleep loss adversely affects performance. Performance may be pushed to extremes by a single night of sleep loss, making circadian fluctuations impossible to detect. This is precisely what happens to sleep latency in the widely used multiple sleep latency test (MSLT). Subjects with extremely high homeostatic sleep pressure fall asleep as quickly as possible regardless of their circadian phase. Nevertheless, when sleep pressure is not extremely elevated and the constant routine procedure is used, both circadian and homeostatic influences on performance can be seen.

Sleep–Wake Regulation

The interaction of circadian and homeostatic influences on performance and alertness has prompted efforts to mathematically model the regulatory processes involved. Current theory and mathematical models posit that the homeostatic process represents the drive for sleep which increases linearly during wakefulness and decreases with sleep. When the 'homeostat' reaches a certain threshold, sleep is triggered; when it is sufficiently dissipated, wakefulness is initiated. The circadian process represents daily oscillatory modulation of these threshold levels, but may have an additional role in promoting wakefulness even in the absence of sleep.

These two processes not only determine the timing of sleep and wakefulness, but also interact to determine waking neurobehavioral performance during sustained wakefulness. Sleep-deprivation experiments clearly show that the circadian process continues to influence performance even as the homeostat increases beyond what would normally initiate sleep.

Forced Desynchrony

Like the constant routine procedure, forced desynchrony is an experimental paradigm that unmasks circadian rhythmicity. In this procedure, study participants stay in an isolated laboratory in which the times for sleep and waking are scheduled to deviate from the normal day. Although humans can entrain to days that are not exactly 24 h, artificial days that are extremely short (e.g., 20 h) or very long (e.g., 28 h) are outside the adaptive capacity of the biological clock. When such a routine is kept for several days, the sleep–wake schedule cannot entrain to endogenous circadian rhythms. Studies using this method have clearly demonstrated that both circadian and homeostatic processes influence sleepiness, performance, and mood. The interaction of the two processes is oppositional during natural waking periods such that a relatively stable level of alertness and performance can be maintained throughout the day.

Ultraradian Days

The 'ultradian' day (meaning 'very short day') paradigm was designed to sample waking behavior across the circadian cycle without significantly curtailing the total amount of sleep allowed. Studies using this method allow subjects relatively brief periods of sleep and wakefulness (e.g., 30 min of sleep followed by 60 min of wakefulness). Thus, ultraradian schedules maintain the normal ratio of sleep to wakefulness (i.e., 1:2). In one such experiment using 7 min of sleep to 13 min of wakefulness, a clear circadian rhythm emerged for the response time on a choice reaction time task. A movement time component was also recorded. This study showed that even in conditions of extremely short artificial days both circadian and homeostatic influences still interact to determine performance.

Circadian Disorders

The laboratory is not the only place one finds people divorcing their sleep–wake pattern from the tendencies of the biological clocks. Recent labor statistics estimate that about 6 million members of the American workforce engage in work at night. While emergency services have been offered 24 h a day for many years, nonessential services are increasingly being offered throughout the night as well. As expected from the research reviewed above, those who work at the circadian phase for sleep show an increased

prevalence of sleep disturbance and related medical and psychiatric conditions relative to those who work during the day.

Field studies have estimated that at best, it takes at least 1 week before the biological clock adjusts to shift work. Complicating the problem is the circadian system's sensitivity to both light and dark. Although in a controlled environment the human circadian system can be shifted a full 12 h over a number of days, in natural settings sunlight will disrupt the reentrainment process. Many shift workers therefore never fully adjust physiologically to their schedules. Melatonin has been investigated as a means of facilitating phase shifts, but its ability to shift circadian phase in the presence of natural light has been found to be fairly weak.

Jetlag is also a common chronobiological problem in modern society. At any given time, approximately 500 000 people are traveling in planes, many of whom will cross many time zones. Most individuals experience unpleasant symptoms for several days after arriving at a destination in a new time zone. Because humans have only very recently achieved high-speed transportation, our circadian systems have not been shaped by evolution to entrain quickly to shifted light–dark cycles. As a result, people experience daytime sleepiness, fatigue, impaired alertness, and difficulty initiating and maintaining sleep for several days after crossing time zones.

Accident Risk

Circadian lows in alertness and performance increase the risk of human-error-related accidents. Motor

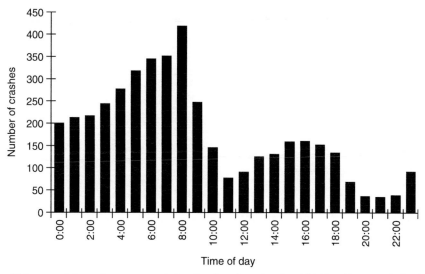

Figure 2 Frequency histogram of time of occurrence during the day of crashes in which the driver was judged to be asleep but not intoxicated. Data for years 1990–92, inclusive. Reprinted from Pack AM, Rodgman E, Cucchiara A, Dinges DF, and Schwab CW (1995) Characteristics of crashes attributed to the driver having fallen asleep. *Accident, Analysis and Prevention* 27: 769–775, with permission from Elsevier.

vehicle accidents caused by drivers falling asleep at the wheel show a circadian rhythm quite similar to laboratory-based findings (see **Figure 2**). Fall-asleep accidents increase throughout the night as sustained attention decreases. Such accidents peak around 8:00 a.m., when vigilant performance is at its lowest, then begin to fall as the circadian system increases alertness and arousal.

Conclusion

Despite the difficulties inherent in identifying circadian influences on performance at various levels of alertness, a large body of research employing diverse methodology has clearly dissociated the effects of oscillating circadian rhythms from homeostatic influences on performance and mood. While controlling for factors that mask circadian rhythms is important for scientific and theoretical purposes, it is important to remember that these factors are also an integral part of the regulation of neurobehavioral functions that allow for adaptation to rhythmic changes in the environment.

Understanding circadian rhythmicity in performance and affective processes is important for understanding circadian disorders (such as shift-work- and jet-lag-induced sleep problems) and for correctly anticipating performance lapses and fatigue-related impairments (e.g., drowsy driving). When sleep is chronically restricted, both homeostatic and circadian influences must be understood in order to correctly predict levels of performance impairment that result from the interaction of these two systems.

See also: Circadian Regulation by the Suprachiasmatic Nucleus; Sleep and Waking in Drosophila.

Further Reading

Boivin DB, Czeisler CA, Dijk DJ, et al. (1997) Complex interaction of the sleep–wake cycle and circadian phase modulates mood in healthy subjects. *Archives of General Psychiatry* 54: 145–152.

Cajochen C, Zeitzer JM, Czeisler CA, and Dijk DJ (2000) Dose–response relationship for light intensity and ocular and electroencephalographic correlates of human alertness. *Behavioral Brain Research* 115: 75–83.

Czeisler CA, Buxton OM, and Khalsa SB (2005) The human circadian timing system and sleep–wake regulation. In: Kryger MH, Roth T, and Dement WC (eds.) *Principles and Practice of Sleep Medicine*, 4th edn., pp. 375–394. Philadelphia: W.B. Saunders.

Czeisler CA, Duffy JF, Shanahan TL, et al. (1999) Stability, precision, and near-24-hour period of the human circadian pacemaker. *Science* 284: 2177–2181.

Dijk DJ and Von Schantz M (2005) Timing and consolidation of human sleep, wakefulness, and performance by a symphony of oscillators. *Journal of Biological Rhythms* 20: 279–290.

Monk TH (ed.) (1991) *Sleep, Sleepiness, and Performance*. Chichester: Wiley.

Phipps-Nelson J, Redman JR, Dijk DJ, and Rajaratnam SM (2003) Daytime exposure to bright light, as compared to dim light, decreases sleepiness and improves psychomotor vigilance performance. *Sleep* 26: 695–700.

Van Dongen HP and Dinges DF (2005) Sleep, circadian rhythms, and psychomotor vigilance. *Clinics in Sports Medicine* 24: 237–249.

Van Dongen HP and Dinges DF (2005) Circadian rhythms in sleepiness, alertness, and performance. In: Kryger MH, Roth T, and Dement WC (eds.) *Principles and Practice of Sleep Medicine*, 4th edn., pp. 435–443. Philadelphia: W.B. Saunders.

Wright KP Jr., Hull JT, and Czeisler CA (2002) Relationship between alertness, performance, and body temperature in humans. *American Journal of Physiology-Regulatory Integrative and Comparative Physiology* 283: 1370–1377.

PHARMACOLOGY OF SLEEP

Modafini, Amphetamines, and Caffeine

C Ballas and D F Dinges, University of Pennsylvania School of Medicine, Philadelphia, PA, USA

Modafinil

Modafinil is a nonamphetamine, wakefulness-promoting drug that has Food and Drug Administration (FDA) approval for the treatment of excessive sleepiness in the following disorders: narcolepsy, residual sleepiness in continuous positive airway pressure-treated obstructive sleep apnea, and shift work sleep disorder. Modafinil has been extensively studied for its effects in these disorders, as well as in sleep-deprived healthy adults. In the latter studies, modafinil was found to reduce physiological sleepiness and improve a wide range of cognitive deficits from sleep deprivation that varied from 36 to 88 h awake. Compared to placebo, modafinil reduced sleep-deprivation deficits on the following cognitive and behavioral tasks: reaction times, vigilant attention, alertness, short-term memory, mathematical calculation speed, logical reasoning, and simulated performance.

Modafinil is designated a controlled substance, Schedule IV, which means it is considered to have less potential for abuse than amphetamines and other schedule I–III stimulants. Although not FDA-approved for attention deficit hyperactivity disorder (ADHD), two clinical trials showed that modafinil resulted in significant improvement in ADHD symptoms, especially attention and concentration outcomes. However, it was associated with Stevens–Johnson syndrome in one child, and approval was rejected pending further investigation.

Modafinil Metabolism and Elimination

Modafinil is the metabolite of adrafinil, lacking adrafinil's terminal amide hydroxyl group. Modafinil exists in racemic form; the I-R-modafinil enantiomer is called armodafinil, and has separate FDA approval. The R enantiomer has a 10- to 14-h half-life, compared to 3 or 4 h for the S enantiomer. Modafinil has a protein binding of 60% and volume of distribution (V_d) of $0.81 kg^{-1}$. The half-life is approximately 12 h. Modafinil is converted primarily to the inactive modafinil acid and modafinil sulfone via CYP3A4/5. The d-form is cleared three times faster than the l-form.

Modafinil Pharmacology

The precise mechanism of action of modafinil is not well understood. It has weak affinity for the dopamine transporter and almost negligible affinity for the norepinephrine transporter. However, in the absence of functional dopamine transporters (e.g., DAT knockout mice), modafinil fails to exert any wakefulness-promoting actions. Medications that inhibit dopaminergic activity, including D1 and D2 blockers and tyrosine hydroxylase inhibitors, fail to inhibit modafinil.

Modafinil inhibits striatal and globus pallidus GABA activity. At doses equivalent to 50 times normal therapeutic doses, it can also enhance glutamate in the striatum. The medial preoptic area and posterior hypothalamus are primarily controlled by tonic GABA inhibition, and consequently it may be that modafinil's effect on glutamate could be blocked with $GABA_A$ antagonists. Modafinil was found at high doses to increase serotonin in the medial preoptic area and posterior hypothalamus, and it was found at lower doses in the cortex and amygdale. Dorsal raphe nuclei receive projections from orexin–hypocretin neurons in the lateral hypothalamus, and they show increases in serotonin with modafinil administration. At therapeutic doses modafinil has no direct effect on serotonin, but it has been found to synergistically augment serotonin increases to fluoxetine and imipramine. Modafinil appears to have minimal effects on the endocrine system, and it has not been associated with significant changes in circadian and sleep modulated hormones such as melatonin, growth hormone, and cortisol.

Loss of orexin–hypocretin cells is associated with narcolepsy, and orexin knockout animals exhibit narcoleptic phenotypes. Because of modafinil's demonstrated efficacy for treating excessive sleepiness in narcolepsy, it was thought that modafinil might exert its effect on orexin–hypocretin cells, which are located only in the lateral hypothalamus but have projections to a wide variety of brain areas. Modafinil was found to increase c-fos immunoreactivity in both orexin–hypocretin neurons and tuberomammilary neurons while having minimal effect in the suprachiasmatic nucleus and anterior hypothalamus. However, modafinil still retains its activity on wakefulness even in orexin–hypocretin receptor knockout animals. In fact, some researchers have found that modafinil has stronger wakefulness-promoting activity in such knockout animals. However, its putative effects via orexin–hypocretin neurons, which are also important in feeding behaviors, have not been associated with effects on food intake. Modafinil appears to also inhibit neurons in the ventrolateral preoptic (VLPO) area, which is the putative sleep switch. This enhances histaminergic activity (i.e., decreases VLPO inhibition of histaminergic activity), which further promotes wakefulness.

Modafinil Adverse Reactions and Safety

Modafinil is generally well tolerated, with the most common reported side effects including headache and insomnia. In phase III trials, three patients with left ventricular hypertrophy and mitral valve prolapse experienced transient ischemic changes on electrocardiogram. As noted previously, one child in a controlled trial had a rash suggestive of Stevens–Johnson syndrome. Modafinil is otherwise relatively safe in overdose, and even chronic usage of high doses has not been found to be associated with significant adverse effects, even after abrupt discontinuation.

Amphetamines

The US Drug Enforcement Agency has listed amphetamines as Schedule II drugs under the Controlled Substances Act because of their high potential for abuse. They are approved by the FDA for the treatment of narcolepsy and ADHD, and several double-blind, placebo-controlled trials in adults and children have found efficacy for this purpose.

Amphetamines have been studied and used for conditions not formally approved by the FDA. These include cognitive dysfunctions associated with brain injury and stroke, and fatigue associated with multiple sclerosis and HIV. Amphetamines have also been studied as augmentation agents in the treatment of depression; however, their efficacy appears to be only moderate, and their effects are concentrated in the expected areas of fatigue and apathy.

Studies on healthy adults have found that amphetamines can improve performance on a variety of tests during sleep deprivation, including visual and auditory reaction times and short-term memory. Subjects in almost all trials have reported subjective feelings of improvements in attention, memory, concentration, and fatigue – even when these were not found on objective outcomes.

Amphetamines have long been known to improve wakefulness and decrease sleep. Whether taken 15 h before sleep or immediately before sleep, they decrease sleep duration. These effects may last into the second night, with further (although lesser) decreases in sleep duration. Amphetamines also affect sleep architecture by decreasing the duration of all sleep stages, and they increase the percentage of light stage 1 sleep. These effects may last into the second night after amphetamine dosing.

Amphetamine Structure–Activity Relationships

In general, amphetamines and their derivatives have three parts that influence their activity: the phenyl, isopropyl, and amine groups. Addition of a CH3 group to the amine carbon creates amphetamines and other chemicals with stimulant properties (e.g., bupropion, ephedrine, cathinone, and fenfluramine). Addition of methoxy groups to the phenyl groups creates hallucinogens (e.g., mescaline and 3,4-methylenedioxy-N-methylamphetamine (MDMA)). For example, 'Ecstasy' (MDMA) is built on a methamphetamine backbone with an aromatic substituted dimethoxy ring. The absence of the methyl group on terminal amine (i.e., 3,4-methylenedioxy-N-amphetamine (MDA), 'Love') further increases the hallucinogenic potency and duration. Collectively, these stimulant/hallucinogen hybrids are referred to as entactogens. Key differences in the activity of amphetamine versus methamphetamine are described later.

The most important structural alterations to the clinically relevant amphetamines consist of the creation of a secondary substituted amine and changes in stereospecificity. The addition of a methyl group to the amine creates levo (L-amphetamine) or dextro (D-amphetamine), which are considerably more potent than amphetamine. However, the L- and D-forms differ in their relative merits and strengths. The dextro form is twice as potent as the levo form in keeping one awake, but both have equivalent effects on rapid eye movement sleep and cataplexy. They do not differ in MAO-A or -B activity, or with regard to norepinephrine efflux. However, the dextro form is markedly more potent in dopamine efflux and reuptake. Importantly, rat experiments have shown that the levo form is considerably less potent (four times less potent) than the dextro form in reinforcement paradigms.

Amphetamine Metabolism and Elimination

Lipid solubility is high for amphetamines and more so for methamphetamine. This allows for rapid absorption and peak levels within 2 h. Its V_d is $5\,l\,kg^{-1}$, with an elimination half-life that varies greatly but is on the order of 20 h. The dextro form is more rapidly eliminated than the levo form. In general, the amount of unchanged amphetamine excreted in urine differs with the pH of the urine. Acidification of urine can double the unchanged fraction to approximately 60%. What is not excreted unchanged is metabolized to phenylacetone, which is then oxidized to benzoic acid and excreted as hippuric acid.

Amphetamine Pharmacology

Amphetamines exert two main effects on the dopaminergic system. Vesicles located inside the presynaptic terminal contain dopamine. They release dopamine into the presynaptic terminal, which is then shuttled out into the synaptic cleft via dopamine transporter (DAT) molecules located on the surface of the presynaptic terminal. Simultaneously, the DAT transports an amount of dopamine back into the terminal. Thus, the transporter serves to release dopamine from the presynaptic terminal into the cleft and/or reuptake dopamine from the cleft into the presynaptic terminal. It is bidirectional, and its kinetics are dependent to a large degree on concentration gradients.

Amphetamine exerts actions in both places. It causes release of dopamine from the vesicles as well as blocking reuptake of dopamine at the DAT. Importantly, amphetamine promotes the outward release of dopamine through the transporter as well. Thus, amphetamine acts on both sides of the bidirectional DAT. Interestingly, amphetamine must first become a substrate of the DAT and be taken into the terminal in order for it to then act to promote efflux. It is this necessary property that has led some to examine the role of blocking the DAT in order to prevent neurotoxins (e.g., MPTP, a model for Parkinson's disease) from entering the terminal and subsequently destroying it.

Additional feedback control of dopamine regulation is provided by norepinephrine. Blockade of the norepinephrine transporter (NET) leads to increased dopamine release and increased amphetamine-related behaviors. Some confusion arises because of the often misleading names of the transporters. The NET and DAT are 70% homologous, and they are both sodium/chloride-dependent transporters. The norepinephrine transporter actually transports norepinephrine and dopamine in the prefrontal cortex, in contrast to the nucleus accumbens, in which the NET has little effect on dopamine. Consequently, NET inhibitors would have a major impact on dopamine levels in the prefrontal cortex but no effect in the nucleus accumbens. This partly explains why amphetamine, a potent NET inhibitor in the prefrontal cortex, raises dopamine there more than methamphetamine, which has minimal effects on the NET. In the nucleus accumbens, both amphetamine and methamphetamine raise dopamine levels equivalently because they are both acting solely through the strength of the DAT.

This same effect is responsible for the often confusing observation that atomoxetine, a 'selective' NET inhibitor, raises dopamine levels as much as clinically equivalent doses of amphetamine in the prefrontal cortex, but it has no effect on dopamine in the nucleus accumbens. From this, it can be seen how atomoxetine may improve cognition (through dopamine and/or norepinephrine in the prefrontal cortex) without causing reinforcement or addiction (the absence of dopamine effects in the nucleus accumbens).

Dopamine efflux is apparently mediated through a change in transporter confirmation. Amphetamine binding to the DAT allows for sodium influx, which in turn causes a calcium influx. This calcium then mediates protein kinase activation that results in the phosphorylation of the N-terminus of the DAT. This appears to relate only to dopamine efflux because deletion of N-terminal amino acids affects dopamine efflux but has no effect on reuptake.

Protein kinase C (PKC) is integral for mediating dopamine efflux and amphetamine's effects; blockade of PKC decreases or abolishes the locomotor activation of amphetamines, and PKC activators cause effects that are similar to those of amphetamine with a similar release of dopamine. PKC activity requires intracellular calcium, further identifying the specific isoform (classical over nonclassical isoforms) of PKC relevant to amphetamine activity.

Behavioral sensitization to the effects of amphetamines is known. However, even before behavioral sensitization, repeated exposure to amphetamine leads to a sensitization of the release of dopamine in the ventral tegmental area and striatal regions of the brain. Investigations have elucidated how dopamine efflux sensitization through the NET is calcium and calcium/calmodulin dependent. Amphetamines increase intracellular calcium through N- and L-type calcium channels. Calcium/calmodulin protein kinase II is activated, which promotes dopamine efflux. Inhibitors of calcium calmodulin protein kinase II block this efflux. This is in contrast to ordinary dopamine efflux after acute doses of amphetamines, which are neither calcium related nor affected by such inhibitors. Additionally, vesicular release of dopamine is not related to calcium or calcium/calmodulin.

Amphetamine Adverse Reactions and Safety

As noted previously, the high abuse potential and risk of dependence and side effects have resulted in the FDA listing amphetamines as Schedule II drugs. Their positive effects on alertness and cognitive functions must be carefully weighed against their unwanted effects on subjective state, disruptions of sleep physiology, cardiovascular and metabolic dysregulation, and abuse liability.

Caffeine

Caffeine is the most commonly used stimulant in the world. Coffee, its primary delivery vehicle, is the second largest commodity traded, after oil. Curiously, despite its popularity, important misconceptions about caffeine persist. The caffeine content in artificial beverages such as soda is generally well controlled, ranging from 80 mg in an 8 oz. energy drink to 38 and 45 mg in two popular colas. The caffeine for these beverages is largely derived from the decaffeination of coffee. The caffeine content in coffee varies for numerous reasons. Each (fresh) coffee bean contains different amounts of caffeine. For example, Arabica – the 'higher quality' bean used in fine coffees and espresso – has half the caffeine of Robusta, the bean used in canned supermarket coffees. Simply roasting the coffee bean (to form dark roast or light roast) can decrease the caffeine content as the volatile caffeine is burned away. However, water is also lost in this process. Thus, by bean weight, dark roast contains more caffeine, but by bean volume (i.e., by measurement with tablespoons) a cup of dark roast will contain less caffeine than light roast. Amounts can vary significantly; light roast Colombia Supremo has a caffeine content of 1.3%, whereas dark roast Sumatras and mocha javas contain 0.02%. Similarly, a typical 1 oz. espresso drink will contain half the caffeine of a regular coffee due to the roasting, as well as the limited amount of water used in the preparation.

Espresso is not a type of bean; it is a fine grind of conceivably any coffee, although typically Arabica. Hot water is forced through the bean at high pressures (130 psi) to extract the oils in the coffee, including the caffeine. Because the 'espresso bean' is generally of the dark roast and Arabica variety, it will contain less caffeine than coffee. However, the primary determinant is the amount of water used and the length of time the water comes in contact with the grounds.

Finally, the caffeine content varies based on the subtleties of preparation. The same coffee from a coffee shop can have significant variability of caffeine content from day to day; differences in the amount of water or coffee used by each preparer can have a great impact. Additionally, home-brewed coffees are generally prepared using more water and consumed in greater volumes.

Caffeine has been shown to improve wakefulness and cognitive performance across a variety of measures (e.g., reaction time, attention, and memory), especially in sleep-deprived subjects. Such effects are found at a variety of caffeine doses. However, doses higher than 200 mg are generally superior to lower doses in improving performance on tasks and sustaining the improvement for longer time periods. Additionally, the longer the duration of sleep deprivation, the greater the caffeine dose necessary to reverse the effects.

Experiments in mice have established caffeine's reinforcing effects, and among prospective trials in healthy adults, a preference was demonstrated for caffeine over decaffeinated controls. Preference appears to be limited to dietary doses, such that increasingly higher doses lead to decreased preference, and avoidance occurs for doses higher than 400 mg. Caffeine reinforcement is also mood and context dependent, with those who prefer the effects of caffeine more likely to choose it. Similarly, negative mood states such as anxiety decrease caffeine preference. It is known that withdrawal effects such as headache or drowsiness reliably predict subsequent caffeine preference. Interestingly, a level of physical dependence (without withdrawal) predicts preference and enjoyment of caffeine, but in the absence of physical dependence this choice is significantly lessened.

There is evidence that caffeine may be protective against Parkinson's disease. It has been observed that caffeine consumption was inversely correlated to Parkinson's disease risk. Using various models of induced Parkinson's disease in mice (e.g., 6-OHDA and MPTP), caffeine was shown to have a protective effect on dopamine levels. This appears to be via its inhibition on A2a adenosine receptors. Decreases in dopamine result in increased glutaminergic activity, which causes destruction of dopaminergic neurons in the substantia nigra. Not inconsequentially, caffeine exerts positive pharmacokinetic effects on Parkinson's disease medications, shortening the time to maximal plasma concentration and the effects of levodopa. The effects of caffeine with levodopa appear to be additive or even synergistic.

An important source of confounding information about caffeine is the tendency to extrapolate acute dosing studies to the effects of chronic use and vice versa. For example, acute administration of 500 mg of caffeine lowers the seizure threshold and is a common maneuver prior to electroconvulsive therapy. However, chronic use (14 days) of caffeine in fact reduces the severity of seizures. Adenosine has anticonvulsant properties, perhaps via inhibition of glutamate or aspartate, and thus it is reasonable to expect that caffeine may alter seizure susceptibility via blockade of adenosine receptors.

Caffeine provides benefits in physical exertion on both performance and endurance. It has even been found that in dietary doses caffeine can reduce by almost half the subjective rating of pain induced by 30 min of exertion. Although its effects on performance are not profound, in competition sports these may be very significant, which may explain why caffeine was once banned by the International Anti-Doping Agency but subsequently taken off the list due to its widespread use. Importantly, the effects of caffeine on performance and endurance are not apparently related to glycogen use or muscle lactate (i.e., anaerobic metabolism). Whether these gains are the direct result of increased sympathetic activity or through some other mechanism is not clear.

Caffeine Metabolism and Elimination

The solubility of caffeine varies substantially with temperature and is approximately $2.2 \, \text{mg ml}^{-1}$ ($66 \, \text{mg oz}^{-1}$.) at $25 \, ^{\circ}\text{C}$ and $180 \, \text{mg ml}^{-1}$ at $80 \, ^{\circ}\text{C}$. Up to 90% of caffeine is hepatically metabolized, mostly to paraxanthine (80%) and theobromine (10%) and trivial theophylline (5%). Paraxanthine is generally further metabolized to inactive compounds before excretion. Importantly, this metabolism is species dependent and has some bearing on the interpretation of pharmacologic data. Rats metabolize less (30–40% excreted unchanged) and have markedly shorter half-lives (1 h for rat, and 3–5 h for human). Paraxanthine has similar effects as caffeine in terms of both its stimulant and its immune system effects. Theophylline is similarly active, but theobromine appears to be considerably less active.

Caffeine Pharmacology

Caffeine's clinical effect on alertness is mediated by nonselective antagonism of adenosine receptors (A1 more so than A2), which are widely distributed throughout the brain. Selectivity for the A2 receptor can be obtained by altering one methyl group with a larger alkyl group. Additionally, the potency can be increased by changing all of the methyls to larger groups.

Although generally not obtained in dietary doses, caffeine has a potent antagonistic effect on phosphodiesterase, which has the result of increasing cAMP and activating protein kinase A. This in turn blocks proinflammatory processes, such as lipopolysaccharide-induced production of tumor necrosis factor-α and interleukin (IL)-12, and promotes anti-inflammatory cytokines including IL-10 and mononuclear cells.

Caffeine's effects through adenosine antagonism are more predictable. Adenosine is released in response to tissue injury and induces collagen production and fibrosis in affected tissue (e.g., liver). Although caffeine is known to inhibit A1 adenosine receptors, data indicate that this may have little relationship to its wakefulness-promoting activity. Pure A1 antagonists will decrease sleep in mice, but A1 knockout mice do not respond to A1 antagonists. Although it may therefore seem logical to conclude that A1 is necessary for sleep, A1 knockout mice do not sleep any longer or shorter than wild-type mice; it is only their sensitivity to A1 antagonists that is different.

Caffeine promoted wakefulness in wild-type mice, but in contrast to the expectation, it also promoted wakefulness in A1 knockouts. It had no effect, however, on A2a knockouts. Taken together, these findings suggest that A1 and A2 receptors work in tandem to regulate wakefulness and sleep, and that a degree of compensation exists, especially for A1 by A2a. Additionally, it suggests that caffeine's wakefulness effects are not related to its majority effect on A1 but, rather, its 'minor' pharmacologic effect – which becomes the main clinical effect – on A2a receptors. This is further supported by location-specific effects of caffeine's stimulant properties in striatal neurons, in which A2a receptors are abundant.

Caffeine Adverse Reactions and Safety

The safety of caffeine as a food and beverage additive has been evaluated, with the conclusion that caffeine added to beverages at a level of 0.02% or less does not present a health risk. Caffeine has been shown to produce a transient increase in blood pressure, but a number of studies have failed to link caffeine with cardiovascular disease. Caffeine may be associated with a small increase in spontaneous abortion in the first trimester of pregnancy, and caffeine can significantly increase 24 h urine output. Because caffeine is not a prescription medication, standard dosing and side effect information is not available. The information in the *Physicians Desk Reference for Nonprescription Drugs and Dietary Supplements* for the product Nō-Dōz, which contains 200 mg of caffeine, indicates that the product may cause "nervousness, irritability, sleeplessness and, occasionally, rapid heart beat." Few research studies report carefully controlled side effect data from matched caffeine and placebo groups. Some investigators estimate that as many as 10% of adults develop the syndrome of caffeinism, which is defined as the daily consumption of large amounts of caffeine (usually in excess of $500 \, \text{mg day}^{-1}$). Symptoms of caffeinism or caffeine intoxication include restlessness, nervousness, excitement, insomnia, flushed face, diuresis, gastrointestinal disturbance, muscle twitching, rambling speech, tachycardia, and agitation. The lethal acute adult dose of caffeine is approximately 5–10 g (approximately 75 cups of coffee).

See also: Behavior and Parasomnias (RSBD); Behavioral Change with Sleep Deprivation; History of Sleep Research; Narcolepsy; Phylogeny and Ontogeny of Sleep; Sleep Apnea; Sleep Architecture; Sleep Deprivation and Brain Function.

Further Reading

Bonnet MH, Balkin TJ, Dinges DF, et al. (2005) The use of stimulants to modify performance during sleep loss: A review by the Sleep Deprivation and Stimulant Task Force of the American Academy of Sleep Medicine. *Sleep* 28(9): 1163–1187.

Dinges DF, Arora S, Darwish M, and Niebler GE (2006) Pharmacodynamic effects on alertness of single doses of armodafinil in healthy subjects during a nocturnal period of acute sleep loss. *Current Medical Research and Opinion* 22: 159–167.

Kantor L, Zhang M, Guptaroy B, Park YH, and Gnegy ME (2004) Repeated amphetamine couples norepinephrine transporter and calcium channel activities in PC12 cells. *Journal of Pharmacology and Experimental Therapeutics* 311(3): 1044–1051.

Mobbs C (2005) Caffeine and the adenosine receptor: Genetics trumps pharmacology in understanding pharmacology. *Cell Science Reviews* 2.

Vanderveen JEC, Armstrong LE, Butterfield GE, et al. (2001) *Caffeine for the Sustainment of Mental Task Performance. Formulations for Military Operations.* Washington, DC: National Academy Press.

Adenosine

R Basheer, J T McKenna, and R W McCarley,
Boston VA Healthcare System and Harvard Medical School, Brockton, MA, USA

Approximately one-third of the human lifetime is spent in sleep. Undoubtedly, sleep is essential for the physical and mental well-being of the individual. Sleep disorders often lead to impaired daytime performance, with at times dire social and financial consequences. Approximately 40 million Americans suffer from chronic sleep disorders each year, and approximately 20 million experience occasional sleeping difficulties. Sleep pathologies such as insomnia, hypersomnia, or parasomnia occur either as a primary disorder or as a secondary disorder in conjunction with another illness. Research has focused on developing new therapeutic strategies, utilizing an increasing knowledge of the brain regions, neurotransmitters, and related molecular/cellular mechanisms that regulate sleep–wake behavior. In this article, we review the current status of pharmacologic agents and their molecular targets used in sleep therapy, followed by an extensive description of developments concerning adenosinergic mechanisms of sleep regulation. Finally, we discuss the prospects of using adenosine receptors or transporters as therapeutic targets for treating sleep disorders.

General Pharmacology of Sleep

Significant advancements have been made in understanding the brain mechanisms involved in sleep–wake regulation (**Figure 1**). The three behavioral states, also termed vigilance states, are waking, non-rapid eye movement (NREM) sleep, and rapid eye movement (REM) sleep (also known as paradoxical sleep or dream sleep). The transitions between these states involve different neurotransmitters, their receptor subtypes, and brain regions. Consequently, the pharmacological advances in sleep-promoting and wake or alertness-promoting drugs target a wide range of these neurotransmitter systems.

The first generation of sleep-promoting drugs composed of hypnotics or sedatives are known to enhance the action of the neurotransmitter GABA at the ionotropic $GABA_A$ receptor. Sedatives such as benzodiazepines (diazepam and valium), barbiturates, neurosteroids, and general anesthetics (isoflurane and propofol) target the $GABA_A$ receptor, potentiating the action of endogenous GABA. These drugs specifically bind to $GABA_A$ receptors that contain the γ subunit, in addition to one of the α subunits (1, 2, 3, or 5). These $GABA_A$ receptors are widely distributed in the cortex and other wake-related areas, such as the thalamus, tuberomammillary nucleus, basal forebrain, and hypothalamus. These drugs may act on targets of the sleep-promoting ventrolateral preoptic area (e.g., the tuberomammillary nucleus) by means of enhancing the inhibitory action of GABA to promote sleep.

A second generation of 'non-benzodiazepine' compounds has been developed which targets the $GABA_A$ receptor but is more selective for the α_1 subunit. Such compounds, including zolpidem, zopiclone, and zalplon, are termed the 'z' drugs. A newer class of compounds with similar actions, such as indiplon, is currently being evaluated. Also, a direct-acting $GABA_A$ agonist ($GABA_A$ receptor with α_4–δ subunits), gaboxadol, has been shown to improve sleep induction, maintenance, and sleep architecture. Interestingly, the selective $GABA_A$ receptor types that bind gaboxadol do not respond to benzodiazepines and are present in high density in the cortex, thalamus, and limbic system. At the cellular level, gaboxadol induces c-Fos expression in VLPO neurons, indicating activation of sleep-inducing neurons.

Drugs that act as antagonists of neurotransmitter receptors on wake-active neurons also exert sedative effects. For example, the histamine blockers dipenhydramine, eplivanserin, and pruvanserin are being developed as potential hypnotic agents. Since it is known that activation of the $5HT_{2A}$ receptor induces arousal, the specific receptor antagonist ketanserin has been developed.

Several recently developed compounds have been used to explore new avenues of pharmacological treatment of sleep disorders. GHB (γ-hydroxybutyrate) has been used to treat cataplexy/narcolepsy (sodium oxybate), and this compound is being considered for treatment of insomnia, although its use as a recreational drug has led to strict controls over prescriptions. Melatonin, secreted by the brain's pineal gland, has been considered a possible dietary supplemental treatment for insomnia, and it may help reduce jet lag. The M1 and M2 melatonin receptor agonist ramelteon has received much attention as a novel therapeutic agent for the treatment of insomnia.

Several drugs are targeted toward enhancing the activities of wake-active neuronal circuitry in order to treat conditions of hypersomnia. The most widely accepted drugs for enhancing wake and alertness are dopamine transport blockers, such as methylphenidale and modafinil. The histamine H_3 autoreceptor

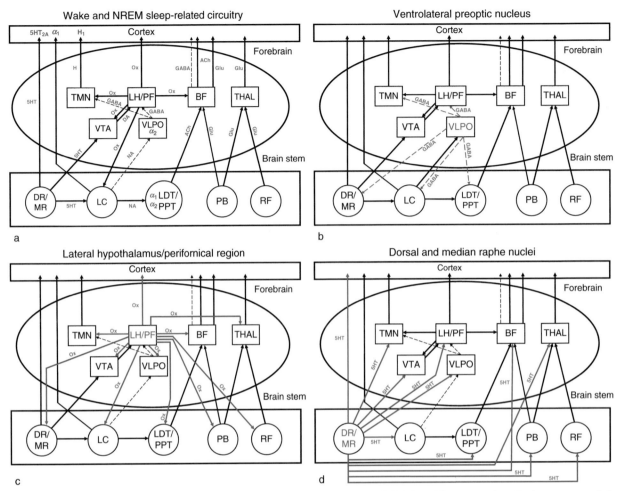

Figure 1 (a) Schematic diagram of the major brain circuitry involved in wake and NREM sleep. The arrows show the known connectivity between the brain regions. Solid lines denote excitatory neuronal outputs, and dashed lines denote inhibitory neuronal outputs. The neurotransmitters (red) shown include the following: ACh, acetylcholine; DA, dopamine; GABA, γ-aminobutyric acid; Glu, glutamate; H, histamine; 5HT, serotonin; NA, noradrenaline; Ox, orexin. Forebrain region: BF, basal forebrain. Cortex: LH/PF, lateral hypothalamus, periformical region; THAL, thalamus; TMN, tuberomammillary nucleus; VLPO, ventrolateral preoptic nucleus; VTA, ventral tegmental area. Brain stem regions: DR/MR, dorsal and median raphe nuclei; LC, locus coeruleus; LDT/PPT, laterodorsal and pedunculopontine tegmental nuclei; PB, parabrachial nucleus; RF, reticular formation. The receptors known to mediate the effects in specific regions (blue) include the following: α_1, excitatory α_1 adrenoceptors; α_2, inhibitory α_2 adrenoceptors; 5HT2A, excitatory serotonin receptors; H1, excitatory histamine receptor. (b) Schematic diagram of proposed inhibitory projections (blue dashed lines) from the ventrolateral preoptic nucleus to proposed wake and NREM sleep-related brain regions. Note that these GABAergic projections are also galaninergic. (c) Schematic diagram of proposed orexinergic excitatory projections (blue solid lines) from the lateral hypothalamus/periformical region to proposed wake and NREM sleep-related brain regions. (d) Schematic diagram of proposed serotonergic excitatory projections (blue solid lines) from the dorsal and median raphe nuclei to proposed wake and NREM sleep-related brain regions. Adapted from Szabadi E (2006) Drugs for sleep disorders: Mechanisms and therapeutic prospects. *British Journal of Clinical Pharmacology* 61: 761–766; and Saper CB, Scammell TE, and Lu J (2005) Hypothalamic regulation of sleep and circadian rhythms. *Nature* 437: 1257–1263.

antagonists ciproxifan and thioperamide can enhance histamine release and increase waking.

In addition to the previously mentioned neuro-transmitters, much research has also focused on endogenous homeostatic factors that integrate sleep regulation with feeding and metabolism, such as the delta sleep factor, cytokines, growth factors, and peptides (orexin, cholecystokine octapeptide, neuropeptide Y, and ghrelin). For example, antagonists of the orexin receptors, such as compound GW649868, are being tested for use as a sedative. Overall, therapeutic strategies that aim to design sleep-promoting and wake-promoting drugs utilize a multitude of approaches developed through knowledge of brain circuitry and manipulation of the different known neurotransmitter systems.

Adenosine as a Potential Target of Hypnotics

The nucleoside adenosine is an endogenous neuromodulator and a homeostatic sleep factor. Developments with regard to adenosine's role in sleep regulation and the delineation of underlying adenosinergic molecular mechanisms have presented several possibilities for treatment of insomnia and hypersomnia. In the following sections, we discuss the brain regions, receptors, and transporters of adenosine that are associated with sleep regulation, and the pharmacological possibilities in designing therapeutic compounds are also discussed.

Adenosine is a ubiquitous molecule, being an integral constituent of several macromolecules such as DNA, RNA, coenzymes, and the high-energy molecule adenosine triphosphate (ATP). The first description of the hypnogenic effects of adenosine was published in 1954 on cats, followed by studies on dogs. Since then, several laboratories have shown that systemic or intracerebroventricular administration of adenosine or agonists of adenosine receptors induces sleep. In 1981, researchers showed that the often-consumed compound caffeine (a theophylline derivative) is behaviorally stimulating due to its structural similarity to adenosine and its ability to block adenosine receptors.

Investigators originally postulated that adenosine levels would increase throughout the brain during wake due to higher neuronal metabolic activity related to ATP breakdown. Thus, in the brain, in which energy metabolism is tightly regulated, adenosine acts as a 'fatigue factor' indicating the need for rest/sleep. In cats, levels of extracellular adenosine, measured by microdialysis, increased 15–20% during wake compared to NREM sleep in brain regions associated with sleep–wake regulation, such as the basal forebrain, cerebral cortex, thalamus, preoptic area of hypothalamus, dorsal raphe nucleus, and pedunculopontine tegmental nucleus. However, prolonged waking achieved by gentle handling revealed a site-specific significant increase in extracellular adenosine in only the magnocellular basal forebrain (140% of the baseline) and, to a lesser extent, in the cingulate cortex. Infusion of adenosine into the rodent or feline magnocellular basal forebrain increased sleep, thus indicating that this region is a major site for the sleep-inducing effects of adenosine. Although adenosine perfusion in other brain regions, such as the pedunculopontine tegmental nucleus and the subarachnoid space ventral to the rostral forebrain, has also been shown to induce sleep, the sleep deprivation-induced increase in extracellular adenosine was rather selective to the magnocellular basal forebrain, suggesting that this region is one of the major sites for the homeostatic regulation of sleep by adenosine.

Four subtypes of adenosine receptors – A1, A2a, A2b, and A3 – have been identified based on their specific order of potency for agonists and antagonists and their particular structure ascertained by gene cloning. Of these, the A1 receptor is widespread throughout the brain, whereas A2a receptors are far more abundant in the basal ganglia and olfactory tubercle, and they are less abundant in the cortex. A2b and A3 receptors are less abundant and their role in sleep is not clear. A1 and A2a receptors are known to regulate adenylate cyclase and cellular activity in opposite ways. The A1 receptor, coupled to the G_i subtype of the G-protein, exerts inhibitory effects, whereas the A2a receptor, coupled to the G_s subtype of the G-protein, exerts excitatory effects on cellular activity. Both A1 and A2a receptors have been implicated in mediating the sleep-inducing effects of adenosine. The effect of adenosine on sleep varies depending on the receptor and the brain region of interest, as described later (**Figure 2**).

Subtype-specific activation of receptors has varying intracellular consequences depending on the receptor-coupling proteins inside the cell membrane (**Figure 3**). The A1 receptor couples to the G_i subtype heterotrimeric G-protein and is recognized for its dual signaling capabilities. It can either inhibit adenylate cyclase activity via the $G_{\alpha i}$ subunit, resulting in reduced cAMP production, or activate phospholipase C (PLC) via the $G_{\beta \gamma}$ subunits, which can lead to inositol trisphosphate (IP_3) production and subsequent release of intracellular calcium by activating IP_3 receptors. In neurons, A1 receptors can activate pertussis toxin-sensitive K^+ channels, as well as K_{ATP} channels, and inhibit Q-, P-, and N-type Ca^{2+} channels. The selectivity of activation and intracellular signaling is also influenced by homo- or heterodimerization. For example, adenosine A1 receptor function can be influenced by the homodimer or heterodimer or by interaction with ectoadenosine deaminase.

A2a receptors are coupled to the G_s-subtype G-protein, and hence increase adenylate cyclase activity. The A2b receptors are positively coupled to adenylate cyclase and PLC. The A3 receptor, like A1, inhibits adenylate cyclase, as well as stimulates PLC and induces intracellular calcium release.

Adenosine has been shown to inhibit neuronal activity by means of A1 adenosine receptors in many of the arousal-related brain regions. In the basal forebrain, where predominantly wake-active neurons have been described, the A1 adenosine receptor has been localized to both cholinergic and noncholinergic neurons. Perfusion of the A1 receptor

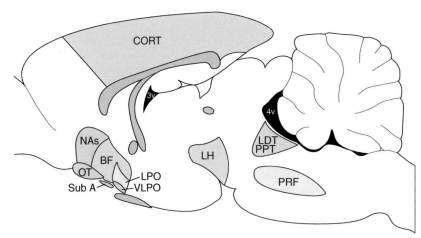

Figure 2 Sagittal representation of the rodent brain showing brain regions with documented adenosine effects on sleep–wake behavior. The brain regions shaded in pink are recognized for A1 adenosine receptor-, blue for A2a receptor-, and green for both A1 and A2a receptor-mediated effects on sleep. BF, basal forebrain; CORT, cortex; LDT/PPT, laterodorsal and pedunculopontine tegmental nuclei; LH, lateral hypothalamus, perifornical region; LPO, lateral preoptic area; NAs, shell of the nucleus accumbens; OT, olfactory tubercle; PRF, pontine reticular formation; SubA, subarachnoid space; VLPO, ventrolateral preoptic area; 3V, third ventricle; 4V, fourth ventricle.

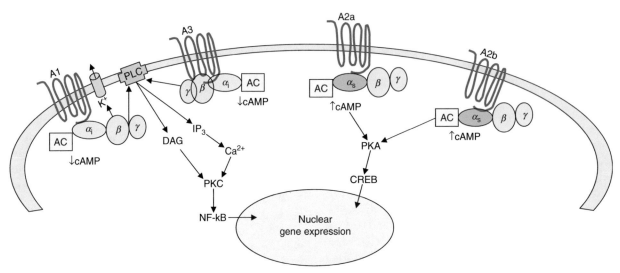

Figure 3 Intracellular signaling pathways of adenosine receptors: A1 and A3 receptors are coupled to pertussis toxin-sensitive G_i-subtype G-protein. Their activation either inhibits adenylate cyclase activity, preventing cAMP production, or activates phospholipase C, releasing diacylglycerol and inositol trisphosphate, which leads to protein kinase C activation. A2a and A2b receptors are coupled to cholera toxin-sensitive G_s-subtype G-protein. Activation of these receptors can increase cAMP production and activation of protein kinase A. The activation kinases can lead to phosphorylation of cellular proteins, including transcription factors such as NF-κB and CREB, leading to changes in gene expression. AC, adenylate cyclase; CREB, cAMP response element-binding protein; DAG, diacylglycerol; IP3, inositol trisphosphate; PKA, protein kinase A; PKC, protein kinase C; PLC, phospholipase C.

antagonist, cyclo-pentyl-1, 3-dimethylxanthine, increased waking and decreased sleep. Reduction of the A1 receptor by infusing antisense for the A1 receptor into the rodent basal forebrain resulted in a significant decrease in slow-wave delta activity during NREM sleep following 6 h of sleep deprivation.

Actions of adenosine at the A1 receptor vary according to its localization, but almost all promote sleep. In cortex, the target of wake-active cholinergic neurons, activation of presynaptic A1 receptors inhibited acetylcholine release, thereby promoting sleep. Presynaptic A1 receptor activation in the ventrolateral preoptic area, known to be sleep active, disinhibited sleep-active GABAergic neurons and led to sleep induction. Also, activation of the somatic A1 receptor on wake-active orexinergic neurons may induce sleep. The wake and REM-active cholinergic neurons of the brain stem laterodorsal and pedunculopontine tegmental nuclei are also directly inhibited by adenosine acting via the A1 receptor. Lastly, in the pontine reticular formation, both A1 and A2a receptors have been implicated in sleep regulation.

A2a receptors are located in select brain nuclei/ regions involved in sleep regulation, such as the subarachnoid space, which is located below the rostral basal forebrain and the shell of the nucleus accumbens. Administration of adenosine or the A2a receptor agonist CGS21680 into the subarachnoid space below the rostral basal forebrain or the shell of the nucleus accumbens near the olfactory tubercle has been shown to promote sleep. Data suggest that prostaglandin D2 may induce the release of adenosine and, via A2a receptors, promote sleep. The sleep-inducing effects of both prostaglandin D and the A2a agonist CGS21680 can be blocked by the A2a receptor antagonist KF17937. In the lateral preoptic area, which contains numerous sleep-active neurons, a clear distinction between the behavioral effects of A1 and A2a receptor activation has been reported. A1 receptor activation of the sleep-active neurons in this area resulted in waking, possibly by inhibition, whereas A2a receptor activation induced sleep by stimulating the activity of sleep-active neurons. The olfactory tubercle contains a high density of A2a receptors, but their role in sleep is unknown.

In addition to activation or inhibition of adenosine receptors, changes in the adenosine metabolizing enzymes or adenosine transporters could potentially contribute to extracellular adenosine levels and promote sleep. For example, administration of the adenosine kinase blocker ABT-702 has been shown to increase sleep. Binding of the equilibrative transporter ENT1 was reduced in the basal forebrain following sleep deprivation in rats, suggesting that a decrease in adenosine transport into cells may contribute to the sleep deprivation-induced increase in extracellular adenosine and subsequent sleep.

Thus, research on the effect of adenosine on sleep has revealed a large number of potential targets for the development of pharmacological drugs. The agonists of adenosine receptors may have potential for treating insomnia, and the agents that antagonize adenosine receptors may be useful for treating hypersomnia and increasing alertness.

Adenosine Receptors as Therapeutic Targets

Adenosine receptor modulation has been utilized in the development of useful therapeutics for treating a variety of pathologies, including inflammatory, cardiovascular, and neurodegenerative diseases. A large number of compounds have been generated as agonists or antagonists that are essentially the derivatives of adenosine and methylxanthine molecules, respectively (**Figure 4**). Advances in molecular modeling have allowed definition of interactions between potential agonists/antagonists and adenosine receptors. Adenosine receptors, like all G-protein-coupled receptors, have a central core consisting of seven transmembrane helices connected with three intracellular and three extracellular loops. The selectivity of these receptors and their respective ligands is often coded by the varying lengths of the N-terminal extracellular domain and the C-terminal intracellular domain, as well as the intracellular loops. In the case of adenosine receptors' consensus sites, N-linked glycosylation has been identified in the extracellular regions, but the location of these sites within the extracellular region and the mode of posttranslational modification vary for each receptor subtype. Site-directed mutagenesis of single amino acids, critical for ligand recognition and responsible for discerning between agonists and antagonists, has been central to the medicinal chemistry of the drug–receptor interaction.

Several potent and selective adenosine receptor agonists have been utilized for treatment, and many are being tested for use in inflammation, analgesia, benign prostatic hypertrophy, and cardiovascular disease, in contrast to the relative paucity of development for use in sleep and wakefulness. Some adenosine antagonists are derivatives of xanthines, but these compounds have limited water solubility and bioavailability. Consequently, nonxanthine compounds with increased water solubility and bioavailability have been synthesized. Examples of A1, A2, and A3 selective agonists/antagonists are presented in **Tables 1–3**, respectively.

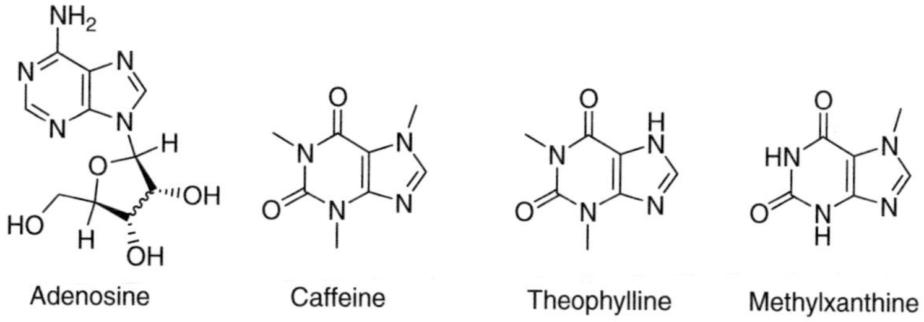

Figure 4 Chemical structures of adenosine, caffeine, theophylline, and methylxanthine.

Table 1 Select human A1 receptor agonists and antagonists

Adenosine A_1 receptor subtype	Compound	K_i value for AR (nM)			
		A_1AR^*	$A_{2A}AR^*$	$A_{2B}AR^*$	A_3AR^*
Agonists					
	CPA	2.3	794	18 600	72
	CCPA	0.83	2270	18 800	38
	S(−)-ENBA	0.38	>10 000	>10 000	915
	ADAC	0.85	210	N.D.	13.3
	AMP579	5.1	56	N.D.	N.D.
	NNC-21–0136	10	630	N.D.	N.D.
	GR79236	3.1	1300	N.D.	N.D.
	CVT-510 (Tecadenoson)	6.5	2315	N.D.	N.D.
	SCZWAG 994	23	25 000	>10 000	N.D.
	Selodenoson	1.1	306	N.D.	N.D.
Antagonists					
	DPCPX	3.9	129	56	3980
	WRC-0571	1.7	105	N.D.	7940
	BG 9719	0.43	1051	172	3870
	BG 9928	29	4720	690	42 110
	FK453	18	1300	980	>10 000
	FR194921	2.9	10 000	N.D.	>10 000
	KW3902	1.3	380	N.D.	N.D.

N.D., not determined.
Reprinted by permission from Macmillan Publishers Ltd: *Nature Reviews Drug Discovery* (Jacobson KA and Gao Z-G (2006) Adenosine receptors as therapeutic targets. *Nature Reviews Drug Discovery* 5: 247–264), copyright (2006).

Table 2 Select human A2 receptor agonists and antagonists

Adenosine A_2 receptor subtypes	Compound	K_i value for AR (nM)			
		A_1AR^*	$A_{2A}AR^*$	$A_{2B}AR^*$	A_3AR^*
Agonists					
A_{2A}	NECA	14	20	140	25
	CGS21680	289	27	>10 000	67
	DPMA	168	153	>10 000	106
	Binodenoson	48 000	270	430 000	903
	ATL-146e	77	0.5	N.D.	45
	CV-3146	>10 000	290	>10 000	>10 000
A_{2B}	LUF5835	4.4	21	10	104
Antagonists					
A_{2A}	KW6002	2830	36	1800	>3000
	CSC	28 000	54	N.D.	N.D.
	SCH 58261	725	5	1 110	1 200
	SCH 442416	1110	0.048	>10 000	>10 000
	ZM241,385	774	1.6	75	743
	VER 6947	17	1.1	112	1 470
	VER 7835	170	1.7	141	1 931
	'Schering compound'	82	0.8	N.D.	N.D.
A_{2B}	MRS1754	403	503	2	570
	MRE 2029-F20	245	>1000	3	>1000
	OSIP-339391	N.D.	N.D.	0.5	N.D.

N.D., not determined.
Reprinted by permission from Macmillan Publishers Ltd: *Nature Reviews Drug Discovery* (Jacobson KA and Gao Z-G (2006) Adenosine receptors as therapeutic targets. *Nature Reviews Drug Discovery* 5: 247–264), copyright (2006).

Nucleoside Transporters as Therapeutic Targets

Adenosine is hydrophyllic and its transport across the cell membrane is carried out by two kinds of nucleoside transporters: the equilibrative nucleoside transporters (ENTs) and the concentrative nucleoside transporters (CNTs). The ENTs allow facilitated diffusion of nucleosides, whereas CNTs are cation symporters. ENT1–4 vary in their tissue distribution pattern. In the brain, ENT1 and ENT4 are

Table 3 Select human A3 receptor agonists and antagonists

A_3 receptor subtype	Compound	K_i value for AR (nM)			
		A_1AR^*	$A_{2A}AR^*$	$A_{2B}AR^*$	A_3AR^*
Agonists					
A_3	IB-MECA	51	2900	11 000	1.8
	Cl-IB-MECA	220	5360	>100 000	1.4
	LJ568	193	223	N.D.	0.38
	CP-608039	7200	N.D.	N.D.	5.8
	MRS3558	260	2300	>10 000	0.29
	MRS1898	136	784	N.D.	1.5
Antagonists					
A_3	OT-7999	>10 000	>10 000	>10 000	0.61
	MRS1292	12 100	29 800	N.D.	29.3
	PSB-11	1640	1280	2 100	4
	MRS3777	>10 000	>10 000	>10 000	47
	MRS1334	>100 000	>100 000	N.D.	2.7
	MRE 3008-F20	1200	141	2100	0.82
	MRS 1220	305	52	N.D.	0.65
	MRS1523	15 600	2050	N.D.	18.9
	Novartis compound	197	1670	3	10

N.D., not determined.

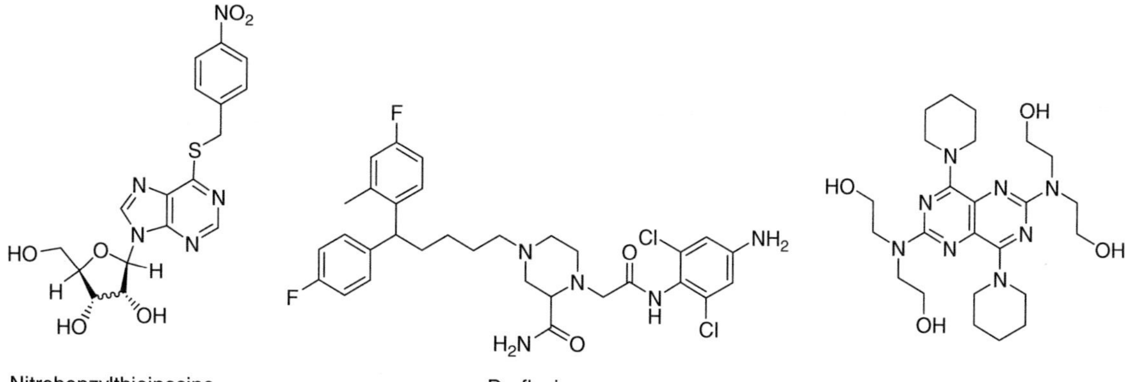

Figure 5 Chemical structures of compounds that block nucleoside transporters.

more abundant. There are three subtypes of CNTs (CNT1–3), and CNT2 is most abundant in brain. Although the affinity for adenosine is much greater for CNT2 compared to ENT1 (K_m of 6 µM for CNT2 and 40 µM for ENT1), the adenosine transport rate is much slower with CNT2. In the cortex, CNT2 mRNA levels and, in the magnocellular basal forebrain, ENT1 binding to NBMPR were attenuated after 24 and 6 h of sleep deprivation, respectively. The downregulation of these transporters was suggested to contribute to the observed increase in extracellular adenosine in these two brain regions following sleep deprivation. Consequently, inhibitors of adenosine transporters that increase the levels of extracellular adenosine may be promising compounds for treatment of such sleep disorders as insomnia. Experimental proof of this phenomenon

has been obtained in the rat with infusion of the ENT1 inhibitor NBMPR into the basal forebrain, where levels of extracellular adenosine increased and sleep was induced. Several such ENT1 inhibitors, including dilazep, dipyridamole, draflazine, and lidoflazine, all with different molecular structures, have been identified for their ability to enhance the extracellular concentration of adenosine by blocking adenosine reuptake (**Figure 5**).

Allosteric Modulators as Therapeutic Agents

Adenosinergic receptor activity may also be influenced by compounds that bind to receptor allosteric sites that are topologically distinct from the orthosteric sites to which endogenous ligands, as well as exogenous agonists and antagonists, bind. Allosteric ligand binding does not overlap with the orthosteric site, thus allowing simultaneous binding of compounds at both sites. These compounds can thus selectively enhance the response to adenosine in only those organs or a localized area of a given organ in which the production of adenosine is increased. At the functional level, molecules that bind to allosteric sites modulate receptor activity by causing conformational changes, leading to either decreased (positive regulators or allosteric enhancers) or increased (negative regulators) dissociation kinetics of the orthosteric ligand. Therefore, the use of allosteric modulators as therapeutic agents has become very attractive. Allosteric modulators exhibit greater subtype selectivity and fewer side effects. Moreover, the neurotransmitter receptors are less sensitive to desensitization or downregulation by allosteric enhancers compared to exogenous agonists. Typical examples are the benzodiazepines, such as diazepam, acting as allosteric enhancers of GABA$_A$ receptors with limited side effects, whereas direct agonists of GABA$_A$ receptors have widespread side effects and are therefore unsuitable for treatment. Similarly, an allosteric modulator of the nicotinic receptor, the alkaloid galanthamine, potentiates nicotinic cholinergic neurotransmission without major side effects and is used for treatment of Alzheimer's disease.

Allosteric Modulators of the A1 Adenosine Receptor Subtype

Allosteric modulators have also been described for adenosine receptors and transporters and are potential therapeutic agents (**Figure 6**). The benzoylthiophene derivatives, such as PD 81,723 (2-amino-4,5-dimethyl-3-thienyl-[3[(trifluromethyl)phenyl]-methanone) and related analogs, were the first compounds to be identified as A1 receptor-selective allosteric enhancers due to their ability to increase the levels of agonist binding. In the past decade, multiple derivatives of PD 81,723, with variable potency for enhancing the agonist binding to the A1 receptor, have been synthesized. Among them, PD 71,605 (T-62) proved to be a potent enhancer of

agonist binding to membranes of Chinese hamster ovary cells expressing the human A1 receptor. 2-Aminothiazlium salts are a new class of allosteric enhancers of the A1 receptor that are suggested to be more stable than benzoylthiophenes. Most of the tests for the potency of this compound have involved *in vitro* systems, however, and clinical use of these compounds has not been reported. On the other hand, tests in animal studies showed that PD 81,723 exhibited some degree of tissue selectivity, modulating neuronal but not adipocyte adenosine receptors. This compound's effects were also species specific, showing greater enhancement of agonist binding in the guinea pig and dog brain compared to the rodent brain. T-62 has been shown to reduce mechanical hypersensitivity in a rat model of neuropathic pain, thus demonstrating its potential use in pain therapy.

Allosteric Modulators of the A2a and A2b Adenosine Receptor Subtypes

The allosteric modulators of the A2a receptor thus far described lack absolute specificity. For example, PD 120918 (4-methyl-7([methylamno]carbonyl)oxy)-2H-1-benzopyran-2-one) has been shown to slow the dissociation of the adenosine agonist from the rat brain A1 receptor, therefore appearing to be not very selective for the A2a receptor. Other compounds, such as amiloride analogs and SCH-202676, increased the antagonist dissociation from the A2a receptor but did not show any effect on the dissociation rate of agonist CGS21680. To our knowledge, no known allosteric compound modulates the A2b adenosine receptor subtype. It is suggested, however, that SCH-20676 can also function as an allosteric modulator of the A2b subtype receptor (**Figure 7**).

Allosteric Modulators of the A3 Adenosine Receptor Subtype

Two distinct compounds and their derivatives have been identified to act as allosteric modulators of the A3 adenosine receptor subtype. 3-(2-pyridinyl) isoquinoline derivatives were very selective for the

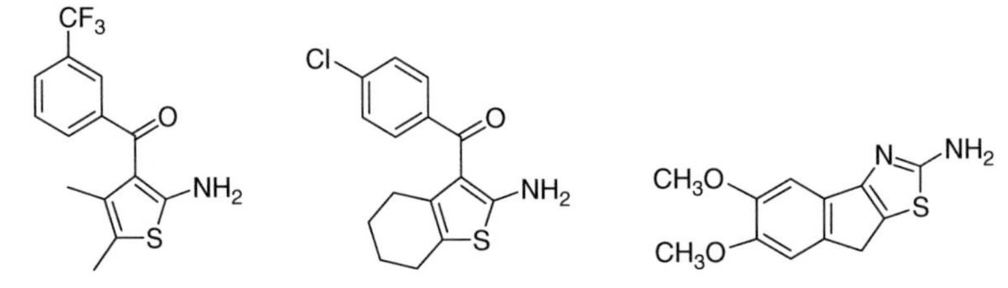

PD 81,723   T-62 (1)   2-Aminothiazolium

Figure 6 Chemical structures of allosteric modulators of A1 receptors.

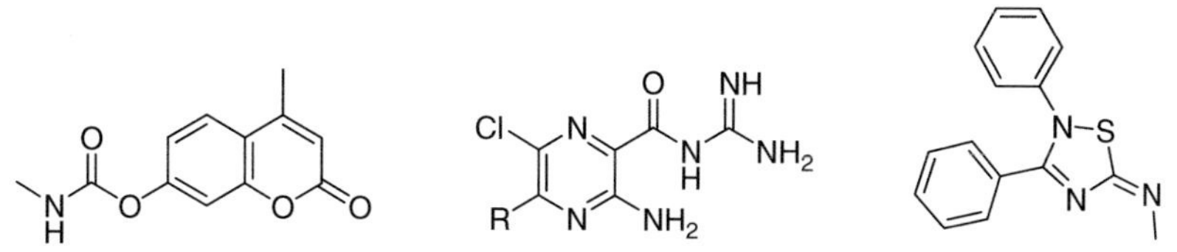

Figure 7 Chemical structures of allosteric modulators of A2a receptors.

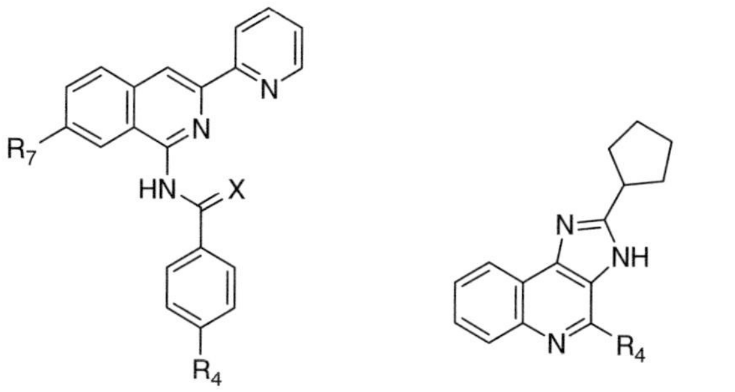

Pyridyl isoquinoline derivatives   Imidazoquinoline derivatives

Figure 8 Chemical structures of allosteric modulators of A3 receptors.

A3 receptor, and they have been shown to enhance binding of the agonist [^{125}I]-N6(4-amino-3-iodobenzyl)-5'-N-methyl carboxamidoadenosine (I-AB-MECA, also known as IB-MECA) in a concentration-dependent manner. These derivatives not only enhanced agonist binding to the A3 receptor but also enhanced A3 receptor function (i.e., inhibition of forskolin-induced cAMP production). However, these compounds also demonstrated antagonist-like activity. Further studies have shown that the removal of a 7-methyl group decreased the antagonistic function without any loss in the allosteric enhancing activity, suggesting that a structural requirement for allosteric enhancement might be distinct from those for competitive antagonism (**Figure 8**).

Another set of compounds, 1H-imidazo-[4,5-c] quinoline and its derivatives, exhibited A3 receptor-selective allosteric enhancement of agonist binding, specific to the human A3 receptors. A few derivatives of this compound decreased the dissociation of the agonist IB-MECA without any effect on A3 receptor antagonist binding or A1/A2a receptor agonist binding. A few of these derivatives, however, tested positive for antagonistic binding to the receptor at the orthosteric site. These compounds demonstrated enhanced inhibition of forskolin-stimulated cAMP production in cells that expressed human A3 receptors.

In addition, another subset of the previously mentioned compounds for the A2a receptor, the amiloride analogs, may be more selective for the A3 receptor. Studies have shown that the former was more effective as an allosteric enhancer for the A3 receptor and acted as allosteric inhibitors of the antagonist binding to A1,

A2a, and A3 receptors. Specifically, it was shown that the dissociation rates for IB-MECA binding of the human A3 receptor, as well as forskolin-stimulated cAMP production, were significantly reduced without any effect on the binding of A1 and A2a agonists to their respective receptors, suggesting a higher selectivity of amiloride compounds for the A3 receptor.

Adenosine as a Therapeutic Agent for Other Diseases

In the drug discovery arena, the development of compounds with receptor subtype-specific pharmacodynamics without undesired side effects has been a constant challenge. Several adenosinergic drugs have been described for the treatment of both central nervous system (CNS) and non-CNS-related illnesses. Some of these compounds are already in use, others are in later stages of clinical trials, and many more are being tested in animals. In patients with paroxymal supraventricular tachycardia, a condition defined as an abnormally fast heart rate due to the rapid firing of electrical impulses from the atrioventricular node, intravenous infusion of adenosine has been found to activate A1 receptors and restore a normal heart rhythm. Another A1 receptor-selective agonist, called selodenoson, tecadenoson, or CVT 2759, in clinical trials, has been developed for its capacity to slow heart rate without lowering blood pressure. However, it has been found to have the detrimental side effect of producing renal toxicity.

Activation of A1 receptors has also been recognized to provide cytoprotection during ischemia. The A1 receptor antagonist BG9719 is undergoing phase II clinical trials for improvement of renal function and treatment of congestive heart failure. Adenosine is being used in imaging for the evaluation of coronary artery disease. Another A2a receptor agonist, regadenoson, is also being used for myocardial perfusion imaging. The A2a receptor agonist GW328267 is undergoing phase II clinical trials for the treatment of asthma, a chronic obstructive pulmonary disease.

Other adenosinergic compounds have been developed to treat other pathologies. Some CNS drugs, such as cabamazepine, exhibit pain-relieving effects which are mostly mediated through A1 and, partially, through A2 receptors. Another A1 receptor-selective agonist, GW-493838, is in clinical trials for the treatment of pain and migraine headaches. An allosteric enhancer of the A1 receptor, T-62, is in clinical trials for treating neuropathic pain. An A1 receptor antagonist, FR194921, is being investigated for use as a potential treatment of dementia and anxiety disorders. NNC-21–0136, another A1 receptor selective agonist, has been shown to be selectively neuroprotective during global ischemia, with significantly few cardiovascular effects. Lastly, the A2a receptor antagonist istradefylline (KW-6002) is in phase III clinical trials for the treatment of Parkinson's disease.

Conclusion

Research has strongly supported the role of adenosine as a potent somnogenic factor. Although no drugs targeting adenosine receptors or transporters have been utilized as compounds for treating insomnia or hypersomnia, there appears to be potential for such therapeutic strategies in the future. With improved understanding of the brain regions and the receptor subtypes mediating the effects of adenosine, it is time to test or design specific modulators of adenosine receptors or transporters for treating sleep disorders. However, it is important to note that one of the basic challenges in using adenosine receptor or transporter modulators as therapeutic agents is the ubiquitous distribution of these molecules and the possibility of undesirable side effects. Strategies that include mediation of effects specific to the brain and, better still, in a localized area of the brain will be important. For example, allosteric modulators may potentiate an adenosine receptor agonist's action in an agonist concentration-dependent manner, which could hold great promise for enhancement of the localized effects of adenosine. The concept of engineering adenosine receptors (neoceptors) in concert with uniquely modified/designed ligands (neoligands) is being explored for use in organ-targeted gene therapy. Early studies have demonstrated successful synthesis of neoceptors for the A2a and A3 adenosine receptors. This approach may hold promise for the future of targeted drug therapy. In summary, the adenosine receptors present a promising target for pharmacological intervention, one heretofore not utilized in the treatment of insomnia and hypersomnia.

See also: Behavioral Change with Sleep Deprivation; Modafini, Amphetamines, and Caffeine; Nightmares; Reticular Activating System; Sleep Architecture; Sleep Deprivation and Brain Function; Sleep in Aging.

Further Reading

Basheer R, Strecker RE, Thakkar MM, and McCarley RW (2004) Adenosine and sleep–wake regulation. *Progress in Neurobiology* 73: 379–396.

Boutrel B and Koob GF (2004) What keeps us awake: The neuropharmacology of stimulants and wakefulness promoting medications. *Sleep* 27: 1181–1194.

Fredholm BB, Chen J-F, Masino SA, and Vaugeois J-M (2005) Actions of adenosine at its receptors in the CNS: Insights from knockouts and drugs. *Annual Review of Pharmacology and Toxicology* 45: 385–412.

Fredholm BB, Ijzerman AP, Jacobson KA, Klotz KN, and Linden J (2001) International Union of Pharmacology: XXV. Nomenclature and classification of adenosine receptors. *Pharmacological Reviews* 53: 527–552.

Gao Z-G, Kim S-K, Ijzerman AP, and Jacobson KA (2005) Allosteric modulation of the adenosine family of receptors. *Mini Reviews in Medicinal Chemistry* 5: 545–553.

Huang Z-L, Urade Y, and Hayaishi O (2007) Prostaglandins and adenosine in the regulation of sleep and wakefulness. *Current Opinion in Pharmacology* 7: 33–38.

Jacobson KA and Gao Z-G (2006) Adenosine receptors as therapeutic targets. *Nature Reviews Drug Discovery* 5: 247–264.

King AE, Ackley MA, Cass CE, Young JD, and Baldwin SA (2006) Nucleoside transporters: From scavengers to novel therapeutic targets. *Trends in Pharmacological Sciences* 27: 416–425.

Moro S, Gao Z-G, Jacobson KA, and Spalluto G (2005) Progress in the pursuit of therapeutic adenosine receptor antagonists. *Medicinal Research Reviews* 26: 131–159.

Porkka-Heiskanen T, Strecker RE, Thakkar M, et al. (1997) Adenosine: A mediator of the sleep-inducing effects of prolonged wakefulness. *Science* 276: 1265–1268.

Saper CB, Scammell TE, and Lu J (2005) Hypothalamic regulation of sleep and circadian rhythms. *Nature* 437: 1257–1263.

Steriade M and McCarley RW (2005) *Brain Control of Wakefulness and Sleep*. New York: Kluwer/Plenum.

Strecker RE, Morairty S, Thakkar MM, et al. (2000) Adenosinergic modulation of basal forebrain and preoptic/anterior hypothalamic neuronal activity in the control of behavioral state. *Behavioural Brain Research* 115: 183–204.

Szabadi E (2006) Drugs for sleep disorders: Mechanisms and therapeutic prospects. *British Journal of Clinical Pharmacology* 61: 761–766.

SLEEP FUNCTION

Endocrine Function During Sleep and Sleep Deprivation

J M Mullington, Harvard Medical School, Boston, MA, USA

Overview

Frequently measured hormonal profiles in healthy volunteers have included the chronobiological pineal hormone melatonin; hypothalamic anterior pituitary hormones, including cortisol, growth hormone (GH), prolactin, and thyroid-stimulating hormone (TSH); and more recently metabolic hormones related to nutrient homeostasis, leptin and ghrelin. The diurnal patterns of some important and frequently measured hypothalamic anterior pituitary hormones along with the rhythm of core body (rectal) temperature and a hypnogram of a typical normal sleep period are presented **Figure 1**.

Melatonin: The Chronobiological Pineal Hormone

Melatonin, the hormonal timekeeper, is produced by pinealocytes in the pineal gland. The pineal is situated in the depression between the colliculi of the midbrain on the dorsal aspect, projecting into the subarachnoid, outside the blood–brain barrier. It receives a large blood supply originating from the posterior cerebral artery and eventually drains into the internal cerebral veins, which flow in turn into the great cerebral vein. The human retinal ganglion cells make bilateral monosynaptic connection with the suprachiasmatic nucleus of the anterior hypothalamic nucleus via the retinohypothalamic tract. These fibers continue and synapse at the paraventricular nucleus of the hypothalamus, project through the medial forebrain bundle, and synapse at the intermediolateral cell column of the spinal cord. This fiber tract then synapses on preganglionic cells that synapse at the superior cervical ganglion.

In humans, light inhibits melatonin secretion during the day. Melatonin is a strongly circadian (from the Latin phrase *circa diem*, meaning about a day) hormone that increases regularly at the dim light-onset time of day, peaking during the nocturnal sleep period, and diminishing again with the onset of bright light. In humans, melatonin increases at night independently of sleep and is inversely related to the rectal temperature rhythm. During early sleep deprivation studies, it was thought that at night light could not suppress melatonin in humans, but this was shown to be dependent on the light level. Later studies that applied bright light (>2000 lux) when melatonin levels were high during the sleep period were able to reduce melatonin to daytime or near-daytime levels, although they returned to presuppression levels within 40 min after resumption of darkness. Studies that followed found that the application of light is more or less effective at delaying or advancing the onset of melatonin, depending on the biological time (or circadian phase) at which it is applied. Bright light applied late in the day delays the melatonin onset, and light applied early in the morning advances it. Although studies have not shown melatonin to have clinically significant hypnotic qualities, the presence of melatonin may facilitate the coordination of sleep-onset processes.

Studies of the responsiveness of the melatonin rhythm to different intensities, spectrums, and timing of light administration have laid the foundation for therapeutic approaches to the treatment of jet lag, chronobiologically related mood disorder (seasonal affective disorder), and sleep disorders such as delayed sleep-phase syndrome, in which individuals are unable to keep a consistent sleep-onset time and instead go to bed later and later each night.

Key Hormones of the Anterior Pituitary

Anterior pituitary hormones are released in a pulsatile fashion and vary in peripheral concentration throughout the day. The determining factors are in part circadian (sleep-independent biological rhythms) and in part related to sleep and other behavioral processes. Cortisol and TSH are more strongly circadian in their diurnal secretion profiles and maintain their diurnal rhythms in the absence of sleep. GH and prolactin are more sleep-dependent and are generally suppressed in the absence of sleep, and they move with the shift of the sleep period in experimental studies of the effects of shifting sleep opportunity. As can be seen in **Figure 1(a)**, cortisol reaches its normal daily low point, or nadir, in the nocturnal sleep period, typically an hour or two after sleep onset, corresponding to the time of maximal nocturnal GH release. Cortisol increases in the early hours of the morning, peaking shortly after the time of rising.

The paraventricular nucleus of the hypothalamus synthesizes and releases corticotropin-releasing hormone (CRH), which in turn regulates the release of adrenocorticotropin (ACTH) in the anterior pituitary, leading to the production and secretion of cortisol by

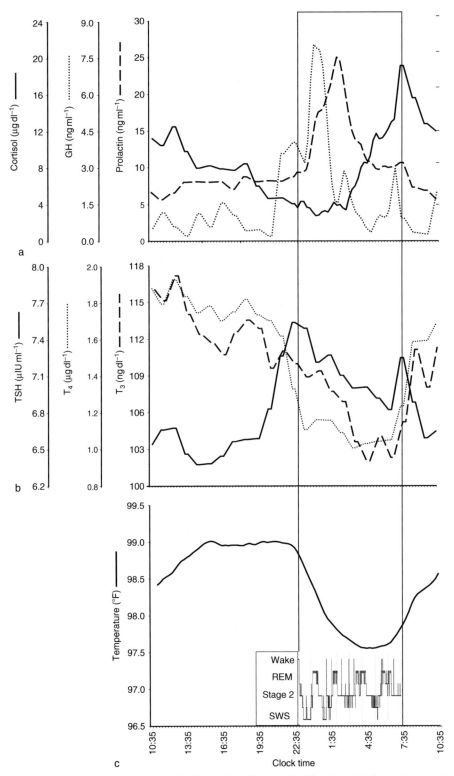

Figure 1 Data from 40 healthy men and women ages 21–40 combined in smoothed tracings: (a) diurnal hormonal profiles for cortisol (based on hourly samples) and for growth hormone (GH) and prolactin (based on half-hourly samples) through the night; (b) normal diurnal rhythm of thyroid-stimulating hormone (TSH), thyroxine (T_4), and triiodothyronine (T_3); (c) normal diurnal rhythm of temperature. The inset in (c) shows a representative sleep hypnogram from one of the subjects in the group of 40, illustrating the descent into slow-wave sleep (SWS) followed by a rapid eye movement (REM) period and subsequent cycling through non-REM and REM periods of sleep for the sleep period. which spans 11 p.m.–7 a.m.

the adrenal cortex. Growth hormone releasing hormone (GHRH) is produced by the arcuate nucleus, which induces the production and release of GH by the anterior pituitary. Somatostatin downregulates the production of GHRH. In addition, insulin-like growth factor (IGF-)I is stimulated by GH. IGF-I indirectly inhibits GH through its stimulation of somatostatin in the hypothalamus.

Cortisol

Cortisol, the most important and abundant glucocorticoid, is essential for proper water, carbohydrate, protein, and fat metabolism, and without this hormone, animals die within a week. In addition to these functions, cortisol is also important for the regulation of vascular reactivity, immune cells and inflammatory mediators, and cellular transcription processes. In addition to having a diurnal rhythm under pleasant relaxed conditions, the hypothalamic anterior pituitary system acts to increase circulating cortisol in response to unpleasant or potentially unpleasant stimuli. Although the reason is not well understood, the ability to withstand noxious stimuli is critically reliant on cortisol. In humans, there is far more cortisol than ACTH measured in the periphery, with a ratio of 7:1. Stimuli which activate ACTH secretion also elevate cortisol, and it is cortisol which is more readily measured in the periphery.

Although ACTH and cortisol are considered to be under strong circadian control, sleep loss and other behaviors do influence portions of the amplitude and shape of the curve. For example, waking causes a surge of ACTH and cortisol. This increase in ACTH is not simply a response to rising, but may also play a role in preparing for rising in that it has been shown that the anticipation of an earlier than normal awakening causes ACTH to rise earlier. Cortisol is also influenced by meals, the first morning exposure to light, stress, and noxious and other stimuli.

During sustained extended wakefulness under controlled laboratory conditions, an increase in the nadir of cortisol during total sleep deprivation has been observed. An afternoon elevation of cortisol has also been described under conditions of sustained reduced sleep. Although the significance of these modifications to the cortisol rhythm under conditions of insufficient sleep is not completely understood, it is known that during aging the nadir of cortisol increases, leading some to consider sleep loss a model of aging.

Cortisol levels are increased in depression and also have been seen to be elevated in individuals suffering from sleep disturbance. Sleep that is disturbed, either by sleep disorders such as sleep apnea or insomnia

or disrupted by experimental means that cause fragmentation of the nocturnal sleep period, may lead to blunted GH and an elevated afternoon and early nighttime cortisol secretion.

Growth Hormone

GH is critical for growth and development, in which it is responsible for linear bone formation during early development, as well as growth and maintenance of tissues through ontogeny. During normal sleep, slow-wave sleep is predominant in the first half of the night with the hormonal release of GH at its highest level of the day. In fact, 50–70% of GH is released during the first half of the nocturnal sleep period in men, although the tightness of this association is somewhat lower in men younger than 25 years of age and higher in men older than middle age. Of particular note, the associative relationship between slow-wave sleep and GH release is also lower in women than it is in men.

Prolactin

Prolactin is usually secreted early after sleep onset and continues to rise through the night, peaking around the middle of the sleep period. Similar to GH, prolactin is suppressed during sleep deprivation and shifts with a delayed bed period. Brief awakenings during nocturnal sleep are associated with decreasing prolactin levels.

Rapid eye movement (REM) sleep predominates in the second half of sleep period and is associated with and accompanied by a rise in cortisol. A reciprocal interaction between the CRH and GHRH systems is thought to be important for the regulation of sleep, with the GHRH system promoting sleep and the CRH system stimulating ACTH and cortisol production and promoting arousal and wakefulness. In men, there is a positive relationship between the amount of slow-wave sleep and GH measured through the sleep period. Men with lower cortisol nadirs have been found to have higher amounts of REM sleep.

Thyroid-Stimulating Hormone

TSH is produced by parvocellular neurons in the paraventricular and anterior hypothalamic nuclei. TSH is increased during the nocturnal bed period and the production of thyroxine (T_4) and triiodothyronine (T_3) by the thyroid gland is concomitantly decreased. During acute total sleep deprivation of one night, TSH and free T_3 are increased above sleep control conditions and free T_4 is unchanged. In contrast to the elevation seen during total acute sleep deprivation, nocturnal TSH has been reported to be reduced in the context of prolonged periods of reduced sleep amount relative to TSH levels seen during extended recovery sleep.

Consolidated nocturnal sleep, initiated at a regular time, when melatonin is rising and ended after the sleep need is satiated, also at a regular time, should facilitate optimal hormonal release and support homeostasis most efficiently. Consistent with this assumption, there is evidence that disruption of sleep, whether due to experimental manipulation or to sleep disorders, leads to nocturnal elevation of cortisol and TSH and a decrease in prolactin; all patterns that are in contrast with the entrained healthy-sleep hormonal profile.

Energy Metabolism and Adipose Tissue Hormones

Because hormonal systems play key roles in the maintenance of energy balance and metabolism and because consolidated sleep is essential for the appropriate timing and secretion amplitude of these hormonal systems, it is perhaps not surprising that sleep itself appears to play a role in metabolism. Under conditions of acute total sleep deprivation and also under conditions of prolonged shortened sleep, glucose metabolism is slowed and young healthy research participants begin to respond in a prediabetic fashion to glucose tolerance testing.

Leptin

Leptin is a hormone produced by adiposites; it signals energy sufficiency and satiety to the brain. It shows a regular diurnal rhythm, reaching maximal levels at night during sleep. Leptin's primary site of action is located in the arcuate nucleus of the hypothalamus. Leptin levels drop during fasting, and it is a hormone known to be important for reproductive capacity. In women, menses ceases when leptin levels drop below normal levels and resume when leptin is supplemented as treatment. Leptin is also integrally involved in energy homeostasis and appetite regulation. Under conditions of total acute and prolonged partial sleep deprivation, leptin amplitude is reduced, with a lowered nocturnal peak.

Ghrelin

Ghrelin is primarily produced by the stomach; it signals hunger to the brain. During reduced sleep, ghrelin levels and appetite are elevated. Insufficient or disrupted sleep leads to a negative energy balance and increased appetite, leading to a higher caloric intake. Furthermore, there is evidence that the food preferences under these conditions favor a high-carbohydrate diet.

Conclusion

The field of sleep and hormonal regulation is a rapidly developing one. New peptides and hormones are currently being discovered that are important for integrating metabolic function. Sleep is critical for the homeostasis of energy systems, and much remains to be learned about the mechanisms involved in translating sleep into health maintenance.

See also: Autonomic Dysregulation During REM Sleep; Behavioral Change with Sleep Deprivation; Immune Function During Sleep and Sleep Deprivation; Sleep Deprivation and Brain Function; Thermoregulation During Sleep and Sleep Deprivation.

Further Reading

Born J, Hansen K, Marshall L, Mölle M, and Fehm HL (1999) Timing the end of nocturnal sleep. *Nature* 397: 29–30.

Czeisler CA, Allan JS, Strogatz SH, et al. (1986) Bright light resets the human circadian pacemaker independent of the timing of the sleep-wake cycle. *Science* 223: 667–671.

Holsboer F, von Bardeleben U, and Steiger A (1988) Effects of intravenous corticotropin-releasing hormone upon sleep-related growth hormone surge and sleep EEG in man. *Neuroendocrinology* 48: 32–38.

Lewey AJ, Wehr TA, Goodwin FK, Newsome DA, and Markey SP (1980) Light suppresses melatonin secretion in humans. *Science* 210: 1267–1269.

Obál F Jr., Payne L, Opp M, Alföldi P, Kapás L, and Krueger JM (1992) Growth-hormone releasing hormone antibodies suppress sleep and prevent enhancement of sleep after sleep deprivation. *American Journal of Physiology* 263: R1078–R1085.

Parker DC and Rossman LG (1971) Human growth hormone release in sleep: Nonsuppression by acute hyperglycemia. *Journal of Clinical Endocrinology* 32: 65–69.

Sassin JF, Parker DC, Mace JW, Gotlin RW, Johnson LC, and Rossman LG (1969) Human growth hormone release: Relation to slow wave sleep and sleep-waking cycles. *Science* 165: 513–515.

Spiegel K, Leproult R, and Van Cauter E (1999) Impact of sleep debt on metabolic and endocrine function. *Lancet* 354: 1435–1439.

Weitzman ED, Nogeire C, Perlow M, et al. (1974) Effects of a prolonged 3-hour sleep-wake cycle on sleep stages, plasma cortisol, growth hormone and body temperature in man. *Journal of Clinical Endocrinology & Metabolism* 38: 1018–1030.

Zee PC and Turek FW (eds.) (1999) *Lung Biology in Health and Disease, Vol. 133: Regulation of Sleep and Circadian Rhythms.* New York: Dekker.

Immune Function During Sleep and Sleep Deprivation

J M Mullington, Harvard Medical School, Boston,
MA, USA

Context and Questions

Physicians through the ages have recommended rest
and sleep to assist in the healing and recovery process.
Most of us have experienced a time when we were
overcome by sleepiness due to an illness and were
unable to function, leaving little alternative but sleep.
Once infection is established, it seems that sleep aids in
the healing process, but does it actually help to clear the
body of pathogens and, if so, how? It is also common
experience that when we do not get adequate sleep for
a prolonged period, we become 'worn down' and
feel susceptible to common colds and other ailments.
It is clear that sleep responds to challenges to the host
defense system (the body's immune protective and
defensive systems), but does sleep actually serve a function in defending against microbial invasion? Are host
defenses bolstered in some way by sleep? Do adequate
quantity and quality of sleep assist in building defenses
against future attacks? Does poor sleep, or going without sufficient sleep for a period, compromise the early
host defense system or the development of immunological memory? Does sleep support good health in a more
general way, such as supporting homeostatic endocrine
and metabolic systems that facilitate a balanced and
measured host defense? Host defense mechanisms are
highly integrated in physiology, and so the answers to
these questions will not be simple. Controlled studies
designed to test these intuitive assumptions have been
sparse. An overview of the major findings is provided
in this article.

Sleep Responds to Experimental Immune Challenge

Animal Models

Studies carried out in many species have shown that
animals need sleep in order to survive. If deprived of
sleep, rats will die within approximately 3 weeks. In
studies in which infectious agents were administered
to rabbits, the probability of survival was increased in
animals that slept more and whose sleep following
challenge showed enhanced slow-wave activity. The
animals that were able to increase their slow-wave
sleep were not only more likely to survive but also
had higher increases over baseline in their white
blood cell count (immune participating) and a lower

triglyceride response to challenge (a marker of the
acute phase response or early immune response to
infection or trauma) compared to the animals that
died. Sleep, it seems, does help the recovery process.

There are specialized receptors for the outer membrane lipopolysaccharide (LPS) fragment of gram-
negative bacteria, called CD14 receptors, that work
through a JAK-STAT (Janus kinase and signal transdu-
cers and activators of transcription) pathway to activate the production of inflammatory protein mediators
such as interleukin-1β (IL-1β), tumor necrosis factor-α
(TNF-α), interleukin-6 (IL-6), and others within
immune-competent cells such as the monocyte, macrophage, and endothelial cell in the periphery and glial
cells in the brain. In animals, levels of IL-1β and TNF-α
in the brain vary with the sleep–wake cycle and are
increased following host challenge.

Several studies in animal models have shown
that IL-1β and TNF-α are key stimulators of the
central nervous system (CNS) production of sleep
and enhanced slow-wave activity, and they are therefore thought to be important for physiological sleep
regulation. If the production of these proinflammatory cytokines is inhibited, the sleep-inducing and
-enhancing effects are also inhibited. In fact, blocking TNF-α and IL-1β inhibits normal increases in
non-rapid eye movement (NREM) sleep that occur
after sleep deprivation. Anti-inflammatory factors
such as glucocorticoids or counterinflammatory
cytokines such as IL-10 also reduce the normal
NREM rebound following sleep deprivation.

The CNS response to immune challenge is not linear.
Sleep may aid in recovery in animals experimentally
challenged with pathogens, but there are clear dose-
dependent effects in the response to challenge. Rabbits
and rats show increased NREM sleep despite moderate fever. Beyond a certain level of inflammatory system activation and fever, sleep onset is delayed and/or
fragmented. The NREM slow-wave enhancing properties of this CNS response to the acute phase reaction is
not simply a by-product of fever because studies that
have blocked the fever response to challenge have
shown that sleep is nonetheless enhanced. Further evidence comes from species-specific differences – for
example, the human sleep response to LPS.

Human Models

As a species, we are exquisitely sensitive to
gram-negative LPS and achieve fevers after being
challenged with very small amounts of endotoxin.
The preparations used for research in humans and
those used in animal research are very different.
When prepared for human use, endotoxin is stripped

of protein so that real infection does not develop. It is the outer membrane component of the gram-negative bacteria, the LPS fragments, that cause the pyrogenic response. Approximately 1 h after injection, TNF-α is already elevated and temperature begins to increase. Unlike the typical NREM sleep response in rabbits and rats that occurs in the context of moderate fever, humans show increased NREM slow-wave sleep only when the dose of LPS is subpyrogenic and when it is too low to increase cortisol. This subpyrogenic dose of LPS is nonetheless able to cause manyfold increases in key cytokines such as TNF-α and its receptors, p55 and p75. IL-1β is generally not measured in peripherally sampled plasma due to insufficient assay sensitivity. Based on animal studies, TNF effects on sleep appear to be mediated through the TNF type 1 receptor (p55) and not its type 2 receptor (p75).

Effects of Sleep Deprivation on Host Defense

Animal Models

The role of sleep in host defense mechanisms is complex. Although there is ample evidence that sleep is necessary for survival, there is also apparently contrary evidence suggesting it is not so important for adequate host defense. For example, although the likelihood of survival is increased by slow-wave sleep following lethal challenge in rabbits, sleep deprivation in rabbits for 4 h by gentle handling, prior to an endotoxin challenge, leads to higher fevers than found in animals that were permitted to sleep prior to challenge. If animals were similarly deprived after challenge, there were no differences in fever response. In fact, except for the fever response, sleep deprivation did not alter the severity of clinical parameters in response to infection. Blood cells, triglycerides, and stress hormones (cortisol and corticosterone) were not negatively impacted by sleep deprivation. The deprivation used may not have been severe enough to cause the expected changes in host response, and the time window with which to see the changes may not have been ideally placed. However, another study using an influenza challenge also failed to demonstrate a sleep advantage compared with sleep deprivation that preceded and/or followed inoculation. Taking yet another approach, tumors implanted in rats that are subsequently sleep deprived shrink in size, counter to what would be expected if sleep loss was compromising host defense. Furthermore, one study in rats failed to find any positive effects of sleep versus 72 h of sleep deprivation on wound healing. Thus, it seems that sleep deprivation, for at least

variations of relatively plausible duration, do little to compromise the host defense response to various forms of challenge. This does not mean that sleep is not involved in supporting host response systems, but it does indicate that the systems function very well and hold up for reasonably long deprivation periods. This is perhaps not surprising, given the central importance of the host defense system for survival and the evolutionary necessity to have the ability to maintain wakefulness for prolonged periods in order to flee danger.

As noted previously, sleep deprivation in animal models does lead to death, usually within 2 or 3 weeks of total sleep deprivation and up to 6 weeks when REM sleep is selectively deprived. Although the ability of animals to withstand the deprivation varies greatly, the pattern of physiological changes is quite consistent. Norepinephrine levels increase early in both total and REM sleep deprivation. Body temperature is increased during the first half of the survival period during total sleep deprivation, but the second half of REM and total sleep deprivation periods are marked by decreases in temperature. Animals undergoing sleep deprivation lose weight despite increased food consumption. This increased energy expenditure and thereby negative energy balance are accompanied by changes in metabolic markers, including decreases in total and free thyroxine and triiodothyronine and reductions in growth hormone, insulin-like growth factor 1, prolactin, and leptin. Fur becomes dull and unkempt in appearance, although grooming time is maintained; tail and paw lesions form.

Leukocytes and some mediators of the immune system (innate and cellular) have been reported to increase under conditions of total sleep deprivation in rats. In addition, consistent with a model of sepsis, bacterial pathogens have been found in organs and tissues of sleep-deprived animals, and these findings have been interpreted by some as evidence of failure of host defense systems. This interpretation is not without opposition, however, because antibiotic treatment is quite successful at controlling aerobic bacteria during deprivation but the animals proceed to accumulate a negative energy balance and die within the same time frame as untreated animals. However, human sepsis, once begun, is often treatment resistant in spite of antibacterial intervention.

In interpreting the findings, it is important to remember that host defense systems are complex and broad in terms of systems mobilized. Sleep supports very complex homeostatic systems, and therefore short-term deprivation and a long-term sleep deficit may have different effects. Very few data are available to enable comparison of the effects of experimental chronic partial sleep deprivation, in which sleep is reduced to a percentage of normal sleep diet for a prolonged period, and total sleep deprivation, in which no sleep is permitted.

Normal Rhythms of Immune-Active Cells and Modulating Hormones in Humans

A diurnal rhythm in human circulating blood cells has long been established. In **Figure 1**, the diurnal pattern of white blood cells and cortisol is illustrated for 25 healthy male and female volunteers between the ages of 25 and 53 years. Participants slept between 11:00 p.m. and 7:00 a.m. before and during the blood sampling period. Blood was sampled every 4 h through an indwelling forearm catheter and results are shown for a 36 h period. The clearly established rhythm of cortisol can be seen. The time of day of lowest circulating levels is morning to early afternoon for lymphocytes, monocytes, and eosinophils, and through the sleep period for neutrophils. Several hormones that show diurnal variation have been reported to modulate immune activity, most notably cortisol, growth hormone, and prolactin. Studies have reported a diurnal rhythm in several circulating cytokines as well, with IL-6 consistently showing an early morning nadir and a nocturnal peak. TNF-α and its receptors have also been reported to have diurnal rhythms, but the peaks and troughs of these rhythms are less consistent across studies.

Cortisol is a clear modulator of immune response. It is used at pharmacological doses to suppress immune function, and at physiological levels it participates in the modulation of the early host response to endotoxin challenge in humans.

Sleep Deprivation in Humans

An interplay of balance between cellular and humoral immune system responding has been described in which type 1 responses are predominantly cellular and type 2 predominantly humoral. The hormonal and cytokine milieu associated with healthy sleep supports growth, probably including development of immunological memory because monocyte production of IL-12, a key cytokine involved in the development of cellular (type 1) memory, is increased during sleep in humans.

Total and partial sleep deprivation studies have been carried out investigating levels of circulating white blood cells and inflammatory mediators. In humans, total sleep deprivation studies have found monocytes and neutrophils to be increased and lymphocytes decreased. Studies of prolonged partial sleep deprivation have also found increased white blood

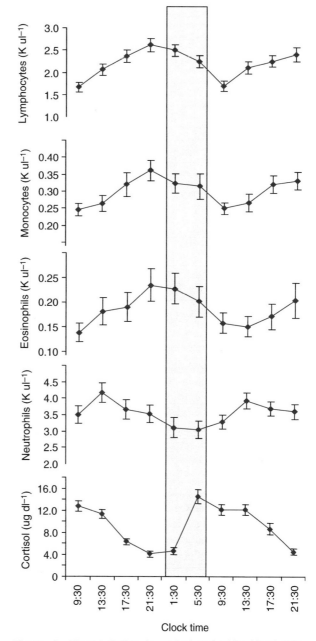

Figure 1 Diurnal rhythm in cortisol and white blood cells. Twenty-five healthy men and women slept between 11:00 p.m. and 7:00 a.m., and blood was drawn through intravenous forearm catheter at 4 h intervals for 36 h. The shaded bar illustrates the time of the sleep period, and the 4 h measurements are shown with standard error bars.

cells due to deprivation. Both total and partial sleep deprivation studies have found that circulating levels of IL-6 and C-reactive protein (RCP) are increased in healthy volunteers. The soluble TNF receptor 1 has been shown to increase during acute total sleep deprivation. These changes may indeed be related to autonomic changes because some studies have shown norepinephrine to be increased in sleep deprivation,

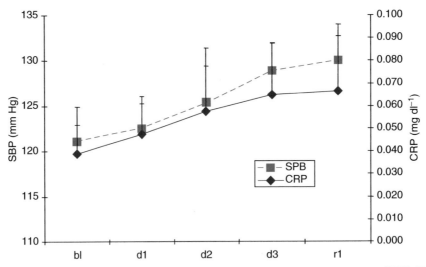

Figure 2 Effects of acute sleep deprivation on C-reactive protein (CRP) and systolic blood pressure (SBP). The average data are shown for the baseline (bl) day, days 1–3 of wakefulness (d1–d3; 88 h total), and the recovery day (r1). Data from Meier-Ewert HK, Ridker PM, Rifai N, et al. (2004) Effect of sleep loss on C-reactive protein, an inflammatory marker of cardiovascular risk. *Journal of the American College of Cardiology* 43: 678–683.

and several studies have shown blood pressure to be increased during sleep deprivation. As can be seen in **Figure 2**, systolic blood pressure has been found to increase with a time course similar to that of CRP. Although the IL-6 and CRP elevations are small, they may be very important in long-term health.

IL-6 and CRP are important acute phase markers of the innate immune system (the type of immunity that humans have without requiring exposure and development of immunological memory) known to be elevated in overweight individuals and which have been linked to development of future cardiovascular disease and diabetes, even independent of their relationship with adiposity. It is important to understand that even very small subclinical elevations in inflammatory mediators have been linked with future risk for the development of disease. It has only been possible to detect these small basal shifts with improvements in assay technology that were made about a decade ago. For this reason, epidemiological studies have only recently been uncovering important predictive relationships that not long ago were undetectable. What used to be thought of as subclinical from the perspective of the acute phase response is now understood to be meaningful from the perspective of understanding risk and disease vulnerability. Do these shifts in inflammatory mediators seen during sleep deprivation translate into increased risk for future disease development? It is too early to know. However, epidemiological studies show that there are risks associated with habitual short sleep duration.

Epidemiological studies show that individuals who sleep less than 6 h per night are at increased risk for developing future cardiovascular disease and

diabetes, independent of other known risk factors such as body mass index. Individuals with sleep-disordered breathing have higher levels of inflammatory mediators, including IL-6 and CRP. Of particular importance in this regard, healthy asymptomatic individuals with very low-level basal increases in CRP have a higher likelihood of developing future cardiovascular disease. This association has also been described for other proinflammatory cytokines such as IL-6, but CRP is particularly useful to examine in population-based studies because it is quite stable and has a long half-life (between 15 and 19 h) and has been shown not to have a diurnal rhythm.

Individuals with disturbed sleep due to disorders such as sleep apnea have been reported to show elevated basal levels of proinflammatory cytokines. It is less clear whether treatment of these disorders resolves the elevated basal inflammatory mediators and to what extent these changes are related to metabolic changes. However, sleep deprivation alters basal metabolic function, which in itself is likely to contribute to changes in immunological homeostasis. Sleep plays a key role in maintaining homeostasis of physiological systems – most important, metabolic and endocrine systems. It is reassuring that host defense mechanisms are intact in the face of acute short-lasting deprivation, but in the long term, sleep is essential for life and both chronic partial sleep deprivation and severe acute sleep deprivation are associated with increased inflammation. Elevations in inflammatory mediators are associated with the development of many diseases, and further investigations are needed to determine the contribution of insufficient sleep to their etiology.

See also: Autonomic Dysregulation During REM Sleep; Behavioral Change with Sleep Deprivation; Endocrine Function During Sleep and Sleep Deprivation; Sleep Apnea; Sleep Deprivation and Brain Function; Thermoregulation During Sleep and Sleep Deprivation.

Further Reading

Born J, Lange T, Hansen K, Mölle M, and Fehm LH (1997) Effects of sleep and circadian rhythm on human circulating immune cells. *Journal of Immunology* 158: 4454–4464.

Dinges DF, Douglas SD, Zuagg L, et al. (1994) Leukocytosis and natural killer cell function parallel neurobehavioral fatigue induced by 64 hours of sleep deprivation. *Journal of Clinical Investigation* 93: 1930–1939.

Everson CA (2005) Clinical assessment of blood leukocytes, serum cytokines, and serum immunoglobulins as responses to sleep deprivation in laboratory rats. *American Journal of Physiology – Regulatory, Integrative and Comparative Physiology* 289: R1054–R1063.

Everson CA and Toth LA (2000) Systemic bacterial invasion induced by sleep deprivation. *American Journal of Physiology – Regulatory, Integrative and Comparative Physiology* 278: R905–R916.

Kuhn E, Brodan V, Brodanova M, and Rysanek K (1969) Metabolic reflection of sleep deprivation. *Activitas Nervosa Superior* 11: 165–175.

Meier-Ewert HK, Ridker PM, Rifai N, et al. (2004) Effect of sleep loss on C-reactive protein, an inflammatory marker of cardiovascular risk. *Journal of the American College of Cardiology* 43: 678–683.

Obal FJ Jr. and Krueger JM (2003) Biochemical regulation of sleep. *Frontiers in Bioscience* 8: 520–550.

Pollmächer T, Mullington J, Korth C, et al. (1996) Diurnal variations in the human host response to endotoxin. *Journal of Infectious Diseases* 174: 1040–1045.

Rechtschaffen A and Bergmann BM (2002) Sleep deprivation in the rat: An update of the 1989 paper. *Sleep* 25: 18–24.

Rechtschaffen A, Bergmann BM, Everson CA, Kushida CA, and Gilliand MA (1989) Sleep deprivation in the rat: X. Integration and discussion of the findings. *Sleep* 12: 68–87.

Renegar KB, Floyd RA, and Krueger JM (1998) Effects of short-term sleep deprivation on murine immunity to influenza virus in young adult and senescent mice. *Sleep* 21: 241–248.

Shearer WT, Reuben JM, Mullington JM, et al. (2001) Soluble TNF-alpha receptor 1 and IL-6 plasma levels in humans subjected to the sleep deprivation model of spaceflight. *Journal of Allergy and Clinical Immunology* 107: 165–170.

Toth LA, Opp MR, and Mao L (1995) Somnogenic effects of sleep deprivation and *Escherichia coli* inoculation in rabbits. *Journal of Sleep Research* 4: 30–40.

Toth LA, Tolley EA, and Krueger JM (1993) Sleep as a prognostic indicator during infectious disease in rabbits. *Proceedings of the Society for Experimental Biology and Medicine* 203: 179–192.

Thermoregulation during Sleep and Sleep Deprivation

R Szymusiak, University of California, Los Angeles, CA, USA

Introduction

The regulation of sleep and the regulation of body temperature are tightly coupled. Sleep onset in mammals is accompanied by reductions in metabolic rate, accelerated heat loss, and an integrated fall in body temperature. Homeostatic responses of the thermoregulatory system to heat or cold challenge are suppressed during sleep, and in some species thermoregulatory adjustments are almost completely absent during rapid eye movement (REM) sleep. Long-term sleep deprivation in animals can be associated with chronically elevated metabolism and food intake, accompanied by significant decreases in core body temperature.

In addition to changes in thermoregulation during sleep, manipulation of environmental temperature or body temperature can alter sleep. In all mammals studied, sleep is suppressed in moderately warm or cold environments. This suppression of sleep by thermal stress can be viewed as adaptive from a homeostatic standpoint because thermoregulatory responses are diminished during sleep. In some circumstances, manipulation of body temperature can promote sleep. For example, in humans, whole-body warming by hot bath immersion a few hours prior to habitual bedtime can reduce sleep latency and increase stage 3/4 sleep amounts.

Neuronal systems that regulate sleep and body temperature are localized in the preoptic and anterior hypothalamus (POAH). Many thermo-sensing neurons in the rostral hypothalamus also exhibit changes in activity across the sleep–wake cycle. Of particular importance may be groups of hypothalamic warm-sensing neurons that are spontaneously activated during sleep onset and sleep. Thus, functional interactions between sleep-regulatory and thermoregulatory neurons in the rostral hypothalamus can explain the close coupling between body temperature and arousal states.

Understanding sleep and thermoregulatory interactions may provide insight into clinical problems. Recent evidence has demonstrated the importance of thermoregulatory adjustments occurring around the time of sleep onset in determining sleep latency, sleep composition, and sleep quality in humans. Therapies that target the thermoregulatory control of arousal states may provide novel approaches to treating disorders of initiating and maintaining sleep.

Thermoregulation during Sleep

Changes in Body Temperature and Metabolism during Sleep

In mammals, the onset of sleep is associated with a fall in core body temperature and a readjustment of thermoregulatory control. The magnitude of the drop in body temperature during sleep varies across species and can be influenced by variables such as ambient temperature and phase of the circadian rhythm. The sleep-related fall in body temperature during sleep is actively regulated and is not simply a passive consequence of behavioral inactivity. Sleep onset in mammals is accompanied by coordinated decreases in metabolic heat production and increases in heat loss through increased peripheral vasodilation and/or sweating.

The decrease in body temperature during sleep can be understood as a resetting of the body's thermostat to a lower level. In other words, sleep appears to be accompanied by a lowering of the thermal set point, the level of body temperature that the thermoregulatory system will defend and maintain. At the transition from waking to sleep, the thermoregulatory system responds as if the set point has been adjusted to a lower level. The transition is accompanied by an integrated series of thermoregulatory responses, including a reduction in metabolic rate and enhanced heat loss to the environment through peripheral vasodilation and sweating. Body temperature falls to a new regulated level which is maintained during sleep.

This lowering of the set point and consequent energy savings can be taken as evidence for a potential energy conservation function of sleep. Energy savings are achieved in two ways: (1) by lowering metabolic heat production and (2) by lowering body temperature. The lowering of body temperature reduces the core temperature-to-ambient temperature gradient, which serves to minimizes heat loss to the environment. This latter factor can be important in small mammals with a high surface-to-volume ratio, which have inherently higher rates of environmental heat exchange. Behavioral aspects of sleep onset such as seeking a thermally comfortable nest and adoption of curled heat-conserving postures can also contribute to energy savings by minimizing heat exchange with the environment.

Thermoregulatory Responses to Thermal Challenge Are Influenced by Sleep

The decline in the thermal set point during sleep onset and non-REM sleep is also accompanied by reduced thermoregulatory responses to thermal challenges. Thermoregulation is not suspended during sleep, but the responsiveness of the thermoregulatory system appears to be diminished; that is, the magnitude of the regulatory response to a given thermal challenge is reduced during sleep compared to waking.

Compared to non-REM, thermoregulatory responses during REM sleep are reduced even further. This is particularly true for responses to cold stress because the inhibition of motor neurons that is characteristic of REM sleep is incompatible with increasing heat production via shivering or elevated muscle tone. As a result of these thermoregulatory changes, animals may be particularly vulnerable to thermal stress during REM sleep. Infants and small mammals with a large surface-to-volume ratio are most vulnerable.

Exposure to hot or cold environments frequently cause sleep disruption and sleep loss. This response may be adaptive in minimizing exposure to thermal stress during sleep when homeostatic compensation is weak. Thermal factors have undoubtedly played a role in the evolution of behavioral strategies for species-specific selection of sleeping environments (e. g., nests, sheltered locations, and adoption of heat conserving postures). Diminished thermoregulatory adjustments to heat and cold stress during sleep dictate that thermal homeostasis can be best achieved over the course of a several-hour sleep episode by the selection of a sleeping environment that poses minimal thermal challenge and ultimately maximizes heat conservation. This is a strategy that has been adopted by several species.

Effects of Environmental and Body Temperature on Sleep

Exposure to moderate hot or cold environments suppresses sleep and/or causes sleep fragmentation in mammals. The most sensitive species are those with large surface-to-volume ratios that experience more rapid environmental heat exchange. In rodents, in which thermoregulatory responses during REM sleep are absent or weak, REM sleep suppression can occur under conditions of minimal thermal stress and in the absence of any significant change in non-REM sleep amounts. From a thermoregulatory perspective, the REM-sleep-suppressing effects of heat and cold exposure in these species can be viewed as adaptive. Increased time spent awake or in non-REM sleep means that lower threshold and stronger thermoregulatory responses are evoked to maintain thermal homeostasis.

Mild increases in ambient temperature or mild warming of the body can promote sleep. An animal's thermoneutral zone is defined as the range of environmental temperatures at which resting metabolic rate is minimal and constant. Within the thermoneutral zone, minimal effort is required to maintain thermal homeostasis because the heat produced by the resting metabolic rate is largely balanced by heat loss to the environment. Within the thermoneutral zone, thermal homeostasis can usually be maintained through peripheral vasomotor adjustments that regulate rates of heat loss from the skin, with vasoconstriction occurring at the low end of the zone and vasodilation at the high end. Exposing rats to an environmental temperature that is at or just above the high end of the thermoneutral zone, at which peripheral blood flow and skin temperatures are high, increases time spent in non-REM sleep and enhances electroencephalogram (EEG) slow-wave activity during sleep.

Whole-body warming, achieved by immersion in a hot bath a few hours prior to normal bedtime, promotes sleep in humans. Hot baths prior to bedtime can reduce sleep latency and increase amounts of stage 3/4 sleep in the first half of the night. Similarly, exercise late in the day reduces sleep latency and increases stage 3/4 sleep. The sleep-enhancing effects of exercise are dependent on an accompanying increase in body temperature. If exercising subjects are prevented from raising their body temperature by being cooled with a fan during exercise, the sleep-promoting effects of exercise are prevented.

Collectively, these findings are consistent with the idea that the mild activation of thermoregulatory heat defense mechanisms is conducive to sleep, particularly sleep onset and slow-wave sleep. The critical feature is that the warming must be mild such that it evokes mild compensatory thermoregulatory responses (e.g., increased peripheral vasodilation). More severe heat challenges that evoke stronger regulatory responses such as intense sweating or panting cause sleep disruption, not sleep promotion.

Brain Circuits That Regulate Sleep and Body Temperature Are Functionally Interrelated

Temperature-Sensing Neurons in the Hypothalamus Control Body Temperature

Neurons in the POAH are critically involved in thermoregulation. Experimental damage to the POAH in animals causes persistent deficits in thermoregulatory defense against heat and cold exposure. Direct infusion of various neurotransmitters or neuromodulators into the POAH can alter body temperature and thermoregulatory function.

Specialized temperature-sensing neurons localized in the POAH play an important role in body temperature control. Warm-sensing neurons are excited by local increases in temperature and inhibited by cooling. Cold-sensing neurons are excited by local decreases in temperature and inhibited by local warming. The functional importance of these temperature-sensing neurons for thermoregulatory control is evidenced by experimental findings that localized warming or cooling of the POAH can evoke fully integrated heat defense or cold defense responses that are qualitatively identical to those evoked by exposing the whole animal to hot or cold environments.

Hypothalamic Temperature-Sensing Neurons Are Involved in Sleep Regulation

The neural control of sleep and thermoregulation intersect in the POAH and adjacent brain structures. Experimental POAH damage that disrupts thermoregulation also causes persistent insomnia, suggesting that thermo- and sleep-regulatory neuronal systems are anatomically related. They appear to be functionally related as well. As previously summarized, local warming or cooling of the POAH can evoke robust whole-body heat and cold defense responses. Milder increases in POAH temperature, in addition to suppressing metabolic rate and augmenting peripheral vasodilation, reduce sleep latency, promote non-REM sleep and increase EEG slow-wave activity during non-REM sleep. Local POAH cooling has the opposite effects, causing sleep suppression and suppression of EEG slow-wave activity during sleep.

Potential mechanisms underlying many of the effects of environmental and body temperature on sleep are revealed by the study of the spontaneous discharge patterns of POAH thermo-sensing neurons during waking and sleep. As documented by single-neuronal recordings in cats and rats, the majority of POAH warm-sensing neurons exhibit an elevation of spontaneous discharge rate during sleep onset and non-REM sleep compared to waking. The majority of cold-sensing neurons are maximally activated during waking and exhibit declining discharge during sleep onset and non-REM sleep. Thus, the changes in POAH thermo-sensing neuronal activity at sleep onset mimic the pattern expected to accompany a lowering of the thermal set point, with the activation of warm sensors and inhibition of cold sensors. This pattern of thermoreceptor activity is consistent with the thermoregulatory responses that normally accompany sleep onset (i.e., reduced metabolic rate and increased heat loss at the periphery). The ability of local POAH warming to evoke non-REM sleep and EEG synchrony indicates that the activation of POAH warm-sensing neurons is sufficient for sleep promotion.

Control of Body Temperature during Sleep Deprivation

Acute Sleep Deprivation: Interactions with Circadian Control of Body Temperature

In healthy individuals, entrained to a light–dark cycle and adhering to a regular sleep–wake schedule, habitual sleep onset reliably occurs a few hours after the 24 h body temperature maxima, on the falling phase of the 24 h body temperature rhythm. Sleep onset is associated with an acceleration in the rate of the decline in body temperature as the activation of heat-loss processes during sleep are expressed. Morning awakenings occur on the rising phase of the temperature rhythm, shortly after the 24 h temperature minimum. Thus, when individuals maintain a regular sleep schedule, sleep–wake rhythms and body temperature rhythms are closely coupled with a characteristic phase relationship.

The fall in body temperature that normally occurs during the sleep period is not a passive consequence of changes in muscle activity and heat production associated with waking versus sleep. The body temperature rhythm persists during sleep deprivation or during constant routine protocols in humans, in whom levels of activity, food intake, light exposure, and so on are maintained at constant levels throughout a 24 h period. Nevertheless, behavioral changes associated with waking and sleep do exert positive and negative masking effects on the circadian temperature rhythms, such that the magnitude of the 24 h peak-to-trough difference in body temperature is reduced during sleep deprivation or constant routine, compared to normal entrained conditions.

An important feature of arousal state regulation is that short-term sleep deprivation or sleep restriction leads to compensatory rebound increases in sleep amount and sleep depth once the opportunity for sleep is provided. Homeostatic increases in sleep following sleep deprivation/restriction involve complex changes in neurophysiological and neurochemical mechanisms that are not yet completely understood. One effect of short-term sleep loss is to change the normal phase relationship between the occurrence of sleep/wakefulness and the 24 h body temperature rhythm. Thus, extending wakefulness through the night and attempting to sleep the next morning results in sleep onset occurring on the rising phase of the body temperature rhythm.

The phase relationship between sleep and body temperature rhythms at sleep onset is an important determinant of subsequent sleep amount and composition. This has been most convincingly demonstrated in human subjects undergoing internal or forced desynchronization in environments without time cues. During internal desynchronization, sleep and body temperature rhythms free run with different periods, and eventually the two rhythms temporally dissociate. In forced desynchrony protocols, subjects adhere to a sleep–wake schedule significantly longer than 24 h (e.g., 28 h), and when this is done in the absence of time cues, body temperature and sleep rhythms dissociate as in internal desynchronization. With both methods, over the course of several days the onset of the major sleep period occurs at all phases of the body temperature rhythm. Sleep episodes that are initiated on the falling phase of the temperature rhythm are associated with shorter sleep latencies, few episodes of waking during the sleep period and maximal amounts of deeper sleep stages. In contrast, sleep episodes initiated on the rising phase of the temperature rhythm are associated with longer sleep latencies, short sleep episode durations, more REM sleep but less stage 3/4 sleep. Thus, depending on the timing of sleep loss and recovery sleep, the adverse effects of short-term sleep deprivation include impaired alertness during sustained waking and impaired depth and quality of recovery sleep. The effects of partial sleep loss are cumulative, and the level of sleepiness will increase over several days of sleep restriction, due, in part, to the inadequate restorative effects of suboptimal recovery sleep.

Chronic Sleep Deprivation Causes Dysregulation of Body Temperature in Rats

The effects of long-term total sleep deprivation on thermoregulation have been studied in rats by Rechtschaffen and colleagues. Total sleep deprivation applied over many days in the rat results in a syndrome that includes increased food intake; increased energy expenditure; weight loss; an initial mild elevation in body temperature, followed by profound decreases in temperature as the duration of deprivation increases; and death within 11–32 days (average ~17 days). Several findings indicate that sleep deprivation in rats causes a profound dysregulation in the homeostatic systems that control metabolism and body temperature. The increased food intake occurring during sleep deprivation is more than offset by increased energy expenditure, such that rats actually lose weight in spite of a near doubling of food intake. The increase in energy expenditure appears, at least

in part, to be an attempt to compensate for increased heat loss. Ultimately, rates of heat loss are so high that the core body temperature falls to well below normal levels, and rats become profoundly hypothermic during the last few days of survival. This fall in body temperature is not regulated; that is, there is no apparent lowering of the set point for body temperature during sleep deprivation. When sleep-deprived rats are given the opportunity to control the temperature of their environment, they select higher temperatures from the beginning of deprivation and select extremely high temperatures later in the deprivation period when they are hypothermic. Thus, they behaviorally try to raise body temperature to combat the hypothermia evoked by unregulated elevations of heat loss. The mechanisms responsible for the changes in metabolism and thermoregulation evoked by sleep deprivation are not understood, but findings do suggest that sleep is necessary for the proper functioning of these regulatory systems.

Thermoregulatory Control of Human Sleep

Thermoregulatory Changes Influence Sleep Onset and Sleep Maintenance

Neurobiological studies in experimental animals have revealed close anatomical and functional interrelationships between sleep and body temperature-control mechanisms in the brain, and they indicate that the activation of hypothalamic neuronal circuits that promote heat loss normally occurs during sleep onset and non-REM sleep. In humans, the magnitude of the heat-loss response around the time of sleep onset is an important determinant of sleep latency, sleep amount, and sleep quality. The magnitude of heat-loss response in humans can be reliably assessed by quantifying the temperature gradient from distal skin sites (the hands and feet) to proximal skin sites (e.g., the forehead, stomach, thigh, or some weighted average of several proximal skin sites). A high distal-to-proximal skin temperature gradient occurs when the temperature of the feet and hands are warm with respect to proximal sites, indicative of elevated blood flow to distal skin and elevated heat exchange at these sites. A low distal-to-proximal gradient reflects peripheral vasoconstriction and diminished peripheral heat loss. A high distal-to-proximal skin temperature gradient is associated with reduced sleep latency, increased stage 3/4 sleep, and reduced number of arousals from sleep. The distal-to-proximal gradient is also positively correlated with subjective sleepiness after morning awakenings and after awakenings from daytime naps.

It can be hypothesized that thermoregulatory responses that normally accompany sleep onset in humans play a role in reinforcing and consolidating circadian and homeostatic aspects of sleep regulation. In an entrained individual, habitual sleep-onset times occur on the descending phase of the body temperature rhythm. Homeostatic sleep drive is also high at this time because of sustained prior waking. As body temperature falls following the nadir of the 24 h rhythm, the activation of heat-loss mechanisms result in elevated skin temperatures. This leads to the activation of peripheral warm sensors and excitation of hypothalamic warm-sensing neurons via peripheral-to-central thermosensory pathways. The activation of central warm-sensors increases sleep propensity via suppression of arousal systems and hasten sleep onset, resulting in further increases in peripheral skin temperature and heat loss. Thus, thermoregulatory changes have the potential to exert feed forward excitatory effects on brain mechanisms that control sleep onset and sleep maintenance, leading to more rapid and seamless transitions from waking to consolidated sleep.

Thermoregulation and Disordered Human Sleep

Insomnia is the most common sleep complaint, and sleep disturbance is a prominent symptom in several psychiatric and neurological disorders. In many instances, insomnia complaints are accompanied by abnormalities in circadian rhythms, including abnormalities in the body temperature rhythm. Sleep-onset insomniacs may exhibit a phase delay in the body temperature rhythm of >2 h, indicating a disturbance of circadian timing or entrainment. This phase delay means that attempts to fall asleep often occur in closer than normal proximity to the body temperature peak and that thermoregulatory adjustments that promote sleep might not be as readily evoked. In the elderly, phase delays in the body temperature rhythm may also be a contributing factor to the increased incidence of insomnia. In affective disorders, the amplitude of the 24 h body temperature rhythm is reduced. The diminished nocturnal peak in the temperature rhythm results in a lower body temperatures at bedtime and a weaker heat-loss response at the time of sleep onset, resulting in more prolonged and disrupted wake-to-sleep transitions.

Although body-temperature abnormalities in disorders associated with insomnia may be secondary to other underlying pathologies, the manipulation of thermoregulatory control, either pharmacologically or behaviorally (e.g., by hot bath immersion), in ways that promote heat loss at sleep onset may help to improve sleep amount and sleep quality.

See also: Autonomic Dysregulation During REM Sleep; Behavioral Change with Sleep Deprivation; Endocrine Function During Sleep and Sleep Deprivation; Hypothalamic Regulation of Sleep; Immune Function During Sleep and Sleep Deprivation; Sleep Deprivation and Brain Function; Sleep in Aging.

Further Reading

Alam MN, McGinty D, and Szymusiak R (1995) Neuronal discharge of preoptic/anterior hypothalamic thermosensitive neurons: Relation to nonrapid eye movement sleep. *American Journal of Physiology* 269: R1240–R1249.

Bach V, Telliez F, and Libert J-P (2002) The interaction between sleep and thermoregulation in adults and neonates. *Sleep Medicine Reviews* 6: 481–492.

Czeisler CA, Weitzman E, Moore-Ede MC, Zimmerman JC, and Knauer RS (1980) Human sleep: Its duration and organization depend on its circadian phase. *Science* 210: 1264–1267.

Gilbert SS, van den Heuvel CJ, and Ferguson SA (2004) Thermoregulation as a sleep signaling system. *Sleep Medicine Reviews* 8: 81–93.

Glotzbach SF and Heller HC (1976) Central nervous regulation of body temperature during sleep. *Science* 194: 537–539.

Heller HC (2005) Temperature, thermoregulation and sleep. In: Kryger M, Roth T, and Dement W (eds.) *Principles and Practice of Sleep Medicine*, 4th edn., pp. 292–304. Philadelphia: Elsevier Saunders.

Horne JA and Reid AJ (1985) Night time sleep EEG changes following body heating in a warm bath. *Electroencephalography and Clinical Neurophysiology* 60: 154–157.

Karuchi K, Cajochen C, Werth E, and Wirz-Justice A (2000) Functional link between distal vasodilation and sleep onset latency? *American Journal of Physiology* 278: R741–R748.

Karuchi K, Cajochen C, and Wirz-Justice A (2004) Waking up properly: Is there a role of thermoregulation in sleep inertia? *Journal of Sleep Research* 13: 121–127.

McGinty D and Szymusiak R (1990) Keeping cool: A hypothesis about the mechanisms and functions of slow-wave sleep. *Trends in Neuroscience* 13: 480–487.

McGinty D and Szymusiak R (2001) Brain structures and mechanisms involved in the generation of NREM sleep: Focus on the preoptic hypothalamus. *Sleep Medicine Reviews* 5: 323–342.

Parmeggiani PL and Rabini C (1970) Sleep and environmental temperature. *Archives of Italian Biology* 108: 369–387.

Rechtschaffen A, Bergman BM, Everson CA, Kushida CA, and Gilliland MA (1989) Sleep deprivation in the rat, X: Integration and discussion of findings. *Sleep* 12: 68–87.

Shaw PJ, Bergman BM, and Rechtschaffen A (1997) Operant control of ambient temperature. *American Journal of Physiology* 272: R682–R690.

Autonomic Dysregulation During REM Sleep

S M Caples and V K Somers, Mayo Clinic, Rochester, MN, USA

Sleep Stage Transitions and the Autonomic Nervous System in Normal Individuals

In humans, wakefulness normally transitions to sleep with the onset of non-rapid eye movement (NREM) sleep, which comprises the majority of total sleep time. Compared with wakefulness, NREM sleep is characterized by cardiovascular stability marked by a reduction in blood pressure (BP) and heart rate (HR). Slowing of the HR is primarily under the influence of vagal activation that is characteristic of NREM sleep. Because of the limited parasympathetic influences on the peripheral vasculature, the decrease in BP is probably a result of concurrent sympathetic withdrawal, as supported by animal models of sympatholysis. During NREM sleep, there is a prominent sinus arrhythmia as cardiac output couples with the respiratory rhythm. With inspiration, as venous return to the heart increases, there is an acceleration of HR followed by a deceleration with exhalation. BP is tightly controlled on account of high baroreceptor gain, which is particularly responsive to BP increments during NREM sleep, serving to ensure the maintenance of stable low BP, referred to as the 'dipping' phenomenon. The absence of this decrement in BP ('nondipping') has, in some studies, been associated with cardiovascular morbidity and mortality – outcomes perhaps mediated by autonomic dysregulation.

Progression of NREM sleep from stage 1 through 4 is characterized by increasingly synchronous cortical and brain stem electrical discharges as exhibited by low-frequency (LF), high-amplitude output on electroencephalography (EEG). This natural sequence is accompanied by further reduction in sympathetic neural traffic such that by stages 3 and 4 (slow wave), the output may be half that encountered in wakefulness. Stage 2 sleep is punctuated by characteristic 'K-complexes' – high-amplitude EEG discharges associated with transient increases in peripheral sympathetic neural activity and an attendant rise in BP (**Figure 1**). Bursts of vagal activity sometimes accompany the transition out of NREM sleep, resulting in HR pauses.

Sharply distinct from NREM sleep, rapid eye movement (REM) sleep occurs cyclically (four to six times) throughout sleep, is more concentrated during the second half of a sleep cycle, and accounts for approximately 25% of total sleep time. With the exception of the respiratory diaphragm and extraocular muscles, REM sleep is heralded by skeletal muscle atonia along with high-frequency (HF), low-amplitude waves on EEG. Because the EEG may otherwise resemble that of wakefulness, REM sleep has been referred to as 'paradoxical sleep.' REM sleep may be subdivided into tonic and phasic stages, the latter of which features characteristic darting eye movements in association with pontogeniculo-occipital EEG spikes, and muscle twitches, predominantly seen in the extremities and face, which break through skeletal muscle atonia.

As discussed later, REM sleep, particularly the phasic component, is considered a state of autonomic instability, marked by surges in both sympathetic and vagal activity with attendant variations in HR, BP, and peripheral vascular resistance. Animal and human studies have demonstrated more concentrated autonomic changes associated with phasic REM sleep, during which intense electrical discharges have been recorded from the brain stem. As a testament to the complexity of the origins of REM sleep and its associated autonomic findings, measurable physiologic parameters representative of autonomic fluctuations have been shown to precede, by several seconds, characteristic eye movements and an active, desynchronized EEG.

Dreaming

Dreaming is thought to occur primarily during REM sleep. Human experiments show that the majority of subjects awoken during REM sleep report a dream experience, compared with a minority reporting the same when awoken from NREM sleep. It has been postulated that the paradoxically active cerebral cortex associated with REM sleep relates to dreams, during which powerful emotions such as anger and fear are frequently generated. Since psychological stress has been associated with cardiac ischemia and rhythm disturbances during wakefulness, it may follow that similar vulnerability could occur during dreaming, as suggested by Mac William in 1923 and also by case reports of patients awakening from dreaming with angina pectoris. There is evidence that brain electrical discharges may directly impact cardiac activity and are perhaps mediated by autonomic tone. Early animal experiments, confirmed by Verrier, have shown that electrical stimulation of certain areas of the brain could induce ventricular arrhythmias and that β-adrenergic blockade, but not vagotomy, reduced vulnerability.

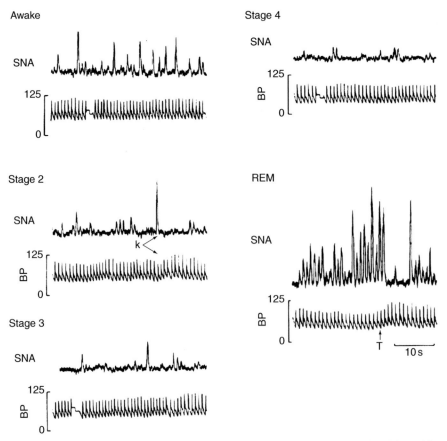

Figure 1 Recordings of sympathetic nerve activity (SNA) and mean blood pressure (BP) in a single subject while awake and while in stages 2, 3, 4, and REM sleep. Note the progressive reduction in SNA and BP as sleep progresses through the stages of NREM sleep. The 'K' during stage 2 corresponds to simultaneous K-complexes noted on the electroencephalogram. In contrast, REM sleep is marked by higher frequency and amplitude of SNA as well as variability in BP. The 'T' during REM sleep refers to a muscle twitch characteristic of phasic REM sleep, associated with abrupt inhibition of SNA and an increase in BP. Reproduced from Somers VK (1993) Sympathetic-nerve activity during sleep in normal subjects. *New England Journal of Medicine* 328(5): 303–307.

Other Physiologic Responses to REM Sleep

In addition to directly observed cardiovascular irregularities, REM sleep is associated with other physiologic perturbations which may indirectly contribute to autonomic dysfunction and, therefore, cardiovascular risk. The respiratory diaphragm usually escapes skeletal muscle atonia, but loss of tone in accessory muscles of respiration and those of the upper airway may pose an increased workload. The ventilatory pattern fluctuates during REM sleep. As a consequence, even though respiratory rate may increase during REM sleep, tidal volumes and minute ventilation usually decrease compared with NREM sleep. There are blunted ventilatory responses to hypoxia and hypercapnia and a heightened arousal threshold, all contributing to prolonged reductions in blood oxygen content, particular in those with impaired respiratory pump function (neuromuscular weakness) or gas

exchange capabilities (chronic obstructive lung disease). Hypoxemia stimulates the chemoreflex, resulting in further sympathetic neural outflow. Increased upper airway resistance related to muscle atonia is operative in the pathogenesis of obstructive sleep apnea (OSA), as detailed later. Partial loss of thermoregulatory control during REM sleep further disrupts homeostasis. Finally, cortical influences related to dream activity during REM sleep may contribute to the irregular breathing pattern, along with increased cortical and brain stem electrical discharges seen in association with dreaming.

Cardiovascular Reflexes Mediating Autonomic Activity

A number of reflexes feed back to the autonomic nervous system to influence cardiac and vascular function during periods of stress or physical exertion.

From the standpoint of sleep, these mechanisms may be particularly important during changes in stage of sleep (NREM to REM), position, or with the occurrence of arousals, all of which may take place repeatedly throughout a sleep cycle and are increasingly common with advancing age, a population vulnerable to cardiovascular disease. Dysregulation of these reflexes may be operative in the pathophysiology of certain disease states, such as sleep disordered breathing.

The Chemoreflexes

The peripheral arterial chemoreceptors, the most important of which are located in the carotid bodies, respond primarily to changes in the blood partial pressure of oxygen. Hypoxemic stimulation elicits a brain stem-mediated increase in respiratory muscle output. By virtue of connections between the carotid bodies and sympathetic ganglia, there is an increase in sympathetic outflow to peripheral blood vessels. Homeostasis is preserved by within-breath activation of pulmonary stretch receptors, which are mediated through the vagus nerve, restraining the overall adrenergic response. This sequence is often part of a normal physiologic response under conditions of hypoxia, such as at high altitude. However, an exaggerated ventilatory response, with attendant further increases in sympathetic tone, has been found in certain disease states, such as heart failure and OSA.

The Arterial Baroreflex

The arterial baroreflex system, composed of sensory receptors in the aortic arch and carotid sinuses that relay signals to the brain stem, provides powerful beat-by-beat negative feedback regulation of arterial BP. Increases in BP stretch the receptors and dampen efferent sympathetic output to cardiac and vascular smooth muscle, thereby relatively increasing cardiac parasympathetic tone. The acute increases in BP associated with peripheral vasoconstriction are partially tempered by the baroreflex. This may occur simultaneously with activation of cardiac vagal drive resulting in bradycardia, collectively referred to as the 'diving reflex,' so-called because of its detailed description in diving mammals and also described in humans. Baroreceptor sensitivity appears to be heightened during NREM sleep, but findings in very small studies that measure REM sleep are inconsistent, with some showing a blunting of baroreceptor sensitivity, no change, or a time-dependent increase in sensitivity. Influences of comorbid illnesses, such as hypertension or heart failure, on baroreceptor function are well documented.

Measures of Autonomic Cardiovascular Regulation

Although there are a number of methods by which to measure autonomic neural output, there have been, with some exceptions, concordant findings of parasympathetic predominance during NREM sleep and sympathetic surges with variable states of parasympathetic tone during REM sleep. A commonly utilized method, heart rate variability (HRV), is defined as the oscillation in the interval between consecutive heartbeats (the R-R interval) as well as the oscillations between consecutive instantaneous heart rates. During NREM sleep, a near sinusoidal modulation of HR variation occurs due to the normally occurring respiratory sinus arrhythmia. Spectral analysis allows quantification of the short-term oscillatory components of cardiovascular variability, which are organized in two frequency bands – the LF (approximately 0.1 Hz) and the HF (>0.15 Hz) respiratory bands. The HF components of R-R variability primarily reflect vagal modulation of cardiac rhythm, whereas the LF rhythm has been attributed, albeit not without controversy, to cardiac sympathetic modulation.

Utilizing HRV analysis, a number of studies have demonstrated increased parasympathetic tone and sympathetic withdrawal with the transition from wakefulness to NREM sleep. These studies also demonstrate evidence for increased sympathetic output with the onset of REM sleep, although not all studies have shown reductions in parasympathetic neural output during REM sleep.

Invasive and technically demanding, microneurographic recording of sympathetic nerve activity (MSNA) involves the direct recording of postganglionic sympathetic nerve traffic which may originate from muscle and skin. MSNA has yielded fairly concordant sleep stage-dependent findings. Peripheral sympathetic neural activity, HR, and BP all decrease in concert from sleep onset and progression of NREM sleep to deep, synchronized slow-wave sleep, such that by stage 4, sympathetic measured output was half that encountered in wakefulness. REM sleep, on the other hand, has been associated with SNA exceeding that encountered during wakefulness, particularly during clusters of eye movements (**Figure 1**). Muscle twitches that break through REM atonia have been associated with abrupt attenuation of sympathetic output, probably due to baroreceptor-mediated inhibition during transient rises in BP. That REM sleep is associated with minimal changes in HR and BP may relate to the dissociation of peripheral muscle and cardiac sympathetic activity and underscores the complexities of cardiac sympathovagal balance

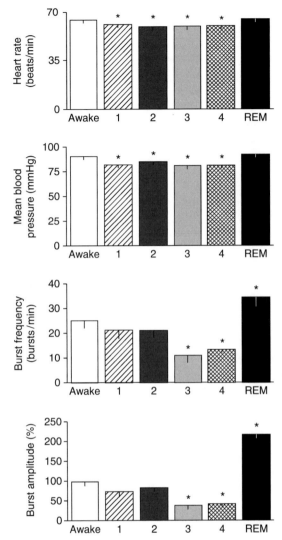

Figure 2 Physiologic indices in a group of normal subjects during wakefulness and NREM and REM sleep. Burst frequency and amplitude refer to sympathetic nerve activity (SNA) measured by intraneural microneurography. Heart rate (HR) and blood pressure (BP) were significantly lower during all stages of NREM sleep than during wakefulness, and SNA was significantly lower during slow-wave sleep (asterisk denotes $p < 0.001$). During REM sleep, sympathetic activity increased significantly, but BP and HR were similar to those recorded during wakefulness. Reproduced from Somers VK (1993) Sympathetic-nerve activity during sleep in normal subjects. *New England Journal of Medicine* 328(5): 303–307.

(**Figure 2**). Thus, this technique provides insight into regional sleep stage-dependent autonomic mechanisms, although it may not necessarily reflect global sympathetic tone.

Other methods for measuring autonomic neural activity, all with their own limitations, include arterial baroreflex function, plasma and urine catecholamine levels, peripheral arterial tonometry, and pulse transit time.

REM Sleep and Heart Rate/Rhythm

Animal experiments have demonstrated marked HR instability during REM sleep, where increases of more than 30% are seen in association with phasic REM sleep, accompanied by an elevation in BP. The frequency of HR surges is higher during phasic REM than tonic REM sleep, with particular concentration during sleep epochs in which an associated muscle twitch was recorded. In animals with coronary artery stenosis, REM sleep causes further decrements in coronary artery blood flow. That these findings were prevented with sympathectomy in cats and dogs suggests an effect attributable to sympathetic overdrive rather than parasympathetic withdrawal. In rats, REM, particularly during phasic episodes, is associated with a higher incidence of bradycardia.

In humans, the average HR is higher during REM than NREM sleep, as is sympathetic neural output, although there is evidence that sympathetic activation is not a global phenomenon during REM sleep. Case reports and series suggest that healthy young adults, sometimes athletes, have periods of exaggerated vagal tone during REM sleep, manifesting as first- or second-degree atrioventricular (AV) block or sinus pauses that may exceed 9 s in duration. Further insight comes from feline experiments demonstrating the highest occurrence of these pauses during the transition from slow-wave sleep to REM sleep as well as during phasic REM sleep. Reversal of the pauses has been reported with atropine or vagotomy. REM sleep therefore appears to be marked by fluctuations in autonomic activity, both to the cardiac conduction system and to peripheral blood vessels.

Patients with established cardiac conduction disease may have distinct pathophysiologic mechanisms during REM sleep, suggested by a small group of subjects with preexisting AV block demonstrating shortening of AV delay during periods of REM sleep. Ambulatory cardiac monitoring (non-polysomnographic) data in healthy elderly populations demonstrate less common occurrences of sinus pauses and AV blocks during sleep episodes, with relatively common supraventricular and ventricular ectopic beats. The circadian variation in paroxysmal atrial fibrillation noted by some suggests a potential influence of the autonomic nervous system. Analysis of HRV in patients with atrial fibrillation demonstrated abrupt shifts in sympathovagal balance preceding the onset of the arrhythmia. Finally, gender-selective effects of REM sleep on cardiac electrical activity were demonstrated by Lanfranchi et al., who measured prolongation of the QT interval exclusive to women.

Arousals/Morning Awakening

Arousal from sleep, either spontaneous or induced by exogenous stimuli, is associated with sympathetic neural surges, leading to increases in HR and BP. Sympathetic tone, however, remains elevated long after return of HR and BP to prearousal levels. Nonspecific arousals and sleep fragmentation increase with advancing age, where the presence of comorbid cardiovascular disease is prevalent and provides a milieu potentially more vulnerable to the effects of sympathoexcitation. Morning awakening induces a stepwise activation of the sympathoadrenal system, with increased HR, BP, and blood catecholamines, and further increases occur with position change and physical activity. Population-based studies have shown that the risk of sudden death from cardiac causes in the general population is significantly greater during the morning hours after waking (i.e., from 6 a.m. to noon) than during other times of the day or night. Because REM is concentrated in the second half of a night's sleep, it is conceivable that autonomic dysregulation associated with REM sleep, among other mechanisms, may be an important contributor, particularly in those with a vulnerable predisposition, such as in the setting of ischemic heart disease. OSA, perhaps due to autonomic dysregulation, has been associated with disruption of this day–night pattern of sudden death.

Obstructive Sleep Apnea

The hallmark of OSA is repetitive upper airway collapse resulting in oxyhemoglobin desaturation. Central nervous system arousals intervene with termination of each breathing event, followed by compensatory hyperventilation. As a result of upper airway/pharyngeal muscle atony as well as reduced lung functional residual capacity, REM sleep usually manifests more upper airway instability and profound deoxygenation than that which occurs during NREM sleep. Further contributing to prolonged hypoxia is the high arousal threshold characteristic of REM sleep. These multiple physiologic stressors all contribute to acute and repetitive surges in sympathetic neural output, the detection of which by measurement of peripheral arterial tone has been used as a diagnostic tool. Some studies report augmented sympathetic responses when central nervous system arousals are accompanied by hypoxia, hypercapnia, and upper airway occlusion. Further evidence for autonomic imbalance in patients with OSA is the finding that the usual decrements in sympathetic activity seen during NREM sleep are lost.

It is possible that there are cumulative effects of long-standing repetitive upper airway events that lead to disruption of autonomic homeostatic mechanisms. In fact, there is abundant evidence for autonomic dysregulation in OSA, in which high levels of MSNA during daytime normoxic wakefulness have been measured (**Figure 3**). HRV is altered in OSA, independent of comorbid illnesses including obesity. Sleep apneics have abnormal baroreceptor function and exaggerated peripheral and central chemoreflex responses, resulting in additive sympathetic stimulation. Deficits in compensatory neural responses to upper airway collapse have been detected in humans with OSA.

Treatment of OSA, typically with the application of continuous positive airway pressure (CPAP) which prevents upper airway collapse, has been shown to acutely attenuate MSNA. Longer term treatment with CPAP reduces daytime sympathetic neural traffic and may also restore, to some degree, autonomic regulatory mechanisms, including baroreceptor function and HRV. Such effects may mediate the salutary effects of CPAP on cardiovascular diseases commonly associated with OSA, such as hypertension and heart failure. Sympathetic neural output may be particularly important in the pathophysiology of heart failure, with OSA having an additive effect and CPAP treatment resulting in acute reductions in sympathetic activity that remain measurable after 3 months of therapy. The long-term implications of OSA treatment on HF outcomes are unknown.

Sleep Deprivation/Short Sleep Duration

National surveys have reported increasingly shorter sleep time per night; as such, there has been intense interest in the systemic effects of short sleep duration. In tightly controlled human experiments, acute restriction of sleep time has been shown to result in metabolic dysregulation, a finding which, in part, could have a physiologic basis in autonomic control. Large population-based studies suggest a higher risk of hypertension in those who self-report shorter sleep duration. Evidence for mediation through autonomic tone has come from experiments demonstrating REM rebound following acute sleep deprivation in young adults and associated increased sympathetic tone by HRV analysis. The blunting of the reduced ventilatory response to hypoxia and hypercapnia associated with sleep deprivation could further contribute to autonomic imbalance.

REM Sleep and Autonomics in the Setting of Cardiovascular Disease

Whereas the acute surges in sympathetic activity associated with REM sleep may be well tolerated in

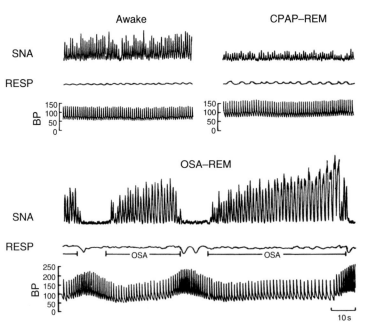

Figure 3 Recording of sympathetic nerve activity (SNA) by microneurography, respiratory excursion (RESP), and intra-arterial blood pressure (BP) in a subject with obstructive sleep apnea (OSA) when awake, during obstructive apneas while in REM sleep, and with elimination of obstructive apnea by continuous positive airway pressure (CPAP) therapy during REM sleep. SNA is high even during normoxic wakefulness, with further increases during REM sleep with obstructive apnea. Note increases in BP as apnea progresses and instantaneous inhibition of SNA with ventilation. Elimination of apneas by CPAP is associated with decreased SNA. Reproduced from Somers VK (1995) Sympathetic neural mechanisms in obstructive sleep apnea. *Journal of Clinical Investigation* 96(4): 1897–1904.

healthy individuals, those with preexisting cardiovascular disease may be vulnerable to adrenergic-mediated vasoconstriction and increments in HR and BP. Polysomnographic recordings of patients with nocturnal angina show ST segment changes on electrocardiography occurring with HR surges associated with REM sleep, as well as with changes in sleeping position that may accompany the transition between NREM and REM sleep. Some patients with diabetes and a recent history of myocardial infarction (MI) have abnormal HRV profiles consistent with higher nocturnal sympathetic tone or reduced vagal tone, which may help explain a higher rate of unfavorable cardiovascular outcomes in these patients. It should also be noted that NREM sleep, which may be associated with hypotension and reduced coronary perfusion pressure, may not be risk-free in susceptible individuals with coronary artery stenosis or after MI. The normal dipping of BP associated with sleep is thought to result from the dominance of parasympathetic tone during NREM sleep over the course of the night. The blunting or absence of dipping suggests a possible role of autonomic dysregulation and has been associated with a higher risk of stroke and congestive heart failure.

Conclusion

There are undoubtedly complex interactions between the autonomic nervous system and REM sleep. The study of cardiovascular function in health and of disease states such as OSA has furthered our understanding of such mechanisms. Additional research will be needed to better characterize these complexities to determine if targeted sleep interventions have any impact on clinical outcomes.

See also: Behavioral Change with Sleep Deprivation; Endocrine Function During Sleep and Sleep Deprivation; Immune Function During Sleep and Sleep Deprivation; REM/NREM Differences in Dream Content; Sleep and Consciousness; Sleep Apnea; Sleep Deprivation and Brain Function; Thermoregulation During Sleep and Sleep Deprivation.

Further Reading

Adlakha A (1998) Cardiac arrhythmias during normal sleep and in obstructive sleep apnea syndrome. *Sleep Medicine Reviews* 2(1): 45–60.

Bonnet MH (1997) Heart rate variability: Sleep stage, time of night, and arousal influences. *Electroencephalography and Clinical Neurophysiology* 102(5): 390–396.

Elsenbruch S (1999) Heart rate variability during waking and sleep in healthy males and females. *Sleep* 22(8): 1067–1071.

Guilleminault C (1984) Sinus arrest during REM sleep in young adults. *New England Journal of Medicine* 311(16): 1006–1010.

Narkiewicz K (1999) Enhanced sympathetic and ventilatory responses to central chemoreflex activation in heart failure. *Circulation* 100(3): 262–267.

Narkiewicz K (1999) Selective potentiation of peripheral chemoreflex sensitivity in obstructive sleep apnea. *Circulation* 99(9): 1183–1189.

O'Driscoll DM (2004) Cardiovascular response to arousal from sleep under controlled conditions of central and peripheral chemoreceptor stimulation in humans. *Journal of Applied Physiology* 96(3): 865–870.

Somers VK (1989) Influence of ventilation and hypocapnia on sympathetic nerve responses to hypoxia in normal humans. *Journal of Applied Physiology* 67(5): 2095–2100.

Somers VK (1993) Sympathetic-nerve activity during sleep in normal subjects. *New England Journal of Medicine* 328(5): 303–307.

Somers VK (1995) Sympathetic neural mechanisms in obstructive sleep apnea. *Journal of Clinical Investigation* 96(4): 1897–1904.

Tank J (2003) Relationship between blood pressure, sleep K-complexes, and muscle sympathetic nerve activity in humans. *American Journal of Physiology – Regulatory, Integrative and Comparative Physiology* 285(1): R208–R214.

Van de Borne P (1994) Effects of wake and sleep stages on the 24-h autonomic control of blood pressure and heart rate in recumbent men. *American Journal of Physiology – Heart and Circulatory Physiology* 266(2): H548–H554.

Verrier RL (1996) Sleep, dreams, and sudden death: The case for sleep as an autonomic stress test for the heart. *Cardiovascular Research* 31(2): 181–211.

Sleep-Dependent Memory Processing

M P Walker, University of California, Berkeley, CA, USA

Introduction

An exciting renaissance is currently under way within the biological sciences, centered on the question of why we sleep, and focusing specifically on the dependence of memory and plasticity on sleep. Although this resurgence is relatively recent in the annals of sleep research, the topic has a surprisingly long history. In the early nineteenth century, the British psychologist David Hartley proposed that dreaming might alter the strength of associative memory links within the brain. Yet it was not until 1924 that Jenkins and Dallenbach performed the first systematic studies of sleep and memory to test Ebbinghaus' theory of memory decay. Their findings showed that memory retention was better following a night of sleep than after an equivalent amount of time awake. However, they concluded that the memory benefit following sleep was passive and resulted from a lack of sensory interference during sleep. They did not consider the possibility that the physiological state of sleep could actively orchestrate these memory modifications.

It is only in the past half century, following the discovery of rapid eye movement (REM) and nonREM (NREM) sleep, that research began testing the hypothesis that sleep, or even specific stages of sleep, actively participated in the process of memory development. In exploring what has become known as sleep-dependent memory processing – and its associated brain basis, sleep-dependent plasticity – we focus herein primarily on evidence in humans (for more extensive reviews covering animal data, see the publications listed in the section 'Further reading').

Delineations and Definitions

We begin our discussion of interactions between sleep and memory by clarifying the complexities that these terms encompass – specifically, sleep states, memory categories, and memory stages.

Sleep States

To begin, it is important to note that the brain does not remain in one single, physiological state across the 24-h day, but instead cycles through periods of differing neural and metabolic activity, associated with distinct biological states, most obviously divided into those of wake and sleep. Sleep has been broadly divided into REM and NREM sleep stages, which alternate across the night in humans in a 90-min cycle (**Figure 1(a)**). In primates and felines, NREM sleep has been further divided into substages 1–4, corresponding to increasingly deeper states of sleep (**Figure 1(a)**). The deepest NREM stages, stages 3 and 4, are collectively referred to as 'slow-wave sleep' (SWS), based on a prevalence of low-frequency cortical oscillations in the electroencephalogram (EEG). Dramatic changes in brain electrophysiology, neurochemistry, and functional anatomy accompany these sleep stages, making them biologically distinct from the waking brain, and dissociable from one another. Thus, sleep cannot be treated as a homogeneous state, which either does or does not affect memory. Instead, each sleep stage possesses a set of physiological and neurochemical mechanisms that may contribute uniquely to memory processing and plasticity.

Memory Categories

In the same way that sleep cannot be considered homogeneous, the spectrum of memory categories believed to exist in the human brain, and the processes that create and sustain memory appear equally diverse. Although often used as a unitary term, 'memory' is not a single entity. Human memory has been subject to several different classification schemes, the most popular based on the distinction between declarative and nondeclarative memory (**Figure 1(b)**). Declarative memory can be considered as the consciously accessible memories of fact-based information (i.e., knowing what). Several subcategories of the declarative system exist, including episodic memory (autobiographical memory for events of one's past) and semantic memory (memory for general knowledge, not tied to specific events). Current neural models of declarative memory formation emphasize the critical importance of structures in the medial temporal lobe, especially the hippocampus, a structure that is thought to form a temporally ordered retrieval code for neocortically stored information, and to bind together disparate perceptual elements of a single event. In contrast, nondeclarative memory is expressed through action and behavior, and includes procedural memory (i.e., knowing how), such as the learning of actions, habits, and skills, as well as implicit learning, and appears to depend on diverse neural anatomies.

Although these categories offer convenient and distinct separations, they rarely operate in isolation in real life. For example, language learning requires a combination of memory sources, ranging from

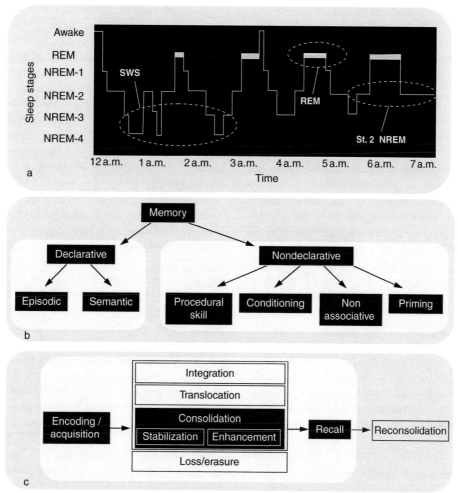

Figure 1 The sleep cycle, memory systems, and memory stages. (a) The human sleep cycle. Across the night, rapid eye movement (REM) and non-REM (NREM) sleep go through a cycle every 90 min in an ultradian manner, while the ratio of NREM to REM sleep shifts. During the first half of the night, NREM stages 3 and 4, slow wave sleep (SWS), dominate, whereas stage 2 NREM and REM sleep prevail in the latter half of the night. Electroencephalogram patterns also differ significantly between sleep stages, with electrical oscillations such as K-complexes and sleep spindles occurring during stage 2 NREM, slow (0.5–4 Hz) delta waves developing in SWS, and theta waves occurring during REM. (b) Memory systems. Human memory is most commonly divided into declarative forms, including episodic and semantic memory, and nondeclarative forms, including an array of different types (e.g., procedural skill memory). (c) Developing stages of memory. Following the initial encoding of a memory, several ensuing stages are proposed, beginning with consolidation and including integration of the memory representation and translocation of the representation, or erasure of the memory. Also, following later recall, the memory representation is believed to become unstable once again, requiring periods of reconsolidation.

nondeclarative memory for procedural motor programs to articulate speech, to memory of grammatical rules and structure, and to aspects of declarative memory for the source of word selection. This too must be kept in mind as we consider the role of sleep in learning and memory.

Memory Stages

Just as memory cannot be considered monolithic, there similarly does not appear to be one sole event that creates or sustains it. Instead, memory appears to develop in several unique stages over time (**Figure 1(c)**). For example, memories can be initially formed or

encoded by engaging with an object or performing an action, leading to the formation of a representation of the object or action within the brain. Following encoding, the memory representation can undergo several subsequent stages of development, the most commonly recognized of which is consolidation. The term 'memory consolidation' classically refers to a process whereby a memory, through the simple passage of time, becomes increasingly resistant to interference from competing or disrupting factors in the absence of further practice . That is to say, the memory becomes more stable.

Recent findings have begun to extend the definition of consolidation. For example, consolidation can be thought of as not only stabilizing memories, but also as enhancing them – two processes that may be mechanistically distinct. While the stabilization phase of consolidation appears to occur largely across time, the enhancement stage appears to occur primarily, if not exclusively, during sleep, either restoring previously lost memories or producing additional learning, both without the need for further practice. From this perspective, the enhancement phase of memory consolidation causes either the active restoration of a memory that had shown behavioral deterioration, or the enhancement of a memory over its simple maintenance. Thus, consolidation can be expanded to include more than one phase of postencoding memory processing, with each phase occurring in specific brain states such as wake or sleep, or even in specific stages of sleep. Finally, following its initial consolidation, a memory can be retained for days to years, during which time it can be recalled.

Although the focus here is primarily on the effects of sleep on encoding and consolidation, it is important to note that additional postencoding stages of memory processing should also be appreciated. These include the integration of recently acquired information with past experiences and knowledge (a process of memory association), the anatomical reorganization of memory representations (memory translocation), reactivation and reconsolidation of memory through recall, and even the active erasure of memory representations, all of which appear to occur outside of awareness and without additional training or exposure to the original stimuli. It is interesting to note that sleep has already been implicated in all of these steps.

Sleep and Memory Encoding

Some of the first studies to investigate the relationship between sleep and human memory examined the influence of sleep on posttraining consolidation (see later), rather than its influence on initial encoding. However, more recent data have described the detrimental consequence of inadequate pretraining sleep on successful memory encoding.

One of the earliest studies to report the effects of sleep deprivation on declarative memory encoding in humans found that 'temporal memory' (memory for when events occur) was significantly disrupted by a night of pretraining deprivation. These findings have been recently revisited in more rigorous studies, again using the temporal memory paradigm. The task comprised photographs of unknown faces, with the temporal memory component involving recency discrimination, together with a confidence judgment. Significant impairments of temporal memory were evident in a group deprived of sleep for 36 h, which

scored significantly lower than did controls. Indeed, significant impairment was evident even in a subgroup that received caffeine to overcome nonspecific effects of lower arousal. Furthermore, the sleep-deprived participants displayed significantly worse insight into their memory-encoding performance.

Neuroimaging reports have begun to investigate the effects of similar periods of total sleep deprivation on verbal memory. As in previous studies, total sleep deprivation resulted in significantly worse acquisition of verbal learning. Surprisingly, however, individuals showed more prefrontal cortex activation during learning when sleep deprived than when not sleep deprived. In contrast, regions of the medial temporal lobe were significantly less activated during learning when sleep deprived. Perhaps most interesting, the parietal lobes, which were not activated during encoding following normal sleep, were significantly activated after sleep deprivation. These findings confirm that sleep deprivation induces a significant behavioral impairment in verbal memory learning, and suggest that these impairments are mediated by a dynamic set of bidirectional changes – overcompensation by prefrontal regions combined with a failure of the medial temporal lobe to engage normally, leading to compensatory activation in the parietal lobes.

We recently investigated the impact of sleep deprivation on declarative memory encoding of both emotional and nonemotional material. Individuals were either sleep deprived for 36 h or allowed to sleep normally prior to an incidental memory encoding session composed of sets of emotionally negative, positive, and neutral words. Following two subsequent nights of sleep, individuals returned for an unexpected recognition task. Overall, individuals in the sleep-deprived condition exhibited a 40% reduction in memory retention relative to individuals who had slept normally prior to encoding (**Figure 2(a)**), results that represent a striking impairment of declarative memory formation under conditions of sleep deprivation. When these data were separated into the three emotional categories (positive, negative, or neutral), the encoding deficit remained, although the magnitude of effect differed across the emotion categories (**Figure 2(b)**). Within the sleep control group, both positive and negative stimuli were associated with superior retention levels relative to the neutral condition, consonant with the notion that emotion facilitates memory encoding. However, there was severe disruption of encoding and hence later memory retention deficit for neutral and especially positive emotional memory in the sleep-deprived group, which exhibited a significant 59% retention deficit relative to the control condition for positive emotional words. Most interesting, however, was the resistance of negative emotional memory

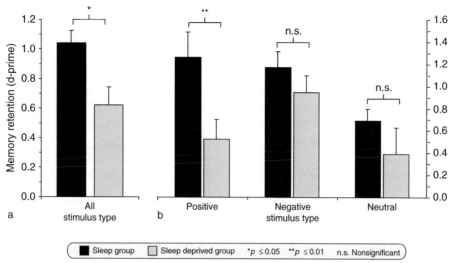

Figure 2   Sleep deprivation and encoding of emotional and nonemotional declarative memory. Effects of 36 h of total sleep deprivation on encoding of human declarative memory (a) when combined across all emotional and nonemotional categories. (b) Effects when separated into emotional (positive and negative valence) and nonemotional (neutral valance) categories.

to sleep deprivation, showing a markedly smaller (19%) and nonsignificant impairment. These data indicate that sleep deprivation severely impairs the encoding of declarative memories, resulting in significantly worse retention 2 days later. Although the effects of sleep deprivation are directionally consistent across subcategories, the most profound impact is on the encoding of positive emotional stimuli and, to a lesser degree, emotionally neutral stimuli, while the encoding of negative stimuli appears more resistant to the effects of prior sleep deprivation.

In summary, sleep deprivation in humans, of varying durations, significantly impairs the ability to effectively form human episodic memories, the effects of which appear be most pronounced for stimuli of positive emotional valence. Furthermore, early evidence suggests that these deficits may be associated with a failure of normal memory encoding networks to respond normally to information acquisition.

Sleep and Memory Consolidation

In addition to the impact of prior sleep deprivation on memory encoding, a plethora of work also demonstrates the critical requirement for sleep after learning for later memory consolidation. Using a variety of behavioral paradigms, evidence of sleep-dependent memory consolidation has now been found in numerous species, including humans, nonhuman primates, cats, rats, mice, and zebra finch.

Declarative Memory

Much of the early work investigating sleep and memory in humans focused on declarative learning tasks.

These studies offered mixed conclusions, some in favor of sleep-dependent memory processing and others against it. For example, significant increases in posttraining REM sleep have been observed after intensive foreign language learning, with the degree of successful learning correlating with the extent of REM sleep increase. Such findings suggest that REM sleep plays an active role in memory consolidation, and that posttraining increases reflect a homeostatic response to the increased demands for such consolidation. However, other studies have reported no alteration of posttraining sleep architecture following learning of a verbal memory task. Similar inconsistencies have been reported both for the degree to which training on declarative tasks alters subsequent sleep-stage properties and for the degree of learning impairment that follows selective sleep deprivation. Recently, several studies by Born and his colleagues have shown improvement on a word-pair associates task after SWS-rich early-night sleep, as well as modification of this posttraining sleep. These findings are striking in the face of earlier studies that showed no effect. However, the discrepancy may reflect the nature of the word pairs used. Whereas older studies used unrelated word pairs, such as dog–leaf, Born used related word pairs, such as dog–bone. The nature of the learning task thus shifts from forming and retaining completely novel associations (dog–leaf) to the strengthening or tagging of well-formed associations (dog–bone) for subsequent recall. Thus, sleep's role in declarative memory consolidation, rather than being absolute, might depend on more specific aspects of the consolidation task, such as the degree of semantic association.

A further example of this subtlety pertains to the emotional nature of the material being learned. For example, it has been shown that late-night sleep, rich in REM sleep, selectively favors retention of previously learned declarative emotional texts relative to neutral texts. We have also investigated the time course of emotional and neutral episodic declarative memory consolidation across the day and overnight. Individuals performed an initial study session containing standardized emotional and neutral pictures, either in the evening or morning. Twelve hours later, after sleep or wake, individuals performed a recognition test, discriminating between the original and novel pictures. When the wake and sleep time periods were combined, significantly better emotional memory recognition was observed, relative to neutral memory – consistent with the notion that emotion enhances consolidation (**Figure 3(a)**). However, when separated according to the wake and sleep periods, a remarkable dissociation was evident (**Figures 3(b)** and **3(c)**). While memory recognition was consistently superior following sleep, consolidation effects were particularly strong for emotional rather than neutral stimuli, with recognition accuracy for emotional pictures improving by 42% overnight, relative to the 12-h waking period. Indeed, emotional memory performance following the wake period was no different from neutral memory in either the wake or sleep conditions. These data indicate the selective facilitation of emotional declarative memory consolidation across sleep, rather than simply across time *per se*, resulting in the enhancement of memory retrieval.

It is important to note that such findings have only begun to test sleep-related effects at the most basic of memory levels – recognition and recall. Several recent studies have now explored more varied measures of memory, demonstrating that the strengths of associative memories are altered in a state-dependent manner. For example, it has been demonstrated that REM sleep provides a brain state in which access to weak associations is selectively facilitated, and flexible, creative processing of new information is enhanced. It has also been demonstrated that, following initial practice on a numeric-sequence problem-solving task, a night of sleep can trigger insight into a hidden rule that can enhance performance strategy the following morning.

Taken as a whole, these studies suggest a rich and multifaceted role for sleep in the processing of human declarative memories. Although contradictory evidence is found for a role of REM sleep in the processing of simple, emotion-free declarative memories, such as the learning of unrelated word pairs, a substantial body of evidence indicates that both SWS and REM sleep contribute to the consolidation of complex, emotionally salient declarative memories, embedded in networks of previously existing associative memories.

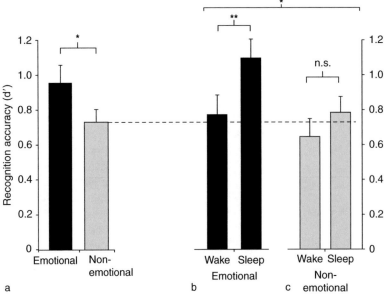

Figure 3 Consolidation of emotional and neutral declarative memory across wake and sleep. Individuals ($n = 14$) entered a repeated-measures design, composed of a sleep phase and a wake phase, presented in counterbalanced cross-over design. During the sleep phase, individuals performed the study session in the evening and 12 h later underwent the recognition test session following a night of sleep. In the wake phase, individuals underwent the study session in the morning and 12 h later performed the recognition test session that evening, without intervening sleep. These two phases were counterbalanced; so that half of the individuals experienced the wake phase first, followed by the sleep phase, while the remaining half experienced the sleep phase first, followed by the wake phase, with the two always being separated by a 1-week interval.

Nondeclarative Procedural Memory

The reliance of procedural, nondeclarative memory on sleep is now a robust finding. These data span a wide variety of functional domains, including both perceptual and motor skills.

Motor learning Motor skills have been broadly classified into two forms – motor adaptation (e.g., learning to use a computer mouse) and motor sequence learning (e.g., learning a piano scale). Beginning with motor sequence learning, we have shown that a night of sleep can trigger significant improvements in speed and accuracy on a sequential finger-tapping task, while equivalent periods of wake provide no significant benefit. These sleep-dependent benefits appear to be specific to both the motor sequence learned and the hand used to perform the task. Furthermore, overnight learning gains correlate with the amount of stage 2 NREM sleep, particularly late in the night (**Figures 4(a)–4(c)**). This sleep window corresponds to a time when sleep spindles – a defining electrophysiological characteristic of stage 2 NREM – reach peak density. Spindles have been proposed to trigger intracellular mechanisms required for synaptic plasticity, and they increase following training on a motor task. Thus, sleep spindles produced in late-night sleep may trigger key cellular events that in turn initiate mechanisms for neural plasticity.

At the behavioral level, the motor sequence task has been dissected to determine where in the motor program this sleep-dependent improvement occurs.

More specifically, differences in transition speeds between each of the separate key-press movements, before and after sleep, were analyzed. For example, in the sequence 4–1–3–2–4, there are four unique key-press transitions: (a) from 4 to 1, (b) from 1 to 3, (c) from 3 to 2, and (d) from 2 to 4 (**Figure 5(a)**). When individual transition-speed profiles of study participants were analyzed before sleep, the speed of individual key-press transitions within the sequence was unequal (**Figure 5(b)**; open circles), with some transitions seemingly easy (fast) and others problematic (slow), as if the entire sequence was being parsed into smaller subsequences during initial learning (a phenomenon termed 'chunking'). Surprisingly, after a night of sleep, the problematic slow transitions (problem points) were preferentially improved, whereas transitions that had already been effectively mastered prior to sleep did not change (**Figure 5(b)**, filled circles). Most remarkable, however, for participants who were trained and retested after an 8-h waking interval across the day, no such improvement in the profile of key-press transitions, at any location within the sequence, was seen (**Figure 5(c)**).

These findings suggest that the sleep-dependent consolidation process involves the unification of smaller motor memory units into one single memory element by selectively improving problem regions of the sequence. This overnight process would therefore offer a greater degree of performance automation, effectively optimizing speed across the motor program, and would explain the sleep-dependent improvements in speed and accuracy previously reported. But more importantly, it again suggests that the role of

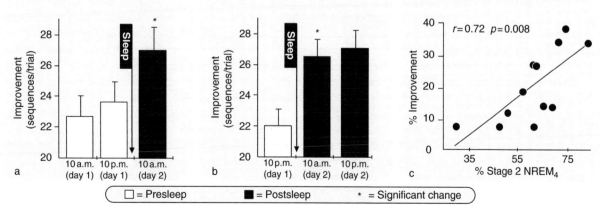

Figure 4 Sleep-dependent motor skill learning. (a) Wake first: after morning training (10 a.m., clear bar), individuals showed no significant change in performance when tested after 12 h of wake time (10 p.m., clear bar). However, when tested again following a night of sleep (10 a.m., filled bar), performance had improved significantly. (b) Sleep first: after evening training (10 p.m., clear bar), individuals displayed significant performance improvements just 12 h after training following a night of sleep (10 a.m., filled bar), yet expressed no further significant change in performance following an additional 12 h of wake time (10 p.m., filled bar). (c) sleep stage correlation; the amount of overnight improvement on the motor skill task correlated with the percentage of stage 2 non-rapid eye movement (NREM) sleep in the last (fourth) quarter of the night (stage 2 NREM$_4$). Asterisks indicate significant improvement relative to training, and error bars indicate standard error of the mean.

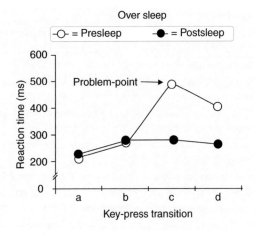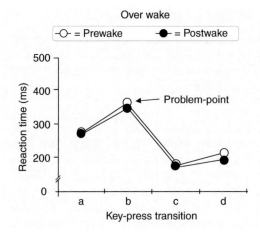

Figure 5 Single-subject examples of changes in transition speeds. Within a five-element motor sequence (e.g., 4–1–3–2–4), there are four unique key-press transitions: (a) from 4 to 1, (b) from 1 to 3, (c) from 3 to 2, and (d) from 2 to 4. (Left) The transition profile at the end of training before sleep (open circles) demonstrated considerable variability, with certain transitions being particularly slow (most difficult; problem points), whereas other transitions appear to be relatively rapid (easy). Following a night of sleep (filled circles), there was a specific reduction (improvement) in the time required for the slowest problem-point transition. (Right) Similarly, at the end of training before a waking interval, transition profiles were uneven (open circles), with some particularly slow transitions (problem points) and other relatively fast transitions (easy). However, in contrast to postsleep changes, no change in transition profile was observed following 8 h of wake (filled circles).

sleep is more subtle and complex than simply increasing the strength of existing memory representations.

Moving from motor sequence learning to motor adaptation learning, it has been demonstrated that selective sleep deprivation impairs retention of a visuomotor adaptation task. All participants were trained and tested on the task and were retested 1 week later. However, some individuals were either completely or selectively deprived of different sleep stages across the first night following memory acquisition. At later retest, individuals deprived of stage 2 NREM sleep showed the most pronounced deficits in motor performance, which again suggests that stage 2 NREM is a crucial determinant of successful motor memory enhancement. Using high-density EEG, more recent studies have been able to show that daytime motor skill practice on a similar task is accompanied by a discrete increase in the subsequent amount of NREM slow-wave EEG activity over the parietal cortex at the start of the night, and that this increase in slow-wave activity is proportional to the amount of delayed learning that developed overnight; individuals showing the greatest increase in slow-wave activity in the parietal cortex that night produced the largest motor skill enhancement the next day.

Visual perceptual learning Learning of a visual texture discrimination task, which does not benefit from 4–12 h of wake following training, has similarly been shown to improve significantly following a night of sleep and appears to require both SWS and REM sleep. For example, overnight improvement has been

shown to be specifically sleep dependent, not time dependent (**Figure 6(a)**), and to correlate positively with the amount of both early-night SWS and late-night REM sleep. Indeed, the product of these two sleep parameters explains more than 80% of interindividual variance (**Figure 6(c)**). Selective disruption of REM sleep, or of either early sleep (normally dominated by SWS) or late-night sleep (normally dominated by REM and stage 2 NREM), results in a loss of overnight improvement, suggesting that consolidation is initiated by SWS-related processes, but that subsequent REM sleep then promotes additional enhancement. In addition, these delayed performance benefits have been shown to dependent absolutely on the first night of sleep following acquisition (**Figure 6(b)**).

Taken together, these reports build a convincing argument in support of sleep-dependent learning across several forms of procedural motor and perceptual skill memory. All of these studies indicate that a night of sleep triggers delayed learning, without the need for further training. In addition, overnight skill improvements consistently display a strong relationship to specific sleep stages, and, in some cases, to specific sleep-stage windows at specific times in the night.

Procedural memory and daytime naps Although the majority of sleep-dependent studies have investigated learning across a night of sleep, several reports have begun to examine the benefits of daytime naps on perceptual and motor skill tasks. Based on evidence that motor learning continues to develop overnight,

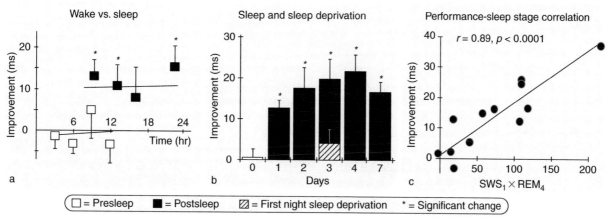

Figure 6 Sleep-dependent visual skill learning. Individuals were trained (during which baseline performance was obtained) and then retested at a later time; the respective improvement in performance (in milliseconds) is illustrated across time. Each individual was retested only once, and each point represents a separate group of individuals. (a) Wake versus sleep: individuals who were trained and then retested on the same day after either 3, 6, 9, or 12 h of subsequent wake (clear squares) showed no significant improvement as a consequence of the passage of waking time across at any of the four time points. In contrast, individuals who were trained and then retested 8, 12, 15, or 23 h after one night of sleep (filled squares) showed a significant improvement occurring as a consequence of sleep. In total, $n = 57$, with $n = 7–9$ for individual points. (b) Sleep and sleep deprivation. Individuals ($n = 89$) who were trained and retested 1–7 days later (filled bars) continued to improve after the first night, without additional practice. Individuals ($n = 11$) sleep deprived the first night after training showed no improvement (crosshatched bar), even after two nights of recovery sleep. (c) Overnight improvement was correlated with the percent of slow-wave sleep (SWS) in the first quarter of the night (SWS_1) and rapid eye movement (REM) sleep in the last quarter of the night (REM_4); $*p < 0.05$; error bars indicate standard error of the mean.

we have explored the influence of daytime naps on motor sequence learning. Two groups trained on the task in the morning. One group subsequently took a 60- to 90-min midday nap, while the other group remained awake. When retested later that same day, individuals who experienced a 60- to 90-min nap displayed a significant learning enhancement of nearly 16%, while individuals who did not nap failed to show any significant improvement in performance speed across the day (**Figure 7**). Interestingly, however, when individuals were retested a second time following a subsequent full night of sleep, those individuals in the nap group showed only an additional 7% overnight increase in speed, while individuals in the control group, who had not napped the previous day, displayed speed enhancements of nearly 24% following the night of sleep. These results demonstrate that as little as 60–90 min of midday sleep is sufficient to produce large significant improvements in motor skill performance, while equivalent periods of wake produce no such enhancement. In addition, these data suggest that there may be a limit to how much sleep-dependent motor skill improvement can occur over the course of 24 h, such that napping changes the time course of when learning occurs, but not how much total delayed learning ultimately accrues. Thus, while both groups improved by approximately the same total amount 24 h later (**Figure 4**), the

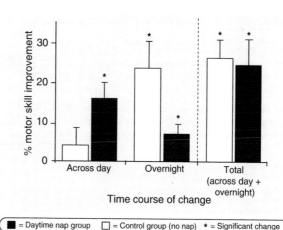

Figure 7 Daytime naps and motor skill learning. Individuals practiced the motor skill task in the morning and either took a 60- to 90-min midday nap or remained awake across the first day. When retested later that same day, individuals who experienced a 60- to 90-min nap (filled bar; across day) displayed significant performance speed improvements of 16%, whereas individuals who did not nap showed no significant enhancements (clear bar; across day). When retested a second time after a full night of sleep, individuals in the nap group showed only an additional 7% increase in speed overnight (filled bar; overnight), whereas individuals in the control group expressed a significant 24% overnight improvement following sleep (clear bar; overnight). Therefore, 24 h later, the groups averaged nearly the same total amount of delayed learning (filled and clear bars; total). Asterisks indicate significant improvement and error bars indicate standard error of the mean.

temporal evolution of this enhancement was modified by a daytime nap.

As with motor skill learning, daytime naps also appear to benefit visual skill learning, although the characteristics of these effects are subtly different. For example, if a visual skill task is repeatedly administered across the day, performance deteriorates rather than remaining stable or improving. This may reflect a selective fatigue of brain regions recruited during task performance, a characteristic not observed in the motor system. However, if a 30- to 60-min daytime nap is introduced among these repeated tests, the performance deterioration is ameliorated. If a longer nap period is introduced, ranging from 60 to 90 min and containing both SWS and REM sleep, performance not only returns to baseline, but is also enhanced. Furthermore, these benefits did not prevent additional significant improvements across the following night of sleep, in contrast to findings for motor skill task performance.

Together these studies build a cohesive argument that daytime naps confer a robust learning benefit to both visual and motor skills and, in the case of visual skill learning, are capable of restoring performance deterioration caused by repeated practice across the day.

Sleep and Brain Plasticity

Memory formation depends on brain plasticity – the brain's lasting structural and/or functional neural changes in response to stimuli (such as experiences). If sleep is to be considered a critical mediator of memory consolidation, then evidence of sleep-dependent plasticity would greatly strengthen this claim. In this section, we consider a mounting wealth of data describing sleep-dependent brain plasticity; our focus here is on neuroimaging studies in humans.

Modification of Posttraining Sleep and Brain Activation

Several studies have investigated whether daytime learning is capable of modifying functional brain activation during subsequent sleep. Based on animal studies, neuroimaging experiments have explored whether the signature pattern of brain activity elicited while practicing a memory task actually reemerges (i.e., is 'replayed') during subsequent sleep. Using positron emission tomography (PET), it has been shown that patterns of brain activity expressed during training on a motor sequence task reappear during subsequent REM sleep, whereas no such change in REM sleep brain activity occur in individuals who received no daytime training. Furthermore, the extent of learning during daytime practice exhibits a positive

relationship to the amount of reactivation during REM sleep. As with previously described animal studies, these findings suggest that it is not simply experiencing the task that modifies subsequent sleep physiology, but rather it is the process of learning that does so. Similar findings have been reported using a virtual maze task. Daytime task learning was initially associated with hippocampal activity. Then, during posttraining sleep, there was a reemergence of hippocampal activation, this time specifically during SWS. The most compelling finding, however, was that the amount of SWS reactivation in the hippocampus was proportional to the amount of next-day task improvement, which suggests that this reactivation is functional, and leads to offline memory improvement. Such sleep-dependent replay may potentially modify synaptic connections established within specific brain networks during practice, strengthening some synaptic circuits while potentially weakening others in the endeavor of refining the memory.

Overnight Reorganization of Memory Representations

A second approach, which more directly examines sleep-dependent plasticity, compares patterns of brain activation before and after a night of sleep. In contrast to approaches that measure changes in functional activity during sleep, this technique aims to determine whether next-day improved performance results from an overnight, sleep-dependent restructuring of the neural representation of the memory. Using a sleep-dependent motor skill task, we have recently investigated differences between patterns of brain activation before and after sleep using functional magnetic resonance imaging (fMRI). Following a night of sleep, and relative to an equivalent intervening period of wake, increased activation was identified in motor control structures of the right primary motor cortex (**Figure 8(a)**) and left cerebellum (**Figure 8(b)**) – changes that allow more precise motor output and faster mapping of intention to key-press. There were also regions of increased activation in the medial prefrontal lobe and hippocampus (**Figures 8(c) and 8(d)**), structures recently identified as supporting improved sequencing of motor movements. In contrast, decreased activity postsleep was identified bilaterally in the parietal cortices (**Figure 8(e)**), possibly reflecting a reduced need for conscious spatial monitoring as a result of improved task automation, together with regions of signal decrease throughout the limbic system (**Figures 8(f)–8(h)**), indicating a decreased emotional task burden. In total, these findings suggest that sleep-dependent motor learning is associated with a large-scale plastic reorganization of memory throughout several brain regions, allowing skilled motor

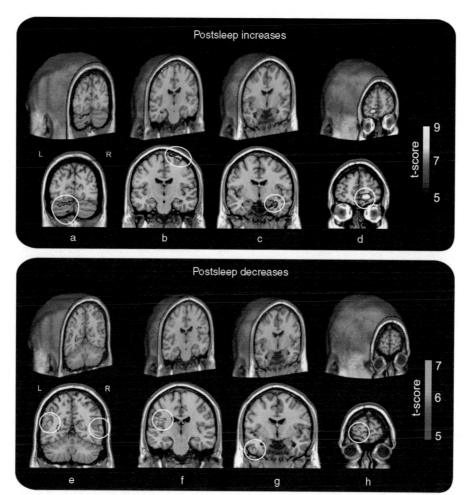

Figure 8 Sleep-dependent motor memory reorganization in the human brain. Individuals were trained on a sleep-dependent motor skill task and then tested 12 h later, either following a night of sleep or following intervening wake, during a functional magnetic resonance imaging brain-scanning session. Scans after sleep and wake were compared (subtracted), resulting in regions showing increased functional magnetic resonance imaging activity postsleep (in red/yellow; a–d) or decreased signal activity (in blue; e–h) postsleep, relative to postwake. Activation patterns are displayed on three-dimensional rendered brains (top row of each graphic), together with corresponding coronal sections (bottom row of each graphic). Following sleep, regions of increased activation were identified in the right primary motor cortex (a), the left cerebellum (b), the right hippocampus (c), and the right medial prefrontal cortex (d). Regions of decreased activity postsleep were expressed bilaterally in the parietal lobes (e), together with the left insula cortex (f), left temporal pole (g), and left frontopolar area (h), all regions of the extended limbic system. All data are displayed at a corrected threshold of $p < 0.05$.

movements to be executed more quickly, more accurately, and more automatically following sleep.

We have also investigated whether overnight reorganization similarly occurs in sensory–perceptual systems using a sleep-dependent visual texture discrimination task, described earlier. Individuals were trained and then tested with or without intervening sleep. Relative to the condition without sleep, testing following sleep was associated with significantly greater activation in an area of primary visual cortex corresponding to the visual target location. However, there were also several other regions of increased postsleep activity throughout both the ventral object recognition (inferior parietal and occipital–temporal junction) and dorsal object location (superior parietal lobe) pathways, together

with corresponding decreases in the right temporal pole, a region involved in emotional visual processing. Thus, a night of sleep appears to reorganize the representation not only of procedural motor but also of visual skill memories. Specifically, greater activation throughout the visual processing streams likely offers improved identification of both the stimulus form and its location in space, with signal decreases in the temporal pole reflecting a reduced emotional task burden resulting from the overnight learning benefits.

While the aforementioned studies examine the benefit of a night of sleep, reports have also investigated the detrimental effects of a lack of sleep on underlying brain activity, using a visuomotor adaptation task. Individuals were trained on the task and tested 3 days later, with half of the individuals deprived of sleep the first night. Controls, who slept all three nights, showed both enhanced behavioral performance at testing and a selective increase in activation in the superior temporal sulcus (a region involved in the evaluation of complex motion patterns), relative to individuals deprived of sleep the first night. In contrast, no such enhancement of either performance or brain activity was observed in these individuals, indicating that sleep deprivation had interfered with a latent process of plasticity and consolidation. These findings indicate that sleep deprivation disrupts not only consolidation, but also the underlying neural mechanisms that support it.

In summary, learning and memory are dependent on processes of brain plasticity, and sleep-dependent learning and memory processing must be mediated by these processes. Many examples of such plasticity across sleep have now been reported over a range of descriptive levels – from whole brain to molecules. As reviewed in this discussion, neuroimaging studies have now identified changes in (1) the functional patterns of brain activity during posttraining sleep periods and (2) the reorganization of newly formed memories following a night of sleep. These plastic brain changes likely contribute to the consolidation of previously formed memory representations, resulting in improved next-day behavioral performance.

Summary

Over the past 25 years, the field of sleep and memory has grown exponentially, with the number of publications per year doubling every 9–10 years, faster than the growth for study of either sleep or memory alone. These reports, ranging from studies of cellular and molecular processes in animals to behavioral studies in humans, have provided converging evidence that pretraining sleep plays a critical role in preparing the brain for initial memory encoding, while posttraining sleep is important in triggering offline memory consolidation and associated neural plasticity. In the end, the question appears not to be whether sleep mediates memory processing, but instead, how it does so. The future of the field is truly exciting, and the challenge to neuroscience will be both to further uncover the mechanisms of brain plasticity that underlie sleep-dependent memory consolidation and to expand our understanding of sleep's role in memory processes beyond simple encoding and consolidation, into the constellation of additional processes which are critical for efficient memory development. Work across the

neurosciences will be necessary to answer these questions, but with the current rate of growth of research in the field, we can expect important advances in our understanding of this critical function of sleep. By way of such a multidisciplinary approach, and with a measured appreciation that sleep plays a fundamental role in forming, consolidating, integrating, and reforming memories, we can look forward to new advances in treating disorders of memory, and perhaps even in improving the capacity of memory in healthy individuals.

See also: Autonomic Dysregulation During REM Sleep; Behavioral Change with Sleep Deprivation; Hippocampal–Neocortical Dialog; Network Reactivation; PET Activation Patterns; Sleep Deprivation and Brain Function; Sleep Oscillations and PGO Waves.

Further Reading

Dudai Y (2004) The neurobiology of consolidations, or, how stable is the engram? *Annual Review of Psychology* 55: 51–86.

Gais S and Born J (2004) Low acetylcholine during slow-wave sleep is critical for declarative memory consolidation. *Proceedings of the National Academy of Sciences of the United States of America* 101: 2140–2144.

Hobson JA and Pace-Schott EF (2002) The cognitive neuroscience of sleep: Neuronal systems, consciousness and learning. *Nature Reviews Neuroscience* 3: 679–693.

Karni A, Tanne D, Rubenstein BS, et al. (1994) Dependence on REM sleep of overnight improvement of a perceptual skill. *Science* 265: 679–682.

Maquet P, Laureys S, Peigneux P, et al. (2000) Experience-dependent changes in cerebral activation during human REM sleep. *Nature Neuroscience* 3: 831–836.

Peigneux P, Laureys S, Fuchs S, et al. (2004) Are spatial memories strengthened in the human hippocampus during slow wave sleep? *Neuron* 44: 535–545.

Stickgold R and Walker MP (2005) Memory consolidation and reconsolidation: What is the role of sleep? *Trends in Neurosciences* 28: 408–415.

Stickgold R, Whidbee D, Schirmer B, et al. (2000) Visual discrimination task improvement: A multi-step process occurring during sleep. *Journal of Cognitive Neuroscience* 12: 246–254.

Wagner U, Gais S, Haider H, et al. (2004) Sleep inspires insight. *Nature* 427: 352–355.

Walker MP, Brakefield T, Morgan A, et al. (2002) Practice with sleep makes perfect: Sleep dependent motor skill learning. *Neuron* 35: 205–211.

Walker MP and Stickgold R (2004) Sleep-dependent learning and memory consolidation. *Neuron* 44: 121–133.

Walker MP and Stickgold R (2006) Sleep, memory and plasticity. *Annual Review of Psychology* 10: 139–166.

Walker MP, Stickgold R, Alsop D, et al. (2005) Sleep-dependent motor memory plasticity in the human brain. *Neuroscience* 133: 911–917.

Walker MP, Stickgold R, Jolesz FA, et al. (2005) The functional anatomy of sleep-dependent visual skill learning. *Cerebral Cortex* 15: 1666–1675.

Behavioral Change with Sleep Deprivation

J D Minkel, S Banks, and D F Dinges, University of Pennsylvania School of Medicine, Philadelphia, PA, USA

Introduction

Sleep restriction is common in modern society, not only as a result of medical and psychiatric disorders but also because of sociocultural influences including occupational and domestic responsibilities. Artificial lighting and around-the-clock services have facilitated lifestyles that do not allow for adequate levels of sleep. The resulting 'sleep debt' is associated with significant daytime performance deficits that carry considerable social, financial, and personal cost. The most well-documented changes in behavior resulting from sleep deprivation include microsleeps intruding into wakefulness and resulting in lapses in cognitive functions that are indicative of state instability. Sleep deprivation experiments have also found deleterious effects on mood, cognitive performance, and motor function due to an increasing sleep propensity and destabilization of the wake state. Vigilant and executive attention, working memory, and divergent higher cognitive functions are particularly vulnerable to sleep loss. These effects have been demonstrated in studies of total sleep deprivation (i.e., those that allow no sleep opportunity) and in studies of chronic partial sleep deprivation.

History of Sleep Research

Although observations on the effects of sleep loss can be found throughout recorded human history, the first systematic investigation of the effects of sleep loss in humans was conducted more than a century ago at the University of Iowa. Observations of three adults across 90 h of sustained wakefulness showed a general decrease in sensory acuity, slowing of reaction times and motor speed, and problems with memorization relative to performance after recovery sleep. This nineteenth-century study identified microsleeps, hypnagogic reverie, lapsing, circadian variation, and recovery sleep depth, all of which have withstood the test of time as behavioral consequences of sleep deprivation in humans.

The first systematic sleep studies were conducted by Dr. Nathaniel Kleitman beginning in the 1920s.

He noted that subjects often seemed to increase their motivation during sleep deprivation to perform as well as they did prior to the sleep loss and that on some tasks, performance appeared normal at the beginning of the test but deteriorated as the test duration increased. These results demonstrated that sleep deprivation did not produce a 'lesion' in performance, which was the conventional wisdom at the time, but reduced neurocognitive capacity and performance.

The next major insight into the consequences of sleep deprivation occurred in the 1930s when Neil Warren and Brant Clark found that performance during sleep deprivation became more variable, despite little change in measures of central tendency of performance. This observation – namely that variability in behavioral output characterized the neurobehavioral performance of sleep-deprived healthy adults – offered a way of reconciling conflicting results among sleep deprivation studies without invoking motivation and duration of measurement. Modern experiments have confirmed the original observation that sleep deprivation increases performance variability rather than simply or solely decreasing average performance.

The physiological basis for the effects of sleep deprivation was first investigated in the late 1940s by Bo Bjerner, who demonstrated that sleep deprivation-induced lapses in performance (which contributed to greater variability) were associated with changes in the human scalp-recorded electroencephalogram (EEG). Specifically, it was observed that alpha-wave (8–12 Hz) depression and high-amplitude slow-wave activity accompanied delayed actions (lapses) on a performance task during deprivation. This EEG profile had been documented 10 years earlier to co-occur with subjective drowsiness.

Thus, by 1950 some of the most fundamental features of performance changes during sleep deprivation had been identified. However, the disparate findings had not yet been integrated into a theoretical model of how sleeplessness affected neurobehavioral functions. Moreover, the behavioral effects of sleep deprivation had not yet been understood relative to the homeostatic drive for sleep. Only in the past 10 years has the full range of consequences from sleep deprivation begun to be understood and a theory that attempts to systematically integrate findings from the neurobiology of sleep deprivation with data on cognitive and neurobehavioral performance changes been developed.

Functional Consequences of Sleep Deprivation

Sleep Deprivation and Accident Risk

Accidents involving sleep-deprived operators have been estimated to have an annual economic impact of $43–$56 billion. Motor vehicle crashes due to drowsy driving contribute substantially to these estimates. Drowsiness markedly increases the risk of crashing, and motor vehicle accidents related to sleep loss are particularly common. Length of time awake, nocturnal circadian phase, reduced sleep duration, prolonged driving duration, and use of sedative–hypnotic medications have all been found to contribute to the occurrence of drowsy-driving and fatigue-related motor vehicle crashes. Studies of shift workers, truck drivers, medical residents, and airline pilots show an increased risk of crashes or near misses due to sleep deprivation. **Table 1** shows the results of a nationwide study of the increased motor vehicle crash risks posed by sleep deprivation from the common practice of scheduling medical residents for extended work shifts of 30 h or more. Sleepiness-related motor vehicle crashes have a fatality rate and injury severity level similar to those of alcohol-related crashes.

Sleep deprivation poses a risk to safety in all modes of transportation and to performance in many other safety-sensitive occupations. For example, studies of interns working prolonged shifts in intensive care units demonstrated that they made 35% more serious medical errors during work schedules that allowed the current practice of ≥30-h shifts compared to work shifts limited to 16 h. Limiting the prolonged

shifts also resulted in the interns sleeping 5.8 h more per week, and they had less than half the rate of attentional failures while working during on-call nights. Thus, sleep deprivation is a preventable cause of accidents in most people.

Sleep Deprivation and the Propensity to Fall Asleep

Under normal conditions, humans experience either stable sleep or wakefulness with relatively little time spent transitioning from one state to another. When sleep is resisted, however, the stability of the sleep–wake system begins to break down. Intrusions of sleep into wakefulness and goal-directed performance become evident through increases in a variety of neurobehavioral phenomena: performance lapses, microsleeps and sleep attacks (i.e., involuntary naps), voluntary naps, shortened sleep latency, slow eyelid closures, and slow rolling eye movements. Wake state instability from inadequate sleep duration or timing can occur while people engage in virtually any type of motivated behavior. In fact, sleep deprivation has been shown to produce psychomotor impairments equivalent to those induced by alcohol consumption at or above the legal limit.

As sleep deprivation increases, the biological initiation of sleep occurs more often and sooner, as evident in sleep propensity/latency tests, which use polysomnography to measure physiological transitions from wake to sleep. Sleep deprivation results in not only a reduction in the latency to sleep onset during the Multiple Sleep Latency Test but also a reduction in latencies from lighter stages of nonrapid eye movement (REM) sleep to deeper slow-wave sleep (SWS). Such rapid transitions from waking to physiologically

Table 1 Risk of motor vehicle crashes and near-miss incidents after extended shifts

Variable	Extended work shifts (>24 h)	Nonextended work shifts (<24 h)
Crashes		
No. reported	58	73
No. of commutes	54 121	180 289
Rate (per 1000 commutes)	1.07	0.40
Odds ratio (95% CI)	2.3 (1.6–3.3)	1.0
Near-miss incidents		
No. reported	1971	1156
No. of commutes	54 121	180 289
Rate (per 1000 commutes)	36.42	6.41
Odds ratio (95% CI)	5.9 (5.4–6.3)	1.0

A within-person case-crossover analysis was used to assess the risks of motor vehicle crashes and near-miss incidents among interns during commutes after extended shifts compared with nonextended shifts. A two-by-two table was constructed for each intern who reported either a crash or a near-miss incident, consisting of the number of crashes or near-miss incidents after an extended shift, the number of crashes or near-miss incidents after a nonextended shift, the number of extended shifts that did not precede a crash or a near-miss incident, and the number of nonextended shifts that did not precede a crash or a near-miss incident. CI, confidence interval. Reproduced from Barger LK, Cade BE, Ayas N, et al. (2005) Extended work shifts and the risk of motor vehicle crashes among interns. *New England Journal of Medicine* 352: 125–134.

deeper stages of sleep reveal the increased neurobiological drive for sleep when wakefulness has been sustained too long, sleep duration has been too brief, and/or sleep consolidation has been abnormal, disrupted, or fragmented.

Whereas the Multiple Sleep Latency Test involves instructing subjects to try to fall asleep while supine in a sleep-conducive environment, the Maintenance of Wakefulness Test involves instructing subjects to remain awake rather than fall asleep in a non-sleep-conducive environment. Like the Multiple Sleep Latency Test, the Maintenance of Wakefulness Test shows reduced sleep latency in response to sleep deprivation. Thus, whether attempting to fall asleep or resist sleep, the latency from waking to sleeping is reduced by sleep deprivation. Although there is some controversy regarding the interpretation of sleep latency measures relative to actual cognitive performance measures during sleep deprivation, sleep latency tends to reflect subjective reports of daytime sleep tendency and the duration of nighttime sleep in the general population.

Microsleeps, Lapses, and Errors

Sleep deprivation not only reduces latencies to enter into normal sleep but also produces microsleeps (i.e., brief episodes of sleep that occur during waking performance). These intrusions result in behavioral lapses or errors of omission (i.e., failure to respond in a timely manner during cognitive performance demands). Lapses contribute to increased moment-to-moment variability in cognitive performance in sleep-deprived adults. Sleep deprivation can also increase errors of commission (i.e., response when no stimulus is present or to the wrong stimulus). Although lapses generally exceed the number of errors of commission on tasks in which the sleep-deprived subject controls the rate of performance, the opposite can be the case when the pace of the cognitive task permits only a very limited time for a response (i.e., tasks done under time pressure). In the latter circumstances, both types of errors are common in sleepy subjects.

Wake State Instability

The consequences of sleep deprivation on human performance were previously understood by the 'lapse hypothesis', which asserted that performance was relatively unchanged except during brief periods of low arousal. Although partially correct, this hypothesis does not explain some of the most robust cognitive changes associated with sleep deprivation, including a general slowing of cognition and reaction time, increased variability in performance, and potentiated decline in performance as time on task increases. To accommodate these effects of sleep deprivation, the wake state instability theory was developed. Rather than suggesting that sleep deprivation simply produces brief periods of impairment in otherwise normal performance, this theory posits that variability in neurocognitive performance increases as homeostatic sleep-initiating mechanisms become progressively more dysregulated with sleep loss. Thus, the theory posits that the co-occurrence of sleep and waking neurobiologies destabilizes performance, making it more erratic, less predictable, and poorly executed toward goal attainment. Not only do behavioral lapses result from sleep deprivation but also even optimal responses can be slowed or made less effective by sleep loss. Higher-order cognitive functions are adversely affected via the impact of sleep loss on basic attention and memory functions, as well as through direct effects on the areas of the cerebrum that mediate executive functions.

Wake state instability is theorized to occur when the brain's capacity to maintain alertness is hindered by the activation of sleep initiation from cortical and subcortical 'sleep switch' mechanisms. In such circumstances, behavioral state instability occurs when sleep pressure is elevated in the presence of goal-directed waking behavior. This conceptualization integrates the neurocognitive effects of sleep deprivation with what is known neurobiologically about sleep–wake regulation. When state instability is high, neurocognitive performance of the sleep-deprived individual is unpredictable and a product of interactive, reciprocally inhibiting neurobiological systems mediating sleep initiation and wake maintenance.

Sleep Deprivation and Basic Cognitive and Psychomotor Performance

Attentional networks, one of the basic building blocks of all cognitive activity, appear to be especially vulnerable to disruption by sleep deprivation. Therefore, any cognitive performance task that relies on stable vigilance and sustained attention with timely responses (e.g., driving) is at risk for deficits induced by sleep loss. Lapses of attention increase in both frequency and duration with increasing sleep deprivation, resulting in a progressive shift toward attentional lability. Individuals who are motivated to remain awake while sleep deprived can temporarily mask their impairment by engaging in compensatory behaviors (e.g., standing, walking, and talking), but the masking effects are typically short-lived. Ultimately, the compensatory effort to resist sleep cannot prevent intrusions of sleep initiation into wakefulness, as evidenced by Kleitman's observations of sleep-deprived subjects 'semi-dreaming' while engaged in verbal cognitive tasks and by first-person reports of sleep-deprived people falling asleep while ambulating in dangerous environments.

There are hundreds of published studies on the cognitive deficits induced by sleep deprivation in humans. Sleep deprivation induces a wide range of effects on cognitive functions, although cognitive tasks vary considerably in their sensitivity to sleep loss. In general, regardless of the task, cognitive performance deteriorates more rapidly in sleep-deprived subjects than in well-rested controls; this is the classic 'fatigue' effect that is exacerbated by sleep loss. However, even performance on very brief cognitive tasks that precisely measure speed of cognitive throughput, and other aspects of attention and memory, consistently has been found to be sensitive to sleep deprivation. In this sense, sleep deprivation adversely affects the cognitive building blocks of attention and memory, on which higher-order cognition depends.

Two often overlooked confounding factors that can obscure the effects of sleep loss on many cognitive tasks are intersubject variability (e.g., aptitude on a task) and intrasubject variability (e.g., rate of learning a task). For example, one person's poorest performance during sleep deprivation may be superior to the best performance of a non-sleep-deprived person. Similarly, someone may be cognitively diminished by sleep loss but continue to improve when measured over time due to continued learning. Moreover, a failure to understand that sleep deprivation increases variability within subjects and between subjects and/or a failure to measure behavior at a frequent rate can result in the effects of sleep deprivation being missed in cognitive measures because less sensitive metrics or inappropriate data analyses are used.

Sleep Deprivation and Higher Cognitive Functions

More complex tasks involving higher cognitive functions have often been reported to be insensitive to sleep deprivation. In general, convergent rule-based deduction and critical thinking (e.g., logic-based tasks) show little change in accuracy during sleep deprivation, although they will reflect cognitive slowing. When tasks are made more divergent, such as multitasking and flexible thinking, sleep deprivation appears to have adverse effects on problem-solving performance. Divergent skills involved in decision making that appear to be affected by sleep loss include assimilation of changing information, updating strategies based on new information, lateral thinking, innovation, risk assessment, maintaining interest in outcomes, insight, communication, and temporal memory skills.

Implicit to divergent thinking abilities is a reliance on executive functions subserved by the prefrontal cortex. Executive function can be defined as the ability to plan and coordinate a willful action in the face of alternatives, monitor and update action as necessary, and suppress distracting material by focusing attention on the task at hand. Many tasks thought to engage different aspects of executive function have been used in studies of sleep-deprived individuals. These tasks also rely on working memory and executive attention, and most show some adverse effects of sleep loss.

Sleep Deprivation and Working Memory

Working memory involves the ability to hold and manipulate information and can involve multiple sensorimotor modalities. Deficits in neurocognitive performance requiring working memory result in difficulty determining the scope of a problem due to changing or distracting information, impairment in remembering temporal order of information, difficulty maintaining focus on relevant cues, taking inappropriate risks, difficulty maintaining flexible thinking, demonstrating poor insight into performance deficits, perseverating on thoughts and actions, and problems making behavioral modifications based on new information. It can therefore be concluded that sleep deprivation adversely affects prefrontal cortex-related executive attention and working memory abilities. Although both neurobehavioral and neuroimaging experiments confirm that sleep deprivation adversely affects neurocognitive functions involving the prefrontal cortex, there is also increasing evidence that subcortical systems involved in sleep initiation – especially hypothalamic, thalamic, and brain stem nuclei – have an important role in altering prefrontal cortical activity during sleep deprivation.

Sleep Deprivation, Mood, and Emotion

In comparison to the effects of sleep deprivation on cognition and performance, relatively little is known about the effects of sleep deprivation on emotional processes, with the exception of subjective mood states. Experimental data have clearly shown that virtually all forms of sleep deprivation result in increased negative mood states, especially feelings of fatigue, loss of vigor, sleepiness, and confusion. These data are limited, however, in that they were obtained without systematic mood induction procedures or other experimental probes designed to influence mood. It is not known if there are systematic differences in how specific events influence mood in sleepy versus rested subjects.

Sleep deprivation is widely believed to be associated with impaired emotional functioning as well. Field studies of sleep deprivation in military and medical personnel have found that reduced sleep is associated with increased negative and decreased positive emotional responses to specific events, but the uncontrolled nature of these studies allows for multiple interpretations of the findings. Other than simple subjective self-report measures of affect, there are few laboratory-based studies on affective responses in sleep-deprived humans. A controlled experiment in rats, however, found that sleep deprivation was associated with increases in aggression and risk-taking behavior. Although data are somewhat limited, the available evidence suggests that emotional processes are sensitive to sleep loss.

Effects of Chronic Partial Sleep Deprivation

Although acute total sleep deprivation is commonly used in research, chronic partial sleep restriction is more frequently experienced in real-world settings. Medical conditions (e.g., pain), sleep disorders (e.g., apnea), work demands (e.g., night shift work), and social and domestic responsibilities generally reduce the amount and quality of sleep that people are able to achieve but rarely eliminate sleep time entirely. Despite the pervasiveness of partial sleep restriction relative to acute total sleep deprivation, only recently has there been substantial experimental focus on its effects.

Daytime Sleep Propensity during Chronic Sleep Restriction

Basic questions concerning how much sleep is required for normal functioning, whether or not there is a cumulative cost to chronically restricted sleep, and to what extent individuals differ in sleep need have only recently been addressed experimentally. Few experiments have studied the cognitive effects of chronic sleep restriction over more than 7 days in a controlled laboratory setting to ensure participants only obtained the sleep durations they were assigned, and that they used no stimulants (e.g., caffeine and nicotine) or other techniques to reduce sleepiness. The two most extensive investigations found adverse effects of chronic sleep restriction on measures of vigilant attention, reaction time, cognitive throughput, and working memory. In one study, truck drivers were randomized to seven nights of 3, 5, 7, or 9 h time in bed for sleep per night. Only the 9-h time in bed condition showed no cognitive impairments across days. In an even larger experiment in which healthy young adults had their

sleep duration restricted to 4, 6, or 8 h time in bed for 14 nights, only the 8-h time in bed condition showed no cognitive impairments across days. Three discoveries in these two experiments were seminal. First, both studies found dose–response cumulative adverse effects of sleep restriction on cognitive functions. That is, deficits in neurocognitive performance accumulated across consecutive days of sleep restriction when time in bed for sleep was chronically reduced below 7 h per night, and the accumulation was more rapid as sleep time became shorter. Second, the cognitive deficits induced by chronic sleep restriction accumulated to levels of impairment equivalent to those observed after one or two nights of acute total sleep deprivation. **Figure 1** illustrates these phenomena for psychomotor vigilance task lapses from the two experiments.

Attenuated Subjective Effects of Chronic Sleep Restriction

The third major finding from the two large-scale controlled dose–response experiments on chronic sleep restriction concerned effects on subjective estimates of sleepiness. Although it is commonly assumed that people know when they are tired from inadequate sleep and can therefore correctly anticipate their ability to perform safely, these and other laboratory studies do not support this assumption. Subjective reports of sleepiness during periods of chronic sleep restriction showed an immediate modest elevation after the first one or two nights of sleep restriction, but this response reached a plateau after a few days and did not demonstrate the cumulative increases that were evident in cognitive performance deficits. Thus, as chronic sleep restriction progressed, subjective perception of alertness became increasingly dissociated from the reality of neurocognitive ability. Subjects often reported that they had adjusted to the reduced sleep durations, but objective performance measures clearly indicated that they had not. This suggests that people frequently underestimate the cognitive impact of sleep restriction and overestimate their performance readiness when sleep deprived.

Polysomnography and Waking EEG and Electrooculography Effects of Chronic Sleep Restriction

Waking EEG changes have also been observed during chronic sleep restriction. These typically involve shifts toward slower EEG frequencies in the theta (4–8 Hz) and delta range (3.75–4.5 Hz). Chronic sleep restriction to 3 or 4 h time in bed has been reported to lead to a decrease in saccadic velocity and slow eye movements (via electrooculography).

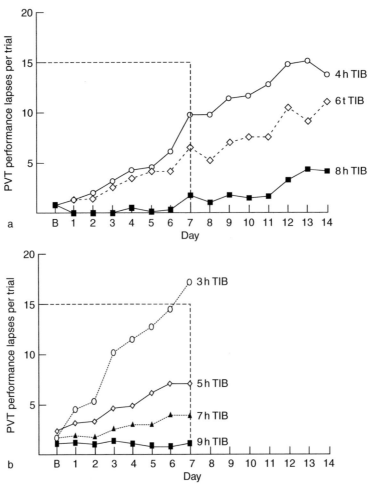

Figure 1 Lapses of attention during a psychomotor vigilance test from two experiments of chronic sleep restriction. (a) In this experiment, sleep was restricted for 14 consecutive nights in 36 healthy adults. Subjects were randomized to 4 h ($n = 13$), 6 h ($n= 13$), or 8 h ($n = 9$) sleep opportunities at night. Performance was assessed every 2 h from 7:30 a.m. to 11:30 p.m. The graph shows cumulative increases in lapses of attention during the psychomotor vigilance task (PVT) per test bout across days. The horizontal dotted line shows the level of lapsing found in a separate experiment when subjects had been awake continuously for 64–88 h. (b) In this experiment, sleep was restricted for seven consecutive nights in 66 healthy adults (mean age, 48 years). Subjects were randomized to 3 h ($n = 13$), 5 h ($n = 13$), 7 h ($n = 13$), or 9 h ($n = 16$) sleep opportunities at night. Performance was assessed 4 times each day from 9 a.m. to 9 p.m. As in a, the graphs show cumulative increases in PVT lapses per test bout across days within the 3 h and 5 h groups ($p = 0.001$). The horizontal dotted line shows the level of lapsing found in a separate experiment by Van Dongen et al. when subjects had been awake continuously for 64–88 h. (a) Adapted from Van Dongen HP, Maislin G, Mullington JM, and Dinges DF (2003) The cumulative cost of additional wakefulness: Dose–response effects on neurobehavioral functions and sleep physiology from chronic sleep restriction and total sleep deprivation. *Sleep* 26: 117–126. (b) Adapted from Belenky G, Wesensten NJ, Thorne DR, et al. (2003) Patterns of performance degradation and restoration during sleep restriction and subsequent recovery: A sleep dose–response study. *Journal of Sleep Research* 12: 1–12. Reprinted from Opp MR (ed.) (2005) *SRS Basics of Sleep Guide*. Westchester, IL: Sleep Research Society.

Slow eyelid closures (which are associated with vigilance lapses) have also been shown to increase with sleep loss and have been found to be a sign of drowsiness while driving.

In addition to affecting Multiple Sleep Latency Test and Maintenance of Wakefulness Test sleep propensity measures, sleep restriction can affect sleep physiology. Following acute total sleep deprivation, rebound increases in non-REM SWS and non-REM EEG slow-wave activity (SWA) – the putative marker of sleep homeostasis – are commonly observed, but these EEG slow-wave rebounds are much less marked in chronic sleep restriction. When sleep duration is 4 h or more, there appears to be modest loss of SWA but marked reductions of stage 2 non-REM and REM sleep time. Consequently, it appears that deficits in SWS and non-REM SWA are not the basis for the cumulative cognitive performance deficits observed during chronic sleep restriction to 4–7 h per night, although they may be more closely related to subjective sleepiness. Other aspects of sleep must be critical to maintaining waking cognitive functions. Total sleep time, rather than non-REM

EEG SWA, therefore remained the major predictor of waking cognitive functions.

Sleep Debt Hypothesis versus Wake Extension Hypothesis

In the chronic sleep restriction experiment involving sleep limited to 4, 6, or 8 h time in bed for 14 consecutive nights, physiological sleep responses to sleep reduction did not mirror waking neurobehavioral responses, but cumulative wakefulness in excess of 15.84 h predicted performance lapses across all sleep deprivation conditions. This suggested that the concept of 'sleep debt' (or cumulative lost sleep) accounting for the cumulative waking neurobehavioral deficits was perhaps a misnomer since it was the additional wake time in excess of 15.84 h per day that appeared to accumulate and have a neurobiological 'cost' over days of sleep restriction. Thus, it is possible that the cumulative cognitive deficits during chronic sleep restriction were not caused by reduction of sleep time *per se* but, rather, by excessive wakefulness beyond a maximum period during which stable neurobehavioral functioning could be maintained. Viewed this way, cumulative wake extension (i.e., excess wakefulness), rather than cumulative loss of sleep (i.e., sleep debt), may be the primary cause of progressively reduced behavioral alertness both across days of chronic sleep restriction and across days of total sleep deprivation. This hypothesis awaits a neurobiological explanation. There has been little basic research published on possible neurobiological mechanisms for the effects of chronic sleep loss on waking cognitive performance, but it has been proposed that an increase in A_1 adenosine receptors in the basal forebrain in response to initial sleep loss may sensitize the brain to subsequent sleep loss.

Physiological Effects of Sleep Deprivation

Endocrine and Metabolic Effects of Chronic Sleep Restriction

Sleep restriction has also been found to affect metabolic responses. Restriction of sleep to 4 h time in bed per night for six nights, compared to extension of sleep to 12 h time in bed per night for six nights, has been associated with elevation in evening cortisol, increased sympathetic activation, decreased thyrotropin activity, and decreased glucose tolerance. Leptin levels have been observed to decrease following sleep restriction to 4 h time in bed. It is believed that these endocrine changes contribute to altered signaling of hunger and appetite, which may promote increased weight gain and obesity. Support for this hypothesis comes from epidemiological evidence of a U-shaped curvilinear association between sleep duration and body mass index (BMI). Those who slept less than 8 h had increased BMI proportional to decreased sleep. In addition, short sleep (5 vs. 8 h) was associated with low leptin (an adipocyte-derived hormone that suppresses appetite) and high ghrelin (predominantly a stomach-derived peptide that stimulates appetite), independent of BMI. The association of sleep duration, especially reduced sleep time, with obesity had also been observed in other epidemiological studies.

Cardiovascular Effects of Chronic Sleep Restriction

An increase in cardiovascular morbidity has been associated with sleep duration in a number of epidemiological studies. For example, there is epidemiological evidence of hypertension and increased risk of coronary events in women reporting less than 7 h of sleep or more than 9 h sleep per day. Similar findings have also been observed in studies examining cardiovascular health in shift workers, who typically experience chronic reductions in sleep duration in addition to circadian disruption. The mechanism underlying the link between chronic sleep restriction and increased cardiovascular risk is unknown. One potential mechanism, however, may be activation of inflammatory processes during sleep loss, which has been reported to occur in laboratory studies of healthy sleep-deprived adults.

Immune and Inflammatory Effects of Chronic Sleep Restriction

A single night of total or partial sleep deprivation can alter natural killer cell activity and circulating levels of leukocytes and cytokines (interleukin-6 and tumor necrosis factor-α) in healthy humans. Although there have been few controlled experiments, it appears that restriction of sleep to 4 h time in bed for six nights acutely reduced antibody titers to influenza vaccination. Attenuation in the febrile response to an endotoxin (*Escherichia coli*) challenge in subjects undergoing chronic sleep restriction to 4 h time in bed for 10 nights has also been observed. The basic mechanisms by which these acute effects of sleep deprivation on immune responses occur remain to be identified, and their clinical significance is unknown.

Differential Neurobehavioral Vulnerability to Sleep Loss

Sleep deprivation (both acute total and partial chronic) not only increases neurobehavioral variability within subjects (i.e., state instability) but also reveals marked cognitive performance differences between subjects. As sleep pressure accumulates over time, intersubject differences also increase markedly. Although most people suffer neurobehavioral deficits when sleep deprived, there are individuals with extreme reactions at both ends of this spectrum. Some people experience very severe impairments even with modest sleep loss, whereas others show few, if any, cognitive deficits until sleep deprivation is very severe. Studies have shown that these responses to sleep deprivation are stable and reliable across subjects, suggesting that they are traitlike and likely genetic rather than the result of situational variability or 'noise'. Thus, some people are much more vulnerable to the neurocognitive effects of sleep deprivation than others. The biological basis of the differential vulnerability to neurocognitive deficits from sleep loss is not known.

Summary

For more than a century, experiments on the effects of total and chronic partial sleep deprivation in healthy human subjects have revealed that sleep cannot be deprived without immediate and often profound changes to neurobehavioral functions, especially those involving attention and memory. In addition to the brain being sensitive to sleep loss, experiments indicate that endocrine, metabolic, immunologic, and inflammatory processes are also affected by sleep deprivation. Despite evidence of the behavioral and physiological risks posed by sleep deprivation, time devoted to sleep has declined in industrialized societies, especially in the past 20 years. Instead of protecting sleep time, there is a tendency in modern cultures to spend resources on ways to safely reduce sleep time to accommodate 24-h operations. These efforts have met with modest success. There is as yet no substitute for sleep, and sleep duration remains the most reliable predictor of wake state stability and neurobehavioral functioning.

See also: Autonomic Dysregulation During REM Sleep; Circadian Rhythms in Sleepiness, Alertness, and Performance; Endocrine Function During Sleep and Sleep Deprivation; History of Sleep Research; Immune Function During Sleep and Sleep Deprivation; Modafini, Amphetamines, and Caffeine; Sleep Deprivation and Brain Function; Sleep-Dependent Memory Processing; Thermoregulation During Sleep and Sleep Deprivation.

Further Reading

Banks S, Barnes M, Tarquinio N, et al. (2004) The maintenance of wakefulness test in normal healthy subjects. *Sleep* 27: 799–802.
Barger LK, Cade BE, Ayas N, et al. (2005) Extended work shifts and the risk of motor vehicle crashes among interns. *New England Journal of Medicine* 352: 125–134.
Belenky G, Wesensten NJ, Thorne DR, et al. (2003) Patterns of performance degradation and restoration during sleep restriction and subsequent recovery: A sleep dose–response study. *Journal of Sleep Research* 12: 1–12.
Carskadon MA, Dement WC, Mitler MM, et al. (1986) Guidelines for the Multiple Sleep Latency Test (MSLT): A standard measure of sleepiness. *Sleep* 4: 519–524.
Chuah YML, Venkatraman V, Dinges DF, and Chee MWL (2006) The neural basis of inter-individual variability in inhibitory efficiency following sleep deprivation. *Journal of Neuroscience* 26: 7156–7162.
Dinges DF, Baynard M, and Rogers NL (2005) Chronic sleep restriction. In: Kryger MH, Roth T, and Dement WC (eds.) *Principles and Practice of Sleep Medicine*, 4th edn., pp. 67–78. Philadelphia: Saunders.
Dinges DF, Pack F, Williams K, et al. (1997) Cumulative sleepiness, mood disturbance, and psychomotor vigilance performance decrements during a week of sleep restricted to 4–5 hours per night. *Sleep* 20: 257–267.
Durmer JS and Dinges DF (2005) Neurocognitive consequences of sleep deprivation. *Seminars in Neurology* 25: 117–129.
Landrigan CP, Rothschild JM, Cronin JW, et al. (2004) Effect of reducing interns' work hours on serious medical errors in intensive care units. *New England Journal of Medicine* 351: 1838–1848.
Saper CB, Scammell TE, and Lu J (2005) Hypothalamic regulation of sleep and circadian rhythms. *Nature* 27: 1257–1263.
Stutts JC, Wilkins JW, Osberg JS, and Vaughn BV (2003) Driver risk factors for sleep-related crashes. *Accident Analysis & Prevention* 35: 321–331.
Taheri S, Lin L, Austin D, Young T, and Mignot E (2004) Short sleep duration is associated with reduced leptin, elevated ghrelin, and increased body mass index. *PLoS Medicine* 1: 210–217.
Thomas M, Sing H, Belenky G, et al. (2000) Neural basis of alertness and cognitive performance impairments during sleepiness: I. Effects of 24 h of sleep deprivation on waking human regional brain activity. *Journal of Sleep Research* 9: 335–352.
Van Dongen HP, Baynard MD, Maislin G, and Dinges DF (2004) Systematic interindividual differences in neurobehavioral impairment from sleep loss: Evidence of trait-like differential vulnerability. *Sleep* 27: 423–433.
Van Dongen HP, Maislin G, Mullington JM, and Dinges DF (2003) The cumulative cost of additional wakefulness: Dose–response effects on neurobehavioral functions and sleep physiology from chronic sleep restriction and total sleep deprivation. *Sleep* 26: 117–126.

Sleep Deprivation and Brain Function

S P A Drummond, University of California at San Diego and Veterans Affairs San Diego Healthcare System, San Diego, CA, USA
B S McKenna, San Diego State University, University of California at San Diego, and Veterans Affairs San Diego Healthcare System, San Diego, CA, USA

Introduction

The behavioral and cognitive consequences of sleep deprivation have been the subject of scientific investigation for over 100 years, since the pioneering work of Patrick and Gilbert in 1896. In general, sleep deprivation produces neurobehavioral deficits in experimental tasks of alertness, attention, memory, cognition, learning, and motor responses, as well as performance deficits in operational settings (e.g., on the highway and in the factory). Despite the wealth of knowledge concerning the behavioral effects of sleep loss, much less work has examined the neurophysiological effects of sleep loss. This article discusses what we have learned about the effects of sleep deprivation on brain function from the perspective of functional neuroimaging studies. First, the definitions of sleep deprivation and brain function are covered. Then we review the literature in which functional neuroimaging is used to study sleep deprivation. Because the widespread application of functional neuroimaging techniques is still relatively new, there are fewer than 20 such studies published. However, rather than providing a detailed review of each one, we highlight the common findings and provide details only when they are illustrative or of particular interest. Finally, we provide a synthesis of the various findings.

Definitions

Sleep deprivation in humans can be broadly classified into three categories: total sleep deprivation (TSD), partial sleep deprivation (PSD), and sleep fragmentation. TSD is the complete lack of sleep for at least one night and often longer. PSD involves restricted sleep for multiple nights, that is, individuals obtaining an inadequate amount of sleep for several consecutive nights. Sleep fragmentation is repeated awakenings from sleep throughout the night. This results in a decreased amount of sleep but a normal time spent in bed. This article focuses on the research on TSD because all published neuroimaging studies in healthy adults to date have used this paradigm.

The term 'brain function' can be used to describe activity ranging from the cellular level (e.g., production of neurotransmitters) to the systems level (e.g., interactions among several brain regions). Functional neuroimaging refers to a set of technologies that allow researchers to examine brain activity with various spatial and temporal resolutions and relate that activity to behavior. Here we focus on two aspects of brain function typically examined in functional neuroimaging studies: (1) that cognitive performance is a behavioral measure often conceptualized as the output of brain function and (2) that functional neuroimaging techniques all provide some measure of brain metabolism. The two functional neuroimaging techniques that have been used most commonly to understand the effects of sleep deprivation are positron emission tomography (PET) and functional magnetic resonance imaging (fMRI). The specific PET technique used measures glucose metabolism over the course of 45 min (although the signal is dominated by activity in the first 20 min). The measure of brain function obtained in the fMRI studies is a complex signal called the blood oxygen-dependent (BOLD) signal. When neurons in a local area increase firing rates (e.g., in response to a cognitive demand), blood flow and blood volume in that area increase. These, in turn, increase oxygen delivery to the area. However, more extra oxygen is delivered than is needed, and the net result is an elevated level of oxyhemoglobin in the local region of activity. It is this elevated level of oxyhemoglobin that produces the BOLD signal. fMRI has a temporal resolution on the order of seconds. Both techniques allow investigators to examine the regions of the brain that are involved in processing a cognitive task being performed by a subject. Brain regions showing positive signal changes when an individual performs the task are said to be activated.

Functional Neuroimaging Studies of Sleep Deprivation

This section reviews the findings from the PET and fMRI studies examining sleep deprivation in healthy young adults. The various reports are categorized by the type of cognitive task examined. Not surprisingly, the type of task has a tremendous influence not only on the brain regions activated when the subject is well rested but also on how brain activation changes with sleep deprivation.

Attention Tasks

There are many types of attention (e.g., sustained, selective, spatial, and divided), but functional neuro-imaging studies have examined only sustained attention and short-term visual attention. Sustained attention is the ability to focus on a single task for an extended period of time. Visual attention refers to the need to process a short stream of visual stimuli and respond to a specific target. In well-rested individuals, sustained attention relies on the right middle frontal gyrus and inferior parietal lobe and, to a lesser extent, on the left inferior parietal lobe and bilateral thalamus. Shorter-term visual attention tasks rely more heavily on the extrastriate cortex, as well as the fusiform, lingual, and superior parietal regions, depending on the specific target of attention.

The first study to use neuroimaging to examine TSD was a PET study involving a sustained attention task in which individuals pressed a button every 3 s for 45 min. Along with a decrease in alertness, this study showed decreased whole-brain metabolism of glucose following approximately 32 h of TSD. In addition, the frontal and temporal lobes showed a decrease in absolute metabolic rates, whereas the parietal lobes showed an increase and the occipital lobes showed no change. The largest decreases, however, were seen in the thalamus, basal ganglia, and cerebellum. These findings are consistent with a later PET study of sustained attention in which subjects performed a serial addition task for 45 min. That study showed a significant drop in cerebral metabolic rates in most brain regions after 24 h of TSD, which continued throughout 72 h of TSD. Both these studies used very long tasks and only examined brain function averaged across the entire task.

Two other studies used fMRI to examine attention after TSD on a finer temporal scale. The first measured visual attention over 40 s and found that increased activation in the ventrolateral thalamus supports intact attention after TSD. The final study used an even finer temporal scale to examine performance on the psychomotor vigilance task (PVT), which measures sustained attention and arousal lability. In this task, an individual presses a button as soon as a milli-second counter starts, which begins randomly within a predetermined interval. The task lasts 10 min and requires sustained attention. Within a testing session, performance on the PVT has been thought to reflect state instability (the variability (waxing and waning) in performance produced by the increased sleep drive during TSD). Findings showed that the frontoparietal

sustained attention network supported fast reaction times whether the subjects were well rested or sleep deprived. These brain regions did not respond during slow reaction times under normal sleep conditions (which is probably why those reaction times were slow). Interestingly, however, these same regions showed increased responses immediately following slow reaction times after TSD. This was thought to reflect individuals reengaging in the task following a slow response because they received feedback regarding this slow reaction time. In essence, then, a slow reaction time prompts compensatory effort from an individual, and this compensatory response is captured in fMRI as an increase in activation in the frontoparietal sustained attention network following the slowest reaction times. A final noteworthy finding is that the study also suggests that the mechanism underlying poor performance after TSD is either (1) the impaired ability of the brain to appropriately allocate resources to task-specific brain regions or (2) the inability to inhibit activation within brain regions that actively interfere with performance.

Overall, it appears that TSD results in decreased brain responses in regions underlying sustained attention demands, as well as diminished performance on sustained attention tasks. Nonetheless, it appears that sleep-deprived individuals can indeed engage these attention regions for brief periods while sleep deprived, but only when accompanied by compensatory activation within the thalamus. This brief engagement allows for intermittent normal levels of performance, despite the fact that average performance across the entire task is diminished with TSD. This set of findings illustrates the importance of the temporal scale at which brain function is measured during TSD.

Learning Tasks

Learning is an essential task for humans, and there are multiple types of learning and memory that rely on various brain systems. Verbal learning (i.e., memorization of language-based material) is the only type of learning studied with functional neuroimaging techniques in the context of sleep deprivation. Studies of TSD have examined brain function during the encoding of verbal material, but none has directly studied recall or recognition of learning materials. Cortical regions associated with verbal encoding are almost exclusively in the left hemisphere and include the medial temporal structures, inferior frontal gyrus, premotor areas, middle temporal gyrus, and cerebellum.

Unlike sustained attention tasks, in which brain function generally decreases with TSD, verbal encoding tasks elicit increased activation in several brain regions after TSD. The most reliable increases are seen in the left and right inferior frontal gyri, as well as the left-hemisphere inferior and superior parietal lobes. This increased activation is typically associated with intact, or only minimally impacted, performance on the learning tasks. In fact, the increased activation within the parietal lobes is directly related to performance after TSD; individuals showing the greatest activation within the left inferior parietal regions after TSD show the best learning performance after TSD.

In addition, the difficulty level of the verbal learning task also influences brain function with TSD. This was demonstrated in a study comparing learning an easy word list versus learning a hard word list. During the memorization of easy words, brain activation did not differ between well rested and TSD states. Memorization of hard words, however, elicited strong increases in brain activation after TSD, and those increases were directly related to performance. The influence of task difficulty in verbal learning has also been examined by employing a distracter task to interfere with learning. That study showed a similar result, with the learning-plus-distraction condition showing greater cerebral responses after TSD than after a normal night of sleep. Once again, the greater the increase in activation with TSD, the better the subjects performed on the learning task. Interestingly, in all the studies that examined verbal learning, the left parietal lobes showed increased activation following TSD and individuals showing the greatest increases in this region also showed the best performance after TSD. This suggests that the left inferior parietal lobe plays a key role in maintaining verbal learning performance during TSD.

Finally, one recent study examined cerebral responses during a verbal learning task from a network perspective using functional connectivity analyses. This study showed that memorization of easy words and hard words each activated the same brain regions (the left and right inferior frontal gyrus and left inferior and superior parietal lobes) but that the interconnections among those regions differed. Following TSD, this study found that the interactions among brain regions within these networks changes. Specifically, in both cases, the interhemispheric connection within the frontal lobe weakened, whereas the intrahemispheric connection between the inferior and superior parietal lobes increased. Thus, this study suggests that changes, positive or negative, in cerebral activation with TSD may reflect alterations in communication among brain regions rather than simply increased or decreased activation within discrete brain regions.

Working Memory

Working memory, also called executive function, has long been a popular cognitive domain in TSD research. This is because (1) working memory tasks typically rely, at least in part, on brain function within the prefrontal cortex; (2) the prefrontal cortex has traditionally been thought of as the brain region most vulnerable to TSD; and (3) electroencephalography (EEG) studies have consistently demonstrated that changes in the frontal lobes following TSD contribute to cognitive performance deficits. The term 'working memory' refers to a large number of cognitive functions, however. It is likely that the specific aspect of working memory examined in any given study influences the measured effect of TSD on brain function. The types of working memory tasks used in functional neuroimaging studies of TSD include those requiring arithmetic, maintenance of information, and manipulation of information.

Arithmetic working memory relies on activation within the bilateral dorsolateral prefrontal cortex, anterior cingulate gyrus, and bilateral parietal lobes. Two studies examined arithmetic working memory following TSD. One of those is the PET study previously discussed in the section on attention tasks. This study showed widespread decreases in glucose metabolism during task performance starting after 24 h of TSD. However, activation in this study was confounded by the sustained attention demands of the task. The second study was an fMRI study examining serial subtraction after 35 h of TSD. Even though this study measured brain activation over 40 s intervals (thereby reducing the sustained attention demands), it confirmed the decreased responses within task-related brain regions following TSD. In both studies, subjects showed diminished performance with TSD.

Tasks requiring maintenance of information in working memory elicit bilateral activation in the prefrontal, precentral, and inferior parietal regions, as well as in the insula and thalamus. Tasks requiring the manipulation of information in working memory show more extensive activation in those same areas, reflecting the fact that manipulation requires more cognitive resources than simple maintenance. The studies examining brain function after TSD during tasks requiring maintenance and/or manipulation of information in working memory have reported both increased and decreased activation with TSD. Some of that inconsistency can be attributed to the task difficulty, similar to the verbal learning studies. More specifically, tasks requiring only the maintenance of information in working memory are more likely to show deceased activation, whereas those requiring the manipulation of information are more likely to show increased activation with TSD. Furthermore, increasing the difficulty level within either a maintenance or manipulation task also results in greater activation during TSD. This is evident from a study that used a grammatical transformation task to manipulate task difficulty and found a stronger modulation of activation in task-related regions as difficulty increased following TSD, compared to the degree of modulation after normal sleep.

Overall, the effect of TSD on the cerebral response to working memory demands remains unclear. It appears that arithmetic demands (at least those not overlearned) produce decreased activation after TSD. The brain response to other working memory tasks probably depends on the specific nature of the task demands and the specific subregions with the prefrontal cortex underlying performance of those specific demands.

Visuospatial Functioning

Visuospatial functioning refers to those cognitive processes involved in visual perception of spatial relationships among objects. One example of visuospatial functioning involves the recognition of objects, such as circular line configurations, that have no verbal properties. In imaging experiments using recognition of nonverbal objects, visual association areas in the ventral visual stream are commonly activated along with the precuneus, an area associated with the dorsal visual stream. As with working memory tasks, studies using visuospatial tasks have shown both increases and decreases in cerebral activation following TSD. Most commonly, activation in the visual association cortex and precuneus decreases after TSD, and this correlates with decreased behavioral performance. Regions showing increased activation during recognition of nonverbal objects after TSD include the basal ganglia and insula. Interestingly, the one study that attempted to use a more real-world spatial navigation task (a piloting task) found exclusively increased activation after TSD, including in both visuospatial processing regions and dorsolateral executive controls regions.

Summary and Conclusion

Research using functional neuroimaging techniques to better understand the effects of sleep deprivation on brain function is still in the early stages. Most studies have taken a relatively descriptive approach; they administered various tasks to examine the differences in brain function after normal sleep and after TSD. A few studies have used more sophisticated manipulations such as parametrically manipulating task

demands or functional connectivity analyses. Despite the fairly small number of studies in this area, some generalizations can be made about the results. There appear to be two distinct brain responses to cognitive challenges during TSD. The first is characterized by decreased activation relative to the well-rested state and diminished performance; this can be thought of as an impaired brain response. The second type of response is characterized by increased activation and relatively intact performance; this has been interpreted as a compensatory recruitment response.

Compensatory Recruitment Hypothesis

The compensatory recruitment hypothesis states that the brain has the ability to recruit additional cognitive resources during task performance following TSD that it typically does not recruit to perform the same task when the subject is well rested. These resources are then engaged in performing the task, and the more an individual engages these cognitive resources following TSD, the better the individual performs the task. It is this correlation between recruitment (measured as activation) and performance that makes the process compensatory. Clearly, however, such recruitment does not always occur because several studies reported decreased activation with TSD. Three broad classes of factors are thought to influence whether, and where in the brain, recruitment occurs following TSD: (1) cognitive task-related factors (What is the subject asking the brain do to?); (2) the length and type of sleep deprivation; and (3) individual difference variables (i.e., some individuals are able to recruit new resources and others are less able).

The studies reviewed here suggest that cognitive task factors certainly play a role in brain function following TSD. For example, sustained attention tasks seem to show impaired brain responses following TSD, whereas verbal learning tasks consistently show increased activation. In addition, each study that systematically manipulated the task difficulty showed that increasing the task difficulty facilitated the compensatory recruitment response. We might think that there is a functional limit to this difficulty-influenced recruitment; however, no study to date has effectively demonstrated such a limit. A similar argument could be made with respect to the length of sleep deprivation. After enough sleep loss, even the verbal learning tasks that consistently show recruitment with TSD should show decreased activation; however, only one published study took functional neuroimaging measurements at multiple time points during TSD. This was the PET study showing decreased glucose metabolism after 24, 48, and 72 h of TSD. Because this study did not even show recruitment at the first time point, it cannot serve as a test of this aspect of the hypothesis. Likewise, no study has examined any length of PSD. The final factor thought to affect recruitment is individual differences, which has only been investigated in a few preliminary studies and, thus, also remains largely untested.

Future Directions

Clearly, much work needs to be done before we will fully understand the impact of sleep deprivation on brain function. All three factors proposed to affect cerebral activation following TSD need more research. Cognitive task-related factors are the most well-studied area. Nonetheless, we need to design cognitive tasks better so as to focus on very specific aspects of cognition rather than on general domains of cognition. For example, instead of examining working memory generically, investigators should employ event-related designs to examine the various components of working memory (e.g., attention, working memory buffer, manipulation, and encoding/retrieval). To the extent this has been done already, the results suggest that some components are more sensitive to TSD than others. Furthermore, future studies should use more real-world tasks when possible. For the second factor, models of PSD need investigation. This is a more realistic model of sleep loss in day-to-day life. Behavioral studies have suggested that a week of 4 or 6 h in bed per night produces cognitive deficits on some tasks similar to 24 or 48 h, respectively, of TSD. However, the time course of the deterioration and whether the brain can show compensatory recruitment during PSD remain unknown. Finally, behavioral data show that some individuals are relatively resilient to the effects of TSD on specific tasks, whereas others are more vulnerable. Understanding the source of those differences could lead to important advances in job selection, safety training, and effective countermeasures to sleep deprivation. Currently, we do not even know if demographic measures such as age, education, IQ, and gender play a role in the brain's response to TSD. Other variables to consider include the subject's mastery of a given task, genomic and gene expression differences, and prior experience with sleep deprivation. Work in this area is growing exponentially. The promise is that a better understanding of the consequences of sleep deprivation, and the factors that protect against the negative consequences, will help us better understand the function of sleep itself.

See also: Autonomic Dysregulation During REM Sleep; Behavioral Change with Sleep Deprivation; Endocrine Function During Sleep and Sleep Deprivation; Immune Function During Sleep and Sleep Deprivation; Modafini, Amphetamines, and Caffeine; PET Activation Patterns; Sleep-Dependent Memory Processing; Thermoregulation During Sleep and Sleep Deprivation.

Further Reading

Chee MWL and Choo WC (2004) Functional imaging of working memory after 24 hr of total sleep deprivation. *Journal of Neuroscience* 24: 4560–4567.

Chee MWL, Chuah LYM, Venkatraman V, et al. (2006) Functional imaging of working memory following normal sleep and after 24 and 35 h of sleep deprivation: Correlations of fronto-parietal activation with performance. *NeuroImage* 31(1): 419–428.

Choo WC, Lee WW, Venkatraman V, Sheu FS, and Chee MWL (2005) Dissociation of cortical regions modulated by both working memory load and sleep deprivation and by sleep deprivation alone. *NeuroImage* 25: 579–587.

Drummond SPA, Bischoff-Grethe A, Dinges DF, et al. (2005) The neural basis of the psychomotor vigilance task. *Sleep* 28: 1059–1068.

Drummond SPA, Brown GG, Gillin JC, et al. (2000) Altered brain response to verbal learning following sleep deprivation. *Nature* 403: 655–657.

Drummond SPA, Brown GG, Stricker JL, et al. (1999) Sleep deprivation-induced reduction in cortical functional response to serial subtraction. *NeuroReport* 10: 3745–3748.

Drummond SPA, Meloy MJ, Yanagi MA, Orff HJ, and Brown GG (2005) Compensatory recruitment after sleep deprivation and the relationship with performance. *Psychiatry Research: Neuroimaging* 140: 211–223.

Drummond SPA, Smith MT, Chengazi V, and Perlis ML (2004) Functional imaging of the sleeping brain: Review of findings and implications for the study of insomnia. *Sleep Medicine Reviews* 8: 227–242.

Mu Q, Nahas Z, Johnson KA, Yamanaka K, et al. (2005) Decreased cortical response to verbal working memory following sleep deprivation. *Sleep* 28: 55–67.

Portas CM, Rees G, Howseman AM, et al. (1998) A specific role for the thalamus in mediating the interaction of attention and arousal in humans. *Journal of Neuroscience* 18: 8979–8989.

Stricker JL, Brown GG, Wetherell LA, and Drummond SPA (2006) The impact of sleep deprivation and task difficulty on networks of fMRI brain response. *Journal of the International Neuropsychological Society* 12: 591–597.

Thomas ML, Sing HC, Belenky G, et al. (2000) Neural basis of alertness and cognitive performance impairments during sleepiness. I: Effects of 24 h of sleep deprivation on waking human regional brain activity. *Journal of Sleep Research* 9: 335–352.

Wu JC, Gillin JC, Buchsbaum MS, et al. (1991) The effect of sleep deprivation on cerebral glucose metabolic rate in normal humans assessed with positron emission tomography. *Sleep* 14: 155–162.

Napping

S C Mednick and S P A Drummond, University of
California at San Diego, La Jolla, CA, USA

Introduction

To answer the question of what the function of
napping may be, we must first define 'function.'
Though this article concentrates mostly on the function of napping for behavior, it would be possible
also to write about the function of napping across
cultures and history. For example, there is a long
history of napping in southern regions of the globe,
in which it is practiced in the form of a siesta. By the
first century BC, the Romans had coined a word for
the afternoon break, *meridiari*, derived from the
Latin word for 'midday.' Later, the church divided
the day into periods designated for specific activities,
such as meals, prayer, and rest. Midday became known
as *sexta*, as in the sixth hour (noon, by their way of
counting), a time when everyone would take rest and
pray. The word has survived as the familiar term,
'siesta.' This postprandial sleep custom disappeared in
northern Europe with the industrial revolution, but
remained a strong tradition in southern Europe and
many Latin American cultures, where people would
shutter their businesses for a few hours to return
home, eat a meal, sleep, and then return to work from
4 p.m. until 9 p.m. While the 'siesta' culture continues
today in these regions, enthusiasm for the siesta has
cooled substantially.

Presently, napping is most popular in highly indus-
trious cultures such as Japan and Germany, whereas
countries such as Spain or Italy have begun phasing
out the practice. Current beliefs about midday rest
are being reassessed from an empirical perspective,
as evidenced by the fact that this topic is enjoying
the spotlight of an entire entry in an encyclopedia
of neuroscience for the first time. One reason for
this resurgence of interest in the function of naps
is due to the changing needs of modern culture.
An increase in 24 h work cycles, the regularity of
international travel and communication with global
markets, and long work days and longer commutes
have brought on a host of secondary problems that
require attention – reports show continual decreases
in nocturnal sleep, increases in sleep disorders, and
increases in sleepiness-related accidents. The nap is
thus being increasingly investigated as an inexpen-
sive, noninvasive, short period of sleep strategically
implemented for sleepiness and fatigue management,
sustained productivity and alertness, and optimal
cognitive processing.

This emerging body of research is summarized in
the following sections, with emphasis on the possible
biological basis of effects of napping, implementa-
tion during night shift and sleep deprivation studies,
and promising new directions investigating the dose-
and sleep stage-dependent benefits possible through
targeted napping schedules designed to improve performance on a wide variety of behavioral measures.

Definitions and Demographics of Napping

What Is a Nap?

Naps, at least as discussed herein, are defined as
intended periods of sleep that can last anywhere from
3 min to 3 h and can be taken anytime during the day or
night. Meta-analysis of surveys and diary studies of
American populations show that naps usually range
between 0.5 and 1.6 h. Naps should be distinguished
from microsleep, which is a brief, involuntary period of
sleep lasting from seconds to minutes. In addition, naps
should be distinguished from 'major' sleep periods that
typically occur overnight and typically last at least 5 h
or more (of course, in night shift workers, these major
sleep periods occur during the day).

Changes in Napping Behaviors across the Life Span

Napping patterns shift throughout life. In infancy,
two basic types of sleep emerge: quiet sleep and active
sleep. These are the infant equivalent, respectively, of
nonrapid eye movement (NREM) and rapid eye
movement (REM) sleep seen in adults. Over 50%
of neonate sleep is active (though premature babies
can achieve levels as high as 80%), but that number
will drop to 30% by the end of the first year. Infants have shorter sleep cycles than adults do,
with the typical cycle lasting 50–60 min as opposed
to 80–100 min in adults. Interestingly, infants also
commonly show sleep-onset active sleep especially
after feeding, a sleep pattern considered pathological
in adults. By the age of 2 years of age, REM occupies
20% of total sleep, a figure that remains relatively
common throughout the rest of life. Napping emerges
a little before the first birthday, when sleep coalesces
into a nocturnal sleep period, a shorter nap (30 min
to 1 h) in the morning around 10 a.m., and a longer
nap (1–3 h) in the afternoon. Eventually napping
consolidates into one long afternoon nap and then
the nap disappears in most children around age

four, reappearing in teenage years and into college years, when up to 60% of students report regular napping habits.

Recent evidence suggests that men take more naps than women do, even though nocturnal sleep hours average about the same for both sexes. On average, men's nocturnal sleep appears to be less efficient, so men may simply need a nap more than women do. Specifically, studies of young adults show that males not only spend more time in bed awake compared to females, but also enjoy less slow-wave sleep (SWS) and REM compared to females. Troubled sleep occurs in women as well, of course, but typically begins later in life, resulting in part from hormone fluctuations that occur across a woman's lifetime.

Napping behavior also changes as a function of aging. Older adults nap more frequently and later in the day compared with younger adults. There may be multiple underlying causes for these differences, including nocturnal sleep deterioration, weaker circadian rhythm and circadian phase advance, or simply that the elderly have more time to practice elective napping. Some studies have shown that napping in older adults is related to decreases in slow-wave activity and reduced sleep efficiency, whereas napping is not related to nocturnal sleep parameters in normal adults or insomniacs.

A number of studies have investigated how napping interferes with nocturnal sleep parameters; effects are likely due to the relieving of sleep pressure, the biological drive to go to sleep that increases as a function of time awake. In general, naps are reported to not interfere with nocturnal sleep. Exceptions to this are that (1) naps can decrease slow-wave activity in subsequent sleep episodes when naps are taken within a 2–3 h window from nocturnal sleep (bedtime) and that (2) when naps are taken during a night shift, the subsequent daytime sleep can show decreased slow-wave activity.

Is Napping a Natural Part of Our Circadian Rhythm?

Timing of Naps

When napping is examined in the laboratory, the consistent finding is that daytime sleepiness is a regularly occurring phenomenon. The afternoon 'nap zone,' first proposed by Broughton, is a period between 14:00 and 16:00, when daytime sleep propensity is highest. Such a propensity for diurnal sleep has been demonstrated in a variety of different experimental milieus, including with removal of all temporal constraints or 'free-running' conditions during ultrashort routines, in which sleep–wake schedules occur over a 90 min period, and as evidenced in the classic 'M-shaped' time-of-day function in studies of sleep propensity using the Multiple Sleep Latency Test (MSLT), a test in which the time it takes to fall asleep, sleep onset latency, is measured at regular intervals across the day. Even in studies when individuals are specifically asked not to nap, resistance to daytime sleep has been most weak during these afternoon hours. Due to the increased sleep propensity because of the time of day, as well as the historical develop of napping behaviors, it has been (perhaps misleadingly) termed 'the postprandial dip.' Studies have shown, however, that the energy slump occurs even in the absence of lunch and/or without knowledge of the time of day.

Physiological Evidence for 'Nap Zone'

A fluctuation in core body temperature (CBT), a fundamental measure of circadian rhythms, represents the best physiological marker to correlate with increased afternoon sleep propensity. Generally speaking, there is a rise in temperature across the daytime and a decrease during the night. The falling temperature traditionally has been considered one important trigger for onset and method of sustainment for nocturnal sleep. CBT starts to fall prior to habitual bedtime and reaches its lowest point approximately two hours prior to habitual wake time (typically between 03:00 and 05:00 in most adults). Though circadian fluctuation of CBT can be fit with a simple sinusoid function spread across the 24 h period, further investigations have found that CBT is better described by adding a 12 h bicircadian component to the model, which corresponds to a robust finding of a dip in temperature in the afternoon (**Figure 1**).

The afternoon dip in temperature corresponds to the time when individuals show a greater sleep propensity. Although decreases in CBT are temporally correlated with increases in sleepiness, a direct mechanistic link has yet to be discovered. Sedative-hypnotics such as melatonin and benzodiazpines decrease CBT and increase peripheral heat loss, which has been directly related to sleep onset latency. In contrast, agents such as caffeine, amphetamines, nicotine, and cocaine decrease sleep propensity and increase CBT. Further, studies have shown that the best predictor of sleep onset (better than melatonin) was the distal–proximal skin gradient, an index of peripheral heat loss. Thus, it is likely that a decrease in temperature (CBT and peripheral heat) is a trigger for sleep in general, and possibly also for afternoon naps. The duration of a sleep episode may also be related to the direction of change in CBT – that is, long nocturnal sleep occurs during an extended period of decreased temperature while short sleep occurs

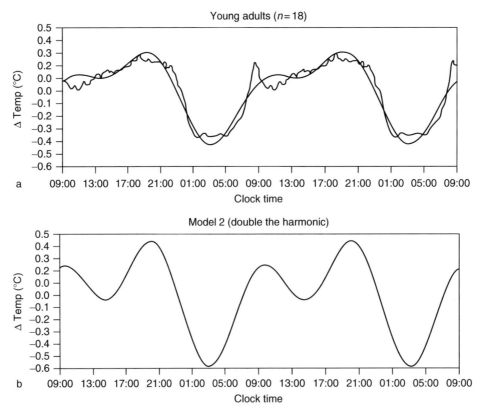

Figure 1 (a) Double-plotted 24 h rectal temperature rhythms from 18 young adults on a nycthemeral routine, together with best fitting composite 24 and 12 h sinusoids (see text). (b) The same fitted model, but with the 12 h amplitude raised from 0.152 to 0.304 °C. Copyright (1996) from 'Circadian determinants of the postlunch dip in performance' by Monk TH, Buysse DJ, Reynolds CF 3rd, and Kupfer DJ. Reproduced by permission of Taylor & Francis Group, LLC.

when CBT increases. Studies have found that increases in CBT are related to more frequent awakenings. Further research is needed to disentangle the causal relationship between changes in CBT and daytime sleep onset.

Behavioral Markers for the 'Nap' Zone

Anecdotal evidence for behavioral measures of the 'nap' zone can be easily found in any typical workplace, with decreased productivity and increased caffeine consumption during the afternoon, as well as an increase in traffic accidents during this time of day (even after taking into account the increased number of cars on the road). On the other hand, it has been more difficult to find consistent laboratory evidence for behavioral markers. Two factors are most likely the cause. First, there may be specific cognitive processes that are vulnerable to circadian peaks and troughs, while others are not. For example, tasks that rely on continuous performance show dips in the afternoon, such as card sorting, serial search, a variety of signal detection tasks, and physical exertion such as sprinting. Memory tasks and perceptual tasks, on the other hand, have not shown a strong circadian component.

Second, individual differences are likely to produce a large source of variation. Intriguing findings of individual differences in vulnerability to a midafternoon performance deficit on a monotonous visual vigilance task demonstrate that only half of the individuals show the 14:00 performance dip. Monk and colleagues compared the CBT of the 'dippers' to the 'nondippers' and showed that performance decreases coincided with a flattening of the CBT in the dippers, whereas CBT in the nondippers continued to increase in a linear manner during this period. Monk proposed that an individual's propensity for midday decreases in performance may be predicted by a combination of the individual's endogenous circadian pacemaker and the length of time the individual has been awake (i.e., magnitude of sleep pressure). Specifically, Monk stated that "the size (or timing) of the 12 h temperature rhythm component might be predictive of the size (or presence) of a post-lunch dip in performance." Further research investigating other biological and genetic determinants of midday dip vulnerability, such as morningness and eveningness measures proposed by Horne, will be extremely interesting pieces of information for answering these questions.

What Is the Function of Naps?

Now that we have provided evidence in favor of naps being a natural part of sleep–wake cycles, at least in some individuals, we turn our attention to the possible function of naps. As already stated, the concept of 'function' can take many forms. For example, one can ask whether the function of a nap is the same as the function of sleep in general, or if it serves a purpose separate from that of the major sleep period. While we do not yet fully understand why we sleep, we do know that a variety of molecular, genetic, and physiological processes occur exclusively, or primarily, during sleep. For the most part, research has not examined whether these same changes occur during naps. What has been examined, though, are the behavioral and cognitive benefits of a nap under a variety of conditions. In other words, research has examined how naps help individuals function better. Thus, it is this more operational definition of function that we discuss here.

Sleepiness versus Fatigue Countermeasures

In the broadest sense, naps are used operationally as either sleepiness countermeasures or fatigue countermeasures. The distinction between sleepiness and fatigue is one that is, unfortunately, often blurred or confused, but is an important distinction to keep in mind. Sleepiness is the physiological propensity to fall asleep, either intentionally or unintentionally. As discussed earlier, the circadian rhythm of the CBT is one traditional biomarker of sleepiness. When naps are used as a sleepiness countermeasure, the intended function is to increase arousal and alertness. This increased alertness level may, in turn, produce better performance. In contrast, fatigue refers to a decrease in physical or cognitive efficiency related to time-on-task and workload, independent of whether someone has a propensity to fall asleep. When naps are used as a fatigue countermeasure, the intended function is to directly boost performance. Most often, the term 'fatigue countermeasure' is incorrectly used to cover both functions (of not only naps, but also interventions such as caffeine) when used in operational settings.

Napping in Sleep-Deprived Conditions

Until recently, napping research has focused primarily on treating sleepiness during extended work periods, such as long-haul truck driving, transatlantic airplane routes, and NASA space flight, as well as during nontraditional work schedules such as night shifts. In both of these circumstances sleepiness due to sleep deprivation is a common danger and has been implicated as contributing to accidents during work

and transit from work to home, as well as to increased health problems in these workers.

Night shift work is particularly vulnerable to extreme performance decrements due to unintentional sleep, increased sleepiness, and decreased performance for most skills, including vigilance, reaction time, serial addition/subtraction, spatial orientation, and flight simulator operation. Critical hours for increased errors and slowness are between 03:00 and 05:00 (coinciding with CBT nadir). The two main sources of reduced alertness and performance during night work are (1) the circadian rhythm of sleepiness and alertness (as discussed earlier) and (2) increased homeostatic sleep pressure. Contributing to this difficulty with night shift work is the poor adaptability of circadian regulated processes such as endocrine, sleep, CBT, adrenaline, alertness, and other physiological rhythms to even long-term (2–3 months) reversal of sleep–wake cycles. Even workers on permanent night shift schedules do not show a change in the timing of the circadian system and continue to show reduced performance and increased mistakes after years on the night shift. Thus, the need to find sleepiness and fatigue management solutions for this population is an imperative.

Work-related napping strategies have categorized three types of naps: prophylactic napping (taken in anticipation of sleep deprivation), compensatory napping (taken after sleep deprivation has begun), and operational napping (napping during working hours). Overall, prophylactic napping taken just before the work night seems to best enhance performance overnight, although a combination of prophylactic napping and caffeine may work even better. With respect to operational napping, it has been found that both pilots and truck drivers unofficially nap during night shifts and long-haul transportation trips. Research has shown that a 20 min operational nap between 01:00 and 03:00 significantly improves speed of response on a vigilance task measured at the end of the shift, compared with a control condition. The potential problem with napping during the night is that there is increased risk for waking with sleep inertia. Sleep inertia is the feeling of slowness, irritability, and poor decision-making ability that can occur during the first 20–30 min after waking from deep sleep, although this period may be shorter with naps than with longer sleep periods. Methods for combating sleep inertia, however, have been proved successful, including exposure to bright light, washing the face, exercise, and, of course, a dose of caffeine. Apart from performance on specific measures, napping has been shown generally to improve alertness, productivity, and mood, and this may be especially so under sleep-deprived conditions, during night shift work, and during prolonged periods of driving.

Cognitive Benefits of Napping Linked to Specific Stages of Sleep

It is perhaps unsurprising that napping has been shown to be an effective sleepiness countermeasure, and that improved alertness can, in turn, have a positive impact on performance. What may be even more interesting to explore is whether napping can be a fatigue counter-measure independent of alerting effects, and whether naps can even enhance normal performance when fatigue and/or sleepiness are not an issue. Recent stud-ies investigating the impact of napping on a variety of memory consolidation measurements have in fact provided evidence for both of these functions. Studies on the benefit of napping on performance have demon-strated that short daytime sleep episodes not only can decrease sleepiness but also show selective enhance-ment of various forms of memory, as well as increase alertness and physical and mental stamina. Further-more, these studies have related these napping-related improvements to specific sleep stages.

An important methodological aspect of napping that has been exploited by these studies is that sleep during naps can be titrated to have specific stages of sleep without disturbing the napper's sleep. This is accom-plished primarily by manipulating the duration and timing of the nap. Such manipulations have been attempted during nocturnal sleep studies by depriving individuals of sleep during the first or the second half of the night, in order to isolate SWS or REM sleep. Although selective cognitive benefits have been shown, this method does not actually isolate stage two, SWS, or REM. Naps, on the other hand, have shown a high degree of specificity in performance between stage one alone versus stage one and two combined, and naps with SWS alone versus SWS and REM combined.

Hayashi et al. demonstrated the recuperative power of napping for sleepiness and alertness measurements even with a 5-min nap limited to stage one sleep. They report that performance improvement on a visual detection task and digit symbol task, as well as decreases in slow eye movements during testing, required a nap with stage two. Walker demonstrated that a nap can improve performance on a motor learning task to the same degree as can a full night of sleep, with stage two sleep playing an important role.

Naps rich in SWS have been shown to improve declarative memory for pictures or word pairs after a nap, as well as prevent deterioration in perform-ance that develops across the day. Mednick and col-leagues have reported a series of studies establishing the efficacy of naps in combating performance

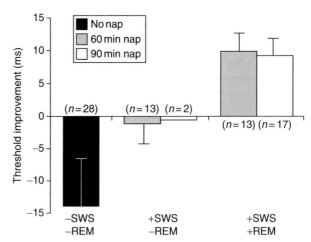

Figure 2 Same-day improvement in no nap, 60 min nap, and 90 min nap groups, with and without rapid eye movement (REM) and slow-wave sleep (SWS). (Left) No-nap group shows deterioration at 19:00 from baseline test at 09:00. (Center) Perfor-mance after naps with SWS but without REM shows neither deteri-oration nor improvement. (Right) Naps with SWS and REM led to significant improvement. Only two individuals in the 90 min nap group showed no REM. Reprinted by permission from Macmillan Publishers Ltd: *Nature Neuroscience* (Mednick S, Nakayama K, and Stickgold R (2003) Sleep-dependent learning: A nap is as good as a night. *Nature Neuroscience* 6(7): 697–698.), copyright (2003).

deterioration. These studies utilized a visual percep-tual task in which individuals reliably show signifi-cant decreases in performance with repeated testing across the day, even when the test is only given twice. Importantly, they found that a 60 min midday nap rich in SWS can reverse perceptual deterioration and restore performance to baseline, with long-lasting benefits to performance (**Figure 2**).

Naps including both SWS and REM actually led to an improvement in perceptual performance equiva-lent to that following a full night of sleep. Further-more, when individuals are tested after a nap and a full night of sleep, they demonstrate as much bene-fit as with two nights of sleep, indicating that sleep-dependent learning is similarly effective whether it is from daytime naps or nocturnal sleep. Also, the benefits from napping and nocturnal sleep are additive (**Figure 3**). It should be emphasized that these sets of studies examined performance after a normal night of sleep, rather than following a period of sleep depriva-tion. Thus they showed that naps can enhance perfor-mance beyond even 'normal' levels.

Summary

In examining the function of naps it appears that modern culture has redefined the functionality to suit the needs of an increasingly 24 h society. The amount of nocturnal sleep continues to decrease as

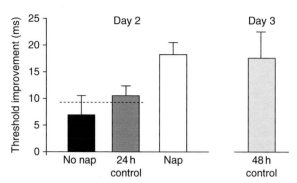

Figure 3 Improvement for nap and no-nap groups. (Left) Improvements 24 h after training for the no-nap group's second retest, the 24 h control group's first retest, and the 90 min nap group's second retest, all at 09:00 on day 2. Dashed line shows nap group's improvement on day 1. (Right) Improvement 48 h posttraining with no nap. Reprinted by permission from Macmillan Publishers Ltd: *Nature Neuroscience* (Mednick SC, Nakayama K, and Stickgold R (2003) Sleep-dependent learning: A nap is as good as a night. *Nature Neuroscience* 6(7): 697–698.), Copyright (2003).

labor demands increase in duration, as well as with the increase in around-the-clock work schedules. Thus the nap serves as both a fatigue and sleepiness countermeasure. Some studies demonstrate that napping in the afternoon may in fact be an inherent part of our natural sleep–wake cycle, as evidenced by physiological and performance decreases during the afternoon that temporally coincide with increased propensity to sleep. Most research on napping has examined either alertness or specific cognitive benefits. An important consideration that has emerged from this body of research is that specific stages of sleep can confer specific benefits to performance. Overall, naps seem to confer a number of possible benefits, but this area of investigation is still young. Present directions of research are investigating the cognitive benefits of napping and the possibility of fitting a nap to an individual's needs by adjusting the duration and time of day of the nap. Important areas for future research concerning the function of naps will include the medical, physiological, and psychological benefits of napping that have been reported to occur with nocturnal sleep.

See also: Autonomic Dysregulation During REM Sleep; Behavioral Change with Sleep Deprivation; Circadian Rhythms in Sleepiness, Alertness, and Performance; History of Sleep Research; Immune Function During Sleep and Sleep Deprivation; Phylogeny and Ontogeny of Sleep; Sleep Deprivation and Brain Function; Sleep: Development and Circadian Control; Sleep in Adolescents; Sleep in Aging; Sleep-Dependent Memory Processing; Thermoregulation During Sleep and Sleep Deprivation.

Further Reading

Carskadon MA and Dement WC (1986) Effects of a daytime nap on sleepiness during sleep restriction. *Sleep Research* 15: 69.

Dinges DF and Broughton RJ (eds.) (1989) *Sleep and Alertness: Chronobiological, Behavioral and Medical Aspects of Napping,* pp. 171–204. New York: Raven Press.

Hayashi M, Motoyoshi N, and Hori T (2005) Recuperative power of a short daytime nap with or without stage 2 sleep. *Sleep* 28(7): 829–836.

Mednick SC, Nakayama K, Cantero JL, et al. (2002) The restorative benefit of naps on perceptual deterioration. *Nature Neuroscience* 5(7): 677–681.

Mednick SC, Nakayama K, and Stickgold R (2003) Sleep-dependent learning: A nap is as good as a night. *Nature Neuroscience* 6(7): 697–698.

Monk TH, Buysse DJ, Carrier J, et al. (2001) Effects of afternoon "siesta" naps on sleep, alertness, performance, and circadian rhythms in the elderly. *Sleep* 24(6): 680–687.

Monk TH, Buysse DJ, Reynolds CF 3rd, et al. (1996) Circadian determinants of the postlunch dip in performance. *Chronobiology International* 13(2): 123–133.

Rosekind MR, Gander PH, and Dinges DF (1991) Alertness management in flight operations: Strategic napping. SAE Technical Paper Series 912138.

Stampi C (ed.) (1992) *Why We Nap: Evolution, Chronobiology, and Functions of Polyphasic and Ultrashort Sleep.* New York: Springer.

Walker MP and Stickgold R (2005) It's practice, with sleep, that makes perfect: Implications of sleep-dependent learning and plasticity for skill performance. *Clinics in Sports Medicine* 24(2): 301–317, ix.

Relevant Website

http://www.nationalsleepfoundation.org – List of Organizations Involved in Sleep Research.

SLEEP DISORDERS

Sleep Apnea

S C Veasey, University of Pennsylvania School of Medicine, Philadelphia, PA, USA

Sleep Apneas Are Neurally Mediated

Distinguishing the types of sleep-disordered breathing events and sleep apnea syndromes is of utmost importance in gauging patient risk and guiding optimal therapy. The term apnea is defined in the adult as cessation in ventilatory airflow for 10 s or longer. In children, cessation of airflow for any duration qualifies as an apnea. Sleep apnea syndromes have been divided into two categories: obstructive and central sleep apnea. This terminology has led to confusion because it implies neural and mechanical processes for central and obstructive apneas, respectively. Indeed, a common misperception is that obstructive sleep apnea (OSA) has a primarily physical etiology (a fixed obstruction in the upper airway), whereas central sleep apnea (CSA) involves alterations in neural input. It is essential to understand, however, that all sleep-disordered breathing syndromes, including OSA, occur only during sleep and thus involve sleep-dependent changes in neural drive to various ventilatory and airway stabilizer muscles. During wakefulness, individuals with OSA have normal ventilation and no significant upper airway obstruction. In wakefulness, these individuals effectively use upper airway muscle to stent open the upper airway. A normal consequence of sleep is reduced neural drive to many muscle groups, including upper airway dilator muscles (e.g., muscles of the tongue and soft palate reduce muscle activity). The termination of an obstructive apnea typically requires wakefulness with resumption of upper airway muscle activity. In other words, the obstructive process in OSA is dynamic and neurally mediated. The neural mechanisms by which OSA occurs may differ for nonrapid eye movement (NREM) and rapid eye movement (REM) sleep. There are differences in neurochmeical drive to the upper airway dilator motor neurons in NREM and REM sleep, and subsets of individuals with OSA are affected differentially in the two distinct sleep states. Specifically, a fraction of patients with OSA have apneas exclusively in REM sleep, whereas others have events in both sleep states but may have more frequent events in NREM sleep than in REM sleep. In most patients, reductions in arterial oxygen levels induced by apneas are more pronounced in REM sleep because REM sleep atonia of chest wall muscles compromises lung volume and oxygenation.

Individuals with CSA do not have increased collapsibility of the upper airway and therefore do not rely on upper airway muscles to maintain patency. There are two major neural mechanisms underlying central apneas. First, in one group of individuals with CSA, ventilatory drive is markedly less in sleep than in wakefulness, and thus a transition into sleep can induce an apnea. These patients are most likely to have apneic events in stages 1 and 2 NREM sleep, when a sigh at the termination of one apnea can lower the $PaCO_2$ sufficiently to initiate the next apnea. In the second group of patients with CSA, there is a reliance on respiratory muscles other than the diaphragm for ventilation. In NREM sleep, these accessory muscles are somewhat suppressed, and in REM sleep the muscles may have atonia. Thus, in these individuals, CSA is expected to be most pronounced in REM sleep.

A second basic concept regarding sleep apnea and its terminology is that not all significant sleep-disordered breathing events are apneas. This concept is most relevant for OSA, for which epidemiological studies justify inclusion of other ventilatory disruptions. In fact, a partial reduction in ventilation occurring with a drop in the oxyhemoglobin desaturation or with an arousal may result in as much sleepiness as apnea events of the same frequency. Clinically significant OSA events occur along a spectrum from what is termed respiratory effort-related arousals (an airflow limitation with an arousal but normal ventilation) to hypopneas (commonly defined as 30–90% reductions in ventilation, measured as airflow, with a 4% decline in the oxyhemoglobin saturation) to apneas (as defined previously). These clinical definitions are relevant because respiratory effort-related arousals and hypopneas can cause significant sleepiness, whereas hypopneas may contribute to cardiovascular disease. These types of disordered breathing events are categorized and defined in **Figure 1**. Disease severity for OSA is defined by the frequency of apnea and hypopnea events per hour of sleep. The American Academy of Sleep Medicine defines mild, moderate, and severe sleep apnea as 5–15, 16–30, and more than 30 apneas and/or hypopneas per hour, respectively. The mathematical modeling for OSA disease severity and morbidity risk, however, is not complete, and it is anticipated that future definitions of severity of OSA will include a factor for the magnitude of oxygenation changes, the time spent hypoxic, and/or the degree to which sleep is fragmented, as described later.

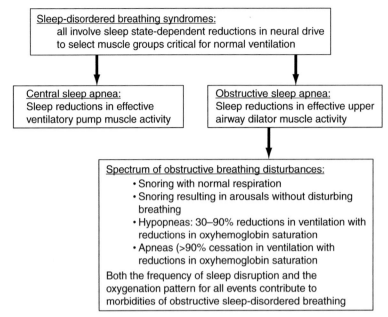

Figure 1 Delineation of sleep apnea syndromes and definitions of sleep-disordered breathing events.

Epidemiology and Genetics of Sleep Apnea

Beyond the common mechanistic theme of sleep-related reductions in neural drive to ventilatory muscles, the pathogeneses for OSA and CSA are remarkably distinct. Consequently, the various sleep apnea syndromes have minimal overlap in epidemiology and genetic bases.

Obstructive Sleep Apnea

The prevalence and/or severity of OSA vary with gender, age, family history, and ethnicity. Specifically, the prevalence of symptomatic OSA (OSA with sleepiness or fatigue) in adults in North America, Europe, and in select regions of Asia is 4–7% for males and 2% or 3% for females. In prepubertal children, there are no apparent gender differences. Prevalence increases with age, rising to more than 40% in elderly individuals. Although some studies have shown increased prevalence of OSA in females after menopause, longitudinal studies are needed to better delineate hormonal and aging effects on OSA. The prevalence of OSA in subjects in the Sleep Heart Health Study appears not to vary with race. However, it is clear that the severity of OSA is greater in age-matched African American versus Caucasian individuals, and overall, OSA develops at an earlier age in African Americans than in Caucasian adults. Familial aggregation is clear for OSA. Approximately 40% of the variance in the apnea hypopnea frequency may be explained by familial factors, and a positive family history increases the relative risk by two- to fourfold. Craniofacial anomalies that

compromise upper airway space and stability provide additional risk factors for OSA. At least two endocrine conditions predispose to OSA: hypothyroidism and acromegaly. The most significant risk factor for OSA, however, is obesity, which is best characterized using the body mass index ($kg\,m^{-2}$) and/or the neck circumference. With the prevalence of obesity increasing in North America, prevalence rates determined 10 years ago will soon require reexamination. In children, the major risk factor has been adenoid and tonsil hypertrophy, although the increase in obesity in children is threatening to make it a major risk factor. A small number of patients diagnosed with OSA are thin. In this group, chronic nasal obstruction from allergies or septal deviation, retrognathia, or chronically enlarged tonsils may predispose patients to increased collapsibility of the upper airway and OSA. In summary, the most important risk factors for OSA appear to be obesity (body mass index and neck circumference), age, family history of OSA, male gender, craniofacial and/or upper airway soft tissue anomalies, hypothyroidism, and acromegaly. However, because OSA is so prevalent, it is essential that the possibility of OSA is considered in all individuals presenting with fatigue and/or unrefreshed sleep.

Although it is clear that OSA has a significant genetic basis, it is less clear whether the genetic basis rests entirely with the intermediate phenotypes (e.g., obesity and craniofacial morphology) or whether there are also genes that determine the severity of OSA and the clinical sequelae of the disease severity. It is important to acknowledge that the many risk factors for OSA and the varied clinical presentations and comorbidities render this syndrome a complex disorder, determined not only by numerous genetic factors but also by environmental and cultural factors. The relative importance of body mass index versus craniofacial structure influences is expected to vary substantially across races with craniofacial differences. Genetics of OSA will require reexamination in large populations in India and China, where obesity may increase with economic expansion. Although there are many candidate genes for the intermediate phenotypes and for ventilatory control, very few have been explored and fewer still in large-scale clinical studies. In adults with OSA, polymorphisms of angiotensin converting enzyme, APOε4, and tumor necrosis factor-α have been identified. Whether these have a direct role in OSA risk or in the severity of comorbidities will require further investigation. Several studies on nonhuman mammals support genetic variation in apnea and hypoxia responses. These studies suggest that genetics may play a very important role in individual susceptibility to complications of OSA.

Central Sleep Apnea

It is estimated that CSA represents just 4% of the sleep apnea diagnoses made in clinical sleep laboratories. CSA occurs in many vastly different conditions. These CSA predispositions may be classified as disorders associated with (1) heightened ventilatory drive, (2) reduced ventilatory drive, and (3) insufficient and abnormal ventilatory pump function. Because of these diverse predispositions, the prevalences of CSA and for many of its subtypes are not known. However, it is estimated that 25–40% of individuals with congestive heart failure with systolic dysfunction may have CSA, and many of these individuals will have Cheyne–Stokes respiration. An additional risk factor for CSA associated with heightened respiratory drive includes periodic breathing or CSA associated with ascent to high altitude (acute hypoxia-induced increased ventilatory drive). The predisposition to CSA at altitude is associated with several angiotensin converting enzyme polymorphisms. Neuromuscular disorders with pump muscle weakness also predispose to CSA. Thus, CSA may be observed in individuals with Duchenne's or myotonic muscular dystrophy or amyotropic lateral sclerosis, myasthenia gravis, multiple sclerosis, and other disorders involving lower ventilatory pump muscle function. In this collection of neuromuscular disorders, the predisposition to sleep-disordered breathing arises from reliance on pump muscles other than the diaphragm. These accessory pump muscles normally have greater reductions in activity in both NREM and REM sleep compared to the diaphragm. Thus, when these accessory muscles are required for normal ventilation, it is expected that

CSA will be present. A very rare form of CSA is congenital central hypoventilation syndrome (CCHS), with apneas occurring in 1 of every 200 000 births. Mutations in the *Phox2B* gene occur in most, if not all, individuals with CCHS. This gene is essential in the development of the autonomic nervous system. Sleep and breathing studies in mice with transgenic absence of the gene have confirmed the importance of the *Phox2B* gene in the development of autonomic ventilatory control and CSA by demonstrating impaired autonomic neuronal development CSA in mice lacking this gene. Whether specific mutations in *Phox2B* or in other gene mutations identified in children with CCHS (e.g., *Ret*, *Mash-1*, and *hash-1*) may contribute to adult forms of idiopathic CSA is not known.

Pathophysiology

Obstructive Sleep Apnea

Three decades ago, Sauerland and Harper measured the electromyographic activity of a major upper airway dilator muscle, the genioglossus, across sleep in subjects with OSA. Coincident with each apneic event, they observed reductions in genioglossus muscle activity, and upon arousal they observed resumption of genioglossus activity and resolution of the apnea. These landmark observations set the stage for further elucidation of mechanisms underlying the pathogenesis of OSA. People with OSA may have normal or reduced upper airway dimensions in wakefulness and may have increased soft tissue surrounding the upper airway, but the universal feature of OSA is increased collapsibility of the upper airway, particularly when upper airway muscle activity is reduced.

Collapse of the upper airway is a dynamic process. Inspiration creates negative pressures in the upper airway. Counterpressure must be generated to prevent collapse of the upper airway in individuals with severe OSA. Collapsibility of the upper airway in snorers is greater than collapsibility in individuals without sleep-disordered breathing but far less than in people with OSA. This increased collapsibility parallels the severity of sleep-disordered breathing events, where the most collapsible airways are present in individuals with OSA who have predominantly apneas.

With the understanding that sleep-dependent reductions in upper airway muscle activity are central to the collapse of the upper airway in OSA, a significant effort has been put forth to determine the neurochemical mechanisms underlying reduced muscle activity. It is now evident that sleep-dependent reductions in serotonin and noradrenaline and increases in GABA delivery to motor neurons innervating the genioglossus contribute to the reduced muscle activity in sleep. It is unfortunate that serotonergic and noradrenergic receptor subtypes on upper airway motor neurons do not represent safe pharmacological targets. It is hoped that novel G-protein-coupled receptors will be identified on upper airway motor neurons and that these may be pharmacotherapeutic targets.

Central Sleep Apnea

As a condition with diverse causes, including congestive heart failure, CCHS, and many neuromuscular disorders, the pathophysiology of CSA varies with the cause of the condition. In deep NREM sleep (stages 3 and 4), ventilation is controlled almost exclusively by autonomic factors and is therefore extremely regular. In lighter NREM sleep (stages 1 and 2), there are major differences between the ventilatory response to low $PaCO_2$. A low $PaCO_2$ in wakefulness will not result in a cessation of ventilation. In contrast, a low $PaCO_2$ in light NREM sleep can suppress ventilation, causing an apnea until the $PaCO_2$ rises enough to stimulate ventilation. Typically, there is an overshoot so that upon arousal, the ventilatory response is so robust that the $PaCO_2$ falls again to a level that induces an apnea if the patient falls quickly back to sleep. This principle underlies the pathophysiology in several forms of CSA. If ventilatory drive is high in waking and subjects enter into NREM sleep with a sufficiently low $PaCO_2$, below the apneic threshold in NREM sleep, apnea will ensue until the CO_2 rises sufficiently to augment ventilatory drive in sleep or induce an arousal. Through this same sleep-dependent reduced ventilatory drive mechanism, a healthy individual may have a post-sigh apnea in NREM sleep. High-altitude periodic breathing also occurs through a similar change in ventilatory drive. At high altitude, there is some hypoxic drive to increase ventilation and sufficiently lower $PaCO_2$. Upon entering sleep, again if the $PaCO_2$ is below the apnea threshold, an apnea will occur until an increased $PaCO_2$ is detected resulting in an arousal that, in turn, greatly augments ventilation, blowing the $PaCO_2$ back down below the apnea threshold, where the process repeats itself throughout the night.

This same sleep-dependent change in ventilatory response to $PaCO_2$ underlies Cheyne–Stokes breathing. In this condition, the heightened ventilatory drive in wakefulness is, in part, a consequence of increased sympathetic drive that drives down the wake $PaCO_2$. Cheyne–Stokes respiration has been successfully treated pharmacologically with adrenergic blockade.

The remainder of conditions predisposing to CSA are those with impaired neuromuscular function. Individuals with neuromuscular weakness may have impaired diaphragm function yet can have normal waking ventilation by using accessory muscles (e.g., the sternocleidomastoid or internal intercostals and abdominal muscles). Sleep-related central apneas occur when sleep-dependent reductions in accessory muscle activity occur. In these conditions, sleep apnea will be worse in REM sleep when atonia is present in all accessory muscles of ventilation.

Clinical Features of Sleep Apnea

Obstructive Sleep Apnea

The clinical presentation of OSA varies across subsets of individuals. Adults with severe OSA, particularly obese males, present with a remarkably consistent constellation of: loud snoring, witnessed pauses in breathing at night, unrefreshed sleep, and daytime sleepiness or fatigue. Dyspepsia, rhinitis, and nocturia are also commonly associated symptoms. Many individuals with severe OSA will also report clumsiness, poor concentration, and impaired dexterity and/or memory impairments. Recognition of mild to moderate OSA in thin males, premenopausal females, and children presents a greater challenge. Patients with less severe OSA are more likely to have complaints of insomnia, fatigue, and quiet or minimal snoring. Therefore, of significant concern is the potential for underdiagnosis. Somatic complaints may predominate and may be attributed to chronic fatigue syndrome, depression, and fibromyalgia. The presentation in children differs greatly from the presentation in adults. Specifically, parents and other caregivers of children with OSA may recognize increased distractibility, labile mood, and hyperactivity in these children, without appreciating sleepiness, sleep disturbances, or even snoring.

The physical exam in many, but not all, cases will provide clues of upper airway compromise, such as a large tongue, a large neck circumference (>17 in. in adult males or >15 in. in adult females), a low-lying soft palate, a posteriorly displaced soft palate (1–4 mm anterior–posterior velopharyngeal space), large tonsils, a small mandible, and, rarely, a high arched palate. However, it is possible to have a normal head and neck exam and yet have enlarged lingual tonsils or collapse of the pharynx that is evident only in sleep. The physical exam should also always include an exploration for signs of hypothyroidism and acromegaly because these would be treated differently, as indicated later.

Central Sleep Apnea

The clinical presentation in CSA varies with the underlying cause, and as with the different pathophysiologies, depending on the cause, the clinical presentation may be classified into two categories: individuals with lower arterial $PaCO_2$ values in wakefulness and those with normal or increased arterial $PaCO_2$. Individuals with CSA and low arterial $PaCO_2$ values may describe sleep as disturbed and unrefreshing, or they may have daytime hypersomnolence or fatigue. Increased fatigue with sedentary behaviors compared to exertional behaviors should alert physicians to examine the possibility of sleep-disordered breathing. Some patients may present with paroxysmal nocturnal dyspnea or may present with quiet snoring and witnessed apneas. However, the witnessed apneas and snoring are less obvious than events witnessed in people with severe OSA. Individuals with normal to increased $PaCO_2$ may present with clinical features more consistent with chronic hypoxemia and hypoventilation: polycythemia, cor pulmonale, or morning headaches.

Diagnosis

The gold standard for diagnosing both CSA and OSA has been overnight polysomnography. Recording sleep in parallel with ventilatory pump effort and nasal and oral airflow, the full spectrum of snoring, hypopnea, and apnea events can be examined. In addition, the effects of sleep-disordered breathing events on sleep quality and on arterial oxygenation may be determined. Together, this information distinguishes CSA and OSA and the severity of OSA can be determined. In OSA, night-to-night variability can be high, particularly in individuals with mild to moderate disease. Thus, when clinical suspicion is high and the first polysomnogram does not show OSA, it may be necessary to repeat the study. In females and in people with mild OSA, events may cluster in REM sleep. Thus, it is essential to examine sufficient REM sleep in individuals before excluding the possibility of OSA. Also in these patient populations, it is important to search for frequent arousals ($>15\,h$) associated with increased ventilatory effort. With CSA, polysomnographic features may be instrumental in elucidating causes of sleep-disordered breathing. For example, Cheyne–Stokes presents with a very characteristic crescendo–decrescendo pattern in ventilation, with the arousal occurring at the peak of ventilation. Congenital central hypoventilation syndrome, which has rarely been identified in adults, manifests with most severe apneas occurring in slow-wave sleep. In contrast, because ventilatory motor neuron suppression is most pronounced in REM sleep, many patients with pulmonary disease or neuromuscular disorders (those that rely on accessory muscles for ventilation) will have more severe

sleep-disordered breathing in REM sleep, when these muscle have atonia. Neither the sensitivity of polysomnography for the diagnosis of CSA nor the night-to-night variability in CSA has been established. Thus, if clinical suspicion is high, repeated polysomnography may be indicated in the hypersomnolent patient.

Of significance, OSA progresses with age and weight gain; thus, reconsideration of the severity of disease and its treatments is required periodically.

Therapies

Obstructive Sleep Apnea

The frontline therapy for moderate to severe OSA is the application of positive airway pressure (PAP) to the upper airway to prevent collapse. The pressure must be titrated individually to a level that prevents obstruction (apneas, hypopneas, and snores with arousals) in both NREM and REM sleep and yet is not so high as to interfere with sleep. Levels typically vary between 5 and 15 cm H_2O. Although remarkably effective for OSA, the therapy is an air pump connected to a relatively airtight mask or nasal interface, and as such this system can be cumbersome and uncomfortable. Consequently, less than half of the individuals prescribed PAP regularly use this therapy. Nonetheless, every effort should be made to encourage use of PAP regularly because this is the only therapy for OSA shown to lessen cardiovascular and neurobehavioral morbidity, even in mild to moderate OSA. Masks and nasal interfaces are more comfortable than the first equipment designed, and many patients who thought they could never tolerate a PAP mask are now able to comply with newer interfaces. Patients with claustrophobia and mask intolerance may undergo behavioral therapy for PAP desensitization. Developments in PAP therapy include machines that can self-adjust the level of PAP based on airflow patterns, gradually increasing pressure as needed across sleep stages and different sleeping positions.

Other therapies for OSA should be considered in young individuals with mild sleep apnea and in individuals with mild disease and inability to desensitize to PAP. These alternative therapies include surgical procedures to shorten the soft palate and reduce collapsibility of the pharynx (uvulopalatoplasty), reduce the tongue volume (genioglossectomy), or advance the genioglossus forward (genioglossus advancement hyoid myotomy) as initial surgical therapies and, if clinically indicated, a second phase of surgery to more dramatically increase pharyngeal space (maxillary advancement or maxillary and mandibular osteotomy). Laser-assisted uvulopalatoplasty is not as effective as surgical uvulopalatoplasty for treating sleep apnea. Clinical indications for use of temperature-controlled radio frequency are still being determined. Some patients who do not tolerate PAP or in whom OSA is mild may benefit from oral appliances for sleep-disordered breathing. Although well tolerated, therapy takes weeks to months to reach optimal effectiveness. Factors predicting success with oral appliances include female gender, non-obese body habitus, young age, and positional OSA with more events in the supine position.

Weight loss should be recommended for all obese individuals with sleep-disordered breathing. Dietary counseling should be the first step taken, and all patients should understand that reduced caloric intake is the primary principle for most, if not all, successful weight loss programs. Exercise may help maintain weight, but in most nonathletic individuals, healthy caloric restriction should be the primary strategy for weight loss. Bariatric surgery is typically reserved for individuals with morbid obesity because there is a 10% chance of significant perioperative adverse events and a 1% chance of mortality in the postoperative period. The majority of individuals who have substantial weight loss after bariatric surgery will experience marked reductions in OSA, if not reversal of the disease. Treatment of endocrine disorders and OSA (hypothyroidism and acromegaly) should begin with PAP therapy, but across the treatment of the underlying endocrine disorder, the PAP may need adjusting. Several medical therapies for OSA may be considered as second-line therapies for mild OSA. There may be subsets of individuals who respond to supplemental oxygen, positional therapy, and, rarely, pharmacotherapies such as selective serotonin reuptake drugs in individuals with mild REM sleep-predominant apnea. Because these therapies are only rarely effective, treatment success should be determined with repeated polysomnography during treatment.

Central Sleep Apnea

Treatment of CSA varies with the cause of sleep apnea. PAP initially appeared promising for treating Cheyne–Stokes respiration in heart failure, but with newer medical therapy including treatment of the sympathetic overdrive, indications for PAP are less clear. High-altitude CSA may be treated with descent, acetazolamide, and medications reducing the hypoxia–angiotensin response or with supplemental oxygen. For other forms of CSA, ventilation must be augmented in sleep, and this may be achieved with PAP that assists ventilation rather than acts as an air splint. This bilevel PAP may be delivered with a mask as in OSA or, rarely, by tracheostomy tube if the noninvasive delivery fails.

Morbidities and Disease Interactions

One of the most significant advances during the past decade regarding OSA has been the identification of OSA as a risk factor for numerous and diverse morbidities. OSA is a major risk factor for excessive sleepiness and sleepiness-related impaired performances. Indeed, OSA is an independent risk factor for motor vehicle crashes, raising the relative risk by 2.5-fold. A direct link is supported by the reduction in car crash risk with successful treatment of sleep apnea. Several reports suggest that OSA is an independent risk factor for insulin resistance. This risk remains after controlling for obesity. As with motor vehicle accidents, further support for this disease interaction comes from studies showing improvement in glucose control with successful use of PAP therapy. It will be exciting to determine whether long-term PAP therapy reduces the occurrence of complications of diabetes. Several reports suggest that OSA may impair liver function and might contribute to nonalcoholic fatty liver disease, a major risk factor for liver failure in developed countries. Perhaps the most impressive observation in individuals with OSA has been the elucidation of OSA as an independent risk factor for several cardiovascular diseases, including hypertension, congestive heart failure, and stroke. Here again, obesity was controlled for. Importantly, the relative risk for hypertension increases even at levels of mild OSA (5–15 events per hour). The rates of significant cardiovascular events during 10-year follow-up in a large European trial also increased with mild to moderate OSA, but to a far greater extent (fourfold increase) in individuals with severe untreated OSA. This risk in individuals with severe OSA declines almost to baseline when treated effectively with PAP across the duration of follow-up. The risk of cardiovascular death is also reduced with PAP therapy in people with severe OSA. In light of the seriousness of morbidities and the disease interactions associated with OSA, every effort to improve PAP therapy acceptance and use should be made.

Future Directions

This area of research is extremely vibrant, and consequently significant progress has been made in understanding the pathophysiology, epidemiology, genetics, therapies, and disease interactions of sleep apnea. In light of the serious morbidities associated with OSA, future directions should emphasize disease prevention, improved detection of mild disease, early disease intervention, and development of novel well-tolerated effective therapies. As disorders with diverse systemic effects, furthering of multidisciplinary approaches to patient care will be ever more important in the future.

See also: Cataplexy; Narcolepsy; Sleep in Adolescents; Sleeping Sickness.

Further Reading

Budhiraja R and Quan SF (2005) Sleep-disordered breathing and cardiovascular health. *Current Opinion in Pulmonary Medicine* 11(6): 501–506.

Culebras A (2005) Sleep apnea and stroke. *Reviews in Neurological Diseases* 2(1): 13–19.

Gozal D (2004) New concepts in abnormalities of respiratory control in children. *Current Opinion in Pediatrics* 16(3): 305–308.

Javaheri S (2005) Central sleep apnea in congestive heart failure: Prevalence, mechanisms, impact, and therapeutic options. *Seminars in Respiratory and Critical Care Medicine* 26(1): 44–55.

Lavie L (2005) Sleep-disordered breathing and cerebrovascular disease: A mechanistic approach. *Neurologic Clinics* 23(4): 1059–1067.

Narkiewicz K, Wolf J, Lopez-Jimenez F, and Somers VK (2005) Obstructive sleep apnea and hypertension. *Current Cardiology Reports* 7(6): 435–440.

Pack AI (2006) Advances in sleep-disordered breathing. *American Journal of Respiratory and Critical Care Medicine* 173(1): 7–15.

Remmers JE (2005) A century of control of breathing. *American Journal of Respiratory and Critical Care Medicine* 172(1): 6–11.

Schwab RJ (2005) Genetic determinants of upper airway structures that predispose to obstructive sleep apnea. *Respiratory Physiology & Neurobiology* 147(2–3): 289–298.

Strohl KP (2003) Periodic breathing and genetics. *Respiratory Physiology & Neurobiology* 135(2–3): 179–185.

White DP (2006) The pathogenesis of obstructive sleep apnea: Advances in the past 100 years. *American Journal of Respiratory Cell and Molecular Biology* 34(1): 1–6.

Young T, Palta M, Dempsey J, Skatrud J, Weber S, and Badr S (1993) The occurrence of sleep-disordered breathing among middle-aged adults. *New England Journal of Medicine* 328(17): 1230–1235.

Narcolepsy

E Mignot and L Lin, Stanford University, Palo Alto, CA, USA

What Is Narcolepsy?

The sleep disorder narcolepsy was first recognized, named, and described in the late nineteenth century when an unusual association of excessive daytime sleepiness and episodic muscle weakness that was believed to be more than simply 'epileptoid' phenomenon was noted. The unusual muscle weakness episodes triggered by emotions that occur in this condition were called 'cataplexy' soon thereafter. Until relatively recently, whether or not narcolepsy was a symptom or a disease entity was much debated. The discovery of a human leukocyte antigen (HLA) association in 1980 added credence to the hypothesis of a unique disease entity with a discrete etiology. This was subsequently confirmed by the discovery of hypocretin (orexin) deficiency as the major cause of narcolepsy–cataplexy.

In its typical form, symptoms include sleepiness, cataplexy, sleep paralysis, hypnagogic hallucinations, and disturbed nocturnal sleep. Sleepiness, abnormal sleep onset into rapid eye movement (REM) sleep, and cataplexy are the most important symptoms. Whereas much is known about the pathophysiology of narcolepsy with cataplexy, little is known regarding the cause of narcolepsy without cataplexy. For this reason, international classifications and research studies typically separate narcolepsy into cases with and without cataplexy. As discussed later, narcolepsy without cataplexy is more likely to be etiologically heterogeneous.

Approximately 1 person in 2000 in the United States and Western European countries has narcolepsy with cataplexy. The exact prevalence of narcolepsy without cataplexy (defined as isolated unexplained daytime sleepiness with abnormal REM sleep onset periods) is unknown but likely to be higher. Narcolepsy is typically diagnosed many years after disease onset. Early detection and treatment is critical to ensure normal schooling and to prevent associated loss of social status. The disorder is socially isolating, and many patients have problems holding a job and maintaining an active social and affective life.

Narcolepsy Symptoms

Excessive daytime sleepiness is the most troublesome symptom and most often causes patients to consult. Typical complaints are frequent and overwhelming sleep attacks and the need to nap frequently during the day. A wide range of severity may be seen, from simply needing a nap everyday or nodding when at rest to an almost constant feeling of irresistible drowsiness. Sleepiness in narcolepsy may be confused with fatigue and depression.

Cataplexy, a brief and sudden loss of muscle strength triggered by emotions, is very important clinically for the diagnosis of narcolepsy. It is the only symptom specific of narcolepsy, but unfortunately it is not present in all patients. Typically, patients feel weak in the knees and have to sit down when they are emotionally excited – for example, in the context of a funny joke, a good laugh, or when becoming angry. Other attacks of muscle weakness may affect the head (head dropping), the jaw (jaw dropping), the face, or the arms. Cataplexy is frequently mild. It may occur only a few times per month or several times a day. Strong attacks may escalate to a complete body paralysis episode lasting up to a few minutes. Cataplexy may be confused with epilepsy or other neurological/psychiatric problems. Importantly, patients are awake and conscious during cataplexy. Cataplexy should also not be confused with catalepsy, a symptom characterized by increased muscle tone and body rigidity, most often in the context of schizophrenia.

The other symptoms (sleep paralysis, hypnagogic hallucinations, and insomnia) are only weakly indicative of narcolepsy. They are also observed in normal individuals or in association with other sleep disorders. Sleep paralysis is characterized by an inability to move when waking up or when falling asleep. The first episode of sleep paralysis is often frightening, but the paralysis always ends after a few seconds to a few minutes. Hypnagogic hallucinations ('hypnagogic' means 'when falling asleep') are dreamlike visions or auditory perceptions that occur when patients are tired or actually falling asleep. Patients with vivid dreamlike hallucinations have occasionally been misdiagnosed as schizophrenic.

Contrary to popular belief, patients with narcolepsy do not typically sleep more than normal people. Rather, they are unable to stay awake or asleep for long periods of time. Insomnia is thus common in patients with narcolepsy. Patients with narcolepsy typically fall asleep easily at any time during the day and night but have great difficulties staying asleep at night. The total amount of sleep during the 24-h day is thus often normal.

Dissociated REM Sleep in Narcolepsy

Many of the symptoms in narcolepsy are due to abnormal transitions from wakefulness to REM sleep (called sleep-onset REM periods), a state associated with vivid dreaming, rapid eye movements, and a complete muscle paralysis. In normal subjects, REM sleep first occurs 90–120 min after falling asleep and then reoccurs periodically throughout the night. In narcolepsy, abrupt transitions from wakefulness to REM sleep often occur before the patient is completely asleep. This problem produces 'dissociated' states in which the patient is half awake and half in REM sleep, for example, being paralyzed (sleep paralysis) or dreaming (hypnagogic hallucinations) but awake. Cataplexy is also considered similar to REM sleep paralysis but occurs when the brain is excited by an emotion such as laughing or anger.

Genetic Predisposition to Narcolepsy–Cataplexy

The occurrence of narcolepsy–cataplexy involves both genetic predisposition and environmental triggers. Approximately 75% of reported monozygotic twins are discordant for narcolepsy, suggesting the importance of environmental factors. Multiplex families are rare, but a 10- to 40-fold increase in relative risk is reported in first-degree relatives.

One of the major genetic susceptibility factors for narcolepsy–cataplexy is HLA-DQB1 (**Figure 1(a)**). Almost all patients with typical cataplexy carry DQB1*0602, an HLA subtype found in 12% of Japanese, 25% of Caucasian, and 38% of African American controls. HLA-DQB1*0602 is always associated with DQA1*0102, another HLA allele encoded by the nearby DQA1 gene, located less than 20 kb telomeric to DQB1 (**Figures 1(a) and 1(b)**). The DQB1 association is especially tight in subjects with hypocretin deficiency; only four HLA DQA1*0102–DQB1*0602 negative narcolepsy subjects with low or undetectable cerebrospinal fluid (CSF) hypocretin-1 have been identified to date (out of several hundred subjects with CSF testing).

The DQB1*0602 association in narcolepsy is primary and not due to linkage disequilibrium with other loci. It is mostly due to DQA1*0102–DQB1*0602, but it is not simply a dominant effect. Indeed, DQB1*0602 homozygotes have a two- to fourfold higher risk than DQB1*0602/X heterozygotes, and specific DQB1*0602 heterozygotes are either at increased (e.g., DQB1*0602/DQB1*0301) or decreased (e.g., DQB1*0602/DQB1*0601 or DQB1*0602/DQB1*0501) risk (**Figure 1(c)**). Additional smaller effects have also been suggested for specific DRB1 alleles, another HLA locus located in the same region. The complexity of the HLA allele association in narcolepsy mirrors that reported in various autoimmune diseases, such as type I diabetes.

Genetic factors other than HLA are likely to be involved in the genetic susceptibility to narcolepsy–cataplexy but are not as well established or replicated. They are also likely to contribute far less to overall genetic risk than HLA-DQ. Studies in multiplex families suggest linkage to 4p13–q31 and 21q11.2 and possible associations with tumor necrosis factor (TNF)-α, catechol-O-methyl-transferase, and TNF receptor-2 polymorphisms. Notably absent in this list is the preprohypocretin gene and its two receptors, hypocretin receptor-1 and -2. Indeed, only a single mutation in any of these three loci has been identified. This case, an HLA-DQB1*0602-negative boy with hypocretin deficiency, had an unusually early onset of narcolepsy–cataplexy at 6 months of age. This boy had a potential dominant mutation in the signal peptide of the preprohypocretin gene that produces impaired hypocretin trafficking and, presumably, cell death.

Narcolepsy–Cataplexy Is Tightly Associated with Hypocretin Deficiency

The role of hypocretin in narcolepsy was first demonstrated in familial canine narcolepsy, a disease caused by mutations in the hypocretin receptor-2 gene, and through the study of a mouse preprohypocretin gene knockout model. This was followed by the demonstration of low CSF hypocretin-1 in 90% or more of cases with narcolepsy and cataplexy. Postmortem studies have shown a selective loss of 50 000–100 000 posterior hypothalamic neurons that produce the neuropeptide hypocretin (orexin) (**Figures 2(a) and 2(b)**). Loss, rather than decreased expression, is suggested by the concomitant loss of associated co-localized markers dynorphin and neuronal pentraxin-2. The HLA association and associated hypocretin deficiency is particularly strong (>90%) in patients with definite cataplexy (**Figure 3**). This association lends support to the hypothesis that narcolepsy is an autoimmune disorder.

Neurobiology of the Hypocretin Systems and Relevance to Narcolepsy

The hypocretin peptides (also called orexins) are encoded by the preprohypocretin gene. This gene encodes two related amidated polypeptides, hypocretin-1 and hypocretin-2, with high sequence homology. Hypocretin-1 (orexin-A), a 33-amino-acid peptide, has two disulfide bonds and is remarkably stable in biological fluids. It has an approximately equal affinity for the two known hypocretin receptors, hypocretin receptor-1 (HCRTR1) and hypocretin receptor-2 (HCRTR2). Hypocretin-2 (orexin-B), a

HLA gene location

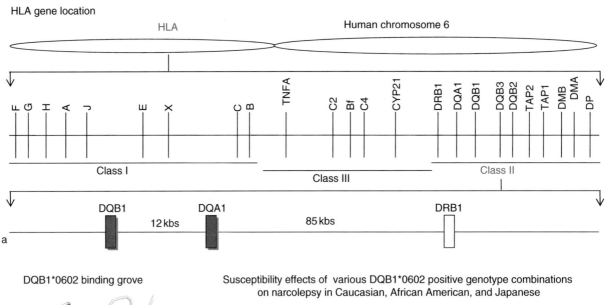

Figure 1 Genetic susceptibility to narcolepsy: human leukocyte antigen effects. (a) Human leukocyte antigen (HLA) genes are numerous and located within the major histocompatibility complex on human chromosome 6. The HLA genes with the tightest genetic association with narcolepsy–cataplexy are the HLA class II region genes DRB1, DQA1 (encoding the DQα chain of the DQ molecule), and DQB1 (encoding the β chain of the DQ molecule). (b) Crystal structure of a DQ molecule associated with a peptide. The HLA-DQ molecule is a heterodimer formed by the polymorphic DQα and DQβ chains (encoded by DQA1 and DQB1 genes, respectively). The two chains form a peptide binding groove in which antigenic peptide (here, the hypocretin-1 peptide) can bind, to be presented to other cells of the immune system for the development of the immune response. (c) Specific DQA1–BQB1 allele combinations and genetic risk for narcolepsy–cataplexy in various ethnic groups. Allele groupings are ranked from the most predisposing to the most protective combinations. Almost all combinations with DQA1*0102–DQB1*0602 are at increased risk; one copy of DQA1*0102–DQB1*0602 is required in almost all cases for developing narcolepsy with cataplexy and hypocretin deficiency. Note that the highest odds ratio (OR) is observed for DQA1*0102–DQB1*0602 homozygotes. Other allele combinations confer various degrees of susceptibility, as often observed in other HLA-associated diseases. (b) Reproduced from Siebold C, Hansen BE, Wyer JR, et al. (2004) Crystal structure of HLA-DQ0602 that protects against type 1 diabetes and confers strong susceptibility to narcolepsy. *Proceedings of the National Academy of Sciences of the United States of America* 101: 1999–2004. (c) Reproduced from Mignot et al. (2001).

smaller peptide of 29 amino acids, has a tenfold higher affinity for HCRTR2.

Hypocretin neurons have tight functional interactions with cholinergic and monoaminergic systems regulating sleep (**Figure 2(c)**). These cell groups are enriched in hypocretin receptors, with a complementary distribution of the two receptors at the anatomical level. HCRTR1 is particularly abundant in the adrenergic locus coeruleus, whereas HCRTR2 is highly expressed on histaminergic cells of the tuberomammillary nucleus. Other regions of interest include hippocampus, paraventricular thalamus, and ventromedial hypothalamic nucleus for HCRTR1,

and cortex, septal nuclei, paraventricular thalamus, paraventricular and arcuate hypothalamic nuclei, and pontine gray area for HCRTR2. In almost all cases, these two receptors have excitatory effects, although there is suggestion of promiscuous coupling of these receptors to multiple G-proteins with occasional inhibitory effects.

It has been hypothesized that the loss of excitatory input to monoaminergic cell groups mediates the sleepiness and short REM sleep in narcolepsy. Indeed, the activity of most monoaminergic cell groups such as locus coeruleus, raphe, and tuberomammillary nuclei, but not the dopaminergic systems in the

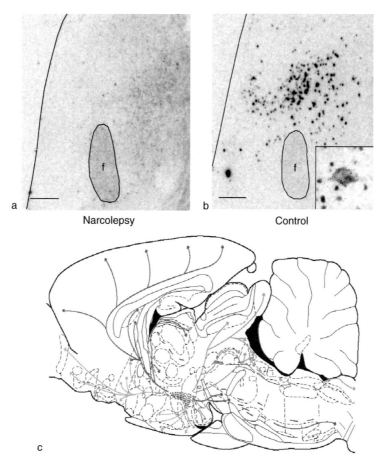

Figure 2 Preprohypocretin *in situ* hybridization in the posterior hypothalamic region in a narcoleptic (a) and control (b) human subject. Note loss of signal in narcolepsy, suggesting neuronal loss of this population. (c) Location and cell bodies of hypocretin-producing cells in a rat. Note the hypothalamic location of cell body and widespread projection through the entire neuroaxis. Scale bar = 1 cm (a, b). Reproduced from Peyron C, Faraco J, Rogers W, et al. (2000) A mutation in a case of early onset narcolepsy and a generalized absence of hypocretin peptides in human narcoleptic brains. *Nature Medicine* 6: 991–997. (c) Reproduced from Peyron C, Tighe DK, van den Pol AN, et al. (1998) Neurons containing hypocretin (orexin) project to multiple neuronal systems. *Journal of Neuroscience* 18: 9996–10015.

ventral tegmental area and substantia nigra, is depressed during sleep and almost silent during REM sleep. The fact that narcolepsy is typically treated with drugs that indirectly stimulate monoaminergic transmission (e.g., amphetamine-like compounds and antidepressants) is also in line with this hypothesis.

Interestingly, hypocretin effects on wakefulness after intracerebroventricular injections are largely blocked or attenuated by monoaminergic blockers, most notably antihistaminergic compounds, suggesting a primary role for downstream histaminergic systems. An important role for histamine is also suggested by findings indicating significantly decreased CSF histamine levels in narcolepsy–cataplexy, a result contrasting with more variable changes in other monoamine and metabolites.

Without hypocretin, narcoleptic patients have sleepiness, inappropriate REM paralysis during wakefulness (cataplexy and sleep paralysis), REM dreaming before falling asleep (hypnagogic hallucinations), and disorganized nighttime sleep. Rapid transitions into REM sleep can be observed during nocturnal sleep and while napping. Finally, narcolepsy is often associated with an increased body mass index and obesity, a possible consequence of sleepiness and decreased energy expenditure.

Little is known regarding the normal function of hypocretins. Hypocretin neurons integrate metabolic and sleep/wake-related inputs. The relative contribution of these various factors may vary across species. In rodents, these cells have receptors for leptin and ghrelin, and they are sensitive to glucose changes, likely facilitating wakefulness in reaction to changes in food availability. In rodents, electrophysiological recording of these neurons indicate maximal activity during wakefulness, especially during purposeful activity. Activity is low during non-REM and REM

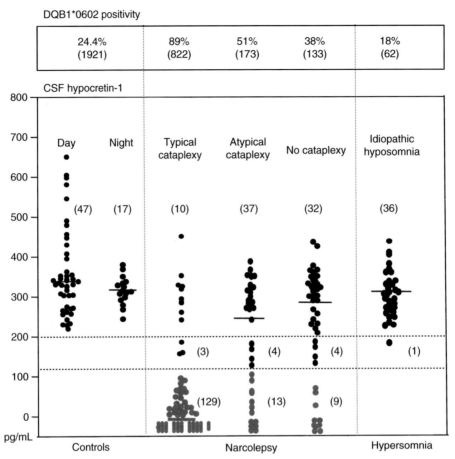

DQB1*0602 positivity

| 24.4% (1921) | 89% (822) | 51% (173) | 38% (133) | 18% (62) |

CSF hypocretin-1

Figure 3 HLA-DQB1*0602 positivity (top) and hypocretin deficiency (bottom), as measured in the cerebrospinal fluid (CSF). (Top) Almost all patients with typical cataplexy are DQB1*0602, suggesting etiological homogeneity. DQB1*0602 positivity decreases in patients with atypical and no cataplexy, but it is still higher than control values. Subjects with idiopathic hypersomnia (isolated sleepiness and no REM transitions) have a normal DQB1*0602 frequency, suggesting no pathophysiological overlap. (Bottom) CSF hypocretin-1 in controls, narcolepsy, and idiopathic hypersomnia. The large majority of subjects with typical cataplexy have low CSF hypocretin-1. Quantitative receiver curve analysis has indicated that values below 110 pg ml^{-1} are most predictive of a narcolepsy diagnosis, although control values are typically above 200 pg ml^{-1}. Only 20–40% of cases without cataplexy have low CSF hypocretin-1. None of the controls or cases with idiopathic hypersomnia have low CSF hypocretin-1. (Top) Reproduced from Lin L and Mignot E (2007) HLA and narcolepsy: Present status and relationship with familial history and hypocretin deficiency. In: Bassetti C, Billiard M, and Mignot E (eds.) *Narcolepsy and Hypersomnia*. New York: Dekker/Taylor & Francis. (Bottom) Reproduced from Mignot E, Lammers GJ, Ripley B, et al. (2002) The role of cerebrospinal fluid hypocretin measurement in the diagnosis of narcolepsy and other hypersomnias. *Archives of Neurology* 59: 1553–1562.

sleep. Hypocretin release is also tightly coupled with locomotor activity in rodents, an effect less evident in wake-consolidated species such as the squirrel monkey. Hypocretin release is also driven by the circadian clock and peaks at the end of the active (wake) period. Finally, hypocretin release is increased by even a brief sleep deprivation, suggesting a role for the hypocretin system in consolidating wakefulness in the face of a mounting sleep debt. This last role of hypocretin may explain the inability of narcolepsy patients to stay awake for periods typically longer than a few hours, as sleep debt is increasing. As expected for a wake-promoting signal, these cells are more active during active wake, decrease their activity during non-REM sleep, and are least active

during REM sleep. The sleep stage-specific activity is consistent with a role in promoting wakefulness and inhibiting REM sleep.

Is Narcolepsy an Autoimmune Disorder?

The tight HLA association and the involvement of both genetic and nongenetic factors in the pathophysiology of narcolepsy suggest an autoimmune basis for the disorder. This hypothesis is also consistent with the peripubertal onset of the disease but not the even sex ratio found in the disorder. The discovery of hypocretin-producing cells as a potential target has given further credence to the autoimmune hypothesis. HLA-DQ is expressed on the surface of B cells, macrophages (including microglia), and activated T cells. It is a heterodimer composed of an α and a β chain encoded by the polymorphic DQA1 and DQB1 genes, two genes in tight linkage disequilibrium. HLA-DQ binds peptide fragments and can present the resulting complex to other cells of the immune system, as recognized by the T cell receptor complex (**Figure 1**).

How HLA molecules in general and HLA-DQ in particular predispose to autoimmune disorders is not well understood. It probably involves the ability of some HLA alleles to bind and consequently present specific autoantigens to the rest of the immune system. In other HLA-associated diseases, however, antibodies directed toward the potential target (e.g., islet cell antibodies in type I diabetes) or T cell clones with primary responsiveness to target antigens (e.g., against myelin basic protein in multiple sclerosis) can be detected. In narcolepsy, however, all attempts to demonstrate T cell reactivity against hypocretin peptides or to detect autoantibodies directed at hypocretin-producing cells have failed.

These generally negative results do not necessarily exclude the possibility of an autoimmune mechanism, but they do raise the possibility of other, more complex neuroimmune mechanisms. Another possibility is an infectious agent with particular tropism toward hypocretin-producing cells, although HLA association in infectious diseases is usually not very tight or it is primarily to a single allele. Alternatively, it has been suggested that these neurons could be unusually sensitive to excitotoxicity. Interest in the autoimmune arena has been rekindled by case reports that intravenous immunoglobulin may reduce the development of narcolepsy severity if administered within a year of onset.

Narcolepsy without Cataplexy

The importance of cataplexy in the diagnosis of narcolepsy has long been suggested by clinical studies. The finding that almost all cases with definite cataplexy (defined as recurrent episodes of muscle paralysis triggered at least sometimes by typical emotions such as laughing or joking) are HLA-DQB1*0602 positive strongly suggests etiological homogeneity in these cases. This was further confirmed by the finding that almost all HLA-positive patients with cataplexy have low CSF hypocretin-1 (**Figure 3**). In contrast, HLA-DQB1*0602 is typically found in 40% of patients without cataplexy or with doubtful or atypical cataplexy (rare events or only events triggered by unusual emotions). This is matched by a much lower percentage of patients without cataplexy who have low CSF hypocretin-1 (5–30% of patients, depending

on the specific case series) (**Figure 3**). Some of these cases (approximately one-third) are children with recent onset (≤ 4 years) who are likely to develop cataplexy within a few years.

These results are consistent with two possible, nonexclusive pathophysiological models for narcolepsy without cataplexy. In the first model, narcolepsy both with and without cataplexy is part of the same disease continuum, with similar pathophysiological effects on hypocretin transmission. Narcolepsy with cataplexy is a generally more severe form of the disease, most commonly associated with an almost complete destruction of the hypocretin system. HLA-DQB1*0602 may be a severity factor more likely to predispose to complete hypocretin destruction. Only in rare cases with cataplexy would hypocretin cell loss be partial but sufficient to produce normal levels of hypocretin and a narcolepsy phenotype. In these cases, compensation by the remaining hypocretin cells likely maintains CSF hypocretin-1 at normal levels. Projections with more effects on CSF hypocretin-1 but of less functional importance for the phenotype (e.g., hypocretin afferents to the spinal cord) also may be more spared in cases with cataplexy and normal CSF levels.

In narcolepsy without cataplexy, cell loss would be less pronounced. The severity of the disease would be generally lower and the cell loss insufficient to result in low CSF hypocretin-1. The HLA-DQB1*0602 association with narcolepsy without cataplexy is weaker. Only in rare cases might neuroanatomical destruction be pronounced, resulting in low CSF hypocretin-1 but still sparing a select hypocretin cell subpopulation projecting to cataplexy-triggering pathways. In support of this model is a study by Thannickal et al. in which 14% of hypocretin cells remained in a case without cataplexy versus 4.4–9.4% remaining cells in five other subjects with cataplexy. Lesion studies in rats also indicate a 50% decrease in CSF hypocretin-1 with a 77% destruction of hypocretin cells, suggesting some degree of compensation.

The fact that HLA-DQB1*0602 is slightly increased in frequency in cases without cataplexy but with normal CSF hypocretin-1 is also consistent with this hypothesis. This finding, however, remains to be confirmed in a population-based sample because HLA positivity may have been increased in clinical samples due to a bias in inclusion (some clinicians use HLA typing to confirm the diagnosis of narcolepsy without cataplexy). The model also predicts that a large number of narcolepsy without cataplexy cases may exist in the general population but have a milder phenotype, consistent with our observation of slightly decreased REM latency in normal population subjects with HLA-DQB1*0602. Similar milder forms of the disease or long latent periods have been suggested for other autoimmune diseases, such as DQ2-associated celiac disease, B27-associated spondylarthropathies, and DR3/DR4-associated type I diabetes. In type I diabetes, for example, a larger number of individuals, especially relatives of affected probands, may be positive for islet cell antibodies without ever developing the disease.

In the second model, the cause of cases without low CSF hypocretin does not involve partial hypocretin cell loss. Other systems downstream of hypocretin, such as hypocretin receptors, histamine, or neuroanatomical systems not associated with hypocretin neurobiology, may be involved. Of note, the two models may not be entirely exclusive. A partial hypocretin cell loss may, for example, not be sufficient in itself to produce symptoms but could lead to narcolepsy when associated with additional defects downstream. A similar model, albeit speculative, has been proposed in some obese-type diabetes children in which both types I and II may coexist. In these cases, subjects with partially reduced islet cell numbers may be asymptomatic when lean but develop insulin resistance if obesity develops and the remaining islet population is unable to produce enough insulin to keep up with increased tissue demand.

To distinguish between these models, neuropathological studies of narcolepsy without cataplexy cases are urgently needed. It is also imperative to study narcolepsy without cataplexy not only in clinical samples but also in the general population because clinical cases may represent a more severe and selected subpopulation. We believe that the most likely explanation will be a combination of these models, including both disease heterogeneity with respect to hypocretin neuropathology and severity gradients for hypocretin cell loss as discussed previously. The extent of this overlap remains to be defined.

The previous discussion is important because the real population-based prevalence of narcolepsy without cataplexy is unknown. Data by our group and others suggest that several percent of the population have unexplained sleepiness and REM transition. Whether a small or large portion of these subjects have a mild degree of hypocretin dysfunction (possibly related pathophysiologically to narcolepsy–cataplexy) is unknown.

Therapies for Narcolepsy

Good sleep hygiene, education, and treatment compliance are important to the management of narcolepsy. Referral to support groups such as the Narcolepsy Network is helpful. Fixed wake-up times, sleep diaries, and regular sleep schedule are recommended. Obesity may develop especially in young children when disease onset is abrupt. It is useful to restrict diet, encourage exercise, and treat sleepiness aggressively at this stage. The risk of driving while sleepy, especially prior to adequate therapy, must be discussed and, if appropriate, regulatory agencies notified. Patients with narcolepsy must avoid jobs that put others in danger and should consider activities that are less sedentary. Jobs that involve repetitive tasks or sitting and looking at a computer all day can be difficult. Employers or teachers can be asked to accommodate 15- to 30-min scheduled naps.

Modafinil is now first-line, standard-of-care treatment for sleepiness associated with narcolepsy. Amphetamine-like stimulants, such as methylphenidate, dextroamphetamine, amphetamine racemic mixture, and methylamphetamine, can also be used but have a number of disadvantages. They are potentially addictive and should preferably be reserved for patients with a well-established diagnosis (e.g., those with cataplexy). Additionally, side effects such as palpitations, hypertension, and nervousness may occur because of their action on the autonomic nervous system. The mode of action of amphetamine-like stimulants and modafinil on wakefulness is thought to involve increased dopaminergic transmission through the inhibition of dopamine reuptake (modafinil and methyphenidate), by increasing dopamine release (methyphenidate), or through more complex effects on release and storage (amphetamine).

Antidepressants are commonly used to treat cataplexy. The rationale for their efficacy is the profound REM suppressant effects most antidepressant share. They are also effective for sleep paralysis and hypnagogic hallucinations, a sometime disabling symptom. Tricyclics are very effective but have anticholinergic side effects. Serotonin reuptake specific inhibitors are efficacious, but relatively high doses are generally needed. Newer medications targeting norepinepherine reuptake, such as venlafaxine and atomoxetine, are most effective. The dual serotoninergic/noradrenergic reuptake inhibitor venlafaxine is a typical first-line treatment. Atomoxetine, a specific adrenergic reuptake inhibitor typically prescribed for attention deficit hyperactivity disorder, can also be used and as it is mildly stimulant. Patients should be warned that rebound cataplexy usually occurs when medications are discontinued, changed, or skipped.

Sodium oxybate (γ-hydroxybutyrate (GHB)) is unique because it is efficacious on all symptoms. This naturally occurring compound is a potent but short-acting hypnotic that consolidates slow-wave and REM sleep, most likely via $GABA_B$ agonistic effect. It has been available since the 1970s but was only recently approved for the treatment of cataplexy. It is thought that the consolidation of nocturnal sleep and REM sleep leads to decreased daytime symptoms. The drug is carefully controlled and can be addictive. It is typically reserved for narcoleptic with both disturbed nocturnal sleep and cataplexy.

Narcolepsy is frequently associated with other sleep disorders. Sleep apnea, a condition in which patients snore and gasp during sleep, is common and must be treated especially if GHB is to be used. Periodic leg movements (repetitive leg twitching occurring during sleep) are also common but very rarely need treatment.

Perspectives for New Treatments

The realization that narcolepsy and related pathologies are not rare is fueling pharmaceutical interest in this area. Novel symptomatic agents are being developed. Agents that act on the histaminergic system, such as H3 antagonists (compounds that increase histamine release through the blockade of the H3 autoreceptor), are being studied. These compounds are wake promoting and effective in animal models of narcolepsy.

New treatments aimed at replacing the missing hypocretin hormone are being developed and will probably be much more effective. These may involve developing hypocretin receptor agonists, preferentially of the HCRTR2 type, using gene therapy or using cell-based replacement therapy. Research aimed at discovering the process that destroys the hypocretin-containing cells in the brain is ongoing. Understanding this process is necessary before truly preventive and curative strategies can be developed.

See also: Cataplexy; Sleep Apnea; Sleep in Adolescents.

Further Reading

Bassetti C, Billiard M, and Mignot E (2007) *Narcolepsy and Hypersomnias of Central Origin.* New York: Dekker/Taylor & Francis.

Blouin AM, Thannickal TC, Worley PF, et al. (2005) Narp immunostaining of human hypocretin (orexin) neurons: Loss in narcolepsy. *Neurology* 65: 1189–1192.

Crocker A, Espana RA, Papadopoulou M, et al. (2005) Concomitant loss of dynorphin, NARP, and orexin in narcolepsy. *Neurology* 65: 1184–1188.

Dauvilliers Y, Carlander B, Rivier F, et al. (2004) Successful management of cataplexy with intravenous immunoglobulins at narcolepsy onset. *Annals of Neurology* 56: 905–908.

Honda Y, Asaka C, Tanimura M, and Furusho T (1983) A genetic study of narcolepsy and excessive daytime sleepiness in 308 families with a narcolepsy or hypersomnia proband. In: Guilleminault C and Lugaresi E (eds.) *Sleep/Wake Disorders: Natural History, Epidemiology and Long-Term Evolution,* pp. 187–199. New York: Raven Press.

Lin L, Faraco J, Li H, et al. (1999) The sleep disorder canine narcolepsy is caused by a mutation in the hypocretin (orexin) receptor 2 gene. *Cell* 98: 365–376.

Lin L and Mignot E (2007) HLA and narcolepsy: Present status and relationship with familial history and hypocretin deficiency. In: Bassetti C, Billiard M, and Mignot E (eds.) *Narcolepsy and Hypersomnia,* ch. 37, pp. 411–426. New York: Dekker/Taylor & Francis.

Mignot E (1998) Genetic and familial aspects of narcolepsy. *Neurology* 50(2 supplement 1): S16–S22.

Mignot E, Lammers GJ, Ripley B, et al. (2002) The role of cerebrospinal fluid hypocretin measurement in the diagnosis of narcolepsy and other hypersomnias. *Archives of Neurology* 59: 1553–1562.

Mignot E, Lin L, Finn L, et al. (2006) Correlates of sleep onset REM periods during the Multiple Sleep Latency Test in community adults. *Brain* 129: 1609–1623.

Mileykovskiy BY, Kiyashchenko LI, and Siegel JM (2005) Behavioral correlates of activity in identified hypocretin/orexin neurons. *Neuron* 46: 787–798.

Nishino S and Mignot E (1997) Pharmacological aspects of human and canine narcolepsy. *Progress in Neurobiology* 52: 27–78.

Peyron C, Faraco J, Rogers W, et al. (2000) A mutation in a case of early onset narcolepsy and a generalized absence of hypocretin peptides in human narcoleptic brains. *Nature Medicine* 6: 991–997.

Peyron C, Tighe DK, van den Pol AN, et al. (1998) Neurons containing hypocretin (orexin) project to multiple neuronal systems. *Journal of Neuroscience* 18: 9996–10015.

Scammell T (2006) The frustrating and mostly fruitless search for an autoimmune cause of narcolepsy. *Sleep* 29: 601–602.

Siebold C, Hansen BE, Wyer JR, et al. (2004) Crystal structure of HLA-DQ0602 that protects against type 1 diabetes and confers strong susceptibility to narcolepsy. *Proceedings of the National Academy of Sciences of the United States of America* 101: 1999–2004.

Thannickal TC, Moore RY, Nienhuis R, et al. (2000) Reduced number of hypocretin neurons in human narcolepsy. *Neuron* 27: 469–474.

Wisor JP, Nishino S, Sora I, et al. (2001) Dopaminergic role in stimulant-induced wakefulness. *Journal of Neuroscience* 21: 1787–1794.

Yoshida Y, Fujiki N, Nakajima T, et al. (2001) Fluctuation of extracellular hypocretin-1 (orexin A) levels in the rat in relation to the light–dark cycle and sleep–wake activities. *European Journal of Neuroscience* 14: 1075–1081.

Cataplexy

S Nishino, Stanford University School of Medicine, Stanford, CA, USA

Cataplexy (one of the symptoms of narcolepsy) is a sudden drop of antigravity muscle tone triggered most often by positive emotional factors (such as laughter) and less frequently by negative emotional factors (such as anger or frustration).

Narcolepsy is a chronic sleep disorder that affects 0.02–0.05% of the general population and has also been reported in dogs (**Figure 1(a)**) and, more recently, in (genetically engineered) mice and rats. Human narcolepsy is caused by an interplay of genetic susceptibility and environmental factors, and the disease is tightly associated with human leukocyte antigen (HLA) DR2 and DQ6 (DQB1*0602). Although many autoimmune diseases are associated with certain HLA types, an involvement of autoimmune mechanisms with narcolepsy has not been demonstrated.

The symptoms of narcolepsy are often classified as the narcolepsy tetrad (excessive daytime sleepiness, cataplexy, hypnagogic hallucination, and sleep paralysis) or pentad (narcolepsy tetrad plus insomnia). Among these symptoms, cataplexy is pathognomonic for narcolepsy, and it occurs almost exclusively in idiopathic narcolepsy and in rare cases of symptomatic (or secondary) narcolepsy (i.e., narcolepsy associated with other neurological conditions such as a brain tumor or stroke). Sleep paralysis and hypnagogic hallucinations also occur in other sleep disorders such as sleep apnea (or even in healthy subjects, especially when their sleep–wake pattern is disturbed). Sixty percent of narcoleptic subjects diagnosed by the current International Classification of Sleep Disorders (ICSD-2) diagnostic criteria developed cataplexy, usually around the onset of sleepiness or within 3–5 years (narcolepsy with cataplexy). The other 40% of narcoleptic subjects do not exhibit cataplexy, but have rapid eye movement (REM) sleep abnormalities (i.e., sleep-onset REM periods) together with excessive daytime sleepiness (classified as narcolepsy without cataplexy). Thus, narcolepsy without cataplexy is also clinically common.

The French physician Gélineau (in 1880) first coined the term narcolepsy and described two cases of excessive daytime sleepiness, discussing how emotions influenced the onset of sleep attacks (a fall astasias was also accompanied with sleep attacks in one of these two cases). However, Lowenfeld (in 1902) is usually considered the first author to characterize cataplexy as part of the narcoleptic syndrome.

Clinical Features of Cataplexy

Severe cataplexy episodes produce a bilateral, generalized weakness of the antigravity muscles sufficient to cause a fall (although usually without injury; a video of cataplexy is available at the Stanford website). Other episodes may be partial, affecting only the face, vocal cords, or limb(s). Consciousness is not impaired unless the patient subsequently falls asleep (a transition to REM sleep often occurs). Calling the person's name or shaking his or her body can interrupt cataplexy, and patients can usually recover without any confusion. Patients can also recall events happening before and during cataplexy if they recover before falling asleep. During cataplectic attacks, the cortical electroencephalogram (EEG) is desynchronized and the EEG pattern is similar to that during waking and REM sleep (**Figure 1(b)** shows the EEG patterns during cataplexy in narcoleptic dogs). In contrast to epileptic seizures, no paroxysmal spikes or burst EEG patterns are observed. Neurological examinations performed at the time of an attack show a suppression of the platellar reflex. The duration of cataplexy varies from a split second to several minutes. Its frequency varies according to the time and individual subject – from less than one episode per year to several episodes per day. Cataplexy is currently treated pharmacologically, most often with tricyclic antidepressants (**Table 1**), but the antiepileptics are ineffective. The anticataplectic effect of tricyclic antidepressants (imipramine) was coincidentally found in the clinical practices in the 1950s. Tricyclic antidepressants are also effective for reducing sleep paralysis and hypnagogic hallucinations, but have little effect on excessive daytime sleepiness. Therefore, most narcolepsy–cataplexy patients need to take additional wake-promoting compounds such as an amphetamine, methylphenidate, or modafinil. Conversely, these wake-promoting compounds used for the treatment of excessive sleepiness have no (modafinil) or only moderate (amphetamine and amphetamine-like compounds) anticataplectic effects.

Cataplexy often worsens with poor sleep and fatigue. Patients may also experience status cataplecticus; this rare manifestation of narcolepsy is characterized by subintrant cataplexy that lasts several hours per day and confines the subject to bed. It can occur spontaneously or more often on withdrawal from anticataplectic drugs.

Figure 1 Continued

Table 1 Drugs commonly used for treatment of cataplexy

Drug	Dosage	Side effects and comments
Imipramine[a]	10–100 mg	Dry mouth, anorexia, sweating, constipation, drowsiness
Protriptyline[a]	5–60 mg	Dry mouth, sweating, anxiety, disturbed nighttime sleep
Clomipramine[a]	10–150 mg	Dry mouth, sweating
Fluoxetine[b]	20–80 mg	Few side effects, nausea, dry mouth
Venlafaxine[b]	75–150 mg	Few side effects, nausea
Sodium oxybate[c]	1.5–4.5 g bedtime and 2–3 h later	Morning sedation, nausea, dizziness, urinary incontinence

[a]Imipramine, protriptyline, and clomipramine are classical tricyclics.
[b]Fluoxetine is a serotonin selective reuptake inhibitor (SSRI) and venlafaxine is serotonin/norepinephrine selective reuptake inhibitor, which have less side effects.
[c]Sodium oxybate is not an antidepressant; it is a short-acting sedative. Taking sodium oxybate at night improves sleepiness and cataplexy the following day.

Physiological and Pathophysiological Aspects of Cataplexy

In humans, the first REM sleep episode occurs approximately 90 min after sleep onset and reappears every 90 min (every 30 min in dogs and cats). A total of four to five REM sleep episodes may occur during the night. Shortly after the discovery of REM sleep, it was found that in narcolepsy REM sleep often occurs at sleep-onset (sleep-onset REM periods) or even during short daytime naps (**Figure 2**). Ever since, cataplexy, sleep paralysis, and hypnagogic hallucinations have been often considered a disassociated manifestation of REM sleep; the occurrence of these symptoms is often explained by an abnormal rapid transition to REM sleep during active wake or sleep-onset. During REM sleep, complete and tonic inhibition of muscle tonus, together with phasic bursts of REMs and swift muscle twitching, occurs physiologically, and the amplitude recorded in the electromyogram (typically monitored in neck or chin muscles, combined with sleep EEG recordings) is the lowest. It is, however, the activity of pyramidal tract neurons in the motor cortex that mediate the limb movements, which are as high during REM sleep as during active waking. It is thought that the tonic inhibitory signal for muscle tonus originating in the dorsal pons overcomes this pyramidal motor activation system during REM sleep and that this results in complete muscle atonia during this sleep state. This notion is supported by brain lesion studies in animals. When the lesions are made in the dorsal pons, the phenomenon

called REM sleep without atonia is observed, and the animals move their limbs during REM sleep. Similar phenomena (i.e., REM sleep behavior disorders) are also reported in humans associated with various neurological conditions (including stroke in the pons).

A series of experiments suggested that an executive mechanism of atonia during cataplexy shares a common pathway with atonia during REM sleep.

1. Electrophysiological experiments using familial narcoleptic Dobermans demonstrated that a population of cell groups in the brain stem that are responsible for loss of muscle tone during REM sleep are also active during cataplexy.
2. Experiments have also recently demonstrated that REM-off cells in the adrenergic locus coeruleus (LC) in the pons cease to discharge during cataplexy.
3. Experiments have shown that H-reflex activity (an electronically stimulated muscle response with monosynaptic latency due to excitation of Ia afferents in the spinal cord) profoundly diminishes or disappears during both REM sleep and cataplexy in humans.
4. The major mechanism of action of anticataplectics is thought to be the suppression of REM sleep by activating the central adrenergic system.

Therefore, the motor inhibitory components of REM sleep at the level of the lower brain stem and the spinal cord are also operative during cataplexy.

In contrast, mechanisms of the induction of cataplexy are not well understood. However, several

---

Figure 1 Cataplectic attacks in Doberman pinschers: (a) photos of attacks; (b) polygraphic recordings of cataplectic attacks observed in a narcoleptic Dobermans. Emotional excitations, appetizing food, or playing easily elicits multiple cataplectic attacks in these animals, and most attacks are bilateral (97.9%). Atonia was initiated partially in the hind legs (79.8%), front legs (7.8%), neck/face (6.2%), or whole-body/complete attacks (6.2%), and a progression of attacks was also seen (49% of all attacks). As shown in (b), a rapid transition from the waking state (desynchronized electroencephalogram (EEG) with high electromyogram (EMG)) to cataplexy (sudden reduction of EMG) is observed (left). During cataplexy (right), the EEG consists of low-voltage fast-wave, occasionally accompanied by rapid eye movement. EOG, electrooculogram.

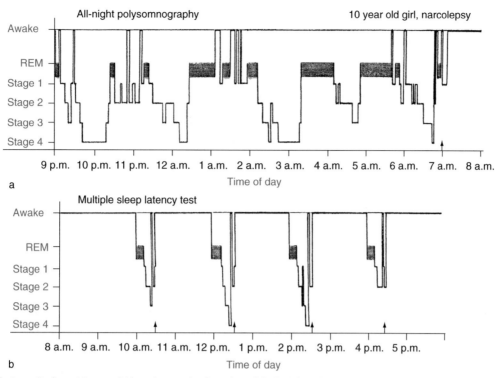

Figure 2 Test results for a 10-year-old female narcoleptic patient: (a) all-night polysomnogram; (b) multiple sleep latency test. In (a), note the appearance of a sleep-onset REM period in the night recordings and marked sleep tendency. In (b), note a sleep-onset REM period at each daytime nap test. Vertical arrows indicate a verbal calls to awaken the patient at the end of each nap test. REM, rapid eye movement. Adapted from Honda Y (1988) Clinical features of narcolepsy: Japanese experiences. In: Honda Y and Juji T (eds.) *HLA in Narcolepsy*, pp. 24–57. Berlin: Springer.

experiments using narcoleptic Dobermans suggested that mechanisms for induction of cataplexy are different from those for REM sleep.

1. Cataplexy in narcoleptic dogs can be elicited anytime by emotional excitation, whereas occurrences of REM sleep in these animals are regulated by a normal 30 min cyclicity.
2. Injection of cholinergic agonist (carbachol) into the basal forebrain (anterior hypothalamus) induces long-lasting cataplexy-like atonia in narcoleptic dogs, but this manipulation does not affect the 30 min cyclicity of the occurrence of the phasic REM sleep phenomena (i.e., REMs).
3. Dopamine D2/3 receptor agonists potently aggravate and D2/3 receptor antagonists potently inhibit cataplexy in narcoleptic dogs, but these classes of compounds have little effect on the occurrences of REM sleep.

Many normal individuals report slight muscle weakness with intense emotion, and during laughter, the H-reflex is markedly reduced in controls and also in narcoleptic subjects. Cataplexy may thus result from excessive activation of the same descending motor inhibitory pathways that are active during strong emotions, but these conceptual (emotion-triggered)

inhibitory pathways and mechanisms are not yet identified.

A series of pharmacological experiments using canine narcolepsy pointed out the importance of the cholinergic system and monoaminergic system for the regulation of cataplexy. Various manipulations and pharmacological compounds reduce cataplexy, but some may be nonspecific; if animals are afraid or do not feel comfortable, cataplexy is generally reduced. In contrast, only a limited class of compounds aggravates cataplexy. These include compounds that activate cholinergic neurotransmission (especially by cholinergic M2/3 receptor agonists) and compounds that reduce monoaminergic neurotransmission (especially by adrenergic α_2 agonists (presynaptic), α_1 antagonists (postsynaptic), and dopamine D2/3 agonists (presynaptic)). Thus, there is a similarity between the pharmacological control of cataplexy and that of REM sleep, but some discrepancies have also been noted with D2/3 compounds.

Cataplexy and Hypocretin/Orexin Deficiency

Cataplexy is also a hallmark symptom of narcolepsy in animals, such as dogs and mice. The identification

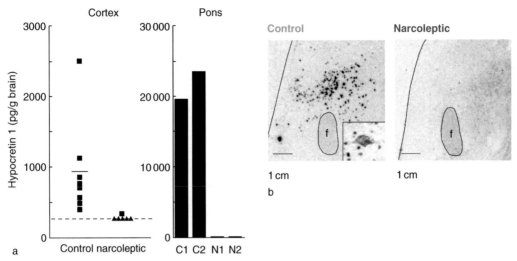

Figure 3 Absence of hypocretin signals and peptide in the hypothalamus of narcoleptic patients: (a) brain hypocretin levels in narcoleptic and control subjects; (b) hypocretin mRNA signal in the lateral hypothalamic area of control and narcoleptic subjects. In (a), cortical hypocretin-1 and hypocretin-2 (data not shown) concentrations were detectable in all controls (— indicates mean) and undetectable (— indicates the detection limit, 332 pg/g brain) in five of the six narcoleptic subjects. The difference in hypocretin concentrations (two narcoleptic subjects vs. two controls) was more pronounced in the pons where high concentrations are observed in the control subjects. In (b) preprohypocretin transcripts are detected in the hypothalamus of control (left) but not narcoleptic (right) subjects. C1/2, controls; f, fornix; N1/2, narcopleptic patients. Adapted from Peyron C, Faraco J, Rogers W, et al. (2000) A mutation in a case of early onset narcolepsy and a generalized absence of hypocretin peptides in human narcoleptic brains. *Nature Medicine* 6: 991–997.

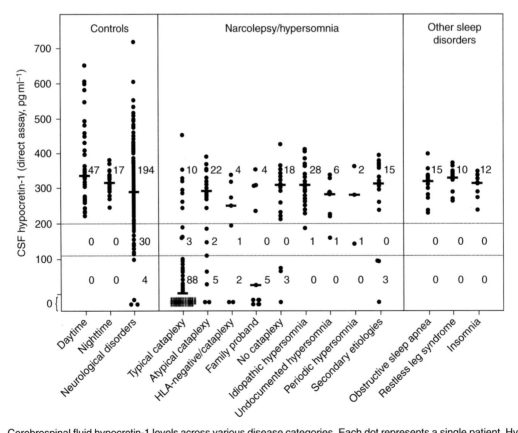

Figure 4 Cerebrospinal fluid hypocretin-1 levels across various disease categories. Each dot represents a single patient. Hypocretin-1 values below 110 pg ml^{-1} were determined to be diagnostic for narcolepsy. Concentrations above 200 pg ml^{-1} best determined healthy control values. The number of patients with hypocretin values below or equal to 110 pg ml^{-1}, above 200 pg ml^{-1}, and between these two values is indicated for each category. CSF, cerebrospinal fluid; HLA, human leukocyte antigen. Adapted from Mignot E, Lammers GJ, Ripley B, et al. (2002) The role of cerebrospinal fluid hypocretin measurement in the diagnosis of narcolepsy and other hypersomnias. *Archives of Neurology* 59: 1553–1562.

of cataplexy in these animals led to the discovery of narcolepsy in these animal species. Studies in these animals (familial canine narcolepsy in narcoleptic Dobermans and gene-knockout mice) led to the discovery of the etiology of narcolepsy in animals. Mutation of one of the two hypocretin/orexin receptors and knockout of the hypocretin/orexin gene (preprohypocretin) causes narcolepsy in dogs and mice, respectively. The hypocretin/orexin system is a recently discovered hypothalamic neuropeptidic system. Hypocretin neurons are exclusively located in the lateral hypothalmic area, but project to most brain areas. Subsequent human studies revealed that the mutation in the hypocretin-related gene is extremely rare, and only one case of early-onset narcolepsy to date was identified as having a mutation in the preprohypocretin gene. However, cerebrospinal fluid (CSF) measures (and a small number of postmortem brain studies) found hypocretin ligand deficiency in most idiopathic narcolepsy–cataplexy cases (~90%) (**Figures 3** and **4**).

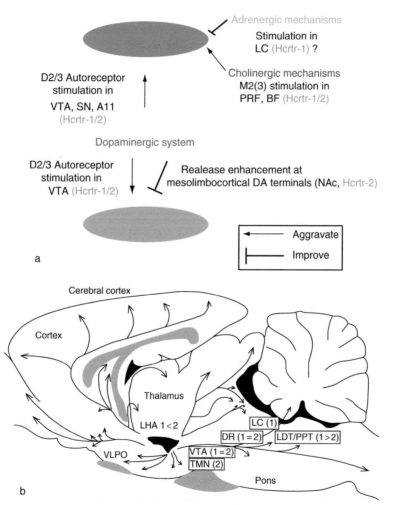

Figure 5 Monoaminergic and cholinergic control of sleepiness and cataplexy in relation to hypocretin input: (a) schematic; (b) hypocretin projections. As shown in (a), the stimulation of adrenergic transmission by adrenergic uptake inhibitors potently reduces cataplexy; this pharmacological property is probably involved in the mode of action of currently used anticataplectic agents (e.g., tricyclic antidepressants). The fact that both the presynaptic α2 autoreceptor stimulation and postsynaptic α1 blockade aggravate cataplexy is consistent with an inhibitory role of adrenergic transmission in the control of cataplexy. D2/3 autoreceptor stimulation aggravates both cataplexy and sleepiness, and amphetamines stimulate DA release at DA terminals and mediate EEG arousal. Muscarinic M2 stimulation induces behavioral wakefulness and cortical desynchrony in control dogs, but it induces cataplexy in narcoleptic dogs. As shown in (b), hypocretin-containing neurons project to these previously identified monoaminergic and cholinergic regions where hypocretin receptors (Hcrtr1/2) are enriched. Impairments of hypocretin input may result in cholinergic and monoaminergic imbalance and the generation of narcoleptic symptoms. BF, basal forebrain; D2/3, dopamine receptors; DA, dopamine; DR, dorsal raphe; EEG, electroencephalogram; LC, locus coeruleus; LDT, laterodorsal tegmentum; LHA, lateral hypothalamic area; M2, muscarinic receptor; NAc, nucleus accumbens; PPT, peduncle pontine tegmentum; PRF, pontine reticular formation; SN, substantia nigra; TMN, tuberomamillary nucleus; VLPO, ventrolateral preoptic area; VTA, ventral tegmental area.

Extended human studies further suggested that hypocretin deficiency is tightly associated with the occurrence of cataplexy and that most patients with narcolepsy without cataplexy and idiopathic hypersomnia and secondary excessive daytime sleep disorders (such as obstructive sleep apnea) have normal CSF hypocretin levels (**Figure 4**). A positive reading for HLA is also tightly associated with hypocretin deficiency. Hypocretin replacements are expected to be a new treatment for hypocretin-deficient narcolepsy, but this is not available due to low penetration (of the brain) of hypocretin peptides; the development of synthetic agonists are likely to be needed.

A significant degree of hypocretin deficiency already occurs around the disease onset and, in some case, before the onset of cataplexy. Moderate hypocretin deficiency is also reported in cataplexy associated with some secondary narcolepsy cases with brain tumors or vascular diseases. In these conditions, the degree of hypocretin deficiency is modest compared to most idiopathic cases (undetectably low), but projection and projection sites of the hypocretin system may also be damaged in these secondary cases. Thus, cataplexy may now appear to be a hypocretin-deficient pathological phenomenon. However, it is not known if impairments of the hypocretin neurotransmission instantly lead to occurrences of cataplexy or if any secondary changes are required to develop cataplexy. If some of these secondary changes are irreversible, hypocretin supplements (when they become available) may not be effective.

Cataplexy is also reported to occur in rare neurological conditions, such as the Nieman-pick type C (without sleepiness) disease, Prader–Willi syndrome, and myotonic dystrophy (with sleepiness). The involvements of the hypocretin system in cataplexy in these conditions have not been systematically studied.

Hypocretins are, in general, excitatory neuropeptides, and they project to most monoaminergic systems (such as the LC, dopaminergic ventral tegmental area and substantia nigra, serotonergic raphe, and histaminergic tuberomammillary nucleus) and cholinergic systems (such as the basal forebrain and brain stem cholinergic nuclei) and their terminal areas, and modulate their neurotransmissions. Hypocretin deficiency is thus likely to induce abnormalities in monoaminergic and cholinergic neurotransmissions (**Figure 5**). Indeed catecholamine contents (dopamine and norepinephrine) are increased in several brain regions, and this may be a compensatory increase due to the lack of excitatory inputs to these systems. This upregulation alone is not likely to prevent the occurrence of the symptoms of narcolepsy, and compounds that enhance the catecholaminergic neurotransmission are needed to control sleepiness and cataplexy. Considering the fact that compounds which shut down the catecholaminergic system (autoreceptor agonists, α_2 and D2/3 agonists) are very potent for inducing cataplexy in the canine model, a sudden decline of catecholamine tonus may be one of the most important mechanisms for triggering cataplexy; however, the precise mechanisms for this are still not understood. We can also assume that the disinhibition of executive components of REM sleep atonia and/or disfacilitation of tonic muscle activity due to the lack of the hypocretin signaling are likely to be involved in the occurrences of cataplexy. The link between this mechanism and the emotional component is still not known.

See also: Acetylcholine; Autonomic Dysregulation During REM Sleep; Narcolepsy; Sleep Apnea.

Further Reading

Honda Y (1998) Clinical features of narcolepsy: Japanese experiences. In: Honda Y and Juji T (eds.) *HLA in Narcolepsy* pp. 24–57. Berlin: Springer.

Lammers GJ, Overeem S, Tijssen MA, and van Dijk JG (2000) Effects of startle and laughter in cataplectic subjects: A neurophysiological study between attacks. *Clinical Neurophysiology* 111: 1276–1281.

Mignot E, Lammers GJ, Ripley B, et al. (2002) The role of cerebrospinal fluid hypocretin measurement in the diagnosis of narcolepsy and other hypersomnias. *Archives of Neurology* 59: 1553–1562.

Nishino S and Mignot E (1997) Pharmacological aspects of human and canine narcolepsy. *Progress in Neurobiology* 52(1): 27–78.

Nishino S, Riehl J, Hong J, Kwan M, Reid M, and Mignot E (2000) Is narcolepsy a REM sleep disorder? Analysis of sleep abnormalities in narcoleptic Dobermans. *Neuroscience Research* 38(4): 437–446.

Nishino S and Sakurai T (eds.) (2005) *The Orexin/Hypocretin System: Its Physiology and Pathophysiology.* Totowa, NJ: Humana Press.

Peyron C, Faraco J, Rogers W, et al. (2000) A mutation in a case of early onset narcolepsy and a generalized absence of hypocretin peptides in human narcoleptic brains. *Nature Medicine* 6: 991–997.

Sakurai T (2005) Roles of orexin/hypocretin in regulation of sleep/wakefulness and energy homeostasis. *Sleep Medicine Reviews* 9(4): 231–241.

Scammell TE (2003) The neurobiology, diagnosis, and treatment of narcolepsy. *Annals of Neurology* 53(2): 154–166.

Relevant Website

http://www-med.stanford.edu – Stanford University School of Medical (animal cataleptic attacks video).

Sleeping Sickness

M Bentivoglio, University of Verona, Verona, Italy
K Kristensson, Karolinska Institutet, Stockholm, Sweden

Introduction

Sleeping sickness or human African trypanosomiasis (HAT) is a re-emergent disease prevalent in sub-Saharan Africa. It is caused by the hemoflagellate protozoan *Trypanosoma brucei* (*T.b.*, also called African trypanosome), and it is transmitted by bites of flies of the genus *Glossina* (tsetse flies).

There are two clinically distinguishable forms of HAT caused by two different subspecies of African trypanosome and localized to separate geographical areas, although they overlap in many foci of central Africa. The Gambian or West African form of the disease occurs west to the Rift Valley and is caused by *T.b. gambiense*, with humans as primary host and main reservoir. The Rhodesian or East African form of sleeping sickness occurs east to the Rift Valley and is caused by *T.b. rhodesiense*. Rhodesian sleeping sickness is a zoonosis, for which cattle or wild game animals are primary hosts. Both forms are lethal if untreated, but their temporal profile differs. The infection caused by *T.b. gambiense* lasts for several months or even years, whereas that caused by *T.b. rhodesiense* can progress more rapidly in humans. This difference in temporal profile may reflect a relative inefficiency of the tsetse flies as vectors, which renders a long-term infection in the primary host necessary to increase the chances for transmission.

A third subspecies of the parasite, *T.b. brucei*, produces a relatively mild form of disease in native game animals but a severe condition in most domesticated animals. *T.b. brucei* is noninfective for humans, who have a serum protein that inhibits this parasite, but produces a lethal disease in rodents and is therefore used in experimental studies.

Although drugs for treatment of both forms of HAT early during the infection are relatively effective, treatment later during infection, when most of the patients may seek medical advice, still widely employs arsenical-based preparations, which have a high degree of toxicity. Better diagnostic tools for surveillance, early detection of HAT in the field, diagnostic tools for the distinction between early and late stages of the infection, as well as less toxic drugs for the cure of brain infection are therefore of high priority. Understanding the mechanisms by which the severe neurological disturbances develop during the disease is essential to prevent, diagnose, and cure neuroinvasion of the parasites. These disturbances are complex but hallmarked by alterations of the sleep pattern. Such alterations and experimental data on their pathogenesis are the focus of this article.

Historical Notes and Current Prevalence

Sizeable epidemics of sleeping sickness in humans were not identified until the turn of the twentieth century, although the disease had been recognized in West Africa for more than 600 years. A plausible explanation for these outbreaks of the disease is the migration of people caused by the European colonization of the continent. By this migration, individuals carrying the parasites were moved from areas with widely spread infections into novel areas. It has been estimated that half a million people died of sleeping sickness along the Congo River at the time of its exploration between 1896 and 1908. The most renowned outbreak of the disease occurred on the northern shore of Lake Victoria between 1898 and 1908. This epidemic may have killed two-thirds of the population in Uganda and the hunt for its causative agent was therefore intensified. In 1902, Robert M. Forde discovered in the blood of a patient with 'Gambian fever' the presence of parasites that were identified as trypanosomes by Joseph E. Dutton. Aldo Castellani found in 1903 the parasites in the cerebrospinal fluid (CSF) of cases of sleeping sickness in Uganda. This finally determined that the etiological agents of the disease were African trypanosomes. David Bruce and his wife Mary discovered in 1895 that tsetse flies were vectors for infections caused by these parasites in horses and cattle (*nagana*, from the Zulu language meaning 'feeble, useless').

In the following decades, the incidence of sleeping sickness gradually decreased and the disease had almost disappeared from several areas by the 1960s. This was due to surveillance and treatment of infected patients, as well as to tsetse fly control programs. Thus, the disease was almost forgotten during the first postcolonial period when other, and more urgent, problems came into focus. Because of the low level of awareness and due to political instability and civil unrest, a recrudescence of the disease occurred during the late twentieth century, most notably in Angola, R-D Congo, Uganda, and southern Sudan. In certain regions of these countries, the incidence of infection reached the level of the early decades of the century. This recrudescence reflects shortcomings in surveillance and treatment, increased migration of human populations due to streams of refugees, as well as disruption of tsetse fly control programs.

Sleeping sickness is certainly the most dreadful of sleep diseases in terms of victims. HAT now occurs in approximately 250 foci in 36 sub-Saharan countries. Approximately 50–60 million people live in endemic areas and are therefore at risk, but only approximately 10% are under surveillance. The estimated number of 70 000 new cases detected and treated per year is therefore underscored. According to the World Health Organization (WHO), the actual annual number of cases may be in the range of 300 000–500 000. The disease was estimated to cause a loss of 2.1 million disability-adjusted life-years in 1999.

The Gambian form of the disease presents the major human health problem, whereas the Rhodesian form only occurs in a minority of patients. During the past few years, the situation has stabilized in areas with less political unrest and improved socioeconomic conditions, although the reemergence of sleeping sickness in many foci in Africa in the past few years is alarming. Cases of infections by *T.b. rhodesiense* and *T.b. gambiense* in travelers have been repeatedly reported in recent years, focusing attention on the increased risk of infection and on the need to consider this diagnostic possibility for imported cases.

With the recent awareness of the situation, eradication of the disease during the coming decades should become a realistic objective. To accomplish this, accurate and cost-effective diagnostic methods have to be developed and need to be implemented in remote and impoverished regions, in addition to re-institution of efficient programs for the vector control.

The Parasite and Human Infections

The Parasite

T.b. gambiense is transmitted mostly by the tsetse fly species *Glossina palpalis* and, as mentioned previously, humans are the major host, although domestic animals such as pigs may also be a reservoir. *T.b. rhodesiense* is transmitted mostly by *Glossina morsitans*, with wild game and domestic livestock as reservoir.

The single-celled *T.b.* parasites belong to the Salivarian group of the genus *Trypanosoma* and they complete their development in the proboscis/salivary glands of the tsetse flies. These flies attack and bite moving animals during the day and have a phototactic behavior, being attracted by blue and black surfaces. The blood-sucking female tsetse flies become infected when encountering a host animal with parasites circulating in the blood. In the tsetse fly, the parasites migrate from the hindgut to the salivary glands, in which they are transformed into metacyclic trypanosomes that are ready to be inoculated into a new mammalian host animal during a blood meal. Approximately 5000 metacyclic trypanosomes are inoculated into the dermal connective tissue of the host with the tsetse fly saliva during a feed. At the site of inoculation, an inflammatory reaction will ensue, sometimes forming a 'chancre' (a skin lesion with edema and also cell infiltration and proliferation). From the inoculation site, trypanosomes enter the lymphatics and bloodstream of the host, where the metacyclic forms are transformed into the slender bloodstream form. The slender parasites divide, resulting in high levels of parasitemia which fluctuate in the host in waves lasting in rodents for approximately 3 or 4 days. Troughs in the parasitemia are caused by the humoral immune response of the host.

The three strains of *T.b.* parasite are morphologically and mostly serologically indistinguishable. They are covered on their surface by a single species of variant surface glycoproteins (VSGs) and express a new species of VSG on their surface after each immune attack. By this mechanism, the parasites evade the humoral immune response and can avoid being eliminated in mammals. The VSG genes occupy 10% of the trypanosome genome and the mechanisms for the process of VSG switching well known. The sequence of the 26-megabase genome of *T.b.* was completed in 2005.

Stages of Sleeping Sickness, Treatment, and Neurological Signs

Sleeping sickness progresses through two clinical stages. The parasites first multiply in the lymph and blood to invade peripheral organs. This first (hemolymphatic) stage is followed by spread to the central nervous system. The second (meningoencephalitic) stage leads to the most severe, and ultimately fatal, consequences of the infection, and it requires treatment with drugs which pass the blood–brain barrier (BBB).

According to current WHO criteria, diagnosis of the second stage relies on the occurrence of trypanosomes, more than five white blood cells, and/or increased levels of protein in the CSF. However, these criteria are debated and need to be better validated considering the important therapeutic implications.

Early during infection, pentamidine is the drug of choice for clearance of *T.b. gambiense*, whereas suramin is used against *T.b. rhodesiense*. Both drugs are given parenterally, but formulas for oral drug delivery are currently being sought. Treatment of the second stage with an arsenical derivative (melarsoprol, the trypanocide introduced in 1949), which crosses the BBB, kills the parasites but in 5–10% of cases causes a reactive arsenical encephalopathy which can be fatal in approximately half of the patients. Another problem is that melarsoprol resistance is increasingly reported, in certain areas in as many as 30% of patients. The relatively new drug difluoromethylornithine (DFMO) is used as an alternative against the West African form of HAT but requires long-term intravenous injections at high doses. Because of this, and also due to side effects and insufficient availability, the utility of DFMO in HAT treatment is limited.

The neurological symptoms and signs are most prominent in the Gambian form of the disease due to its more protracted course. There is considerable variation between afflicted individuals in the neurological picture of the disease, which involves a constellation of sensory, motor, and mental disturbances. Increase in muscle tone and abnormal movements, which include trembling of the hands and fingers, as well as choreiform and athetoid movements, are common. There are no objective sensory defects, but patients usually develop a peculiar pain syndrome in which a slight touch is sufficient to provoke an intense pain episode (Kerandel's sign).

States of mental confusion are common and patients may develop psychiatric disturbances, including emotional instability, indifference, aggression, asocial and stereotypic behavior, impulsive actions, fugue states, manic episodes, melancholia, and delirium. Dementia is common at the end stage, and patients will finally become comatose and die in a state of wasting. The most characteristic sign of sleeping sickness, however, is represented by sleep alterations.

Sleep and Circadian Rhythmicity Disturbances

Interestingly, sleeping sickness is not a hypersomnia, as its name may suggest. The name of the disease derives from the observation that patients develop frequent and sudden sleep episodes during the day. On the other hand, patients have wakefulness episodes during the night. Polysomnographic recordings in HAT patients have demonstrated that the total sleep time and total amount of each sleep stage during 24 h are not increased. Instead, the main alternations during sleeping sickness involve the sleep–wake cycle and the sleep pattern. The sleep–wake alteration tends to occur in short cycles, with sleeping and waking occurring equally during the day and night (**Figure 1**). The internal structure of sleep is altered, with frequent onset of rapid eye movement (REM) sleep periods and REM sleep episodes occurring soon after long wakefulness transitions with latency shorter than 15 min. Such sleep onset of REM sleep

(SOREM) periods without any intermediary non-REM sleep, and with the occurrence of SOREM episodes during wakefulness, are considered characteristic of the disease. However, such alterations have also been documented in narcolepsy. In addition, abnormal electroencephalographic (EEG) events have been recorded in sleeping sickness patients, with slowing of EEG or periodic slow waves during wakefulness and sleep stages.

It has been reported that the sleep–wake disturbances in sleeping sickness are related to the severity of the disease, and that the occurrence of sleep alterations can be used as diagnostic criterion for the second stage. However, it has not been determined experimentally whether the onset of sleep changes corresponds to the brain parenchyma invasion by the parasites or stage I drug treatment failure.

The alterations in sleep and sleep–wake cycle generally ameliorate with arsenical drug treatment, but the recovery of these parameters in surviving patients is extremely slow and can occur months after the interruption of treatment. This indicates that sleep-regulatory structures are not destroyed by neurodegenerative phenomena during the disease, as also assessed experimentally, and that sleep changes are not directly related to the presence of the parasites in the brain parenchyma. Further potential long-term consequences have not been thoroughly investigated.

Loss in the circadian rhythmicity of cortisol and prolactin secretion, as well as loss of growth hormone release, has been recorded. The relationship between plasma hormone pulses and sleep–wake stages is preserved, however. Melatonin secretion maintains its circadian rhythmicity but exhibits a phase advance. Together, these findings, in the presence of a severe alteration of the sleep–wake cycle, point to the complexity of the interrelationships between different circadian rhythms.

Neuropathological Findings

The pathological changes in both forms of sleeping sickness are similar, and the Rhodesian form may be regarded as a compressed Gambian form. The neuropathological picture of the disease is relatively mild compared to its clinical severity. The alterations are mainly inflammatory, without marked neurodegenerative changes, and they are diffuse. The inflammatory changes are predominantly seen in the white matter of the cerebral hemispheres (leucoencephalitis), but areas of demyelination, when present, are confined to perivascular sites. Perivascular infiltration of monocytic inflammatory cells also occurs at several sites, including basal ganglia, brain stem, and cerebellar white matter. There is also an infiltration in the brain of peculiar cells, the so-called Mott cells,

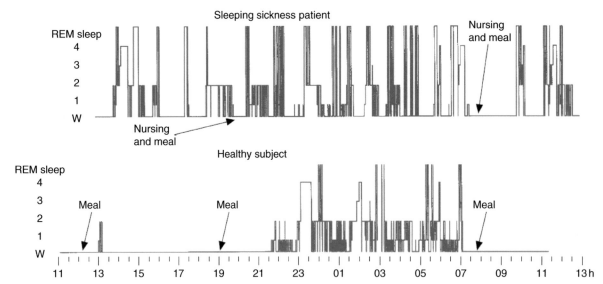

Figure 1 Hypnograms showing the 24 h distribution of wakefulness (W), rapid eye movement (REM) sleep, and stages 1–4 of non-REM sleep in an African patient affected by sleeping sickness (top) and in a healthy African subject (bottom). Hypnograms were recorded in both subjects at bed rest and with an intravenous catheter (due to which the sleep of the healthy subject had many interruptions). Note the disruption of the circadian distribution of sleep and wakefulness in the patient, with frequent occurrence of sleep-onset REM sleep episodes. Reproduced from Buguet A, Bourdon L, Bouteille B, et al. (2001) The duality of sleeping sickness: Focusing on sleep. *Sleep Medicine Reviews* 5: 139–153, with permission.

which are morular shaped and contain IgM; they are most numerous in long-standing cases. Marked activation of microglial cells and widespread astrocytosis in the brain of human victims have been reported in early descriptions of sleeping sickness.

In most of these studies, trypanosomes were not detected in the brain parenchyma. This may reflect a high degree of sensitivity of the parasites to post-mortem autolysis, or it could be due to arsenical drug treatment that may kill the parasites in the brain before patients die.

Experimental Studies and Hypotheses on Pathogenesis of Sleep Disruption during the Disease

The pathogenesis of sleeping sickness is complex and poorly understood. Because the causative agent is an extracellular parasite, neuronal dysregulation causing neurological symptoms should be due to factors released by the parasite, its interactions with the immune system of the host, or both.

In view of laboratory safety conditions, as mentioned previously, most experimental studies have employed *T.b. brucei*.

Neuroinflammation in *Trypanosoma brucei*-Infected Rodents

Experimental studies have revealed a temporal profile of *T.b.* neuroinvasion, which seems to occur stepwise.

Early after infection in experimental rat and mouse models, *T.b. brucei* localizes to the choroid plexus and circumventricular organs, which lack a BBB. Later during infection, the parasites can penetrate into the brain parenchyma across the BBB. However, the parasites can cross the BBB with preserved endothelial cell tight junction proteins. This finding supports the clinical observation that drugs that cannot penetrate the BBB are inefficient in the treatment of HAT late during infection. As described later, a major unsolved pathogenetic problem is the identification of factors which, in the parasite and/or the host or their interaction, determine *T.b.* passage through the BBB.

The pathological picture in experimental infections in rodents is neuroinflammatory, as in humans, with only minor signs of neurodegeneration, and the reaction is predominant in the white matter, into which the parasites penetrate. There is a marked activation of astrocytes and microglia, which shows a predilection for certain areas including the hypothalamus, as well as production of cytokines in the brain.

The protagonists of the immune response to the infection (in experimental animals, as well as in humans) are proinflammatory cytokines, including interleukin (IL)-1, tumor necrosis factor (TNF)-α, and interfron (IFN)-γ. TNF-α was isolated in 1985 from *T.b.*-infected rabbits in the search for the cause of cachexia in trypanosomiasis, and it was initially named 'cachectin.' In the brain, cytokines can have an endogenous origin, being released by activated glial cells. The action of proinflammatory molecules, however, can be counteracted by the release of anti-inflammatory cytokines such as IL-6 and IL-10 and also other extra- and intracellular factors that control inflammation. Free radicals, including nitric oxide, and prostaglandins are also induced in the brain during the infection. In particular, high levels of prostaglandin D2 have been reported in late-stage patients. A prostaglandin released by *T.b.* has also been identified, but its eventual role in sleep changes during the disease has not been examined.

Alterations of Sleep and Endogenous Rhythms in Rodent Models

In rats infected with *T.b. brucei*, the sleep pattern has been found to be markedly fragmented, with major dysregulation of the REM/non-REM sequence of sleep episodes on EEG recordings. Frequent awakenings during slow-wave sleep (SWS) episodes, a reduction of the average length of SWS episodes, and a reduced REM sleep latency have been observed. The sleep structure is marked by abnormal transitions and especially the appearance of SOREM-like episodes. However, the duration of REM sleep episodes is unaffected.

The experimental infections in rodents do not reproduce the full spectrum of the human disease, which mostly develops over a long period of time, whereas the rodent infections allow only a limited survival of a few weeks. However, 24 h EEG recordings have revealed that, in addition to sleep fragmentation, disruption of the sleep–wake cycle can also occur in late stages of the experimental infection in rats.

It should be emphasized that, as in humans, despite the pronounced secretion of somnogenic and pyrogenic cytokines (i.e., IL-1 and TNF-α), neither hypersomnia nor fever are characteristic of the infection in experimental animals. The reason for this paradox is unknown.

Experimental data also point to a disturbance in the melatonin-generating system in the pathogenesis of the disease, with decreased amplitude of the nocturnal peak of melatonin secretion in the urine of *T.b.*-infected rats. The EEG alterations are also ameliorated by treating infected animals with melatonin. However, the locomotor activity and body core temperature rhythms are preserved in *T.b.*-infected rats, but the amplitude of locomotor activity is markedly reduced even early during infection, and a phase advance in the temperature cycles has been noted.

Involvement of Neural Centers That Regulate Circadian Rhythms and Sleep

The sleep disturbances during *T.b.* infection in rodents are in part reminiscent of those observed after lesions of the hypothalamic suprachiasmatic nucleus (SCN), which is the master circadian pacemaker in the mammalian brain. Therefore, functional parameters of this nucleus have been investigated during experimental infections.

The SCN is entrained by environmental light through the retinohypothalamic tract, which utilizes glutamate as a neurotransmitter. The circadian timing system can interact with diencephalic sleep-regulatory centers, whereby circadian information can be transmitted to arousal- and sleep-promoting centers. In the posterior hypothalamus, the histamine-containing tuberomammillary nucleus and the dorsolateral group of neurons containing the peptide orexin/hypocretin (**Figure 2**) have been identified as key relay centers in the regulation of wakefulness. The ventrolateral preoptic nucleus of the anterior hypothalamus (**Figure 2**) has been identified as a sleep-promoting center. These regions have been proposed to be involved in a bistable 'flip-flop' model by which sleep and wakefulness can be switched by reciprocal inhibitory circuits. A decreased resistance in this circuit would weaken the stability of the switch and could therefore result in disruption of the sleep–wakefulness pattern. In particular, the orexin/hypocretin-containing neurons have been suggested to stabilize the switch, causing arousal through an excitatory effect on tuberomammillary neurons directly and by disinhibition.

Experimental data in rodents suggest an involvement of sleep–wakefulness-regulating hypothalamic cell groups during trypanosome infections (**Figure 2**). Marked impairment of the SCN response to a photic stimulus, in the absence of structural changes of retinohypothalamic fibers, was documented in infected rats, and this was accompanied by altered expression of α-amino-3-hydroxy-5-methyl-4-isoxazole propionic acid (AMPA) glutamate receptor subunits. The amplitude of spontaneous firing in the SCN was also found to be severely altered in *T.b.*-infected rats. Alterations in the synaptic machinery of SCN neurons, particularly disturbances in excitatory neurotransmission, were found in slice preparations of the SCN from infected rats, and these were also caused by treatment with IFN-γ in combination with the endotoxin lipopolysaccharide and TNF-α. Data also indicate a dysregulation of orexin/hypocretin-containing neurons in infected rodents. Furthermore, effects of trypanosome infections on the brain stem serotoninergic raphe nuclei (**Figure 2**), which is also a major sleep-modulatory center, have been suggested.

Together, the findings indicate that due to targeting of trypanosomes in the brain to circumventricular organs close to diencephalic centers and prolonged release of proinflammatory molecules during the infection, trypanosomes could interfere with the function

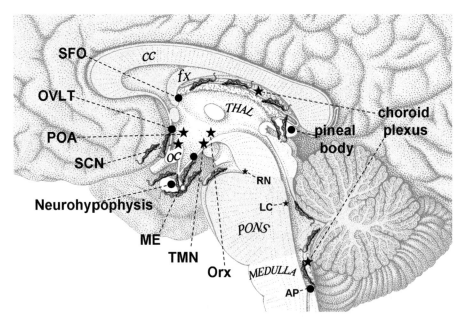

Figure 2 This drawing depicts on the medial surface of the human brain the localization of African trypanosomes at early stages of central nervous system infection, as derived from experimental studies on rodent models. Sitting in circumventricular organs and in the choroid plexus before penetration into the brain parenchyma, the parasites release factors and elicit response of the immune system of the host, with release of inflammatory molecules, including proinflammatory cytokines. Long-term exposure to these molecules could affect neural centers which regulate the sleep–wake alternation and organization. Over time, this could cause dysfunction of circadian timing and sleep-regulatory neuronal assemblies and/or their coupling. The solid circles indicate circumventricular organs; the stars indicate the choroid plexus in the lateral and fourth ventricles and also circadian timing and sleep-regulatory neural centers. AP, area postrema; cc, corpus callosum; fx, fornix; LC, locus coeruleus; ME, median eminence; oc, optic chiasm; Orx, orexin/hypocretin-containing cell groups; OVLT, vascular organ of lamina terminalis; POA, preoptic area; RN, raphe nuclei; SCN, suprachiasmatic nucleus; SFO, subfornical organ; THAL, thalamus; TMN, tuberomammillary nucleus.

of the master circadian pacemaker and sleep-regulatory centers and/or with the coupling between these structures. This neural functional derangement could be built up over time in the molecular interaction between the parasite and the host and/or reach a threshold revealed by the onset of sleep changes in the host.

The Blood–Brain Barrier and Implications for Treatment

As indicated previously, a major problem with regard to therapy of the encephalitis that evolves during sleeping sickness is that the only available drugs which penetrate the BBB are very toxic. Important questions to address are how the parasites cross the BBB, whether there are clinical correlates or markers for parasite penetration into the brain parenchyma, and whether the parasites can hide behind the BBB from the immune response or whether a continuous influx of parasites into the brain is needed to maintain the cerebral infection.

In experimental rodent models, trypanosomes do not move passively into the parenchyma through a disrupted BBB but, rather, follow processes similar to the infiltration of white blood cells (WBCs) into the brain during inflammation. Infiltration of WBCs into the brain can be modulated by cytokines, which

also influence *T.b. brucei* neuroinvasion. Thus, the migration of trypanosomes from the lumen of cerebral blood vessels into the brain parenchyma is impeded in mutant mice deficient in IFN-γ, IFN-γ receptor, IL-12p40, and recombinant activating gene-1, despite higher levels of parasitemia in the mutant strains compared to wild-type mice. In the mutant mice, there is a striking accumulation of parasites in the perivascular compartment, confined between the endothelial and the parenchymal basement membranes. These data also indicate that a host immune response molecule, IFN-γ, paradoxically may facilitate penetration of trypanosomes across cerebral blood vessels.

That mechanisms for penetration of parasites into the brain parenchyma are related to those of WBCs is further supported by studies on the effect of the drug minocycline on parasite neuroinvasion. This tetracycline antibiotic impedes the penetration of WBCs into the brain in experimental allergic encephalomyelitis models and, similarly, it impedes neuroinvasion of trypanosomes in *T.b. brucei*-infected mice. Weight loss occurring during infection with *T.b. brucei* is not observed after minocycline treatment of the mice, which also survive longer than nontreated mice.

Due to the treatment hazards of the disease at a late stage, there is an urgent need for better staging criteria of the infection. For instance, since experimental animals develop sleep pattern changes, a correlation between changes in sleep parameters and passage of trypanosomes across the BBB should be sought. Until there are novel drug discoveries for late-stage disease, early diagnosis and treatment are mandatory to control the infection.

See also: Cataplexy; Narcolepsy; Sleep Apnea; Sleep: Development and Circadian Control.

Further Reading

Bentivoglio M and Kristensson K (2007) Neural-immune interactions in disorders of sleep-wakefulness organization. *Trends in Neuroscienses* 30: 645–652.

Buguet A, Bert J, Tapie P, et al. (1993) Sleep–wake cycle in human African trypanosomiasis. *Journal of Clinical Neurophysiology* 10: 190–196.

Buguet A, Bourdon L, Bouteille B, et al. (2001) The duality of sleeping sickness: Focusing on sleep. *Sleep Medicine Reviews* 5: 139–153.

Dumas M and Bisser S (1999) Clinical aspects of human African trypanosomiasis. In: Dumas M, Bouteille B, and Buguet A (eds.) *Progress in Human African Trypanosomiasis, Sleeping Sickness*, pp. 215–233. Paris: Springer.

Kennedy PGE (2006) Human African trypanosomiasis – neurological aspects. *Journal of Neurology* 253: 411–416.

Kristensson K and Bentivoglio M (1999) Pathology of African trypanosomiasis. In: Dumas M, Bouteille B, and Buguet A (eds.) *Progress in Human African trypanosomiasis, Sleeping Sickness*, pp. 157–181. Paris: Springer.

Lundkvist GB, Kristensson K, and Bentivoglio M (2004) Why trypanosomes cause sleeping sickness. *Physiology* 19: 198–206.

Masocha W, Robertson B, Rottenberg ME, Mhlanga JDM, Sorokin L, and Kristensson K (2004) Cerebral vessel laminins and IFN-γ define *Trypanosoma brucei brucei* penetration of the blood–brain barrier. *Journal of Clinical Investigation* 114: 689–694.

Mhlanga JDM (1996) Sleeping sickness: Perspectives in African trypanosomiasis. *Science Progress* 79: 183–214.

Wang CC (1995) Molecular mechanisms and therapeutic approaches to the treatment of African trypanosomiasis. *Annual Review of Pharmacology & Toxicology* 35: 93–127.

Welburn SC, Coleman PG, Maudlin I, Fèvre EM, Odiit M, and Eisler MC (2006) Crisis, what crisis? Control of Rhodesian sleeping sickness. *Trends in Parasitology* 22: 123–128.

DREAMING

Theories of Dream Function

D Kuiken, University of Alberta, Edmonton, AB, Canada

Introduction

Theories of dream function posit a correspondence between the effects of dreaming and what enables the dreamer, in some sense, to thrive. Perhaps the most common proposal is that dreams metaphorically represent disturbing life events and allow the covert review of self-protective or self-restorative responses. Another long-standing hypothesis is that dreaming distinctively reorganizes memories for newly learned tasks and enhances the retrieval of related knowledge and skills. And, still another conception is that dreaming directly alters attention, evokes feeling, and enhances the flexibility and emotional relevance of postdream thought. As is evident from these broad themes, models of dream function vary considerably in their scope and in the nature of the evidence required to substantiate them.

Contrasting Conceptions of 'Function'

Theories of dream function differ markedly in their articulation of the temporal locus of what is beneficial about dreaming. Researchers have sometimes taken a broad perspective, hypothesizing that dreaming facilitates long-term adaptation (e.g., species survival). In contrast, some investigators have identified immediate and transient dream effects (e.g., postawakening attentional flexibility). Also, theories of dream function are usually presented as though all dreams fulfill a common function. Perhaps dreaming is a sufficiently uniform phenomenon to serve consistently some function (or functions), but, by abandoning strict assumptions about highly general dream effects, the contrasting effects of different dream types – for example, nightmares versus night terrors, or rapid eye movement (REM) dreams versus non-REM (NREM) dreams – have also been fruitfully considered.

Contrasting conceptions of the temporal locus and uniformity of dream effects became important when evaluating the efficacy of REM sleep research protocols during the past half century. Although researchers originally posited an isomorphism between REM sleep and dreaming, it is now widely acknowledged that dreaming in some form occurs elsewhere in the sleep cycle, most notably during sleep onset and stage 2 NREM sleep. The implications of these observations remain controversial. Even though dreaming occurs during NREM sleep, it is still unclear whether dreams reported upon awakening from REM sleep are qualitatively or merely quantitatively different from those reported upon awakening from NREM sleep. If REM dreams are qualitatively different from NREM dreams, comparison of the consequences of REM and NREM sleep deprivation may help to establish the contrasting functions of REM and NREM dreams. On the other hand, if NREM dreams are only somewhat more intense but of the same basic type as REM dreams, the psychological functions of 'typical' dreaming can only be weakly clarified by comparing the effects of REM and NREM sleep deprivation.

Moreover, even if the differences between REM and NREM dreams were clearly established, and REM sleep physiology could confidently be regarded as a useful index of REM dreaming, the use of REM sleep research protocols in studies of dream function would leave unresolved an even more basic issue: the psychological function of REM dreaming is not reducible to the physiological function of REM sleep. Thus, even if REM sleep were reliably associated with a distinct form of dreaming, the challenge of differentiating the physiological effects of REM sleep from the psychological effects of REM dreaming would remain. This issue is far from resolved and contributes to very different perspectives on the significance of REM sleep research protocols for studies of dream function.

Dream Carryover Effects

Awakenings from REM and NREM sleep are followed by a brief period (≈20 min) during which thoughts and feelings are systematically altered. Observations during this period are often thought to reflect the nature of mental activity during the sleep stage that immediately precedes awakening. However, these observations are also appropriately considered as the direct influence of the preceding sleep stage on subsequent waking experience.

Carryover Effects of REM Dreaming

In cats, it has been observed that (1) high-amplitude pontogeniculooccipital (PGO) waveforms accompany the orienting response during waking, (2) comparable high-amplitude PGO waves occur spontaneously during REM sleep, and (3) novel auditory stimuli that elicit the orienting response during waking reliably induce PGO waves in both REM and NREM sleep. Such observations suggest that REM sleep is characterized by persistent and spontaneous activation of the orienting response. Correspondingly, in humans, there is evidence that an automatic, stimulus-induced shift of attention toward unexpected stimuli occurs more quickly during the period immediately following awakening from REM sleep than it does following awakenings from NREM sleep.

Fluidity of thought   Such attentional flexibility helps to explain dreamers' post-REM readiness to entertain infrequently considered possibilities. For example, individuals awakened from REM sleep show greater priming by weakly associated words (e.g., thief–wrong) than by strongly associated words (e.g., hot–cold), whereas individuals awakened from NREM sleep show greater priming by strongly associated than by weakly associated words. Also, individuals awakened from REM sleep use strategies that differ from those used during wakefulness to solve simple problems (e.g., anagrams), whereas, after awakening from NREM sleep, strategies for solving these problems more nearly resemble those used during wakefulness. Finally, after REM deprivation, individuals awakened from REM sleep score higher on fluidity and flexibility in traditional tests of divergent thinking than do those deprived of NREM sleep. In sum, numerous findings support the contention that awakenings from REM sleep are more likely than awakenings from NREM sleep to be followed by a brief period of fluid thinking. Whether this fluidity reflects the lingering effects of REM dreaming, rather than the lingering effects of REM physiology, remains, as indicated earlier, unresolved.

Felt presence   The fluid thinking that follows REM dreaming acquires additional significance when it is considered that carryover effects also involve the readiness to see motion where it objectively is not. Individuals awakened from REM sleep are more likely than those awakened from NREM sleep to (1) perceive one dot moving between two positions when two stationary dots are alternately presented (the beta phenomenon) and (2) perceive movement in a static figure after a period of concentration on a rotating spiral (the spiral aftereffect). This 'feeling' of motion is complemented by enhanced 'feeling for' objects. After awakenings from REM sleep, performance on a left-handed tactile-kinesthetic recognition task is better than after NREM awakenings (even though analogous differences are not evident with the enumeration of dot patterns). Together these results suggest that the felt presence of objects – rather than simply their visual appearance – is intensified during the period that immediately follows REM dreaming.

Felt position   There is also evidence that REM dreaming accentuates the felt sense of the dreamer's position

in relation to the external world. It has been proposed that, during REM sleep, the initiation of bodily movement without corollary proprioceptive feedback (a pattern associated with intensification of the orienting response) precipitates perceived adjustments in bodily position. This possibility is consistent with evidence that REM deprivation increases REM density, the associated sense of self-participation (i.e., active participation in the dream scenario), and also the perception of human movement in ambiguous figures after awakening.

Emotional feeling   The carryover of emotional dream feelings might seem to follow directly from the preceding findings. However, emotional feeling is attenuated immediately following awakening from a typical REM sleep period, either because of the lingering atonia of that sleep stage or, more likely, because of the mood-regulating function of REM dreams. Regardless of its origin, imaginative expression (e.g., creating stories in response to ambiguous stimuli) immediately after REM awakenings seems to evoke emotional feeling more than does straightforward dream reflection.

Auxiliary effects   Carryover effects of typical REM dreaming are subtle and transient. Nonetheless, post-REM fluidity of thought is sometimes not only greater than after NREM sleep but is also greater than during ordinary wakefulness. Correspondingly, there is evidence that reflection on remembered REM dreams is more likely than reflection on remembered NREM dreams to facilitate feeling expression during insight-oriented therapy. Although the carryover effects of typical REM dreaming may exert their adaptive influence especially through timely and vivid recollection, this auxiliary function of dreaming has considerable practical import.

Carryover Effects of Intensified Dreaming

Sometimes dream carryover effects are intensified during the transition from REM sleep to wakefulness and during the period immediately following REM dreaming. These extreme forms of dreaming sometimes have lasting significance.

Sleep paralysis episodes   In sleep paralysis episodes, the general atonia of REM sleep persists despite the capacity for external perception and the presence of consciousness. In this liminal state, the accompanying hallucinatory experiences are of three types. One type involves the felt presence of a spatially distant but threatening figure (accompanied by assorted noises, humanoid apparitions, and sensations of being touched), a second type involves the felt presence of a spatially near and disturbing figure (accompanied by breathing difficulties, feelings of suffocation, and thoughts of impending death), and a third type involves sensations of linear and angular acceleration (e.g., flying, floating), fictive motor movements (ranging from simple arm movements to apparent locomotion), and feelings of bliss. Consistent with laboratory research on REM sleep carryover effects, these three types of transitional dream experience involve the distinctly felt but nonvisual presence of an 'other' and a keenly felt but anomalous sense of 'self.'

The effects of sleep paralysis episodes on waking thoughts and feelings are often direct and persistent. The lingering 'reality' of these dream experiences, traceable to the compellingly felt presence of external objects and the experience of an unembodied but uniquely positioned self, contribute to beliefs about spirit figures and self-transcendence. Whether these effects are lasting and whether they are beneficial depend upon the cultural context – even when these dreams challenge, rather than merely reflect, the surrounding cultural milieu.

Impactful dreams   Significant features of sleep paralysis episodes correspond to empirically established types of impactful dreams. Impactful dreams are defined as those that directly influence postdream thoughts and feelings, most likely through the intensification of REM carryover effects. One type, the nightmare, with features such as intense fear, the avoidance of threatening others, and physical metamorphoses, is followed by externally directed vigilance and apprehension. A second type, the existential dream, with features such as intense sadness, rejection by or separation from significant others, and spontaneous feeling change, is followed by sadness, autobiographical review, and changed self-perception. A third type, the transcendent dream, with features such as surprise, magical accomplishment, and shifts in visual–spatial orientation (floating, flying), is followed by astonishment and reported spiritual change. In forms that are analogous to sleep paralysis episodes, nightmares involve the threatening presence of distant external objects, existential dreams involve the disturbing presence of near external objects, and transcendent dreams involve an anomalously positioned self.

The effects of impactful dreams also seem direct. Dreamers are often reluctant to have them interpreted, and there is evidence that, although they may alter the way dreamers understand their daily lives, they do not provide optimal material for therapeutic dream reflection. However, the auxiliary processes by which such intense dream experiences do exert influence on the dreamer's personal, social, or cultural life have not been systematically investigated.

Memory Consolidation

The long-standing hypothesis that memories are reorganized and stabilized during sleep became more focused during the 1970s when researchers began to examine specifically whether REM sleep was the psychobiological state within which memory consolidation occurs. At that time, the apparent isomorphism between REM sleep and dreaming suggested that the processes by which dreams are formed are also the processes by which memories are consolidated. So, REM deprivation became a common means by which memory consolidation was hypothetically disrupted, and assessing the effects of prior learning on REM sleep parameters was often used to assess increased pressure for REM-dependent consolidation. Although research initially supported the memory consolidation hypothesis among animals but not humans, in recent years, studies of human memory during sleep support the role of REM sleep, and by implication REM dreaming, in memory consolidation among humans.

Memory Consolidation and REM Sleep

A turning point in this research occurred with the realization that REM sleep may play a particularly important role in the consolidation of procedural, as opposed to declarative, memory. Review of the literature indicated that, when a task requires the development of perceptual-procedural, verbal-procedural, or cognitive-procedural skills, REM sleep facilitates learning; when the task requires the retrieval or recognition of generic semantic information, REM sleep does not facilitate learning.

Effects of learning on REM sleep   In one commonly used research paradigm, a learning task is presented prior to sleep and then REM sleep parameters (e.g., REM density, percentage of sleep in stage REM) are assessed. For example, when people are required to adapt to lenses that distort their visual field, there is a subsequent increase in the percentage of stage REM sleep. Comparable effects have been found after learning verbal-procedural skills (e.g., second language learning) and cognitive-procedural skills (e.g., solving a conceptual problem that involves recursive steps).

Effects of REM deprivation on consolidation   In the frequently used REM deprivation paradigm, people are awakened whenever they enter REM sleep, or, in comparison conditions, whenever they enter NREM sleep. In general, as hypothesized, when people are deprived of REM sleep after learning perceptual-procedural skills (e.g., a complex visual discrimination task) or cognitive-procedural skills (e.g., the recursive cognitive problem), their performance during retesting is impaired. These findings have sometimes been challenged because deprivation procedures are often confounded with the stress precipitated by repeated awakenings from REM sleep. However, some studies of REM deprivation have demonstrated that performance disruption is still evident several days after the termination of REM deprivation and, hence, after recovery from the related stress or fatigue. Also, REM deprivation during specific temporal windows, sometimes days after the learning task, sometimes disrupts performance even more than does REM deprivation immediately after learning, precluding the possibility that stress associated with REM deprivation accounts for impaired performance.

Convergent findings from other paradigms Other research methods have provided converging evidence that REM sleep facilitates perceptual-procedural and cognitive-procedural memory. Research using functional neuroimaging procedures indicates that specific brain areas activated while learning a serial reaction time task are significantly more active during REM sleep among those who have previously trained on the task than among those who have not. Also, when people who have learned a recursive cognitive skill in the presence of a clicking sound are presented clicks coincident with their rapid eye movements during REM sleep, they perform better on retesting than do people who have learned the skill and then are presented clicks during a period of REM sleep without rapid eye movements or during a period of NREM sleep. These findings suggest a special role for the phasic events of REM sleep in memory consolidation, a possibility bolstered by evidence that hippocampal theta rhythms, which have been implicated in memory consolidation during wakefulness, are not only prominent during REM sleep but also are precipitated by the PGO waveforms that generate the REMs, twitches, and other phasic features of that sleep stage.

However, because hippocampal theta rhythms are especially implicated in the consolidation of declarative rather than procedural memory, the association between hippocampal theta and REM sleep is inconsistent with the notion that REM-mediated memory consolidation is specific to procedural learning. Moreover, there is evidence from a partial sleep paradigm that declarative memory involving emotional information is enhanced by REM sleep. Specifically, when memory is assessed after periods of undisturbed sleep during either the first half of the night (when NREM sleep predominates) or the second half of the night (when REM sleep predominates), the consolidation of emotional declarative memories (e.g., for emotional stories), but not of neutral declarative

memories (e.g., for nonemotional stories), is enhanced by late-night (REM-dominated) sleep. These findings suggest that declarative memory for emotional events, although dependent upon hippocampal theta activity, is enhanced during REM sleep by amygdalar modulation of hippocampal function. Since, within any particular learning task, declarative and procedural learning are often mixed, there clearly is need for more refined characterization of the particular kinds of memories that are consolidated during REM sleep.

Memory Consolidation and NREM Sleep

Although there is evidence that sleep stages other than stage REM are implicated in memory consolidation, identification of their distinctive contributions has only begun. There are replicable findings indicating that NREM sleep facilitates performance on motor tasks (e.g., a pursuit rotor task). And, there is evidence that a word-pair association task influences some parameters of early-night (NREM-dominated) sleep (e.g., the number of sleep spindles) and that the disruption of early-night (NREM-dominated) sleep interferes with this type of declarative memory. This might be expected from dual-process theories, which suggest that REM and NREM sleep have specialized memory consolidation functions. However, other evidence (e.g., correlations between retesting performance and time spent in both NREM and REM sleep) suggest that NREM and REM sleep independently contribute to memory consolidation. According to sequential process theories, the orderly succession of NREM and REM sleep is required for memory consolidation, regardless of the nature of the task.

Dream Content and Memory Consolidation

On a few occasions, researchers have directly linked REM-dependent memory consolidation to dream content. For example, individuals who have successfully adapted to lenses that distort their visual field report dreams including visual and motor difficulties, as well as misfortune and confusion. Similarly, among people learning a second language, those performing successfully are among the first to incorporate that second language into their own dream communications. However, there is ample room in this literature for closer examination of the relations between the incorporation of the presleep task (into both REM and NREM dreams) and subsequent task performance.

Such research must take into account the extended temporal reach of dream formation processes. The incorporation of challenging life events into dreams appears to be modulated by a 7-day rhythm. One peak in incorporation occurs for dreams following the event by 1 day (the day-residue effect), and a second peak occurs 6–7 days later (the dream-lag effect). Dream-lag incorporations (1) seem related to interpersonal problem-solving (interpersonal relationships, positive emotions, and resolved personal problems), (2) are most pronounced among individuals who report that their dreams influence waking thoughts and feelings, and (3) are at least compatible with evidence that REM deprivation during specific, delayed temporal windows disrupts REM-dependent memory consolidation. Documenting the complex processes by which presleep tasks and challenging events are incorporated into dreaming may help clarify the forms of memory reorganization and elaboration that support memory consolidation.

Emotional Adaptation

Studies of dream function often are couched in the language of problem solving. Usually, the proposal is that dreams incorporate aspects of a waking problem, the process of dream formation actively reorganizes material related to that problem, and the dream effect is articulation of a novel solution to the problem, either within the dream or during immediate waking reflection upon the dream images. Whereas research supports the notion that REM dreaming, as a form of fluid thinking that briefly persists after awakening, contributes to the formulation of solutions that are not characteristic of waking thought, generalization from transient nonemotional problems (e.g., anagrams) to enduring emotional problems (e.g., bereavement) is common but problematic. Research procedures suited to the investigation of short-term dream effects cannot simply be extended to studies of the influence of dreaming on long-term emotional well-being. Longitudinal (or quasi-longitudinal) research designs are required.

Within longitudinal designs, the contribution of REM sleep research protocols continues to be significant for two reasons. First, laboratory studies of the immediate effects of REM dreaming often guide articulation of the hypotheses examined in longitudinal studies. Second, sleep laboratory procedures remain crucial when the research question concerns changes in dreaming during the course of the night, contrasts between REM dreams and NREM dreams, etc.

Mood Regulation

The hypothesis that REM sleep has a mood regulatory function is consistent with evidence of an overnight attenuation of either positive or negative mood intensity. An extension of this hypothesis is that nightmares reflect the failure of this regulatory mechanism to subdue disturbing emotions, a proposal that is reminiscent of Freud's suggestion that the function of dreaming is to protect sleep. A recent and modified version of the mood-regulation hypothesis has influenced studies of the role of REM dreaming in the alleviation of depression.

REM sleep and mood regulation during depression There is evidence that early-night and late-night REM sleep stages play different roles in mood regulation during depression. It is well established that depressed individuals show reduced latency to the first REM sleep episode of the night and an increase in REM density during the first REM sleep period, indicating a phase advance of the diurnal cycle. Also, during depression, REM density during the first REM period is negatively correlated with the attenuation of negative mood during the night, whereas REM density during the last REM period is positively correlated with the attenuation of negative mood. Finally, depressed individuals reporting more negative dreams at the beginning of sleep, and fewer negative dreams toward the end of sleep, are more likely to be in remission 1 year later than are those with fewer negative dreams at the beginning of the night and more negative dreams toward the end of the night. Taken together, these findings indicate that, among depressed individuals, the progressive attenuation of negative affect during successive REM periods and the associated attenuation of negative mood in the morning alleviate depression. While it is not clear what normally brings this about, REM deprivation is one intervention that seems to do so.

REM deprivation and depression When acutely depressed individuals (e.g., responding to divorce) undergo selective REM deprivation over several weeks, the phase advance of the diurnal cycle is reversed, leading to sustained remission from depressive symptoms. Preliminary evidence that dreaming is implicated in these improvements includes (1) the attenuation of negative mood on REM deprivation nights, (2) the maintenance of this attenuation during the following day, (3) the vividly dreamlike character of REM mentation early in the treatment period, and (4) dream reports that incorporate emotionally significant aspects of the life event that precipitated the depressive episode (e.g., the former spouse; the children). The preceding pattern is best described as emergent; its various components remain somewhat unstable from study to study. However, the prevalence of depression (including as a complement to other distressing life events, e.g., trauma) makes clarification of this pattern especially important.

This research is a reminder of the procedural complexity that is required in studies of long-term dream function. Among those who are depressed, it is demonstrably important to (1) differentiate the types of dreaming that occur early in the night from those that occur late in the night, (2) examine the blending of and variations in dream themes between early-night and late-night REM periods, (3) monitor not only the immediate mood carryover effects, but also the waking events that support or obstruct stabilization of these mood changes during wakefulness, and (4) assess individual differences in the patterns of dreaming and dream effects.

Conflict Resolution

In recent years a number of investigators have focused on repeated themes, and their variations, in individual dream series. These repeated themes vary on a continuum, from the reappearance of particular images in a dream series, to the recurrence of analogous dream scenarios, to the repeated replication of overwhelming traumatic moments. In a variety of theories, these repeated themes are understood as indicating the presence of unrecognized and unresolved personal conflicts. Correspondingly, the repetition and variation of these themes are understood as fostering their recognition and resolution.

There is some evidence from research using quasi-longitudinal research designs that is consistent with this formulation. When adults who are presently experiencing recurrent dreams are compared with those who experienced recurrent dreams in the past and with those who have not experienced recurrent dreams, recurrent dreamers report dreams that are more anxious, dysphoric, and conflicted laden; they also have lower scores on several indices of personal well-being (e.g., anxiety, depression, life stress). Past recurrent dreamers report greater personal well-being, as well as more positive dream content, than do recurrent dreamers or nonrecurrent dreamers. These results have been replicated with younger adults among whom past recurrent dreamers were defined as those who had recurrent dreams during late childhood. In this case, past recurrent dreamers did not differ from nonrecurrent dreamers. These studies indicate that the cessation of adult – but not childhood – recurrent dreams is associated with enhanced well-being.

Vestigial Defense

There are two conceptions of dream function that posit their historical rather than current adaptive significance. In an ontogenetic version of this possibility, the capacity of REM dreaming to enhance the consolidation of memories for emotional events may fix traumatic memories in a way that is not adaptive in the dreamer's current life. A prototypic example occurs in posttraumatic stress disorder (PTSD), with its hyperemotional memories, flashbacks, and nightmares. Hypothetically, the enhanced consolidation of declarative memory for overwhelming events (through amygdalar modulation of hippocampal function) consolidates those memories so that they remain easily evoked long after the threat associated with trauma has passed (e.g., in posttraumatic nightmares). Generalization of this hypothesis to more mundane emotional memories may explain the repetition principle – that is, the persistence in dreams of a variety of recurrent themes that are unrelated to current concerns.

In an analogous evolutionary model, dreaming purportedly simulates threatening events of the kind that early humans were likely to encounter frequently (e.g., aggressive interaction with animals) but that contemporary humans typically do not. In hunter–gatherer societies, such dreamed threat simulations may have provided covert rehearsal of a defensive response to actual threatening circumstances, even though they have no contemporary ecological validity. So, when contemporary threats evoke dreams with those primordial characteristics, the threat simulation is likely to be unrelated to, and perhaps even incompatible with, adaptive responses. These conceptions of vestigial dream function may help to explain why dreams related to trauma, or even nightmares that do not have origins in identifiable trauma, often disrupt sleep, prompt hypervigilance, and create lingering distress.

Conclusion

It sometimes seems that theories of dream function are strongly conflicting – as though if one is valid then the others cannot be. However, since short-term dream effects may serve auxiliary functions within more obviously 'adaptive' activities, long-term dream function may be understood as the capacity of a complex self-organizing system that depends upon integration of numerous simpler capacities. From this perspective, the short-term functions of dreaming, when coordinated according to the system's superordinate integrative principles, may subserve a long-term function that is irreducible to any particular dream effect. This integrative possibility remains undeveloped but increasingly plausible.

See also: Dreams and Nightmares in PTSD; Incorporation of Waking Events into Dreams; REM/NREM Differences in Dream Content; Sleep and Consciousness; Sleep-Dependent Memory Processing; Theories and Correlates of Nightmares.

Further Reading

Blagrove M (1992) Dreams as the reflection of our waking concerns and abilities: A critique of the problem-solving paradigm in dream research. *Dreaming* 2: 205–220.

Cartwright R, Baehr E, Kirkby J, et al. (2003) REM sleep reduction, mood regulation and remission in untreated depression. *Psychiatry Research* 121: 159–167.

Cheyne JA, Rueffer SD, and Newby-Clark IR (1999) Hypnagogic and hypnopompic hallucinations during sleep paralysis: Neurological and cultural construction of the night-mare. *Consciousness and Cognition* 8: 319–337.

Greenberg R and Pearlman C (1993) An integrated approach to dream theory: Contributions from sleep research and clinical practice. In: Moffitt A, Kramer M, and Hoffman R (eds.) *Functions of Dreaming*, pp. 363–377. Albany: SUNY Press.

Kramer M (1993) The selective mood regulatory function of dreaming: An update and revision. In: Moffitt A, Kramer M, and Hoffman R (eds.) *Functions of Dreaming*, pp. 139–195. Albany: SUNY Press.

Kuiken D and Sikora S (1993) The impact of dreams on waking thoughts and feelings. In: Moffitt A, Kramer M, and Hoffman R (eds.) *Functions of Dreaming*, pp. 419–476. Albany: SUNY Press.

Maquet P, Laureys S, Peigneux P, et al. (2000) Experience-dependent changes in cerebral activation during human REM sleep. *Nature Neuroscience* 3: 831–836.

Rauchs G, Desgranges B, Foret J, et al. (2005) The relationships between memory systems and sleep stages. *Journal of Sleep Research* 14: 123–140.

Revonsuo A (2003) The reinterpretation of dreams: An evolutionary hypothesis of the function of dreaming. In: Pace-Schott EF, Solms M, Blagrove M, et al. (eds.) *Sleep and Dreaming: Scientific Advances and Reconsiderations*, pp. 85–109. New York: Cambridge University Press.

Smith C (2001) Sleep states and memory processes in humans: Procedural versus declarative memory systems. *Sleep Medicine Review* 5: 491–506.

Stickgold R (2005) Sleep-dependent memory consolidation. *Nature* 437: 1272–1278.

Wagner U, Degirmenci M, Drosopoulos S, et al. (2005) Effects of cortisol suppression on sleep-associated consolidation of neutral and emotional memory. *Biological Psychiatry* 58: 885–893.

Sleep and Consciousnesss

J A Hobson, Harvard Medical School, Boston,
MA, USA

Introduction

When we go to sleep at night and dream, our conscious
state changes radically. When we first become drowsy,
fleeting dream images may occur. We then become
oblivious and hard to rouse from deep sleep. Later in
the night, when sleep lightens, we may experience
vivid, bizarre dreams from which spontaneous or easily induced awakenings may occur. By studying the
changes in the brain which are associated with these
changes in our state of consciousness, we may learn
more about the physiological mechanisms underlying
our states of mind.

During waking we can attend to events in the outside world, form perceptions of events, and/or direct
our thoughts to those percepts or to other matters,
such as the review of our behavior, our plans, or our
feelings. When we go to sleep, we become unaware of
the outside world and lose the ability to construct
thoughts. These perceptual and cognitive defects persist even when we dream: perceptions are then almost
entirely internally generated and we have difficulty
reasoning about them; we almost always believe that
we are awake instead of dreaming.

From the vantage point of the cognitive neuroscience of consciousness, we can tentatively explain
these phenomena as a function of three factors:

1. Factor A: the level of electrical activation in the
   brain.
2. Factor I: the status of gating information flow to
   and from the brain.
3. Factor M: the nature of the mode of information
   processing within the brain.

The level of consciousness is thus low or high, depending upon the level of activation (A). The focus or
direction of consciousness is either strongly external
or strongly internal, depending on whether the input–
output gates of the brain are open or closed to the flow
of information (I). The mode (M) of information processing within the brain is set by the ratio of aminergic
to cholinergic modulation. Since these three factors
can be quantified using experimental data, it is possible to construct a three-dimensional state space model
(**Figure 1**). The state of consciousness, a point with
values A, I, and M, varies over time, the fourth dimension of the model. Normal sleep cycles appear as
elliptical trajectories in the resulting state space.

Definition of Consciousness

For the purpose of the conscious state paradigm,
consciousness can be simply defined as awareness of
the outside world, our bodies, and ourselves. Consciousness can be further distinguished in two ways.
The first distinction is between primary consciousness
(which does not depend upon language and may well
be shared with many other mammals) and secondary
consciousness (which does depend upon language
and may be a uniquely human capability). The second
distinction is between the components of consciousness, which are shown in **Table 1**. These components
of consciousness vary significantly in strength over
the normal wake–sleep–dream cycle.

The functions listed in **Table 1** are the aspects of
consciousness which are commonly studied by cognitive neuroscientists. Here we investigate the common
mechanisms by which all of these aspects are changed
as a function of the changes in conscious state experienced by us as waking, sleeping, and dreaming.

Characteristics of Consciousness

Consciousness is graded and its components vary in
strength as the brain changes state over the course of
each day of each life and over a lifetime. In other
words, consciousness is more or less intense and its
components change in relative strength as a function
of the sleep–wake cycle. For example, in waking, perception can be exteroceptive, thought can be logical,
and memory can be good. By contrast, in dreaming, all
three of those cognitive functions are altered; internal perception is enhanced, emotion is intensified
and confabulation runs wild, and logical thought and
memory are greatly impaired. **Table 2** lists some of
the state-dependent consciousness components.

Changes in the Brain and in the Body during Sleep; Changes in Consciousness

As shown in **Figure 2**, sleep onset is associated with
thalamocortical deactivation. This deactivation is signaled, first, by electroencephalogram (EEG) slowing
(stage I) and next by definitive blockage of thalamic
transmission with the appearance of EEG spindles
(stage II). After that there is a further slowing in
frequency and an increase in amplitude of cortical
slow waves (stages III and IV). Muscle tone decreases
passively, postural shifts stop, and eye movements
are greatly reduced as this pattern evolves. During
the evolution of these stages of so-called nonrapid
eye movement (NREM) sleep, individuals may have

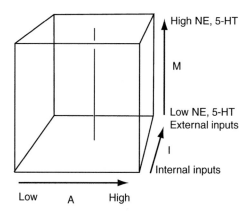

Model factor	Psychological	Neurobiological
A-Activation: energy level of processing capacity	• Word count • Cognitive complexity, e.g., perceptual vividness, emotional intensity, narrative	• EEG activation • Firing level and synchrony of reticular, thalamic and cortical neurons
I-Information: source internal or external	• Real world space, time and person referents and their stability • Real vs. imagined action	• Level of presynaptic and postsynaptic inhibition • Excitability of sensorimotor pattern generators
M-Mode: organization of data	• Internal consistency? • Physical possibility? • Linear logic?	• Activity level of aminergic neurons

Figure 1 The activation/input source/neuromodulation (activation/information/mode; AIM) model. Illustration of three-dimensional state space and the psychological neurobiological correlates of each dimension.

Table 1 Components of consciousness

Component	Definition
Attention	Selection of input data
Perception	Representation of input data
Memory	Retrieval of stored representations
Orientation	Representation of time, place, and person
Thought	Reflection upon representations
Narrative	Linguistic symbolization of representations
Emotion	Feelings about representations
Instinct	Innate propensities to act
Intention	Representations of goals
Volition	Decisions to act

fleeting dreams as their brains are disconnected from sensory input, but they soon become oblivious with little or no memorable mental activity. Measures of this deactivation process yield low values of factor A in the AIM model. At the same time that factor A decreases, factors I and M are also decreasing as the sensory gates close and aminergic modulation decreases.

After 70–80 min, often signaled by a posture shift, the EEG is reactivated and reverses its downward path, moving rapidly up through stages III and II to enter stage I. At the same time the electromyogram (EMG) is actively inhibited and the electooculogram (EOG) shows the increasingly clustered and intense eye movements that give REM sleep its name. The values of factor A activation increase to waking levels (and beyond), but sleep persists. But both factors I and M go to their lowest values in REM as the activated brain is actively put offline and further demodulated. The alternation of NREM and REM sleep continues throughout the night at 90- to 100-min intervals. As the night progresses the NREM periods become shorter and less deep while the REM periods become longer and more active. In the last two cycles the alternation is usually between stage II and stage I REM. This increasing brain activation is associated with an increase in dreaming. Although dreaming is always more intense in REM, it also occurs in NREM sleep, especially in the second half of the night.

Formal Analysis of Dream Content

In order to compare with waking the nature of conscious experience in REM and light NREM sleep, it has

Table 2 State-dependent consciousness components

Function	Nature of difference	Causal hypothesis
Sensory input	Blocked	Presynaptic inhibition
Perception (external)	Diminished	Blockade of sensory input
Perception (internal)	Enhanced	Disinhibition of networks storing sensory representations
Attention	Lost	Decreased aminergic modulation causes a (decrease in) signal-to-noise ratio
Memory (recent)	Diminished	Because of aminergic demodulation, activated representations are not restored in memory
Memory (remote)	Enhanced	Disinhibition of networks storing mnemonic representations increases access to consciousness
Orientation	Unstable	Internally inconsistent orienting signals are generated by cholinergic system
Thought	Reasoning *ad hoc*; logical rigor weak; processing hyperassociative	Loss of attention, memory, and volition leads to failure of sequencing and rule inconstancy; analogy replaces analysis
Insight	Self-reflection lost (failure to recognize state as dreaming)	Failure of attention, logic, and memory weakens second- and third-order representations
Language (internal)	Confabulatory	Aminergic demodulation frees narrative synthesis from logical restraints
Emotion	Episodically strong	Cholinergic hyperstimulation of amygdala and related temporal lobe structures triggers emotional storms, which are unmodulated by aminergic restraint
Instinct	Episodically strong	Cholinergic hyperstimulation of hypothalamus and limbic forebrain triggers fixed-action motor programs, which are experienced fictively but not enacted
Volition	Weak	Top-down motor control and frontal executive power cannot compete with disinhibited subcortical network activation
Output	Blocked	Postsynaptic inhibition

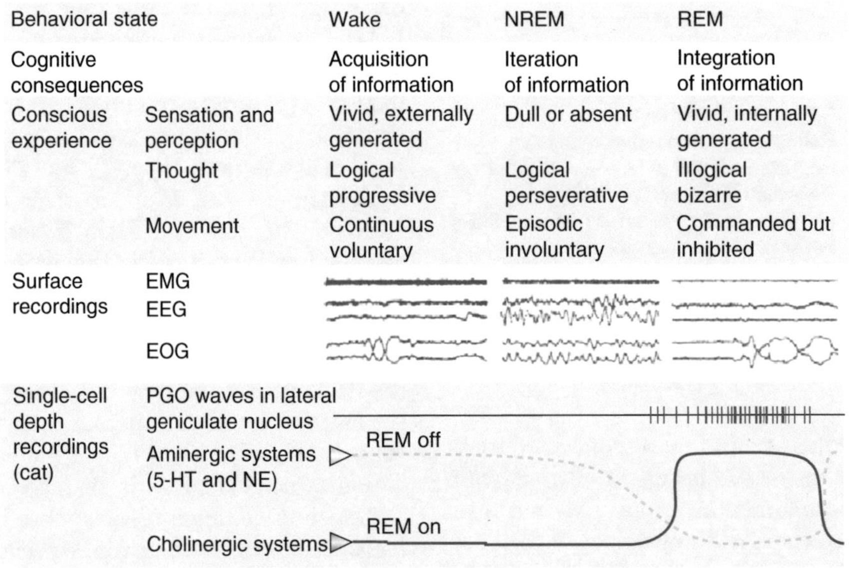

Figure 2 Behavioral states of humans. States of waking, nonrapid eye movement (NREM) sleep, and REM sleep have behavioral, polygraphic, and psychological manifestations. The sequence of these stages is represented in the polygraph channel. Sample tracings of three variables used to distinguish state are also shown: the electromyogram (EMG) is at its highest in waking, intermediate in NREM sleep, and lowest in REM sleep; the electroencephalogram (EEG) and electooculogram (EOG), which are both activated in waking and REM sleep and inactivated in NREM sleep. PGO, pontogeniculooccipital; 5-HT, 5-hydroxytryptamine (serotonin); NE, norepinephrine.

proved propitious to quantify such aspects of consciousness as (1) the sense modalities that are represented and the intensity of the internally generated percepts, (2) the occurrence and strength of self-reflective awareness and executive thought, (3) the dreamer's orientation in time and space, and dream personnel, (4) the dreamer's emotion profile (presence and strength of emotions such as elation, anxiety/fear, anger, shame, guilt, sadness, and eroticism), and (5) the identification by the dreamer of the possible memory sources of the dream content. This approach has led to the characterization of dream consciousness as having (1) strong sensorimotor perceptions (with a preponderance of vivid visual imagery as the dreamer moves through a fictive reality); (2) impoverishment of executive ego function (with the delusional belief that one is awake and a failure to develop or sustain logical thinking); (3) discontinuity and incongruity of plot features as the basis of bizarre cognition; (4) potentiation of anxiety, anger, and elation together with a weakening of shame, guilt, and sadness; and (5) inability to specify the memory source of the confabulated dream content in about 80% of the cases.

Dreaming is thus characterized by visual hallucinations, by disorientation, by lack of insight and judgment, and by memory loss (with compensatory confabulation). These formal similarities are so strong as to lead to the hypothesis that dreaming, although normal, is a state akin to delirium. Since delirium is caused by diseases and by drugs which interfere with normal neuromodulation, it is natural to wonder if, in normal sleep, there are changes in neuromodulation during the wake–sleep cycle.

Cellular and Molecule Mechanisms of Mammalian Sleep

All mammals share with humans the regular periods of NREM and REM sleep. This is so whether the animals are aquatic, terrestrial, or arboreal, and whether they are large surface-dwelling carnivores (e.g., lions) or small nest-dwelling herbivores (e.g., mice or rabbits). As an experimental model, the domestic cat has appealed to neurobiologists because it is so docile and so prodigious a sleeper. As experimental animals, cats are more expensive and more ethically problematic than are rats, but they are large enough to carry the microelectrode, microinjection, and microdialysis array, the equipment necessary to perform physiological and pharmacological experiments. They are well studied neuroanatomically so that stereotaxis is facilitated.

Using cats, neuroscientists have shown that the two brain-activated states, waking and REM sleep (with high values of A), are at opposite poles with respect to both input–output gating (high I in waking, low in REM) and neuromodulation (high M in waking, low M in REM). On all these dimensions, NREM sleep is quite different (low A and intermediate I and M). In addition to demonstrating that aminergic levels decline in NREM sleep and reach their nadir in REM, it has been shown that the cholinergic level declines as animals go from waking to NREM, but rises again and reaches its highest level, in REM sleep, especially in the brain stem. This gives factor M (defined as the ratio of aminergic to cholinergic modulation) a highly differentiating power for distinguishing the three states: M is highest in waking, intermediate in NREM sleep, and lowest in REM sleep. This finding is potentially relevant to our understanding of the brain basis of the many phenomenologic distinctions between waking, sleeping, and dreaming. The mode of information processing has changed dramatically.

The importance to REM of the pontine brain stem is demonstrated by the presence of the cell bodies of both cholinergic and aminergic neurons there. Furthermore, it has now been conclusively demonstrated that the microinjection into the pontine tegmentum of both cholinergic agonists and cholinesterase antagonists produces dramatic increases in REM sleep. All of these effects are blocked by pretreatment of the injection sites by atropine. There is no longer any doubt that compared to waking and NREM sleep, REM sleep is hypoaminergic and hypercholinergic. This robust physiological difference correlates with – and hypothetically causes – the psychological differences.

Based upon new studies in rats, a flip-flop switch for REM sleep triggering has recently been proposed. Each side of the switch contains γ-aminobutyric acid (GABA)ergic neurons, and this resulting mutual inhibition guarantees that when one side is on, the other is off. The REM-on population contains two populations of glutaminergic neurons: one population projects to the basal forebrain and produces the forebrain activation while the other projects to the medulla and spinal cord, producing atonia. The cholinergic system is activated as part of the REM-on switch and can, in turn, activate that side of the switch. The REM-off population includes the dorsal raphe nuclei. At a still deeper level, the activation of protein kinase in the pontine tegmentum has been shown to accompany REM sleep, suggesting that pedunculopontine intracellular protein kinase A (PKA) activation is involved in REM sleep generation.

Input–Output Gating

Michel Jouvet called REM sleep 'paradoxical' because sleep persists despite the EEG evidence for strong brain activation. A partial explanation of the paradox is the simultaneous blockade of sensory input and motor output, the neurophysiological mechanisms of which are now well understood. Ottavio Pompeiano used classical Sherrington reflexology techniques to show that motor output is blocked in REM by descending inhibition from the pontomedullary brain stem to the anterior bone cells of the spinal cord. That this inhibition is induced by powerful and active postsynaptic inhibition of the final common path motor neurons was shown by Michael Chase. Chase also demonstrated that the spinal inhibition of REM is mediated by glycine.

As for the inhibition of sensory input, it was also Pompeiano who demonstrated that presynaptic inhibition is associated with the clusters of eye movement during REM sleep. An associated occlusion of sensory input excitation also accompanied the REMs. This co-option of input-processing channels (including the all-important thalamocortical system) is effected by the internally generated pontogeniculooccipital (PGO) waves, first described by Jouvet. These cortical activation waves were later shown to be of entirely internal origin and were enhanced by cholinergic microstimulation. PGO wave generation is inhibited by noradrenaline and serotonin.

The emerging picture of REM sleep is a high level of activation in a brain which is not only off-line by virtue of sensory input and motor output blockade, but is also stimulated by phasic activation signals of its own devising. The coactivated mind is therefore fooled by its own parasitic excitation; the brain-mind then develops the illusion or fictive experience of perceiving and moving through dream space.

Efferent Copy Generation

The PGO waves of REM sleep not only constitute internally generated phasic activation signals for the dreaming brain-mind, but they also convey to the thalamus and cortex feed-forward information about the direction of the upcoming eye movements. As such, they are classical efferent copy stimuli from the oculomotor circuits to the visual forebrain. Such eye movement activity can also be recorded in waking when it forms an integral part of the startle response. In waking, the PGO waves rapidly habituate as the startle response diminishes. The second and third stimuli of a train evoke a much weaker or no response because they are no longer novel. The mechanism of this attenuation is thought to be aminergic inhibition of the cholinergic-modulated PGO generator neurons of the lateral pons.

The relative continuity of waking consciousness is, in part, a function of PGO wave damping, whereas the bizarre discontinuity of dream consciousness is a function of PGO wave disinhibition. Dream consciousness reflects this dishabituation by the increase in the emotions of surprise and anxiety. These dream features could be mediated by excitation of the amygdala, a major target of the PGO waves. In REM sleep, the efferent copy information about the direction of eye movement that is conveyed by the PGO system is not integrated with changes in visual input, as occurs in waking, because in sleep there is no visual input. It is tempting to speculate that the efferent copy information that is generated in REM sleep is not only a physiological cause of dream surprise, but is also used in perceptual aspects of dream plot construction.

The oculomotor system of the pons is only one of many motor program generators that are activated in REM sleep. Locomotion is commanded but, because the anterior horn cells are hyperpolarized, real motor behavior is not generated and REM sleep dreams reflect this by the intensity of fictive motoric action in dreams. There is hardly a sentence in the REM sleep dream report that does not contain an action verb. We are always walking, running, flying, or swimming in our dreams. Even if we lead sedentary waking lives, stationary activity is not represented in our dreams.

The finding of motor pattern generator activation in REM sleep and its correlation with the fictive motility of dreaming is a good example of how the close study of neurophysiology can provide dream theory with a new paradigm and new hypotheses regarding the origin of dream plots. Of course, the approach does not allow us to propose why a given dreamer has a given dream or what that dream might mean. But it does provide encouragement for an alternative way of approaching even interpretive questions. For example, our dreams are full of movement because brain motor pattern generators are activated in REM sleep. Further evidence that this theory may be correct is given by experimental and clinical discoveries, both of which reveal unexpected and unwanted enactment of REM sleep motor commands. The experiments of Michel Jouvet and of Adrian Morrison showed that cats with lesions in the vicinity of the locus coeruleus in the pons evinced REM sleep without atonia. In other words, the cats showed sleep with EEG activation REMs but suddenly stood up and evinced a wide variety of attack and defense behaviors. They were then in a dissociated state of REM sleep in which they expressed automatic behaviors that are normally suppressed by the motor inhibition of REM. On the basis of these observations, Jouvet boldly proposed that the cats were acting out their dreams.

That idea gained credence when it was reported by Carlos Schenk and Mark Mahowald that humans did precisely the same thing when they became afflicted by what they called the REM 'sleep behavior disorder.' Schenk's and Mahowald's patients, mostly >50-year-old males, performed motor acts in sleep that corresponded to their dream reports. One patient, dreaming that he was playing football, got up out of bed and tried to tackle a chest of drawers. Another, dreaming that he was diving, stood up on the side of the bed and propelled himself into the baseboard of his bedroom, which he imagined to be a swimming pool. Presumably, these experimental and natural pathological conditions of REM sleep without atonia are due to damage of that subpopulation of glutaminergic REM-on neurons that convey inhibitory commands from the flip-flop switch to the spinal cord.

These examples would seem to make more plausible the suggestion that dream consciousness is a function of brain activation in sleep and that the important differences between the dreaming and waking states of consciousness are determined by specifiable differences between the mechanisms of brain activation in the two states. In particular, it is the presence or absence in input–output gating that determines the relationship between the outside world and internal perceptions, beliefs, and actions.

Differences in Regional Brain Activation

EEG evidence had long suggested that the brain was similarly activated in waking and REM sleep dreaming. The recent advent of PET imaging technology revealed that this was an oversimplified conclusion. Most brain regions are less active in NREM sleep than in either the wake or REM sleep state. But when REM is compared to the wake state, differences emerge which are of great relevance to studies of conscious state determination.

Certain brain regions are more active in REM than in waking. They include the pontine tegmentum (which animal studies have shown to be the probable site of the REM sleep and PGO wave generator), the amygdala (which animal studies have shown to be involved in mediating emotion, especially the anxiety which is so common in dreaming), the parietal operculum (which is thought to mediate visuomotor integration of cardinal perceptual feature of dreaming), and the parahippocampal and deep frontal cortices (which may process emotional data and integrate it with cognitive information).

One brain region is notably less active in REM than in waking. The dorsolateral prefrontal cortex (DLPFC) remains at the same low level seen in NREM sleep. Other human studies have implicated the DLPFC in the executive ego functions that are so conspicuously deficient in dreaming: short-term memory,

self-reflective awareness, intentional decision making, and volitional action. The brain lesion studies done by Mark Solms have revealed that two of the structures selectively activated in REM sleep are, when damaged, associated with a complete cessation of dream recall. They are the parietal operculum (damage to which causes problems with sensorimotor integration) and the deep frontal white matter (damage to which causes emotional apathy). Patients often complain of cessation of dreaming following lobotomy, which consists of severance of the deep frontal white matter.

Field Studies of Waking, Sleeping, and Dreaming Consciousness

The invention of the 'Nightcap', a two-channel event recorder, has made possible the field study of conscious experience in humans as they lead their otherwise normal waking and sleeping lives. In addition to wearing the Nightcap for recording their head and eye movements while asleep, participants carried radio beepers so that they could be contacted throughout the day and night to collect reports of conscious experience (see **Figure 3(a)**). This combination of techniques allowed us, for the first time, to collect reports of consciousness experience in the same persons, from several brain states, around the clock. Over 3000 reports from ten people ages 20–40 years were elicited from active and quiet waking states, from five intervals during sleep onset and from both NREM and REM sleep. The reports were transcribed and scored for dimensions such as word count (REM reports were 6 times as long as NREM reports), internally generated percepts (REM sleep highest, active waking lowest). These data, shown in **Figure 3(b)**, suggest a general psychophysiological law: thinking is incompatible with hallucination and vice versa. The implications of these findings for conscious state control and for psychiatry are loud and clear. One striking finding was a reciprocal increase in internally generated perception and a decrease in thinking across the five states.

Convergence of Physiology, Psychology, and Philosophy

The integration of the fields of psychology and physiology, envisaged by William James and the young Freud over 100 years ago, is now beginning, with sleep and dream science leading the way. The Finnish philosopher of science Antti Revonsuo has recently championed the theory, first put forth by Michael Jouvet, that REM sleep dreaming allows the sleeping brain to run adaptive behavioral programs of fight and flight. According to this view, REM sleep dreaming has a Darwinian function even if dream behavior

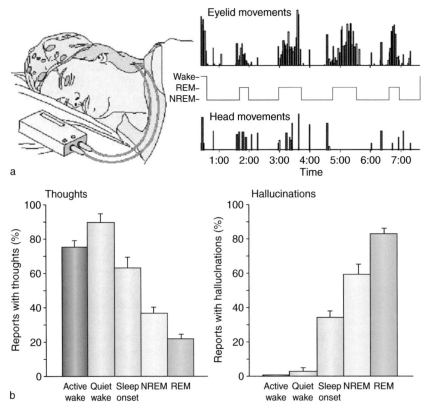

Figure 3 (a) Central arousal accompanying the activated states of rapid eye movement (REM) sleep and waking can be measured using the 'Nightcap,' a simple ambulatory monitor. The Nightcap is a two-channel recording device that distinguishes waking, REM and NREM sleep. One channel of the Nightcap monitors eye movement and the other monitors body movements. The Nightcap eyelid-movement readout is thought to reflect activity in portions of the brain stem oculomotor nucleus that innervate the eyelid and are adjacent to portions of the medial brain stem ascending reticular system, the activity of which, in turn, generates forebrain activation. (b) Decline in directed thought and reciprocal increase in hallucinations during progression from active waking through sleep onset and NREM to REM sleep.

is fictive. For the German philosopher Thomas Metzinger, the occurrence of lucid dreaming, out-of-body experience and dreaming itself, supports his phenomenal self-model of consciousness. The phenomenal self is a constantly renewed functional state of the brain whose vicissitudes can, for the first time in human history, begin to be explained.

AIM: An Integrated Model of the Human Brain Mind

Using physiological measures, it is now possible to create a three-dimensional model of brain-mind state space (see **Figures 1** and **4**). Activation, assessed from EEG and/or thalamocortical cellular activity, gives values and axis A, running from left (low) to right (high) across the front wall of the state space. Input–output gating is measured as EMG activity, values of H-reflex, and PGO waves. It runs from the front to the back of the state space. The vertical axis, M, measures the ratio of the strength of aminergic to cholinergic neuron modulation. Unfortunately, factor

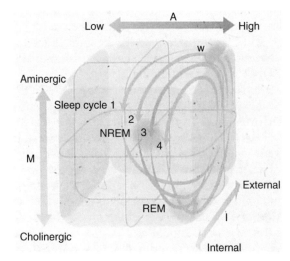

Figure 4 The three-dimensional activation/information/mode (AIM) state space model showing normal transitions within the AIM state space, from wake, to non-rapid eye movement (NREM), and then to REM sleep. REM occupies the lower right-hand front corner, in which activation is high, input is entirely internal, and the forebrain is cholinergically activated and aminergically demodulated. ACh, acetylcholine; NE, norepinephrine; 5-HT, 5-hydroxytryptamine.

M cannot yet be measured in humans, but pharmacological studies indicate that inferences from neurobiological studies in animals are valid.

In the model, time is the fourth dimension and the solution of AIM is constantly changing, even in the waking domain (at the far right, upper, rear corner of the state space) as we change the focus of attention, day dream, or enter altered states such as hypnosis. We may become drowsy, doze, or fall frankly asleep, in which case AIM moves down, forward, and to the left and then occupies the NREM sleep domain in the center of the state space. Finally, AIM moves to the REM sleep domain in the right front lower corner of the state space.

During a normal night of sleep, AIM function follows an elliptical trajectory that repeats itself 4 or 5 times per night. The ellipse shifts sequentially to the right and to deeper parts of the state space in successive cycles. These elliptical orbits in AIM state space conform to the mathematical description of the reciprocal interaction model, giving an internally consistent picture of sleep physiology.

Consciousness is most intense in the right portion of the state space and lowest on the left. It is interoceptive toward the front of the state space and exteroceptive toward the back. The kind of consciousness is given by the mode dimension. It is highest at the top of the state space and lowest at the bottom. Abnormal states of consciousness such as narcolepsy, the REM sleep behavior disorder, delirium, and other psychoses can also be modeled in the normally forbidden zones of AIM state space.

See also: Dreams and Nightmares in PTSD; History of Sleep Research; Incorporation of Waking Events into Dreams; Phylogeny and Ontogeny of Sleep; REM/NREM Differences in Dream Content; Theories and Correlates of Nightmares; Theories of Dream Function.

Further Reading

Amini-Sereshki L and Morrison AR (1986) Effects of pontine tegmental lesions that induce paradoxical sleep without atonia on thermoregulation in cats during wakefulness. *Brain Research* 384(1): 23–28.

Bandyopadhya RS, Datta S, and Saha S (2006) Activation of pedunculopontine tegmental protein kinase A: A mechanism for rapid eye movement sleep generation in the freely moving rat. *Journal of Neuroscience* 26(35): 8931–8942.

Hobson JA, Pace-Schott EF, and Stickgold R (2000) Dreaming and the brain: Toward a cognitive neuroscience of conscious states. *Behavioral and Brain Sciences* 23: 793–842.

Jouvet M (1962) Recherche sur les structures nerveuses et les mechanismes responsables des differentes phases du sommeil physiologique. *Archives Italiennes de Biologie* 100: 125–206.

Jouvet M (1973) Essai sur le reve. *Archives Italiennes de Biologie* 111: 564–576.

Jouvet M, Vimont P, and Delorme F (1965) Elective suppression of paradoxical sleep in the cat by monoamine oxidase inhibitors. *Comptes Rendus des Seances de la Societe de Biologie et de ses Filiales* 159(7): 1595–1599.

Lu J, Sherman D, Devor M, et al. (2006) A putative flip-flop switch for control of REM sleep. *Nature* 441(7093): 589–594.

Metzinger T (2003) *On Being No One.* Cambridge: MIT Press.

Revonsuo A (2005) *Inner Presence: Consciousness as a Biological Marker.* Cambridge, MA: MIT Press.

Schenck CH and Mahowald MW (1996) REM sleep parasomnias. *Neurological Clinics* 14: 697–720.

Solms M (1997) *The Neuropsychology of Dreams: a Clinico-Anatomical Study.* Mahwah, NJ: Lawrence Erlbaum Associates.

Yamuy J, Fung SJ, Xi M, et al. (2004) Hypocretinergic control of spinal cord motor neurons. *Journal of Neuroscience* 24(23): 5336–5345.

REM/NREM Differences in Dream Content

J S Antrobus, The City College of New York, New York, NY, USA
E J Wamsley, Harvard Medical School, Boston, MA, USA

Dreaming in Rapid Eye Movement Sleep

The neurocognitive study of sleep mentation began when Aserinsky, in the first all-night electroencephalograph (EEG) study of sleep, observed periodic intervals of rapid eye movements (REMs) associated with reports of long, visually vivid, and often bizarre dreams. Although Aserinsky quite plausibly assumed that sleepers were scanning their dream imagery, this assumption has not been confirmed. Relative to periods of non-REM (NREM) sleep, REMs occur during periods of low-voltage, high-frequency EEGs so similar to that of waking that European scientists called it 'paradoxical' sleep. The first REM period typically follows about 90 min of NREM sleep and lasts about 10 min. Subsequent REM periods increase in duration and make up a total of 21% of sleep time. Although REMs are most frequent at the onset of each REM period, and are associated with the most visually vivid dream reports, vivid dreaming is reported throughout the period, even after intervals of REM quiescence. Sleepers report some content from about 80% of REM awakenings, as compared to 50% of NREM awakenings. REM reports are, on average, more vivid, emotional, storylike, and bizarre than are NREM reports. Relative to these REM reports, NREM reports are generally shorter and thoughtlike, more often consisting of isolated, static images.

Although the association between REM sleep and dreaming is one of the strongest, nonpathological, neurocognitive associations ever discovered, in 1953 what was known about waking neurophysiological processes was insufficient to instruct a model of dreaming sleep. Consequently, neurological models of dreaming initially emphasized the pontine processes that produced the alternation between REM and NREM sleep, at the expense of describing how cortical structures might generate the thought and imagery of those states. REM/NREM models were also silent about the fact that 6% of NREM reports were more dreamlike than were REM reports, matched by sleeper and time of night, and about the rather dramatic increase in the dreamlike quality of both REM and NREM reports in the late morning.

The Activation–Synthesis Model

Though the study of reported sleep mentation is necessarily limited to humans, the first brain models of REM/NREM sleep were based on the cat. Hobson, McCarley, and their colleagues solved the puzzle of the paradoxical wakelike EEG pattern of REM sleep when they showed that it is produced by reticular–thalamic–cortical circuits similar to those that Moruzzi and Magoun had previously shown to produce general wakefulness. The 90-min periodic alternation between REM and NREM sleep is controlled by pontine circuits. The primary distinction between waking and REM sleep is that the pontine-initiated REM activation process strongly inhibits both afferent input, from the sensory pathways, and efferent motor output. In short, it produces an active brain, with high sensory thresholds and, except for the oculomotor system, motor paralysis. In REM sleep, the dreamer can attempt to run, but efferent commands from the motor cortex are inhibited and the afferents that normally confirm feedback of movement, both kinesthetic and visual, are not received. The dreamer may interpret this to mean that he/she is 'stuck.'

But how, in the absence of sensory input, does the brain produce imagery and thought? In their activation–synthesis model (1977), Hobson and McCarley proposed that the 'input' is oculomotor information initiated in the pons and transmitted via pontogeniculooccipital (PGO) circuits to the cortex. In the attempt to 'synthesize' this information, the cortex creates a semicoherent visual story (i.e., the dream).

The large amplitude of the PGO spikes during REM sleep in cats suggested that they might interrupt the cortical EEG and thereby the coherence of the ongoing story sequence, producing the bizarre characteristics of the dream. But this assumption has yet to receive empirical support. Because of the pontine origin of this circuit, Hobson further argued that the Freudian assumption of a cortical (i.e., higher order) origin of dream process was without merit. The synthesis portion of the model is based on the quite reasonable neurocognitive isomorphic assumption that the cortex interprets subcortical processes similarly during sleep and waking. Thus, a command from the lateral geniculate nucleus to move the eyes rightward in waking should be interpreted similarly in REM sleep. However, the geniculooculomotor left–right orientation pattern of waking is not sustained, at least consistently, in REM sleep – the isomorphism assumption does not hold in this particular case.

Synthesis and Dream 'Incorporation'

The stimulus–response experimental procedures that are so effective in waking cognitive neuroscience are almost useless in sleep, since processing a stimulus generally drives the sleeper out of the biological state being studied. Nevertheless, once it was established that REM sleep is the time when the likelihood of dreaming is high, investigators began to systematically study how the dreamer 'incorporates' external stimuli into the dream. Although a fine spray of water or a soft bell presented during sleep generally elicit a short interval of EEG alpha, sleep often returns quickly to the prior REM state. Using this procedure, Dement showed that the time interval between stimulus presentation and awakening was similar to the length of the subsequently reported dream experience – which discredited the theory that dreams occurred in the momentary transition between sleep and waking.

The content of such incorporations, while fascinating, provided little information about the processes by which they were produced. Their salient characteristic was that they rapidly made remarkably good sense out of one or more features of the external stimulus – within the apparently prior context of the ongoing dream episode. For example, Dement found that spraying individuals with water during REM sleep led one dreamer to report that his children had spilled water on him while he was reading a book, whereas another dreamer reported that while watching an actress walk on stage ahead of him, it began to rain through the ceiling, so the actress immediately opened her umbrella. These quick, apparent 'interpretations' by the sleepers, in response to their wet skin, are remarkably similar to the left-brain interpretations of right-brain input later reported by Gazzaniga in split-brain patients.

Activated Cortex as an Attractor Landscape

A neural network attractor is essentially a network with recurrent connectivity that represents a well-learned feature, object, person, or event. Attractors that share many features ('wet face,' 'wet clothes,' 'rain,' 'shower') lie closer together than do attractors that represent dissimilar objects. A large array of attractors can be represented in an attractor landscape in which each attractor is a basin within a large plane; attractors with similar features are neighbors within a common attractor valley. When information reaches the landscape, it is cycled recurrently until it 'settles' into (i.e., recognizes) one particular attractor. Cortical regions that process different classes of features or events may be represented as different, but interconnected, landscapes. Several characteristics of attractors make them particularly

useful for representing the brain processes that generate dreaming and other forms of sleep mentation. Even when information is extremely 'noisy,' it can settle into the basin of an attractor. Second, the settling process can be strongly biased by prior context, whether the context is the preceding dream narrative, the events of the preceding day, or chronic life stressors. Biases may be represented as temporary deepening of attractor basins – neurally by long-term potentiation (LTP) of the synapses that form their neural networks, particularly in the hippocampal system. Finally, the patterns of regional cortical activation that distinguish REM, NREM, and waking states can be represented by leveling or deepening the basins across different regional attractor landscapes.

The pontine–reticular–cortical activation network described by Hobson and his colleagues provides a basis for amplifying a large cortical attractor network during REM sleep. Other sources of activation capable of supporting attractor networks, thereby sleep mentation, even in NREM sleep, will be described in the following sections. This attractor model is nicely supported by the 'incorporation' studies, whereby a single feature of an external stimulus is attracted to a basin that has previously been deepened by the context of the prior dream. For example, if the dream is located in the living room close to the telephone, when the experimenter, Dement, rings a buzzer, the attractor landscape 'interprets' the sound as the telephone, and the dreamer gets up to answer it.

Incorporation studies of sleep mentation imply that external input is the origin of the sleep mentation content. A neural network attractor model, however, can 'recognize' and produce well-learned sequences of activity even when the input is sheer neural noise (i.e., possesses no information). This suggests that an activated cortical region can produce the same type of output that it was trained to do in the waking state, even when there is no concurrent external stimulus input. It further implies that the higher the level of regional activation, the deeper the attractor basins, and therefore the greater the likelihood that any noisy information will settle into one of the basins of attraction (i.e., produce mentation).

Characteristics of Sleep Mentation

Hallucinatory Quality

The ability of activated cortical attractors to produce vivid, even bizarre, imaginal sequences is characteristic not only of REM sleep but also of waking mentation, under conditions of mild sensory deprivation. The truly unique characteristic of the REM dream is its hallucinatory quality. Because inhibition of sensory input allows minimal processing of external stimuli, the constructed imaginal world is experienced as veridical perception. The dreamer imaginally interacts with the imaginal perceptions. By contrast, though waking daydreams in a dark room may be similarly vivid, the daydreamer reports them as imaginal. We assume that the hippocampal system tends to maintain a background representation of the daydreamer's external environment even when the daydreamer is not actively attending to those stimuli.

Lucid Dreaming

Opposing this hallucinatory, defining characteristic of REM dreams is the relatively infrequent phenomenon of 'lucid' dreams, whereby the 'sleeper' reports being aware that he/she is dreaming, and in some cases is able to deliberately modify the events of his/her ongoing dream. Lucid dreams are commonly reported from the final hours of morning sleep. The phenomenon suggests that the hallucinatory characteristic of dreamlike mentation may have a neurological basis different from that of other properties. Circadian-driven reticular formation/cortical circuits may not produce the deafferentation associated with the REM/NREM cycles, so that sensory, including proprioceptive, thresholds are lower than in earlier REM periods. These lower thresholds may allow the lucid dreamer to process stimuli that identify the external world and distinguish it from the internal dream experience, thereby producing the experience of 'knowing' that one is dreaming.

Personal Motives, Values, and Emotions

Night terrors are associated with the most violent emotions in sleep. These experiences occur upon sudden arousal from what is otherwise the deepest stage of NREM sleep (stage 4), and continue into a sustained hallucination of being attacked. The cry of terror and motor escape from the imagined attacker is preceded by only a few seconds of autonomic arousal, more intense than is seen in waking. The 'cause' is thought to be the loss of normal parasympathetic inhibition of the sympathetic system.

By contrast, emotions reported from REM sleep are more modest and tend not to awaken the sleeper or produce an autonomic response. We assume that the inhibitory afferent–efferent processes that protect sleep during the REM period extend to the autonomic system. Reported emotions provide us with an index of the motives that are active during sleep. This index suggests that emotions are stronger in REM than in NREM sleep, and that negative emotions are more prevalent than positive. Nevertheless, REM reports, especially in the early night, often

describe without emotion events that in the waking state would elicit a strong emotional response.

The abstract nature of motives and the difficulty of determining whether a motive or emotion precedes a visual dream event, or is elicited in response to it, make the study of motivation in sleep very difficult. Fortunately, an increasing understanding of how the amygdala, medial prefrontal, and frontal cortices represent goal or motive structures in the waking state provides us with a neurological basis for exploring how these structures may participate in the production of sleep mentation.

Positron emission tomography (PET) studies show that the amygdala is unusually active in REM sleep, relative to NREM and waking, providing a neural basis for the negative emotions characteristic of that sleep stage. That the amygdala is associated with the experience of fear and pain in the waking state suggests that it plays a similar role in REM dreaming. Following the amygdala–fear model of LeDoux, we suggest that the imaginal events continuously generated by the cortex in REM sleep are also continuously evaluated by the amygdala, via their links to the medial prefrontal cortex, for their potential imagined threat to the dreamer. Positive emotions are reported less frequently than are negative emotions in REM mentation, but, with the exception of night terrors, both are reported less frequently in NREM mentation. We note that this description of affect in REM dreams is inconsistent with the long, vivid, emotional dreams that define most people's experience of their dreams. The latter tend to be produced while sleeping late on a weekend morning, and their neural generators are circadian and homeostatic processes that amplify the generation of mentation in REM and NREM sleep.

Analyzing brain-lesioned patients, Solms demonstrated that forebrain regions with limbic connections – which, in the waking state, participate in estimating the value of stimuli – are in sleep critical for dreaming in both REM and NREM sleep. Arguing against the activation–synthesis model, he finds that dreaming can persist even in the face of massive lesioning of the pontine brain stem mechanisms responsible for REM sleep. By contrast, dreaming can be entirely eliminated by forebrain lesions, particularly in the ventromesial quadrant of the frontal lobes, an area which connects frontal and limbic structures with dopaminergic cells in the ventral tegmentum. This forebrain circuit is often referred to as the 'seeking' or 'wanting' center of the human brain. Administration of L-dopa, stimulating this area, results in excessive and intense dreaming (in addition to psychotic symptoms). Conversely, inhibition of this circuit with haloperidol inhibits dreaming. Solms therefore hypothesizes that dreaming is controlled, not by the cholinergic brain stem mechanisms described by Hobson but rather by this dopaminergic forebrain motivational system. In summary, the construction of sleep mentation is strongly influenced by neural processes that, in waking, determine the value of stimuli.

Bizarreness

There are two primary classes of bizarre mentation. The most common in laboratory reports are sudden changes in scene or location. The second, the juxtaposition of features, objects, or persons that would be improbable or impossible in the waking state, is the hallmark of the popular conception of dreaming, but is relatively infrequent in both REM and NREM sleep, especially in the early night. The neural basis for both classes is a matter of speculation. Certainly, the characteristics of attractors that make sense out of noisy input mitigate against bizarre patterns. Densely connected attractors in the same feature neighborhood tend to actively inhibit features that rarely occur together (i.e., they prevent bizarre juxtaposition). For example, in waking, a person cannot be recognized simultaneously as a boy and a girl. Yet one dreamer reported that "it was my brother," but "he was a girl" in the dream. We assume that the sense of knowing the relationship – "it was my brother" – was accomplished in an attractor field other than the field that saw his visual features as "a girl." Many dreamers do not recognize these incongruities until they awaken, at which time the normal communication between these cortical processing regions is restored. PET and transcranial magnetic stimulation studies have shown that cortical regions tend to be less effectively interconnected during sleep, as compared to wakefulness. Transcollosal communication may be especially impaired during REM sleep. We suggest that this lack of connectivity in sleep allows attractors that normally constrain one another in the waking state to produce bizarre combinations of visual objects, conversation, meanings, and responses, which would otherwise mutually inhibit one another.

The neural basis of sudden changes of topic or scene is even less well understood. We speculate that all thought, sleep or waking, is sustained by the value assigned to it by the amygdala and other limbic structures. We assume that the continuity that we, correctly or not, attribute to waking thought is due largely to the stability of our external environment. Lacking this external basis for continuity in sleep, each dream scenario simply fades away and is replaced by a new scene of significance to the dreamer.

Visual and Auditory Imagery

Visual imagery is much brighter and clearer in REM than in NREM periods (reported as being about 70% of the magnitude of brightness and clarity of waking perception). Although we assume that most of this imagery is generated in the occipital and right parietal cortex, heightened activity in extrastriate visual areas during REM sleep is accompanied by substantial decreases in blood flow to primary visual cortex. Sleepers who have been blind from birth report surprisingly normal dreams, which are apparently constructed from spatial coordinates learned in the waking state from auditory, tactile, and motor cues. Reported auditory imagery in REM sleep is also more frequent than in NREM, as are reports of conversation and the meanings of those conversations. Reports of tactile or pain imagery in sleep are extremely rare.

Total Reported Information

Given that REM reports exceed NREM on almost any variable on which they are compared, one must ask whether the cortical regions that generate mentation simply generate more information in REM sleep. With the exception of emotional intensity, after the effect of total reported information is partialled out, no other variable contributes to the qualitative difference in REM versus NREM mentation. REM reports, for example, tend to contain more bizarreness than NREM reports do, simply by virtue of their greater length. This suggests that aside from the more vivid visual imagery, which we attribute to right parietal cortex activation in REM sleep, the remaining characteristics of REM mentation may be attributed to a widespread increase in cortical activation. The active regions simply do what they were trained to do in the waking state, except that with high sensory thresholds, their processing sequences are not responsive to external stimuli, and decreased connectivity between regions results in less mutual constraint than is operative in waking perception. Many investigators have claimed, and continue to claim, that the unique pattern of cortical activation in REM must produce a unique pattern of mentation, but few have statistically corrected for the total amount of reported information, and some have even discarded reports with no content.

Circadian Influences on Sleep Mentation

Report length, and with it most characteristics of sleep mentation, tends to increase during the final hours of sleep. It is this longer, more dramatic imagery and emotion that people typically remember and identify as dreaming. Because REM periods also increase in length across the night, it has been generally assumed that these longer, more dramatic late-morning dreams are the result of a more active REM process. However, increases in report length and dreamlike quality across the night occur in NREM as well as in REM sleep, correlated with the rising phase of a circadian cortical activation cycle, roughly correlated with the core body temperature rhythm. Propensity for REM sleep, which increases across the night, is controlled by the endogenous circadian clock, as is spindle frequency activity in stage 2 and alpha activity in REM sleep. Levels of the wake-promoting neuromodulator cortisol also rise as the sleeper moves toward morning wakefulness. Each of these activation-related physiological changes depends not on the number of hours of prior wakefulness or sleep, but rather on internal clocklike mechanisms. Although circadian-driven changes in cortical activation are difficult to measure directly, core body temperature is known to begin rising about 2 h before the time of normal waking. Inasmuch as circadian-driven cortical activation does not include the inhibition of sensory input characteristic of the REM generator, REM mentation produced during this late-morning interval may sometimes tend to be lucid in the sense described earlier. The size of the circadian effect on report length can be quite large, equaling the strength of the REM/NREM effect on propensity to report dreaming.

Joint and Unique Patterns of REM and Circadian-Based Mentation

That both REM and circadian generators are associated with increased mentation reporting and dreaming places us in a better position to ask what patterns of sleep mentation are unique to, or common to, the two generators. We note that the late-morning circadian increase in NREM (as well as REM) dreaming accounts, at least in part, for the often disputed source of dreaming in NREM sleep. Consistent with the heightened amygdala activation specific to REM sleep, a study from our lab recently found that circadian-driven activation amplified the emotional intensity of late-morning REM reports, but did not effect emotion in NREM sleep.

Bizarreness Revisited

That bizarreness increases with the active phases of both the circadian and the REM cycles suggests that bizarreness is not produced by an activation process unique to REM sleep and its pontine generator. We suggest that as cortical activation increases, more objects and events are produced, and if the regions that generate these objects and events are weakly interconnected, failing to mutually constrain one another, each will be regarded as improbable in the context of the other upon awakening.

Homeostatic-Driven Processes, Naps, and Dreaming

Circadian-driven cortical activation would be expected to increase steadily throughout the morning hours and into the afternoon. The joint effect of the circadian and REM/NREM generators should therefore produce intense dreaming at this time, compared to during the night. However, these activation effects appear to be opposed by the familiar, but poorly understood, 'sleep need' process. This homeostatic process, commonly measured by propensity for slow-wave activity in the EEG, accumulates as an exponentially saturating function of time awake, and then discharges during sleep according to an exponentially decaying function. It follows that even a short period of waking might suppress the activating effects of the rising phase of the circadian rhythm. Supporting this assumption, a recent study in our laboratory found that NREM reports from early-afternoon naps, when circadian activation is presumably high, yielded less total information than did late-morning NREM reports preceded by sustained sleep. Ultimately, the generators and suppressors of cortical activation responsible for mentation can only be isolated by further studies that employ forced desynchrony protocols in which individuals sleep at various phases of the circadian rhythm.

Sleep State Ambiguity and Mentation Features

Although the vernacular labels 'wake' and 'sleep' imply that the two states are discretely different, the father of sleep research, Nathaniel Kleitman, showed that biological characteristics of the two states can be dissociated. Inasmuch as the feature patterns of night terrors and lucid dreaming, discussed earlier, are incompatible with the defining characteristics of either waking or sleep, it would obfuscate our understanding of those phenomena if we were to force them into one category or the other. This dilemma raises the question of how much weight to give to an individual's own classification of his/her sleep/wake state when it conflicts with neurological and behavioral criteria. Sleep scientists traditionally distinguished sleep from waking by EEG criteria. However, these criteria were brought into question when it was found that many normal sleepers reported having been awake when interrupted from stage 2 NREM sleep during the first 2 h of the night. Interestingly, the latency of some of these 'awake' responses was sometimes over 30 s. One systematic study of this issue found that people use the subjective quality of their preinterruption mentation to make the judgment as to whether they were awake or asleep.

Thus, if mentation were an internal subvocal mono-log common in the waking state, the stage 2 sleeper might say, "I was awake," whereas streams of vivid imagery common to REM dreaming would elicit the judgment of being asleep.

See also: Dreams and Nightmares in PTSD; History of Sleep Research; Incorporation of Waking Events into Dreams; Phylogeny and Ontogeny of Sleep; Sleep and Consciousness; Sleep: Development and Circadian Control; Theories and Correlates of Nightmares; Theories of Dream Function.

Further Reading

Antrobus J, Kondo T, and Reinsel R (1995) Dreaming in the late morning: Summation of REM and diurnal cortical activation. *Consciousness and Cognition* 4: 275–299.

Aserinsky E and Kleitman N (1953) Regularly occurring periods of eye motility and concomitant phenomenon during sleep. *Science* 188: 273–274.

Bertini M, De Gennaro L, Ferrara M, et al. (2004) Reduction of transcallosal inhibition upon awakening from REM sleep in humans as assessed by transcranial magnetic stimulation. *Sleep* 27: 875–882.

Braun AR, Balkin TJ, Wesenten NJ, et al. (1997) Regional cerebral blood flow throughout the sleep–wake cycle: An H2O-O-15 PET study. *Brain* 120: 1173–1197.

Dement W and Wolpert EA (1958) The relation of eye movements, body motility, and external stimuli to dream content. *Journal of Experimental Psychology* 55: 543–53.

Ellman S, and Antrobus J (eds.) (1991) *The Mind in Sleep*, 2nd edn. New York: Wiley.

Hobson J and McCarley R (1977) The brain as a dream state generator: An activation-synthesis hypothesis of the dream process. *American Journal of Psychiatry* 134: 1335–1348.

Hobson JA, Pace-Schott E, and Stickgold R (2000) Dreaming and the brain: Towards a cognitive neuroscience of conscious states. *Behavioral and Brain Sciences* 23: 793–842.

Maquet P, Ruby P, Maudoux A, et al. (2005) Human cognition during REM sleep and the activity profile within frontal and parietal cortices: A reappraisal of functional neuroimaging data. *Progress in Brain Research* 150: 219–227.

Massimini M, Ferrarelli F, Huber R, et al. (2005) Breakdown of cortical effective connectivity during sleep. *Science* 309: 2228–2232.

Moruzzi G and Magoun HW (1949) Brain stem reticular formation and activation of the EEG. *Electroencephalography and Clinical Neurophysiology* 1: 455–473.

Nielsen TA (2000) A review of mentation in REM and NREM sleep: "Covert" REM sleep as a possible reconciliation of two opposing models. *Behavioral and Brain Sciences* 23: 793–1121.

Pivik T and Foulkes D (1968) NREM mentation: Relation to personality, orientation time, and time of night. *Journal of Consulting and Clinical Psychology* 32: 144–151.

Smith M, Antrobus J, Gordon E, et al. (2004) Motivation and affect in REM sleep and the mentation reporting process. *Consciousness and Cognition* 13: 501–511.

Solms M (2000) Dreaming and REM sleep are controlled by different mechanisms. *Behavioral and Brain Sciences* 23: 793–850.

Wamsley EJ, Hirota Y, Tucker MA, et al. (2007) Circadian and ultradian influences on dreaming: A dual rhythm model. *Brain Research Bulletin* 47(4): 347–354.

Dreams and Nightmares in PTSD

A Germain, University of Pittsburgh School
of Medicine, Pittsburgh, PA, USA
A Zadra, Université de Montréal, Montréal, QC, Canada

Clinical Significance of Nightmares in Posttraumatic Stress Disorder

Posttraumatic stress disorder (PTSD) refers to symptoms of reexperiencing, avoidance, and hyperarousal that persist for more than 1 month after exposure to a traumatic event. Violent crimes, including rape, physical assaults, and combat exposure, constitute traumatic events that involve threat to integrity of the self or to others and are accompanied by intense fear, helplessness, or horror. Epidemiological studies indicate that community prevalence estimates of PTSD are in the range of 1–10%, with higher estimates reported for victims of interpersonal violence (20–30%) and combat veterans (15–30%). Recommended first-line treatments of PTSD include selective serotonin reuptake inhibitors (SSRIs) and cognitive-behavioral interventions such as prolonged exposure and cognitive restructuring.

Nightmares are a core feature of PTSD. As many as 90% of trauma-exposed individuals who develop PTSD report disturbing dreams that bear varying degrees of resemblance to the actual traumatic event. Growing evidence suggests that trauma-related nightmares may directly contribute to the pathophysiology of PTSD. The persistence of nightmares and sleep disruption 1 month posttrauma increases the likelihood of developing PTSD at follow-up assessments. Moreover, a personal history of nightmares prior to trauma exposure can predict the severity of PTSD and other posttraumatic psychiatric symptoms. Frequent recurrent nightmares about a traumatic event can persist for years and even decades after the trauma. Trauma-related nightmares disrupt sleep and independently contribute to increased severity of daytime PTSD symptoms and increased alcohol use in trauma survivors, and adversely affect quality of life and perceived physical health.

Nightmares and Sleep following Trauma Exposure and in Chronic PTSD

In healthy persons, elaborate dream recall is more prominent following awakening from rapid eye movement (REM) sleep than from awakening from non-REM (NREM) sleep. In trauma-exposed adults, two preliminary studies suggest that disruption of REM sleep, as indicated by increased sleep stage transitions and arousal, as well as increased sympathovagal tone during REM sleep are associated with increased PTSD symptom severity and may increase the likelihood of developing PTSD at follow-up assessments. Thus, REM sleep disruption and increased arousal during REM sleep may increase one's vulnerability to experience vivid nightmares following trauma exposure and to develop PTSD.

REM sleep disturbances have been most consistently reported in patients suffering from chronic PTSD, but other sleep alterations, such as reduction in slow-wave sleep, have also been documented. More general sleep disturbances, such as increased sleep latency, total sleep time, and duration of and number of nocturnal awakenings, have been reported in some, but not all, studies. Polysomnographic studies based on refined objective measures of central and peripheral physiological arousal, such as quantitative electroencephalography and quantitative electrocardiography, have observed indices of heightened central and peripheral arousal during sleep in patients suffering from PTSD. PTSD in patients who report trauma-related nightmares show increased duration of wakefulness after sleep onset in comparison to those who report non-trauma-related nightmares. However, REM sleep parameters, or more global sleep measures, do not correlate with specific aspects of nightmare content. These observations suggest that in chronic PTSD, trauma-related nightmares may represent expressions of a more global sleep disturbance related to heightened arousal.

Dream Recall and Nightmare-Related PTSD

Studies on dream recall frequency following trauma exposure have yielded inconsistent results. Increases in dream recall frequency following a traumatic event have been reported and the hypothesis put forth is that dream recall is facilitated by the intensity, affective charge, and references to the victims' trauma depicted in their dreams. Conversely, significant decreases in dream recall have also been documented. These data, however, are largely based on retrospective assessments conducted many months or even several years following trauma exposure. It is possible that dream recall increases immediately following an individual's exposure to trauma and subsequently diminishes over time. One comparison of high-functioning Holocaust survivors, average-functioning Holocaust survivors, and controls found that the high-functioning survivors reported fewer dreams from REM sleep (33.7%), compared to the average-functioning survivors (50.5%)

and controls (80.8%). A reduction in dream recall may thus reflect improved posttrauma adaptation in this population.

Repetition of Trauma in Dreams

The recall of dreams that replicate the traumatic event experienced by trauma victims has been described in many populations, including war veterans, adults and children exposed to war, witnesses of violence and abuse, burn victims, and individuals exposed to natural disasters. A classification system for trauma-related dreams has been put forth. Posttraumatic dreams refer to the replication of the traumatic event during sleep or the introduction of the encapsulated memory of the trauma during sleep. Modified dreams present distorted themes, elements, or emotions from the trauma. Finally, disguised dreams represent the traumatic event symbolically or metaphorically. These categories form a continuum of replication (according to the extent to which the traumatic event is repre-sented in the dream) and of repetition (i.e., the dream's recurrence). Empirical studies support this classifica-tion system. For example, in one study of Vietnam War veterans, 304 of 316 veterans reported combat-related nightmares. An examination of their dream narratives revealed that over half of the veterans reported realistic combat dreams, 21% reported plausible war sequences that they nevertheless had not experienced, and 26% reported dreams that alluded to the war, but also included fantastical and everyday elements. In addi-tion, a laboratory study of war veterans found that only 21% of dreams exactly replicated the traumatic event, while the majority contained distortions related to the traumatic event. Similarly, one study of patients hospitalized following an accident or assault found that 46% of reported dreams replicated the traumatic event, 33% were dissimilar to the traumatic event while containing high levels of distress, and 19% showed little similarity to the trauma or distress.

Taken together, the results reveal changes in dream content with the passage of time and with improvement in the posttrauma reaction. Dreams recalled during the initial posttrauma phase tend to include some type of repetition of the trauma while subsequent phases are characterized by more symbolic representations of the trauma and greater integration of the individual's everyday life in recalled dream content.

Trauma and Dream Content

Several studies have reported salient alterations in the content of trauma victims' dreams. Some data suggest that when compared to the dream reports from controls, dreams recalled by trauma victims tend to be more ordinary and realistic and less salient, bizarre, and imaginative. The 'ordinary' quality of trauma victims' dreams may reflect an alteration in the process of dream construction or may be related to a protective mechanism which keeps traumatic images from intruding into everyday dreams.

Intense emotions have been documented in several studies of trauma victims' dreams. Feelings of horror, anxiety, rage, sadness, and frustration are among the most frequently reported. PTSD patients rate their nightmares as being significantly more distressing than do non-PTSD patients. Emotional preoccupations prior to bedtime may also play a role in the affective content experienced during subsequent sleep. The finding that one's mood prior to sleep is inversely related to positively or negatively experienced dream affect supports the idea that dreams may help regulate emotional equilibrium in traumatized individuals.

Some research has been directed at delineating rela-tionships between measures of dream content and specific types of traumas.

Victims of Physical and Sexual Abuse

Themes of attack, pursuit, and of one's own death are more frequently found in the dreams recalled by vic-tims of sexual abuse than in those recalled by control participants. Episodes of verbal and physical abuse are more evident in the dreams of physically abused women when compared to the dreams of controls. Although themes of violence and aggression often char-acterize women's nightmares, the nightmares reported by victims of sexual abuse are more likely to include unique themes of blood and dismemberment as well as a higher occurrence of male strangers. In addition, the dreams and nightmares of sexually abused women contain more frequent references to negative sexual activity, including themes concerning lack of trust and shame, guilt, jealousy, anger, and violence. These dreams often depict situations wherein elements of sex and aggression are confounded, in which sexuality is unpredictable and often results in fear. Victims of sexual abuse also frequently report the presence of serpents and worms in their dreams as well as refer-ences to both sexual and nonsexual body parts.

Victims of War and Violence

Exposure to an environment of war or violence is associated with dreams containing high proportions of aggressive and hostile human interactions. Themes involving immediate dangers to one's life appear more frequently in the dreams of Holocaust survivors, chil-dren that have been kidnapped, and children involved in car accidents than in nontrauma populations. It is not unusual for war veterans to report that their traumatic dreams entail being killed in the place of their compatriots. A study of war-exposed Palestinian children found that traumatized children had more dreams with threatening strangers. Their dreams were often composed of attacks, anxiety, persecution, hostility, and nondesirable endings. In addition, nontraumatized children report more dreams of school and of their peers than do traumatized children.

That said, conflicting results exist regarding the impact of trauma and PTSD on dream content. For example, in addition to having more dreams with threatening strangers, victims of violence also report more dreams referring to their family, house, affiliations, and human connections than do controls.

Frequency of Dream-Related Disorders

Dream-related disorders, such as posttraumatic dreams, nightmares, bad dreams, and recurrent dreams, are the most frequently reported and most persistent symptoms exhibited by trauma victims. Several factors appear to mediate this association. First, a positive relationship exists between the degree and severity of trauma exposure and the frequency of dream-related disorders. Second, the victim's psychological reaction to trauma (e.g., level of anxiety) positively influences the occurrence of nightmares and other dysphoric dream experiences. Third, the time elapsed since trauma exposure appears to play an important role, with reductions in dream-related disorders occurring over time. However, it should be noted that dream disturbances can nevertheless persist for many years following the trauma.

Functional Hypotheses of PTSD-Related Nightmares

Many contemporary dream theorists suggest that dreaming is functionally significant and may subserve a biologically important function, but some argue that dreams are epiphenomenal to neurophysiological activity during REM sleep and have no value in and of themselves. While this overarching question on the function of dreaming is of phenomenological interest, the resolution of this debate is of little clinical relevance for chronic, frequent, and distressing nightmares in PTSD patients. However, understanding the possible role(s) of nightmares as one of the normal initial reactions to trauma exposure, which may also become a perpetuating condition contributing to maladaptive, chronic trauma response, has direct clinical implications regarding the development and refinement of effective nightmare-specific treatments. The following sections describe the main hypotheses put forth to explain the possible dual role of nightmares occurring posttrauma.

Nightmares and Emotional Adaptation to Trauma

Several neuropsychophysiological hypotheses of the function of dreaming have been adapted to explain the occurrence of nightmares after exposure to a trauma or other significant life events, and the role of nightmares in the pathogenesis and maintenance of PTSD. Freud proposed that nightmares reflect attempts to master anxiety and guilt associated with a traumatic experience, and to integrate traumatic experiences into one's psyche during sleep. Contemporary hypotheses are based on a similar premise – that dreaming serves a function of emotional adaptation to emotionally salient or traumatic events – and combine empirical observations supporting the role of REM sleep (and by association, dreaming) in learning, memory consolidation, emotional processing, and adaptation to stress. Given that REM sleep is associated with muscle atonia, and given the lack of detectable physiological activation during REM sleep episodes preceding awakening from nightmares, dreaming and REM sleep may provide a unique psychophysiological milieu during which traumatic memories can be reexperienced in the absence of physiological arousal. This process could facilitate the attenuation of emotional and physiological responses associated with these memories, leading to desensitization to the trauma-related memories, including a reduction of nightmares. Prospective dream collection studies in combination with repeated objective assessment of sleep patterns and other physiological parameters of central and peripheral arousal in people at high risk for trauma exposure (e.g., disaster workers, police officers, firefighters, deployed military personnel) are necessary to assess the role of nightmares in the recovery of trauma response and PTSD.

Stickgold proposed an explicit REM sleep model of the neurobiological substrates potentially subserving emotional processing and integration of trauma-related memories. Specifically, it is suggested that REM sleep provides a unique neurochemical and neurobiological brain state that allows the transfer of hippocampally mediated episodic traumatic memories and amygdala-dependent salient affect into cortically distributed semantic networks. It should be noted that the possibility has not been evaluated that PTSD-related nightmares may also reflect the failure of compensatory mechanisms aimed at preserving processes such as emotional adaptation, memory consolidation, and sleep continuity.

Sensitization Hypothesis of PTSD Nightmares

While the adaptation process may be successful in individuals who recover from trauma exposure, the persistence of nightmares in those who develop PTSD may, instead, reflect a failure of dreaming and REM sleep to fulfill their role of emotional adaptation and memory integration. In this perspective, nightmares are the outcome of unsuccessful processing during dreaming and REM sleep. In addition, nightmares may directly contribute to the failure of the adaptation and integration process by promoting sensitization (rather than desensitization) to trauma-related memories. PTSD-related nightmares are often associated with awakenings and are accompanied by intense emotional and physiological arousal. Trauma-related nightmare content, sleep disruption induced by awakening from a nightmare, and heightened physiological and emotional arousal levels that follow such awakenings may contribute to reinforce the occurrence of nightmares and related distress during both wakefulness and sleep. In other words, the recall of nightmares can expose patients to their prior trauma and may even induce retraumatization, since PTSD nightmares often repeat a traumatic situation in whole or in part, whereas habituation to physiological arousal elicited by these memories does not occur.

There is evidence that frequent and distressing PTSD-related nightmares are associated with increased severity of daytime reexperiencing, arousal, and avoidance PTSD symptoms. Alternatively, some PTSD patients who show clinically significant improvements in PTSD following exposure-based therapy also report improvements in trauma-related nightmares, suggesting that habituation to trauma memories during wakefulness can also attenuate nightmare frequency, intensity, and related distress. Desensitization therapy has also been used successfully for the treatment of trauma-related nightmares. These observations support the hypothesis that chronic nightmares without physiological habituation may indeed contribute to the maintenance of PTSD by sensitizing patients to trauma memories.

Potential Psychophysiological and Neurobiological Underpinnings of PTSD-Related Nightmares

While several polysomnographic studies have been conducted in PTSD samples, these sleep measurement methods do not allow the identification of neurobiological underpinnings of trauma-related nightmares or, more generally, of PTSD during sleep. Such an endeavor requires the use of functional sleep neuroimaging approaches. Sleep neuroimaging has not yet been conducted in PTSD patients. However, animal models of the effects of fear conditioning on sleep, and neuroimaging sleep studies in humans, have provided valuable insights into the potential neurobiological underpinnings of altered REM and NREM sleep mechanisms following stress exposure PTSD,

and into the neurobiology of normal sleep, respectively. While animal models and sleep imaging in healthy persons limit possible extrapolation of neurobiological underpinnings of trauma-related nightmares, findings derived from these two distinct fields provide evidence for functionally significant overlaps between the neurobiology of fear and sleep regulation.

Functional neuroimaging studies conducted during wakefulness in PTSD patients are consistent with animal models of fear conditioning. PTSD is associated with hyperresponsiveness of the amygdala to threat-related stimuli, and/or a with blunted response of medial prefrontal cortical regions, which exert inhibitory control over the amygdala. Other functional findings include increased thalamic activity with trauma-related stimuli in PTSD compared to non-PTSD subjects, and hyperresponsiveness of the noradrenergic system. Of note, neuronal activity in these structures cannot be captured by conventional sleep measurement methods, which further reinforces the need to use state-of-the-science sleep neuroimaging methods in patients with PTSD.

In rodents, cued fear conditioning increases REM sleep latency, decreases REM sleep duration and number of REM bouts, and increases pontogeniculooccipital (PGO) waves, a marker of alerting mechanisms during sleep and wakefulness analogous to REMs in humans. The effects of fear conditioning on REM sleep in rodents are mediated by amygdalar projections to brain stem regions involved in alerting response and REM sleep generation. Anatomically, the interconnections that the amygdala shares with the basal forebrain, hypothalamus, preoptic area of the anterior hypothalamus, brain stem reticular formation, and solitary tract nucleus allow the amygdala to influence both wakefulness-promoting and sleep-promoting areas. Neuronal firing of the amygdala varies across the sleep–wake cycle, and stimulation of the amygdala during REM sleep increases PGO waves in REM and NREM sleep, whereas inactivation of the amygdala with tetrodotoxin decreases sleep latency and increases slow-wave activity during wakefulness, REM sleep, and NREM sleep. Conversely, ablation of the amygdala in rhesus monkeys is associated with increased sleep consolidation and total sleep time. These findings indicate that the amygdala is an important modulator of sleep and wakefulness regulation mechanisms.

Sleep neuroimaging studies conducted in healthy persons have reliably shown that REM sleep is a natural activator of the amygdala and anterior paralimbic areas and of the medial pons and thalamus. REM sleep is also associated with relative deactivation of lateral prefrontal parietal regions and primary sensory cortices. These selective activation and deactivation patterns during REM sleep relative to wakefulness have yielded the hypothesis that dreams may reflect the mental representations of high limbic activations in conjunction with deactivation of high-order cortical regions. NREM sleep is a natural deactivator of arousal-promoting structures.

The functionally significant interconnections between the amygdala and wakefulness- and sleep-promoting areas, the unique limbic activation and cortical deactivation patterns observed during sleep in healthy persons, and the functional neuroanatomical findings in PTSD patients during wakefulness all support a role for a heightened amygdalar activity during sleep in PTSD, and provide preliminary guidance for the investigation of the neurobiology of PTSD during REM and NREM sleep. While the relationship between heightened amygdalar activity and dream content cannot yet be ascertained, the role of the amygdala in acquisition of the fear response and the emotional salience of stimuli raises the possibility that heightened amygdalar activity during REM sleep also contributes to salient, negative aspects of dreams and nightmares.

Treatments of PTSD-Related Nightmares

Nightmares are closely related to sleep disturbances, including increased sleep latency, decreased total sleep time, and increased number and duration of awakenings after sleep onset and restless or fitful sleep. Nightmares often produce an insomnia-like pattern of sleep disturbance, if not outright insomnia. Sleep avoidance and fear of darkness, frequent features associated with PTSD-related nightmares, can further reinforce the maintenance of both nightmares and insomnia.

Recommended (by the Expert Consensus Guideline Series on the treatment of PTSD) first-line treatments include Food and Drug Administration (FDA)-approved SSRIs such as sertraline and paroxetine, and cognitive-behavioral interventions such as prolonged exposure and cognitive therapy. However, nightmares are often resistant to such first-line PTSD treatments, and adjunctive treatments are often required to alleviate PTSD-related nightmares.

Several psychological and pharmacological nightmare-focused treatments have been shown to effectively reduce and eliminate nightmares in PTSD patients. Successfully treated nightmare patients often report improvements in sleep quality, feeling more rested upon awakening and having more daytime energy, and reduction in nightmares is a significant predictor of sleep improvement. In addition, nightmare reduction is associated with marked improvements in daytime PTSD symptoms, quality of life, depression and anxiety symptoms, and overall functioning. These findings suggest that at least in some cases,

chronic nightmares may represent a primary sleep disorder rather than a symptom of a psychiatric disorder, and as such can benefit from targeted, adjunctive cognitive-behavioral or pharmacological treatments.

Imagery Rescripting and Rehearsal

Numerous techniques have been proposed to treat PTSD-related nightmares, including hypnosis, lucid dreaming, eye movement desensitization and reprocessing, desensitization, and imagery rescripting and rehearsal (IRR). However, only desensitization and IRR have been the objects of controlled studies. Only IRR has been shown to effectively reduce nightmare frequency and nightmare-related distress in patients suffering from idiopathic, recurrent, or PTSD-related nightmares.

IRR for nightmares is composed of two general elements, each of which utilizes several steps in the therapeutic process. The first component is an educational/cognitive restructuring element, focused on helping nightmare sufferers to consider their disturbing dreams as a learned behavior. The second component is an imagery education/training element, which teaches nightmare patients about the nature of human imagery and how to implement a specific set of imagery steps to decrease nightmares. Generally, patients are shown how a nightmare can be effectively treated as a problem in and of itself, without any discussion or emphasis on prior traumatic events or nonsleep-related PTSD symptoms. As such, IRR seeks to minimize the use of exposure therapy as an ingredient in its procedures.

When learning IRR, patients are instructed to modify an original nightmare in any manner of their preference, and to create a new dream scenario. The new dream scenario is then rehearsed in session, and practiced each day until the next session. The rationale is that the rescripting and rehearsal of new dream imagery during the day can replace or alter nightmare scenarios experienced during sleep. Generally, patients report reductions in nightmare frequency and intensity within 6–12 weeks, and improvements are maintained for as long as 30 months.

Several variations of IRR have now been described. The distinguishing features between variations of IRR generally revolve around the degree of exposure used during treatment sessions and/or the specific application of the technique during the sessions. IRR can be delivered in group or individual formats. The number of sessions required for IRR can vary from one to six sessions. Since IRR involves minimal exposure and abreaction, its effectiveness is most likely related to the patient's ability to increase mastery over distressing dream elements by generating new dreams. Research on the scripting of new dreams linked to IRR indicates that mastery is in fact a key ingredient in this treatment. The observed increase in mastery is all the more remarkable given that IRR does not include instructions to increase mastery when the technique is described during treatment. Although these data are consistent with the view that a dysfunctional imagery system is involved in traumatic nightmares, more work is required to elucidate the relationship between variables related to patient characteristics (e.g., traumatized vs. nontraumatized, presence of psychiatric and sleep disorders, dispositional factors, degree of distress), nightmare content and type (e.g., replicative, recurrent, lifelong, idiopathic), and therapeutic effects (e.g., enhanced self-efficacy, perceived control, modes of emotion expression and representation).

Prazosin

Prazosin is a central α1 adrenoreceptor blocker that is FDA-approved for the treatment of hypertension. Prazosin is currently the most efficacious treatment for PTSD-related nightmares in military veterans and civilian populations with PTSD. Promising results for the use of prazosin to reduce PTSD-related nightmares became available after the Expert Consensus Guideline Series on the treatment of PTSD were established. Results from controlled trials indicate that prazosin effectively reduces chronic PTSD-related nightmares in military and civilian sample groups with PTSD. Reduction in nightmares with prazosin is associated with clinically meaningful improvements in sleep quality, overall PTSD severity, and daytime functioning in combat veterans with chronic PTSD. Most patients experience a recurrence of insomnia and nightmares upon prazosin discontinuation. The drug is well tolerated by patients, is associated with minimal side effects, and no adverse events have been reported with long-term use.

Serotonin-Potentiating Drugs

Randomized controlled trials on the efficacy of FDA-approved SSRIs for the treatment of PTSD (i.e., sertraline and paroxetine) either did not measure improvements in sleep using validated measures, or reported minimal improvements in nightmares or overall sleep quality. In fact, some SSRIs may increase the number of arousals during sleep and decrease total sleep time. Open-label trials with fluvoxamine suggest moderate improvements in PTSD-related nightmares, but randomized controlled trials have not been conducted and side effects may impede tolerability. Open-label trials and case reports suggest that serotonin-potentiating non-SSRIs, such as nefazodone and trazodone, can be associated with reductions in PTSD-related nightmares and sleep disruption. However, double-blind randomized controlled trials

in PTSD samples are not available. The need for rigorous randomized controlled trials is highlighted by the example of cyproheptadine. Initial case series suggested that cyproheptadine was a potent agent for the alleviation of PTSD-related nightmares. However, a subsequent double-blind, randomized controlled trial in Vietnam veterans with PTSD found that cyproheptadine could in fact exacerbate nightmares and sleep disruption. A few randomized controlled trials have evaluated the efficacy of monoamine oxidase inhibitors and tricyclic antidepressants for PTSD, and slight to moderate improvements were reported in some, but not all, studies. Optimal doses, durability of therapeutic gains, side effect profiles, and tolerability have not been assessed specifically for PTSD-related nightmares and sleep disruption with non-SSRIs, monoamine oxidase inhibitors, or tricyclic antidepressants. It should be noted that very few trials targeting clinically meaningful reductions in PTSD-related nightmares and sleep disturbances have used specific nightmare measures (e.g., questionnaires, sleep or dream logs) as primary outcome measures. Such measures are necessary to determine the effects of targeted treatments on specific aspects of nightmares, such as frequency, intensity, nightmare-induced sleep disruption, and associated distress.

Other Pharmacological Agents

Several medications that have sedative properties have been used in PTSD patients. Antipsychotics such as olanzapine and quetiapine have been tested in PTSD and have been associated with mixed findings as to their potential for reducing PTSD-related nightmares and insomnia. Despite the lack of evidence for their efficacy in reducing both daytime and nighttime PTSD symptoms, benzodiazepines are often prescribed to patients with PTSD. An open trial with zolpidem, a nonbenzodiazepine hypnotic, reported improvements in sleep and some improvements in nightmares in military veterans with PTSD. Case series reports on other pharmacological agents, such as topiramate, respiridone, and gabapentin, among others, are also available. While most cases report some degree of improvement in nightmares and insomnia in PTSD patients, proper randomized controlled trials are necessary to fully assess the efficacy of these agents as adjunctive PTSD treatments for nightmares and insomnia, and to determine associated health risks in light of side effects such as tolerance, weight gain, and glucose dysregulation.

See also: Incorporation of Waking Events into Dreams; Nightmares; PET Activation Patterns; Phylogeny and Ontogeny of Sleep; REM/NREM Differences in Dream Content; Sleep and Consciousness; Theories and Correlates of Nightmares; Theories of Dream Function.

Further Reading

Barrett D (ed.) *Trauma and Dreams*. Cambridge: Harvard University Press.

Foa EB, Davidson JRT, and Frances A (eds.) (1999) The Expert Consensus Guideline Series: Treatment of posttraumatic stress disorder. *Journal of Clinical Psychiatry* 60(supplement 16).

Harvey AG, Jones C, and Schmidt DA (2003) Sleep and posttraumatic stress disorder: A review. *Clinical Psychology Review* 23: 377–407.

Krakow B, Hollifield M, Johnston L, et al. (2001) Imagery rehearsal therapy for chronic nightmares in sexual assault survivors with posttraumatic stress disorder: A randomized controlled trial. *JAMA: Journal of the American Medical Association* 286: 537–545.

Maher MJ, Rego SA, and Asnis GM (2006) Sleep disturbances in patients with post-traumatic stress disorder: Epidemiology, impact and approaches to management. *CNS Drugs* 20: 567–590.

Mellman TA, David D, Bustamante V, et al. (2001) Dreams in the acute aftermath of trauma and their relationship to PTSD. *Journal of Traumatic Stress* 14: 241–247.

Mellman TA and Hipolito MM (2006) Sleep disturbances in the aftermath of trauma and posttraumatic stress disorder. *CNS Spectrums* 11: 611–615.

Raskind MA, Peskind ER, Kanter ED, et al. (2003) Reduction of nightmares and other PTSD symptoms in combat veterans by prazosin: A placebo-controlled study. *American Journal of Psychiatry* 160: 371–373.

Rothbaum BO and Mellman TA (2001) Dreams and exposure therapy in PTSD. *Journal of Traumatic Stress* 14: 481–490.

Stickgold R (2002) EMDR: A putative neurobiological mechanism of action. *Journal of Clinical Psychology* 58: 61–75.

Wood J, Bootzin R, Rosenhan D, et al. (1992) The effects of the 1989 San Francisco earthquake on frequency and content of nightmares. *Journal of Abnormal Psychology* 101: 219–224.

Relevant Websites

http://www.aasmnet.org – American Academy of Sleep Medicine.

http://www.istss.org – International Society for Traumatic Stress Studies.

http://www.ncptsd.org – National Center for PTSD (US Department of Veterans Affairs).

Theories and Correlates of Nightmares

T Nielsen, Université de Montréal, Montreal,
QC, Canada
R Levin, Ferkauf Graduate School of Psychology,
Bronx, NY, USA

Nightmares Defined

Nightmares are currently defined in both the *Diagnostic and Statistical Manual of Psychiatric Disorders, Fourth Edition, Text Revision* (DSM-IV-TR) and the *International Classification of Sleep Disorders, Second Edition* (ICSD-2) (**Table 1**). These definitions are consistent in linking nightmares to abrupt rapid eye movement (REM) sleep awakenings with a clear recall of primarily fearful dream content. The more recent ICSD-2 criteria acknowledge that nightmares may involve dysphoric emotions other than fear and anxiety and distinguish idiopathic nightmares from the more severe and highly distressing nightmares associated with posttraumatic stress disorder (PTSD). However, the ICSD-2 does not subscribe to a criterion of subjective distress as does the DSM-IV-TR, even though converging evidence suggests that one's habitual manner of affective responding (nightmare distress) is an important determinant of nightmares as a clinical problem.

Nightmare Prevalence and Correlates

Large community-based epidemiological studies demonstrate a high prevalence of occasional nightmares: about 85% of adults report at least one within the past year, whereas 8–29% report at least one per month. These same studies indicate that 2–6% of respondents report weekly nightmares, a frequency widely considered to be of moderately severe pathology. The 2–6% range is robust cross-culturally, with similar rates reported in Canada, Europe, Japan, the Middle East, and America. This range may nonetheless underestimate the real prevalence of nightmares for a number of reasons, such as reliance on retrospective measures and undergraduate samples, inconsistent operational definitions, and failure to distinguish traumatic from nontraumatic nightmares.

Frequent nightmares are up to three to four times as prevalent in childhood and adolescence as in adulthood. According to the DSM-IV-TR, 10–50% of children aged 3–5 years of age have disturbing dreams, with a prevalence that increases through early adolescence. Nightmare 'problems' (duration >3

months) occur with prevalences in the same order of magnitude: 24%, 41%, and 22% for children between age ranges of 2–5, 6–10, and 11–12 years, respectively. Evidence from the Finnish Twin Cohort, a community-based sample of 1298 monozygotic and 2419 dizygotic twin pairs aged 33–60 years, suggests that nightmares are a common and stable trait from childhood to middle age and are substantially affected by genetic factors. In contrast, nightmare prevalence and frequency in elderly populations are considerably lower than rates found in younger adults.

Females at all ages consistently report nightmares at significantly higher rates than do males, although some portion of this difference may be attributed to trauma-induced nightmares. This consistent gender gap in nightmares appears to first emerge around early adolescence. Nightmares are more frequent and more prevalent in psychiatric populations and are associated with a diversity of symptoms: anxiety, neuroticism, and global symptom reporting, schizophrenia-spectrum symptoms; heightened risk for suicide; dissociative phenomena; health behavioral problems; and sleep disturbances. Nightmares are also linked to a diversity of psychopathological traits. They are particularly prevalent in PTSD, for which they are considered a hallmark symptom. A broad range of traumatic events may trigger nightmares: combat exposure, motor vehicle accidents, natural disasters, crime victimization, and rape.

Nightmares are also reactive to intense stress. More frequent nightmares are reported during increased life stress and are associated with specific stressors, such as anticipated surgery, experimental pain stimulation, menstruation, pregnancy, miscarriage, new motherhood, preparation for exams, sham intelligence testing, stock market downturns, and the viewing of disturbing movies. Their onset is often immediately preceded by stressful events, such as death of someone close, interpersonal conflict, news of a disaster (e.g., an earthquake), and major life events in general.

Nightmares also accompany several chronic health problems, including migraine, bronchitis/asthma, chronic obstructive airways disorder, cardiac disease, and substance abuse. In addition, individuals who report frequent nightmares are characterized by a number of personality variables strongly associated with waking emotional distress: heightened physical and emotional reactivity, imagery vividness, fantasy proneness, maladaptive coping, and thin boundaries.

The association of nightmares with this wide spectrum of pathological symptoms and conditions, all of which are marked by considerable waking emotional

Table 1 DSM-IV-TR and ICSD-2 diagnostic criteria for nightmare disorder[a]

Component	DSM-IV-TR nightmare disorder (2000)	ICSD-2 nightmare disorder (2005)
Nature of dream recall	A. Repeated awakenings from the major sleep period or naps with detailed recall of extended and extremely frightening dreams, usually involving threats to survival, security, or self-esteem	A. Recurrent episodes of awakenings from sleep with recall of intensely disturbing dream mentation, usually involving fear or anxiety but also anger, sadness, disgust, and other dysphoric emotions
Nature of awakening	B. On awakening from the frightening dreams, the person rapidly becomes oriented and alert (in contrast to sleep terror disorder and some epilepsies)	C. Alertness is full immediately on awakening, with little confusion or disorientation. Recall of sleep mentation is immediate and clear
Nature of associated distress	C. The dream experience, or the sleep disturbance resulting from the awakening, causes clinically significant distress or impairment in social, occupational, or other important areas of function	D. Associated features may include: • Return to sleep after the episodes is typically delayed and not rapid
Timing of event	A. The awakenings generally occur during the second half of the sleep period	D. Associated features may include: • The episodes typically occur in the latter half of the habitual sleep period
Physiological criteria	None	None
Differential diagnosis	D. The nightmares do not occur exclusively during the course of another mental disorder (e.g., a delirium, posttraumatic stress disorder) and are not due to the direct physiological effects of a substance (e.g., a drug of abuse, a medication) or a general medical condition	Nightmares are distinguished from: seizure disorder, arousal disorders (sleep terrors, confusional arousal), REM sleep behavior disorder, isolated sleep paralysis, nocturnal panic, posttraumatic stress disorder, acute stress disorder

[a]DSM-IV-TR, *Diagnostic and Statistical Manual of Mental Disorders, Fourth Edition, Text Revision*; ICSD-2, *International Classification of Sleep Disorders, Second Edition*. Items A–D refer to specific definitional criteria described in the manuals; REM, rapid eye movement. Data from American Psychiatric Association (2000) *DSM-IV-TR: Diagnostic and Statistical Manual of Mental Disorders*, 4th edn. Arlington, VA: American psychiatric association; and American Academy of Sleep Medicine (2005) *International Classification of Sleep Disorders: Diagnostic & Coding Manual*, 2nd edn. Westchester, IL: American Academy of Sleep Medicine.

distress, supports the contention that nightmare production is related to a personality style characterized by intense reactive emotional distress.

Models of Nightmare Production

Despite a literature that has proliferated over the past two decades, relatively little is known about the etiology and pathogenesis of nightmares. There is a surprising paucity of working models of nightmare production, with most existing attempts being based on clinical observation but little experimental verification. Many existing models address nightmares only indirectly within the broader context of theories about dream function or personality. Further, some models are concerned almost exclusively with elucidating a nightmare function but not how nightmares are formed.

Psychoanalytic and Neopsychoanalytic Models

It is widely accepted that Freud's dream theory failed to deal adequately with nightmares. To incorporate them into his theoretical framework, Freud resorted to the notion of masochistic wish-fulfillment dreams, within which anxiety feelings stemmed from the transformation of libido. Despite the unpopularity of this idea, twentieth-century theories of nightmare production were nonetheless largely informed by

Freud's work. For example, various neo-Freudian models proposed that nightmares were expressions of repressed incestuous impulses, were residues of unresolved psychological conflicts, were attempts to transform shame into fear, and so forth. Some of these views (e.g., incestuous impulse theory) have held little currency in contemporary deliberations over processes of nightmare production, whereas others have continued to wield considerable influence. For example, one contemporary model embraces an explicitly psychoanalytic framework in describing a three-stage neurological mechanism responsible for normal dreaming and nonrecurring nightmares. Other, more recent, mastery and adaptation models of dream function (see later) are also broadly consistent with neopsychoanalytic orientations.

Personality and Evolutionary Models

Boundary permeability Based on extensive clinical work with frequent lifelong nightmare sufferers, Hartmann proposed the personality dimension of 'boundary permeability' to explain nightmare pathology. Nightmare sufferers fall to one extreme of this dimension ('thin' boundaries) by virtue of their striking openness, sensitivity, and vulnerability to cognitive and emotional intrusions – and possibly by biologically based differences as well. Hartmann reasoned that thin-boundary individuals were susceptible to internal events not usually perceived by most others as threatening or traumatic, that they had difficulty discerning internal fantasy from external reality, and that they were prone to source monitoring disturbances across various states of consciousness. Their boundary 'thinness' thus consists in a vulnerability to spikes of heightened emotional distress; during dreaming this vulnerability leads to nightmares.

A modest body of empirical work supports the validity of the Boundary Questionnaire created to measure an individual's boundary thinness or thickness. As predicted by the model, thin-boundary persons have high levels of dream recall, dream bizarreness, and nightmare recall. Relationships have been found for adolescents as well as for adults, but not for healthy elderly individuals. However, failures to find relationships between boundary thinness and nightmare recall have also been reported. In general, despite the generally positive findings and the publication of a book on the boundary model, this approach has not been elaborated into a complete theory of nightmares and has not enjoyed widespread application as such by the research community.

Image contextualization   Unlike most other models that consider nightmares to result from a breakdown in emotional regulation processes, Hartmann's contextualization model proposes that (with the exception of nonrepetitive PTSD nightmares) nightmares serve the function of contextualizing, or finding a picture context for an individual's predominant emotional concerns. Contextualization proceeds by establishing a wide swath of new associations to the emotion, the result of which is emotionally therapeutic. Contextualization is presumed to be a function of dreaming more generally, but the contextualizing processes in normal dreams are often difficult to discern when the underlying emotions are too weak or diffuse. However, after exposure to stressful or traumatic events that engender high levels of emotionality, the processes are more evident. Emotion is thus the central instigating force in the formation of dreaming and nightmares, and integration of this emotion, via the establishment of broad memory associations, is the ultimate goal of dreaming.

Contextualization occurs via contextualizing images, which are powerful central images of a dream whose emotions are consistent with a central concern but whose content may differ from this concern. For example, an image of being swept up in a tornado may contextualize an individual's feelings of helplessness, fear, and foreboding that stem from a prior assault. Hartmann's group has demonstrated that such images are, in fact, more frequent after trauma. Research has also revealed that the presence and intensity of contextualizing images are related to thinness on the Boundary Questionnaire, to a history of trauma, and to the impact of a dream on waking life. Independent validation of some features of this model was also recently reported.

Emphasis on a contextualizing mechanism likens this model to connectionist models of memory and emotional processing; both of which are thought to be central to REM sleep function. Recent evidence does support the notion that memory systems become hyperassociative and more flexible during REM sleep. Hartmann's focus on context formation during dreaming also presages recent theoretical trends in neuroscience – specifically, the view that sleep-related alterations in hippocampal context-building functions are key in determining how dreaming transmogrifies episodic memories into oniric worlds and relationships. Thus, the contextualization model anticipates both recent speculations that an REM sleep function is to create contextual memories and recent research demonstrating sleep-related facilitation of the consolidation of implicit, contextual (hippocampal-dependent) learning.

In sum, the contextualization model is a relatively new proposal which is still in the process of validation but which is consistent with several emerging trends in the neuroscience of memory.

Threat simulation   The threat simulation model of nightmares is an evolutionary theory that considers nightmares to be virtual representations that enable the self to engage in behaviorally realistic responses to subjectively real threatening events. Active 'rehearsal' of responses during such simulations enhances threat-avoidance skills in the waking world and confers behavioral and survival advantages. Thus, Revonsuo suggests that nightmarish dreams are prime examples of dreams that fully realize this biological function. The suggestion that ancestral threats are more apparent in children's dreams than in those of adults underlines the importance of assessing the dreams of these two populations comparatively.

This model has generated substantial interest and empirical test in the scientific community but remains controversial. The evolutionary assumption of the model, that nightmares are heritable, has been supported to some extent. The only available study on this question used structural equation modeling of the responses of subjects from the Finnish Twin Cohort to reveal persistent genetic effects on the disposition to nightmares both in both childhood (44–45% of phenotypic variance) and adulthood (36–38%).

However, heritability *per se* is insufficient to establish function; many psychiatric disorders bear a genetic component. As described earlier, nightmares are associated with increased vulnerability, psychopathology, and dysfunctional adaptation to the environment. A functional role for nightmares is not supported by the finding that sleep disturbances (including nightmares) increase the risk of developing PTSD upon subsequent trauma exposure and that nightmares following trauma are associated with more severe PTSD. PTSD is associated with a host of sleep and waking state abnormalities and does not appear to be an adaptive condition in any sense.

While this model has stimulated little research on the content of nightmares *per se*, assessment of dream content has provided several consistent findings. Revonsuo's group reported evidence that severely traumatized children living in environments of threat report dreams with more threatening events and dream threats that are more severe than do less traumatized or nontraumatized children. Similarly, the earliest dreams remembered by adults (i.e., dreamed when they were children) contain a very large proportion of threat themes, also as predicted by the model. Revonsuo also reported that the dreams of college students contained threats that were frequent (66% of reports), severe (39% nontrivial), realistic (aggression, failures, misfortunes, etc.), and directed toward the self (73%), and to which the self responded with relevant defensive behaviors (56%) – all findings consistent with the model. Mixed support comes from an independent study of recurrent dreams in which six of eight predictions from the model were judged to be supported. However, less than 15% of these dreams contained realistic and probable threats critical for physical survival or reproductive success. Further, less than 2% of the dreams supported all of the predictions. Similarly, only 8% of undergraduate students' most recent dreams contain realistic life-threatening events (and escape is depicted in only a third of these), whereas severe life-threatening events are experienced in real life by 45% of them. It bears noting, however, that the presence of threats *per se* in dream content does not necessarily prove that dreams are adaptive in an evolutionary sense; these threats may simply reflect daily reality, as stipulated by the continuity hypothesis of dreaming.

In sum, the threat simulation model remains a provocative but largely unproved explanation of nightmares. As an evolutionary theory, many of its tenets are difficult, if not impossible, to test empirically. However, it provides a context, which places nightmare studies within the broader field of evolutionary biology, and has generated many novel hypotheses for empirical testing.

Neurobiological Models

Neurotransmitter imbalance Hartmann and colleagues early proposed a model of nightmare etiology based upon the effects of pharmacological treatments for chronic nightmare sufferers. An imbalance in neurotransmitter systems, characterized by diminished levels of norepinephrine and/or serotonin and elevated levels of dopamine (or a combination of these), leads to repeated experiences of nightmares. Hartmann reviewed much of the evidence available at the time but appears to have abandoned this work in favor of the boundary and contextualization models. Evaluation of this complex model today would require an extensive review of the side effects and interactions of several classes of drugs and is not attempted here. Interested readers are directed to the 'Further reading' section for relevant reviews of this literature.

REM sleep desomatization Fisher and colleagues are among very few groups to have recorded polysomnographic variables during spontaneous nightmares. They found nightmares to be associated with smaller than expected levels of autonomic activation during REM sleep and, in some cases, stage 2 sleep. The extent of autonomic activation recorded with measures of heart (HR), respiratory rate (RR), and eye movement (EM) activity was low, was limited to the last few minutes of preawakening sleep, and was, in 60% (12 of 20) of the nightmares recorded, absent altogether. Even lower levels of activation were found in a more recent study. This apparent separation of subjective fear (fearful dream imagery) from its normal psychophysiological concomitants (low or no autonomic arousal) was referred to as 'desomatization,' that is, an REM sleep mechanism for modulating anxiety by abolishing or diminishing its physiological concomitants. Such a mechanism was thought to help preserve REM sleep, to prevent the self-perpetuation of anxiety, and to contribute to the mastery of traumatic experiences. Nightmares result when the anxiety exceeds a certain threshold and the REM desomatization mechanism breaks down, allowing autonomic activation to occur.

Findings from this study, though influential in clinical research, remain somewhat questionable because an indeterminate number of patients with borderline psychosis, prior trauma, and comorbid sleep terrors participated in the study sample. Nonetheless, some of the findings have been replicated with a sample of idiopathic nightmare cases. Little new empirical evidence has been brought to bear on this model, although some laboratory findings are consistent with the possibility that dream emotion is inhibited by REM sleep processes related to the orienting response. Fisher's findings and speculations, however, have had an impact. Several investigators have also suggested that components of REM sleep may be responsible for desomatization or desensitization – that is, the eye movements of REM sleep desensitize affect in a way similar to what occurs in eye movement desensitization and reprocessing (EMDR) therapy.

In sum, the desomatization model is based upon a limited number of polysomnographic recordings and has not been tested systematically, but has generated much further speculation. The specificity of its proposed mechanism remains an intriguing explanation of nightmare function that is compatible with other theories of dream and nightmare function.

Mood regulation Kramer's mood regulatory theory of dreaming is premised on laboratory findings consistent with the claim that REM sleep is characterized by a 'surge' of affective arousal (e.g., a progressive increase and plateau in limbic system, eye movement, and heart and respiratory activity across the REM period). Dream content is proposed to serve the adaptive function of containing these surges by decreasing the intensity and variability of the associated emotion. This is achieved by a particular pattern of dream content that unfolds across the night and is referred to as 'progressive-sequential' (P-S) in nature. P-S dream series enable a form of emotional problem solving that ameliorates mood. P-S dreaming is distinguished from a repetitive-traumatic pattern, during which an emotional conflict is simply stated and restated without evidence of adaptive change. Nightmares presumably occur when the capacity of dreaming to assimilate the emotional surge in this fashion is exceeded.

While the physiological description of REM sleep as surgelike remains debatable, Kramer has marshaled some empirical support for the proposed function of dream content. In general, evidence that dreams are influenced by one's immediate presleep thoughts and emotional experiences, and that one's waking state mood is related to the previous night's dreams, is consistent with the notion that intervening dream activity regulates mood. More specific evidence that dreaming mediates this regulation is that pre- to post-decreases in mood scores, the 'unhappiness subscale' especially, are correlated with intervening dream content scores, scales involving the number of dream characters especially. While one study by Kramer's group failed to replicate this relationship, consistent findings have been reported by independent researchers. In this case, presleep depression scores were found to be correlated negatively with emotional tone in the dreams of the first REM period of the night (higher depression associated with more negative dreaming) but not with sleep physiology variables. There is also evidence for a mood regulatory function of dreaming in studies of persons undergoing marital separation, who report more negative dreams at the beginning and fewer at the end of the night and who prove more likely to be in remission a year later, compared to those with the opposite pattern. These findings may indicate that negative dreams early in sleep reflect a within-sleep mood regulation process, while negative dreams late in sleep reflect a failure of regulation.

The functional claims of the mood regulation model parallel those of several other nightmare models (e.g., image contextualization, REM sleep desomatization) which claim that dreaming functions to adaptively modify emotions. This claim is supported by a small but growing body of experimental evidence. However, there is as yet no convincing evidence that the P-S dream pattern *per se* is the active agent of mood regulation.

Affective network dysfunction Our neurocognitive model of nightmare formation is based upon a synthesis of findings in the areas of sleep physiology, PTSD, and anxiety disorders. It stipulates that nightmares reflect dysfunction in a network of processes that, during normal dreaming, serves the adaptive function of fear memory extinction. At the cognitive level, this extinction function depends upon three imagery processes that operate on the constituent elements of fear memories: (1) element activation, or the increased availability of isolated features of fear memories removed from their episodic (real-world) contexts, (2) element recombination, or the reorganization of these features into novel, virtual, 'here-and-now' simulations of reality, and (3) emotional expression in reaction to the recombined features, which allows new, fear extinction, memories to be formed and maintained.

At the neural level (see **Figure 1**), the fear extinction function is supported by a network of four limbic, paralimbic, and prefrontal regions that constitute the control center for a number of emotional processes, including the perception and representation of emotional stimuli and the expression and regulation of emotional responses. These four brain regions, which we refer to as the 'AMPHAC model,' include the amygdala (Am), the medial prefrontal cortex (mPFC), the hippocampal (Hip) complex, and the anterior cingulate cortex (ACC). The AMPHAC regions operate in a coordinated manner to influence other perceptual, cognitive, memorial, and affective brain events. However, general correspondences between each region and cognitive processes are also postulated (e.g., Am, emotional activation; Hip, control of memory context; mPFC, control of extinction memories; ACC, regulator of affect distress).

The cognitive and neural explanatory levels together define an affective network within which

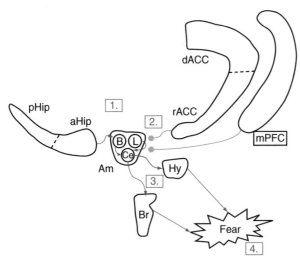

Figure 1 Schematic representation of a network of brain regions hypothesized to be implicated in the production and extinction of fear during normal dreaming; dysfunction in this network is responsible for nightmares. (1) Context is relayed in realistic (virtual) form via anterior hippocampus (aHip) to basal nucleus (B) of the amygdala (Am) and is further processed by the central (Ce) nucleus. (2) Medial prefrontal cortex (mPFC) and dorsal and rostral anterior cingulate cortex (dACC, rACC) afferents to the amygdala regulate the output of Ce neurons to induce extinction and signal distress and maintain appropriate levels of fear. (3) Ce nucleus signals brain stem (Br) and hypothalamus (Hy) to produce (4) the autonomic and behavioral correlates of fear. Excitatory connections are shown in red, inhibitory connections in green. For readability, only connections judged most pertinent to the consolidation and extinction of fear are indicated. Reproduced from Levin R and Nielsen TA (2007) Disturbing dreams, posttraumatic stress disorder, and affect distress: A review and neurocognitive model. *Psychological Bulletin* 133(3): 482–528, with permission from APA.

perturbations may produce a variety of dysphoric dreaming types – from occasional bad dreams to nontraumatic nightmares to recurring posttraumatic nightmares. Further, network activity is 'cross-state,' meaning that it is active during both sleep and wakefulness, which means that its functions – and especially its dysfunctions – should be measurable with various psychophysiological instruments both during sleep and waking states.

Two cross-state factors are proposed for which substantial supporting evidence exists. One is a situational or state factor, affect load, and reflects the combined influence of stressful and emotional negative events (e.g., interpersonal conflicts, daily hassles) on an individual's capacity to effectively regulate emotions. Increases in affect load result in intensified dysphoric dreaming in general. A second factor is a dispositional or trait-like factor, affect distress, and is characterized as a long-standing tendency to experience heightened distress and negative affect in response to increases in affect load, and to react. Affect distress is influenced by events such as prior neglect and trauma, is akin to both the negative affect and negative emotions

(personality dimensions), and is associated with a wide variety of psychopathological conditions (anxiety disorders, health behaviors, PTSD, etc.). Affect distress is related to early development and acts as a risk factor diathesis for emotional dysregulation in the face of subsequent stress.

Thus, during nightmares, high levels of affect load and/or affect distress interact with the neurophysiological state of REM sleep so as to favor the activation of fear memories that are coherent (i.e., that resist feature recombination). These memories are akin to those occurring in waking, fear-based pathological conditions such as phobias or social anxiety. Nightmare-related fear memories are highly resistant to extinction, overly weighted with response elements (usually involving escape or avoidance), and, in more severe instances, corrupted by elevated affect distress. A fuller delineation of this model can be found in recent work by Levin and Nielsen.

As a new model of nightmare formation, the affective network dysfunction model has not yet been tested empirically. However, its tenets are consistent with much literature; its neural propositions, especially, are consistent with recent neuroimaging studies of REM sleep, PTSD, anxiety disorders, and personality factors. To illustrate, several studies indicate that REM sleep is characterized by high levels of activation in Am, mPFC, ACC, and the hippocampal complex. Additionally, PET studies have found that glucose metabolic rates in mPFC during REM sleep are highly correlated with elevated anxiety in the content of dreams sampled during these REM periods. Neuropsychological evidence from brain-lesioned patients also supports the model, demonstrating a link between temporolimbic brain regions and frequent nightmares of both recurring and nonrecurring types. Eight of nine patients who reported recurring nightmares were found to have temporal or frontotemporal lesions, including, in some cases, the Hip or ACC. Nightmares and 'dream–reality confusions' (DRCs) were also habitually associated with limbic lesions (nine of ten patients); six of these patients had lesions affecting the mPFC or ACC or both.

Findings from PTSD patients also support this model. All four of the designated brain regions are affected in PTSD patients. Further, heightened affect distress may be expressed in the form of several sleep-related hyperarousal symptoms, including increased awakenings, wake after sleep onset (WASO), and insomnia, as well as nightmares in stages other than REM sleep and at times other than the habitual last third of the night (e.g., stage 2 nightmares occurring early in the sleep episode). Hyperarousal is also suggested by the expression of motor activity in sleep, including REM-related twitches in leg muscles, more periodic leg movement during sleep in all stages, frequent large body movements, and more REM-related motor activity and vocalization. Explosive motor activity can be elicited from any stage of sleep in some patients with war-related PTSD.

In sum, the affective network dysfunction model is new and still untested but consistent with a large and recent empirical literature. The proposed fear extinction function, while novel, is nonetheless modeled after a well-established research literature on the nature, learning, and unlearning of fear memories. This function is compatible with adaptive emotional functions proposed by other dream and nightmare theorists and may serve as an integrative foundation for resolving these diverse theories.

See also: Behavior and Parasomnias (RSBD); Dreams and Nightmares in PTSD; History of Sleep Research; Incorporation of Waking Events into Dreams; Nightmares; Phylogeny and Ontogeny of Sleep; REM/NREM Differences in Dream Content; Sleep and Consciousness; Theories of Dream Function.

Further Reading

Arkin AM and Steiner SS (1978) The effects of drugs on sleep mentation. In: Arkin AM, Antrobus JS, and Ellman SJ (eds.) *The Mind in Sleep*, pp. 393–415. New York: Lawrence Erlbaum Associates.

Braun AR, Balkin TJ, Wesensten NJ, et al. (1998) Dissociated pattern of activity in visual cortices and their projections during human rapid eye movement sleep. *Science* 279: 91–95.

Cartwright R (2005) Dreaming as a mood regulation system. In: Kryger MH, Roth T, and Dement WC (eds.) *Principles and Practice of Sleep Medicine*, 4th edn., pp. 565–572. Philadelphia, PA: Elsevier Saunders.

Davis M, Myers KM, Chhatwal J, et al. (2006) Pharmacological treatments that facilitate extinction of fear: Relevance to psychotherapy. *NeuroRx* 3: 82–96.

Hublin C, Kaprio J, Partinen M, et al. (1999) Nightmares: Familial aggregation and association with psychiatric disorders in a nationwide twin cohort. *American Journal of Medical Genetics* 88: 329–336.

Hull AM (2002) Neuroimaging findings in post-traumatic stress disorder. Systematic review. *British Journal of Psychiatry* 181: 102–110.

Kim JJ and Jung MW (2006) Neural circuits and mechanisms involved in Pavlovian fear conditioning: A critical review. *Neuroscience and Biobehavioral Reviews* 30: 188–202.

Kramer M (1993) The selective mood regulatory function of dreaming: An update and revision. In: Moffitt A, Kramer M, and Hoffmann R (eds.) *The Functions of Dreaming*, pp. 139–196. Albany, NY: State University of New York.

Levin R and Fireman G (2002) Nightmare prevalence, nightmare distress, and self-reported psychological disturbance. *Sleep* 25: 205–212.

Levin R and Nielsen TA (2007) Disturbing dreams, posttraumatic stress disorder, and affect distress: A review and neurocognitive model. *Psychological Bulletin* 133(3): 482–528.

Maquet P, Ruby P, Maudoux A, et al. (2005) Human cognition during REM sleep and the activity profile within frontal and parietal cortices: A reappraisal of functional neuroimaging data. *Progress in Brain Research* 150: 219–227.

Nielsen TA (2005) Disturbed dreaming in medical conditions. In: Kryger M, Roth N, and Dement WC (eds.) *Principles and Practice of Sleep Medicine*, 4th edn., pp. 936–945. Philadelphia, PA: Elsevier Saunders.

Nielsen TA and Stenstrom P (2005) What are the memory sources of dreaming? *Nature* 437: 34–38.

Nielsen TA and Zadra AL (2005) Nightmares and other common dream disturbances. In: Kryger M, Roth N, and Dement WC (eds.) *Principles and Practice of Sleep Medicine*, 4th edn., pp. 926–935. Philadelphia, PA: Elsevier Saunders.

Nutt DJ and Malizia AL (2004) Structural and functional brain changes in posttraumatic stress disorder. *Journal of Clinical Psychiatry* 65(supplement 1): 11–17.

Pace-Schott EF, Gersh T, Silvestri R, et al. (2001) SSRI treatment suppresses dream recall frequency but increases subjective dream intensity in normal subjects. *Journal of Sleep Research* 10: 129–142.

Revonsuo A (2000) The reinterpretation of dreams: An evolutionary hypothesis of the function of dreaming. *Behavioral and Brain Sciences* 23: 877–901.

Solms M (1997) *The Neuropsychology of Dreams*. Mahwah, NJ: Lawrence Erlbaum Associates.

Stickgold R (2005) Why we dream. In: Kryger M, Roth N, and Dement WC (eds.) *Principles and Practice of Sleep Medicine*, 4th edn., pp. 579–587. Philadelphia, PA: Elsevier Saunders.

Thompson DF and Pierce DR (1999) Drug-induced nightmares. *Annals of Pharmacotherapy* 33: 93–98.

Incorporation of Waking Events into Dreams

E J Wamsley and R Stickgold, Beth Israel Deaconess
Medical Center, and Harvard Medical School, Boston,
MA, USA

Overview

We all have the intuition that our dreams in some way
relate to experiences from waking life. Even in cases
in which no specific event presents itself as the likely
source of a particular dream, still, our nightly reveries
often involve places, persons, or activities which are
familiar to us. Indeed, empirical investigations of the
effects of presleep experience on dream content support the generally held presumption that mentation
during sleep is constructed, at least in part, from
memories of both recent and remote experiences.
But what does this tell us? Despite decades of research
effort, we still have surprisingly little understanding
of the cognitive mechanisms controlling incorporation of waking events into sleep mentation, and virtually no understanding of the neural events underlying
these processes.

What algorithms determine which recent and remote
memories will appear in sleep mentation and what
form incorporation of these memories will take?
'Does the incorporation of waking events into dreaming rely on the engagement of particular memory systems, to the exclusion of others? Is the appearance of
recent events in dreaming related to neural processes
facilitating memory processing during sleep? Might the
reexpression of waking events in sleep mentation serve
some adaptive function?' Answering these questions
not only will provide us with a better understanding
of the neural basis of the dreaming process, but may
also provide us with insights into the nature of human
cognition and memory processing in general.

The definition of 'dreaming' remains a contentious
subject, and one that we do not address here. For our
purposes, we refer to 'dreaming' as being all sleep mentation of which an individual is aware either during
sleep or upon awakening, and we refer to a 'dream' as
being the content of that mentation. In addition, we will
sometimes use the term 'dream' to refer to the waking
report of dream content, and 'dream content' to refer to
the content of such reports. We do so only for the sake
of clarity in reading, and not to stake out a particular
position in regard to the nature of dreaming.

Effects of Presleep Experience on Dream Content

Manipulation of Presleep Experience in the Sleep Laboratory

During the 1960s and 1970s, a considerable amount
of research effort was devoted to understanding the
relation of waking events to dreaming through laboratory studies manipulating participants' presleep experience. Unfortunately, much of this research was
plagued by methodological weaknesses, including
frequent failure to employ blind raters and/or reliance
on subjective interpretation of symbolic relationships
between waking experience and dream content.
Overall, the most consistently observed direct effect
of waking experience on laboratory-collected dream
reports to emerge from these studies was the powerful influence of the laboratory setting on dream
content. In an analysis of 813 rapid eye movement
(REM) dream reports collected across several studies,
Dement and co-workers reported that 22% of reports
unambiguously incorporated either isolated elements
(e.g., the experimenter, the sleep laboratory itself,
electrodes) of the laboratory situation (in 10% of
the reports) or more complete representations (e.g.,
a combination of the previously mentioned elements)
of the experimental setting (in 12%).

In contrast to the strong influence of the laboratory
setting, experimental manipulations of the presleep experience have often failed to demonstrate clear effects
on dream content. In fact, the most striking outcome of
such investigations has been the conspicuously rare
appearance of content directly related to a particular
presleep intervention. Even the use of emotionally
salient stimuli (e.g., violent films) or the manipulation
of basic biological drives (e.g., the induction of thirst)
has rarely resulted in dream content obviously related
to the manipulation. In studies in which unambiguous
incorporations of experimental stimuli have been
reported, the influence of the laboratory environment
often greatly overshadows these effects.

In retrospect, this is hardly surprising, given the
relative salience of sleeping in a strange place and
being awakened during the night to report dreams.
The salience of any particular film, activity, etc. introduced to individuals as a part of an experimental protocol likely pales in comparison. At the same time, these
findings suggest that emotional salience may be a major
determinant of dream content. This conclusion must be

tempered, however, by the fact that the individuals were receiving sensory input from the laboratory environment (e.g., from the recording electrodes) even as they slept, and such concurrent sensations might have dictated their incorporation.

Well-designed experimental investigations utilizing appropriate control groups have had relatively more success in demonstrating statistically significant effects of experimental manipulations when formal or qualitative dream features (e.g., length, affect), rather than content of dreams *per se*, were used as the outcome measure. In short, it has proved to be difficult, if not impossible, to reliably manipulate the content of dreams through presleep interventions administered in a laboratory setting.

Correspondence between Presleep Experience and Dreaming in the Home

Other, nonexperimental approaches, however, clearly show that waking experience is a primary source of sleep mentation content. In an analysis of the possible waking memory sources of 299 home-collected dream reports, Magdalena Fosse and colleagues demonstrated that fragments of recent experience are often seen in dream reports. Of 299 reports, 51% analyzed in this study were judged by individuals to contain at least one feature bearing strong similarity to a recent waking event. In this particular study, participant-identified incorporations were typically direct in nature. As in the case of the aforementioned experimental studies, however, naturally experienced waking events appear to sometimes affect qualitative dream features, as opposed to resulting in direct incorporations of the experience. In an analysis of home dreams collected in the weeks prior to and following the terrorist attacks of 11 September 2001, Ernst Hartmann reported that, although the events of September 11 were not typically incorporated into participants' dreams in the period following the attacks, the emotional intensity of central dream images was significantly increased.

Other lines of research have explored the more general correspondence between sleep mentation and waking life. Calvin Hall, an early cognitive dream theorist, pioneered the use of content analysis to quantitatively assess the content of large sets of mentation reports. This classification system, developed in collaboration with Robert Van de Castle as a response to subjective interpretations of dream content generally employed by psychoanalytic dream theorists, counts the occurrences of various categories of characters, settings, objects, social interactions, etc., as explicitly described in the text of dream reports. Using this system, Hall was able to create surprisingly accurate psychological profiles and histories for psychiatric patients, based solely on blind content analysis of

their dream reports. Later work employing this same system of quantifying dream content has reported consistent, statistically significant differences in the dream content of groups of individuals with divergent waking experience (e.g., between males and females, children and adults, and blind and sighted individuals). These investigations have contributed to our understanding of the dreaming process particularly inasmuch as they remind us that, while attempts to predict or control the waking source of specific dream elements have often failed, a broad correspondence between dream content and waking experience is nonetheless transparently obvious. In short, people clearly tend to dream of the same persons, activities, etc. which dominate their waking lives. While psychoanalysts have long presumed that dreaming conceals a hidden message demanding the application of a complex 'decoding' process, this research, on the contrary, suggests that the relationship between dream content and waking experience may be, if not straightforward, at least discoverable without resorting to symbolic interpretation or reference to hidden 'latent' content.

Incorporation of Presleep Experience into Sleep Onset Dreaming

In contrast to the relative difficulty of manipulating dream content later in the night, experimentally administered tasks appear to more clearly affect mentation during the first few minutes of sleep. In a series of studies employing engaging video games as a presleep task, Stickgold has demonstrated a powerful influence of presleep experience on dreaming at sleep onset. In one such study, participants played the video game Tetris in the morning and evening, several hours prior to retiring for the night. Dream reports were then repeatedly elicited from each participant after just 30 s to 3 min of sleep (in Tetris, the gamer must manipulate falling block shapes and fit them together; problematic subjective interpretations of dream imagery were not necessary in this study, as gamers reported clearly related content elements, such as "Tetris shapes" and "tiles fall[ing] down"). All told, 64% of participants reported isolated, unambiguous, Tetris-related images in at least one sleep onset report. In a related investigation utilizing a downhill skiing arcade game, a surprising 88% of participants reported direct incorporation of game features into at least one mentation report collected during the 3-day study. The frequency of direct incorporation in these sleep onset studies, which is dramatically higher than that observed in any previous overnight investigation, suggests that the first minutes of sleep represent the ideal conditions for reexpression of waking experience in dreams. The instance of direct Alpine Racer incorporation, however, is greatly decreased when reports are collected from a delayed sleep onset period following 2 h of sleep, even when these report collections are matched for time of night with initial sleep onset controls. It may be, then, that direct incorporation of these experiences into initial sleep onset reports is dependent upon the position of the report collection period within the sleep phase, rather than on either simple proximity to waking events or general neurobiological features of the sleep onset state.

Taken together, these data strongly support the common presumption that waking experiences influence dreaming both at sleep onset and later in the night. Yet, the question of 'whether or not' waking experience affects dream content may not be a particularly meaningful one. While empirical verification of the folk psychology presumption that dreams incorporate experience-related content is a necessary step, its confirmation is hardly a revelation, and, more importantly, it tells us little about the neural processes which create dreaming, nor about a possible function for the dreaming process. Instead, a more detailed examination of the nature of the incorporation waking experience, grounded in neurobiological data regarding the engagement of memory systems across states of consciousness, is needed if we are to gain clues to the neural and cognitive processes supporting incorporation of waking experience into sleep mentation.

Form of Waking Experience Incorporations

Even in cases in which mentation reports clearly do include elements originating from a specific spatio-temporal episode, these dreams almost never take the form of an exact replay of a life event. Instead, sleep mentation typically incorporates only isolated elements, or fragments, of a waking episode. These fragments, in turn, may be interleaved with remote and semantic memory material. For example, although Fosse's group found that more than half of the reports in their sample contained individual features with strong similarity to a recent experience, less than 2% of these preserved all of the key elements of the original episode, including the setting, characters, objects, and actions found in the memory of the waking event. Mentation reports following Tetris and Alpine Racer exposure exhibit similar characteristics, with individuals describing seeing clearly identifiable isolated features of the game (e.g., Tetris shapes) without the larger context of the activity (e.g., the computer screen, keyboard, or desk).

The following report from Fosse's group provides an illustrative example of a waking memory source and corresponding report, in which the sleep mentation clearly incorporates fragments of a waking experience, without replicating the original context in which these fragments were embedded:

> Waking Experience: "*When I left Starbucks, we had so many leftover pastries and **muffins** to throw away or take home. **I couldn't decide which muffins to take and which to toss**...*"
>
> Corresponding Sleep Mentation: "*My dad and I leave to go shopping. We go from room to room, store to store. One of the stores is filled with **muffins, muffins, muffins** from floor to ceiling, all different kinds, **I can't decide which one I want**...*"

The various components of a waking episode, then, do not appear to be 'bound' together in the manner that characterizes the 'mental time travel' (as described by psychologist Endel Tulving) of episodic memory recall in waking life. This phenomenological disparity between the appearance of memories in dreams and the recall of recent episodes in waking life suggests that divergent, though likely overlapping, processes underlie these two phenomena. We consider in the following discussion how the fragmented appearance of recent memory in dream reports, interleaved with remote and semantic content, might reflect the consolidation of recent memories during sleep as they become integrated into older cortical memory networks.

Note that it may be precisely the fragmentary nature of event incorporation which frustrated the efforts of so many early presleep-stimulus studies. Investigators have often, for example, exposed study participants to multidimensional film stimuli prior to sleep, expecting to see veridical 'replay' of this stimulus in subsequent mentation reports. The stimuli employed might contain a number of relatively well-learned features (e.g., persons or commonplace objects and activities). If only isolated features of these experiences appeared later in dream reports, their ubiquitous nature may have prevented the recognition as having originated in the presleep manipulation of interest. In contrast, the novel and easily identifiable features of Tetris, for example, provide the optimal conditions for tracking isolated features of waking events in subsequent dreams.

It may be important to note also that, as in the preceding example, isolated dream elements originating from the identified/manipulated presleep experience clearly appear in combination with additional elements derived either from other memory sources or from generalized semantic memory networks (e.g., 'shopping with dad' in the preceding example). Indeed, although participants often cite relatively recent experiences as the source of their mentation content, remote memories from several years or longer ago, as well as generic semantic memories, are also frequently cited as the mnemonic referent of dream material. In sum, the incorporation of recent experiences into dreams is manifested in the intrusion of isolated, salient features of a presleep experience, intermingled with other content.

Sleep Stages, Waking Memory Sources, and Memory Systems

There appear also to be systematic differences across various sleep states in the types of memories incorporated into dream content. This distribution of memory sources, in turn, may inform us about the engagement of particular memory systems across sleep stages and their participation in the dream creation process. In a series of studies, primarily conducted by researchers at the University of Bologna, Italy, participants identified possible waking memory sources of mentation collected from non-REM (NREM), REM, and sleep onset. Blind raters classified each participant-identified memory source as either 'episodic' (autobiographical episodes taking place in a specific spatiotemporal context) or 'semantic' (elements of general knowledge not tied to a particular spatiotemporal context). Perhaps the most interesting result emerging from this work is that NREM reports, rather than REM reports, are associated with the highest proportion of episodic memory sources. This observation fits well with what we know about regional cerebral blood flow (rCBF) patterns during sleep. Acquisition and retrieval of recent episodic memories rely crucially on the engagement of the hippocampal formation, which is more active during NREM than in either waking or REM sleep. In addition, the low levels of acetylcholine characterizing NREM sleep are hypothesized to preferentially facilitate output from the hippocampus to the cortex.

Consistent with the data of Stickgold, however, reports at sleep onset tend to contain an even greater proportion of episodic memory sources than do NREM reports, as well as to contain a very high proportion of 'day residues' (episodic memory sources originating from an experience during the previous day). Comparatively little is known about the activity of the hippocampal system at sleep onset, although at least one investigation has reported that rCBF in the hippocampus is not significantly decreased at sleep onset relative to waking. However, there is cognitive evidence that the hippocampus is not required for incorporation of recent experience into sleep onset mentation (see later).

When participants, returning to sleep following awakening after the first complete REM cycle, provide dream reports from a delayed sleep onset period, episodic sources in general are still proportionally greater than in either REM or NREM reports. But 'day

residues' are significantly less frequent. Though proximity to waking experience has been postulated to account for this decline in incorporation of very recent events into later sleep onset periods, recall that Stickgold et al. observed a similar decline in Alpine Racer game-related content when mentation was collected from a delayed sleep onset phase, even though time of night and time since last game play were both kept constant. Again, this suggests that the disproportionate appearance of recent memories in sleep onset reports is not a function of simple proximity to the daytime experience. Rather, memory processing occurring during the initial hours of sleep may alter the incorporation of recent experiences into subsequent dreams, such as into delayed, as compared to initial, sleep onset periods.

Emotional Value and Waking Experience Incorporation

But what processes determine 'which' memory fragments will appear in subsequent sleep mentation? Though doubtless others operate in tandem, the emotional value of waking experiences appears to be one relevant factor. Analyses of the correspondence between dreamed and waking activities, for example, demonstrate that while some relatively mundane activities consuming a great deal of our waking time (e.g., reading and writing) are underrepresented in dreams, other activities of greater personal significance (e.g., socializing with friends) are overrepresented relative to their frequency during waking. Other lines of research suggest a direct link between the emotional value of specific waking events and their appearance in dream reports. Rosalind Cartwright's work on the adaptive function of dreaming in women experiencing divorce directly addresses personal relevance as a mediator of the incorporation of waking experiences into dream content. Cartwright found that individuals' self-rated concern about their spouse is strongly correlated with the percentage of dream reports containing explicit mention of the spouse. Meanwhile, DeKoninck and Koulack have reported similar findings in a group of individuals viewing a film of work-related accidents prior to sleep. Those individuals reporting increased levels of anxiety following morning film viewing reported significantly more film-related content in their subsequent dream reports. Each of these studies suggests that degree of emotional involvement in waking experiences modulates their incorporation into subsequent dreams.

Recall also that in laboratory studies of the effects of presleep experience, the experimental situation was often more strongly incorporated into mentation reports than any was particular experimentally manipulated stimulus. Presumably, in these situations, the experience of sleeping in a strange place, being wired with electrodes, and participating in an experiment is far more salient and more important than is any specific presleep experience (e.g., a film or other activity) of interest to the investigators. Again, this general set of findings indicates that those experiences in which an individual is more emotionally involved are more likely to appear in subsequent dreams. We also see this effect dramatically, of course, in the dreams of trauma survivors, who tend to replay the traumatic emotional event in their dreams to such a degree that the disturbing dreams themselves become a clinically significant symptom of posttraumatic stress disorder (PTSD).

Time Course of Incorporations

Incorporation of recent experiences into dream content may follow an organized temporal pattern, examination of which could provide clues to a possible function of dreaming. First of all, very recent events may more often be incorporated into mentation occurring early in the sleep phase. Though only a small handful of studies have examined the effect of time of night on such incorporations, both recency of subject-identified memory sources and the similarity of sleep mentation reports to presleep thoughts have been reported to correlate negatively with time since sleep onset. Meanwhile, Stickgold has reported that, during the first few minutes of sleep, incorporation of Tetris imagery into sleep onset dreams declines linearly as a function of time since sleep onset. Recall also that, overall, very recent experiences are more often incorporated into initial sleep onset mentation than into delayed sleep onset mentation or dream reports later in the night. As discussed earlier, simple proximity to waking events appears to be inadequate to explain these effects, as incorporation of recent events declines later in the sleep cycle, even when time of night is held constant.

Examining the incorporation of waking experience into sleep mentation across longer timescales also suggests that mere proximity to a waking event is a relatively poor predictor of incorporation. Although some studies conducted across multiple nights have reported incorporations occurring most frequently on the night immediately following a presleep experience, others have found that frequency of incorporations peaks only after a delay of ≥ 1 day. Stickgold, for example, found that sleep onset reports of novice game players contained far more Tetris imagery on the second night following initiation of game play than on the first. Cartwright has similarly reported increased incorporation of a sexually arousing film on the second night following viewing the film, though these incorporations were most often 'symbolic' in nature.

Roffwarg and colleagues have reported even longer incorporation delays, as in the case of the influence of extended wearing of colored goggles on the perceptual quality of dreams. Here, the incorporation effect was seen most strongly several days after participants began wearing the goggles.

In addition, Tore Nielsen has conducted a series of careful studies that find two distinct postevent time frames in which incorporation is most likely to be identified. In accordance with the findings of several Roffwarg and colleagues, the first of these is on the night immediately following a presleep experience (the 'day residue' effect). Nielsen also finds evidence, however, for a 'dream lag' effect, in which waking experiences are again likely to be incorporated into dreams recalled approximately 7 days after the initial experience. Though broadly consistent with the long incorporation delays reported by Roffwarg et al., this is a relatively novel finding and the cognitive and neurobiological mechanisms that might underlie it remain unknown. Nonetheless, this 'dream lag' effect has remained relatively robust across several studies, and controls have ruled out a number of potential confounds.

On an even broader timescale, the temporal distribution of participant-identified dream memory sources is quite similar to that of experimentally elicited waking autobiographical memory samples. In each case, the largest proportion of memory references is drawn from more recent experiences, with references to more remote memories declining in concert with increased temporal distance. This observation is consistent with the idea that some overlap exists between the neurobiological basis for waking recall of autobiographical episodes and their appearance in dream content, and, contrary to Freudian theories, dreams are relatively infrequently associated with memories from the distant past.

Recent Memories, Dreaming, and Memory Consolidation

A growing body of evidence suggests that sleep, and in particular, NREM sleep, facilitates the offline processing of recent declarative memories (e.g., memories of events and facts). Behavioral studies in humans, for example, suggest that periods of NREM sleep are beneficial for declarative memory performance, relative to periods of wakefulness or REM sleep. Meanwhile, both human and animal studies demonstrate that patterns of neural activity associated with recent hippocampally dependent learning are reexpressed during subsequent NREM sleep. This neural 'replay' of recent episodes during NREM sleep is thought to support consolidation processes that mediate the transformation of recent memories from an initial hippocampus-dependent labile form into more permanent long-term memories in the cortex.

Naturally, the observation that recent declarative memories are reexpressed in the sleeping brain has prompted speculation that these processes may be related to the content of dreams. Although very little work has directly addressed the possible relationship between dreaming and sleep-dependent memory consolidation, a small group of studies do support the notion that such memory processing is reflected in the content of sleep mentation. For example, Carlyle Smith reported preliminary evidence of dream reports related to a mirror-tracing task learned prior to sleep. These task-related dreams were present to a greater degree in individuals 'cued' during sleep using a sound associated with the learning task, compared to noncued individuals. Fiss has provided evidence of a link between overnight changes in memory performance and dreaming, reporting that dreaming of stories presented to individuals prior to sleep was associated with enhanced morning recall of those stories. Thus, the available cognitive evidence is at least consistent with the notion that sleep mentation may reflect consolidation of memories in sleep.

Neural Basis for Dream Incorporation of Recent Events: Connection to Declarative Memory Consolidation?

Research into the neurophysiological basis of sleep-dependent memory consolidation may shed light on neural mechanisms mediating the relationship between recent experience and dreaming. Models of memory consolidation are grounded in evidence that, although declarative memory recall is initially dependent on the hippocampus, over time, repeated memory trace reactivation enables the cortex to support retrieval without the need of hippocampal involvement. Repeated memory reactivation is thought to be crucial to the process of memory consolidation, and, as noted previously, evidence suggests that this occurs optimally during NREM.

Supporting this general model, a number of studies have demonstrated that neural activity initially established during waking experience is 'replayed' during subsequent sleep. These data primarily derive from studies in rodents utilizing multiple single-neuron recordings. Analogous observations have been reported in humans, however, as patterns of regional brain activation associated with specific learning experiences in wakefulness are again seen during subsequent NREM sleep.

Several key features of the replay of learning-associated neural activity in sleep exhibit interesting parallels with the preceding cognitive data on sleep mentation:

1. The neural replay of recent experiences has most often been reported to occur in NREM sleep. As described earlier, NREM is also the sleep stage during which episodic memories are most often incorporated into sleep mentation.

2. Within NREM sleep, neural replay of recent experience is expressed most strongly immediately after learning. The strength of this reactivation effect then tends to decay quickly across time. Similarly, as already described, sleep mentation may be most strongly related to recent experience early in the sleep cycle.

3. Recent learning is not 'replayed' during sleep in its original form. Following learning, firing sequences established during wakefulness are reexpressed only intermittently during rodent NREM, with relatively low fidelity, and on a faster timescale than the original experience. Similarly, as already described, recent experiences are not 'replayed' in dream content in their original form. Instead, intermittent fragments of recent experience are recalled from sleep, intermingled with other content.

Although far from conclusive, this correspondence between features of neuronal-level experience replay during sleep and dream content is consistent with the hypothesis that subjective experience is derived, at least in part, from memory reactivation supporting consolidation processes in sleep. Though these generalizations are largely based on neural replay of recent events observed in the hippocampus, it is important to note that reexpression of learning-related activity has also been observed in cortical structures. Matt Wilson, for example, recently observed sequential replay of a recent experience in the rodent primary visual cortex. Though the inference that an observed pattern of brain activity is associated with conscious awareness always presents a problematic conceptual leap, experimental stimulation of visual cortex in humans does result in associated visual experience.

Despite these arguments, it must be pointed out that little evidence directly links subjective experience recalled from sleep to neurophysiological and cognitive data on memory replay and consolidation. In fact, the study of Tetris imagery in sleep onset dreaming suggests that, at least at sleep onset, this incorporation likely does not rely on neural processes involved in sleep-dependent declarative memory consolidation. Replay and consolidation of recent declarative memory during sleep is presumed to depend crucially on involvement of the hippocampus. Remarkably, however, Stickgold found that amnesiac patients with complete loss of hippocampal function incorporated Tetris images into their sleep onset dream reports at rates similar to normal persons. These images appeared even though the patients could neither identify the source of the Tetris imagery nor recall ever having played the game. This observation suggests that, at least at sleep onset, dream incorporation of recent experiences does not rely on the hippocampal processes presumed to underlie consolidation of declarative memory. It seems likely that this is exclusively a sleep onset phenomenon, as sleep onset is a state with distinct neurophysiological and phenomenological features. Alternatively, as described in early studies of dreaming in hippocampal patients, there may be subtle differences in the quality of memory content seen in the dreams of amnesiacs. Certainly, more research is needed before a definitive statement can be made about the relationship between dreaming and memory consolidation in sleep.

A Function for Dreaming?

To the extent that dreaming does turn out to reflect memory consolidation processes, neural activity supporting the dreaming process may serve an important adaptive function. Presumably, the function of memory consolidation, in sleep or in wakefulness, is not only to facilitate long-term retention of information, but also to identify those memories of recent events worth retaining and then to integrate these with memories of related older experiences and with general semantic information, so that this new knowledge can be most adaptively utilized to guide future action. It may be that dreaming is associated both with the identification of valuable memories and with the integration of these memories with remote and semantic information stored in cortical networks.

Supposing that dreaming is indeed associated with such a function, might there be cases in which this function is disrupted? Although we have characterized the incorporation of recent experiences into dream content as a fragmentary process, with elements of the waking event interleaved with remote and semantic content, there are cases in which more veridical, complete replay of experiences is reported. In patients suffering from PTSD, the traumatic experience is often replayed in an abnormally direct manner, and the patients' frequent, near-veridical replay is a persistent symptom of the disorder, which can continue for years. It is conceivable that this abnormal replay of traumatic experiences in the dreams of PTSD patients is reflective of the dysfunctional consolidation of the traumatic memories, perhaps involving a failure to successfully integrate the traumatic experience with existing cortical memory networks.

In closing, it is important to remember that despite this or any speculation on a possible 'function' for dreaming, the question of whether the conscious experience of dreaming *per se*, as opposed to the neural activity associated with it, supports these processes is essentially unapproachable until more is understood about the neural basis of consciousness in general.

See also: Dreams and Nightmares in PTSD; History of Sleep Research; Phylogeny and Ontogeny of Sleep; REM/NREM Differences in Dream Content; Sleep and Consciousness; Sleep-Dependent Memory Processing; Theories and Correlates of Nightmares; Theories of Dream Function.

Further Reading

Baylor GW and Cavallero C (2001) Memory sources associated with REM and NREM dream reports throughout the night: A new look at the data. *Sleep* 24(2): 165–170.

Buzsaki G (1996) The hippocampo–neocortical dialogue. *Cerebral Cortex* 6(2): 81–92.

Cartwright R, Agargun MY, Kirkby J, et al. (2006) Relation of dreams to waking concerns. *Psychiatry Research* 141(3): 261–270.

De Koninck JM and Koulack D (1975) Dream content and adaptation to a stressful situation. *Journal of Abnormal Psychology* 84(3): 250–260.

De Koninck J and Brunette R (1991) Presleep suggestion related to a phobic object: Successful manipulation of reported dream affect. *Journal of General Psychology* 118(3): 185–200.

Dement WC, Kahn E, and Roffwarg HP (1965) The influence of the laboratory situation on the dreams of the experimental subject. *Journal of Nervous and Mental Disease* 140: 119–131.

Fiss H, Kremer E, and Lichtman J (1977) The mnemonic function of dreaming. *Sleep Research* 6: 122–136.

Fosse MJ, Fosse R, Hobson JA, et al. (2003) Dreaming and episodic memory: A functional dissociation? *Journal of Cognitive Neuroscience* 15(1): 1–9.

Grenier J, Cappeliez P, St-Onge M, et al. (2005) Temporal references in dreams and autobiographical memory. *Memory & Cognition* 33(2): 280–288.

Hall C and Nordby V (1972) *The Individual and His Dreams*. New York: New American Library.

Hartmann E (2000) We do not dream of the 3 R's: Implications for the nature of dream mentation. *Dreaming* 10: 103–110.

Hasselmo ME (1999) Neuromodulation: Acetylcholine and memory consolidation. *Trends in Cognitive Science* 3(9): 351–359.

McClelland JL, McNaughton BL, and O'Reilly RC (1995) Why there are complementary learning systems in the hippocampus and neocortex: Insights from the successes and failures of connectionist models of learning and memory. *Psychological Review* 102(3): 419–457.

Nielsen TA, Kuiken D, Alain G, et al. (2004) Immediate and delayed incorporations of events into dreams: Further replication and implications for dream function. *Journal of Sleep Research* 13(4): 327–336.

Stickgold R, Malia A, Maguire D, et al. (2000) Replaying the game: Hypnagogic images in normals and amnesics. *Science* 290 (5490): 350–353.

Subject Index

Notes

Cross-reference terms in italics are general cross-references, or refer to subentry terms within the main entry (the main entry is not repeated to save space). Readers are also advised to refer to the end of each article for additional cross-references – not all of these cross-references have been included in the index cross-references.

The index is arranged in set-out style with a maximum of three levels of heading. Major discussion of a subject is indicated by bold page numbers. Page numbers suffixed by T and F refer to Tables and Figures respectively. vs. indicates a comparison.

This index is in letter-by-letter order, whereby hyphens and spaces within index headings are ignored in the alphabetization. Prefixes and terms in parentheses are excluded from the initial alphabetization.

Declarative (explicit) memory
  consolidation
    sleep role 233
    see also Memory consolidation; Sleep-dependent memory
        processing
  definition 230
  encoding, sleep role 232
  episodic memory *see* Episodic memory
  neuroanatomy/neural substrates 230
    hippocampus role 230
  procedural (implicit, nondeclarative) *vs.* 230
  semantic memory *see* Semantic memory
  spatial memory *see* Spatial memory
Deep sleep *see* Slow-wave sleep (SWS)
Delayed sleep phase syndrome (DSPS) 177–178
  characteristics 178–179
  diagnosis 178
  genetic basis 181
  light response 179–180, 180*f*
  melatonin treatment 209
  pathophysiology 179
    decreased circadian period 179
    homeostatic sleep regulation and 180–181
    light entrainment and 180
    masking 179–180, 180*f*
    melatonin and 180
    phase relationship 180
  prevalence 179
Delta waves 23–24
  ascending reticular activating system (ARAS) 136
  PET studies 31–32
  thalamus and 87
Dementia with Lewy bodies *see* Lewy body disease (DLB)
Depression *see* Depressive illness
Depressive illness
  sleep disorders
    cortisol and 211
    RAS and 138
  stress response *see* Depressive illness, HPA axis and stress
  winter *see* Seasonal affective disorder (SAD)
Depressive illness, HPA axis and stress
  glucocorticoids and
    cortisol level changes 211
Diagnostic and Statistical Manual of Mental Disorders
  4th edition (DSM-IV)
    Text Revision (DSM-IV-TR), nightmares 323, 324*t*
Diaphragm
  REM sleep and 224
Dipyridamole 203*f*
Disk-over-water method (DOW), sleep deprivation 174
Distal–proximal skin gradient, napping 255–256
Diurnal rhythm(s) *see* Circadian rhythm(s)
Divided brain *see* Split-brain patients
Dog(s)
  narcolepsy 280*f*
    pharmacological experiments 281
Dolphin(s)
  sleep/sleep states 62–63, 63*f*, 112
Domestic cat (*Felis catus*) *see* Cat
Dopamine (DA)
  anatomy *see* Dopaminergic neurons/systems
  arousal and *see* Dopamine, arousal role
  motivational system
    dreaming and 312–313
  neurotransmission *see* Dopaminergic neurons/systems
  pleasure/reward and *see* Dopaminergic neurons,
    reward role
  receptors *see* Dopamine receptor(s)
  reuptake/transport
    stimulant-induced arousal and 128–129
    transporter proteins *see* Dopamine transporter
  reward and *see* Dopaminergic neurons, reward role
  see also Dopaminergic neurons/systems
Dopamine, arousal role **125–130**
  dopaminergic disorders and sleep 125–126, 127
    evidence for 127
    hypersomnia (excessive daytime sleepiness) 127–128
    REM sleep behavior disorder 128
    see also Parkinsonian syndromes/parkinsonism
  dopamine transporter and
    DAT knockout mice and 125–126, 128–129
      modafinil effects 129

stimulant effects 128–129
  dopaminomimetic effects 125–126
  historical perspective 125
  stimulant-induced arousal 128
    anatomical substrates
      basal forebrain cholinergic neurons and 129
      lateral hypothalamus orexin neurons and 129
      VLPO inhibition 129
    DAT transporter and 128–129
  wake-active dopamine neurons 125, 129
    evidence 125–126
    midbrain cell groups 125–126, 126*f*
    REM sleep role 127
    vPAG dopamine neurons 126–127
      as functionally distinct population 126–127
      identification 126–127, 127*f*
      projections 126–127
Dopamine receptor(s)
  antagonists
    arousal effects 128–129
Dopaminergic drugs (dopaminergics)
  arousal effects 125–126
Dopaminergic neurons/systems 125
  ARAS 92
  arousal role *see* Dopamine, arousal role
  diencephalon 125
  midbrain tracts *see* Midbrain dopaminergic neurons
  striatal *see* Striatum
Dopamine transporter (DAT)
  amphetamine effects 193
  arousal role
    knockout mice and 125–126, 128–129
      modafinil effects 129
    norepinephrine transporter *vs.* 128–129
    stimulant-induced arousal and 128–129
  blockers
    arousal effects 128–129
  genetic manipulation
    arousal effects 125–126, 128–129
Dorsolateral prefrontal cortex (DLPFC)
  REM sleep
    PET deactivation during 32–33
Dorsomedial hypothalamic nuclei (DMN)
  circadian rhythm and 159
DQB1*0601, narcolepsy–cataplexy 271
DQB1*0602, narcolepsy–cataplexy 271, 272*f*, 274*f*
Draflazine 203*f*
Dreams/dreaming 295–301, 310–315
  autonomic responses 223, 312
  bad dreams
    coping strategies 142*t*
    see also Nightmares
  blindness and 313
  cardiac activity and 223
  carryover effects 295
    auxiliary effects 296
    emotional feeling 296
    felt position 296
    felt presence 296
    fluidity of thought 296
    of intensified dreaming 296
      impactful dreams 297
      sleep paralysis episodes 296
    REM dreaming 295
  cat studies 310
    PGO waves 28, 310
  characteristics 34, 305, 311
    bizarreness 34, 312, 314
    emotions 34, 312
    hallucinatory quality 311
    imagery 313
    lucid dreaming 312
    motives 312
    total reported information 313
    values 312
  circadian influences 313
    bizarreness and 314
    joint *vs.* unique patterns of circadian-based
      mentation 314
    lucid dreams and 312
  content, trauma 317
  contrasting conceptions 295

Printed in the United States
By Bookmasters